HANDBOOK OF
Computer Vision Algorithms in Image Algebra

SECOND EDITION

HANDBOOK OF
Computer Vision Algorithms in Image Algebra

SECOND EDITION

Gerhard X. Ritter
Joseph N. Wilson

CRC PRESS

Boca Raton   London   New York   Washington, D.C.

Library of Congress Cataloging-in-Publication Data

Ritter, G. X.
Handbook of computer vision algorithms in image algebra / Gerhard X. Ritter, Joseph
N. Wilson.—2nd ed.
   p. cm.
  Includes bibliographical references and index.
  ISBN 0-8493-0075-4 (alk. paper)
   1. Computer vision—Mathematics. 2. Image processing—Mathematics. 3. Computer
algorithms. I. Wilson, Joseph N. II. Title.

TA1634 .R58 2000
006.4′2—dc21
                                                         00-062122

Preface

The present edition differs from the first in several significant aspects. Typographical errors as well as several mathematical errors have been removed. In a number of places the text has been revised to enhance clarity. Several additional algorithms have been included as well as an entire new chapter on geometric image transformations. By popular demand, and in order to provide a better understanding of image algebra, numerous exercises have been added at the end of each chapter. Starred exercises at the end of a chapter depend on knowledge of material from subsequent chapters.

As with the first edition, the principal aim of this book is to acquaint engineers, scientists, and students with the basic concepts of image algebra and its use in the concise representation of computer vision algorithms. In order to achieve this goal we provide a brief survey of commonly used computer vision algorithms that we believe represents a core of knowledge that all computer vision practitioners should have. This survey is not meant to be an encyclopedic summary of computer vision techniques as it is impossible to do justice to the scope and depth of the rapidly expanding field of computer vision.

The arrangement of the book is such that it can serve as a reference for computer vision algorithm developers in general as well as for algorithm developers using the image algebra C++ object library, iac++.[1] The techniques and algorithms presented in a given chapter follow a progression of increasing abstractness. Each technique is introduced by way of a brief discussion of its purpose and methodology. Since the intent of this text is to train the practitioner in formulating his algorithms and ideas in the succinct mathematical language provided by image algebra, an effort has been made to provide the precise mathematical formulation of each methodology. Thus, we suspect that practicing engineers and scientists will find this presentation somewhat more practical and perhaps a bit less esoteric than those found in research publications or various textbooks paraphrasing these publications.

Chapter 1 provides a short introduction to the field of image algebra. Chapters 2–12 are devoted to particular techniques commonly used in computer vision algorithm development, ranging from early processing techniques to such higher level topics as image descriptors and artificial neural networks. Although the chapters on techniques are most naturally studied in succession, they are not tightly interdependent and can be studied according to the reader's particular interest. In the Appendix we present iac++ computer programs of some of the techniques surveyed in this book. These programs reflect the image algebra pseudocode presented in the chapters and serve as examples of how image algebra pseudocode can be converted into efficient computer programs.

---

[1]   The iac++ library supports the use of image algebra in the C++ programming language and is available via anonymous ftp from ftp://ftp.cise.ufl.edu/pub/src/ia/.

Acknowledgments

We wish to take this opportunity to express our thanks to our current and former students who have, in various ways, assisted in the preparation of this text. In particular, we wish to extend our appreciation to Dr. Paul Gader, Dr. Jennifer Davidson, Dr. Hongchi Shi, Ms. Brigitte Pracht, Dr. Mark Schmalz, Mr. Venugopal Subramaniam, Mr. Mike Rowlee, Dr. Dong Li, Dr. Huixia Zhu, Ms. Chuanxue Wang, Dr. Jaime Zapata, and Mr. Liang-Ming Chen. We are most deeply indebted to Dr. David Patching who assisted in the preparation of the text and contributed to the material by developing examples that enhanced the algorithmic exposition. Special thanks are due to Mr. Ralph Jackson, who skillfully implemented many of the algorithms herein, and to Mr. Robert Forsman, the primary implementor of the `iac++` library. We also wish to thank Mr. Jeffrey Palm for preparing the fractal and iterated function system images.

We wish to express our gratitude to those at Wright Laboratory for their encouragement and continuous support of image algebra research and development. This book would not have been written without the vision and support provided by numerous scientists at the Wright Laboratory at Eglin Air Force Base in Florida. These supporters include Dr. Lawrence Ankeney who started it all, Dr. Sam Lambert who championed the image algebra project since its inception, Mr. Neil Urquhart our first program manager, Ms. Karen Norris, and most especially Dr. Patrick Coffield who persuaded us to turn a technical report on computer vision algorithms in image algebra into this book.

Last but not least we would like to thank Dr. Robert Lyjack of ERIM and Dr. Jasper Lupo of DARPA for their friendship and enthusiastic support during the formative stages of Image Algebra.

Notation

The tables presented here provide a brief explantation of the notation used throughout this document. The reader is referred to Ritter [1] for a comprehensive treatise covering the mathematics of image algebra.

Sets Theoretic Notation and Operations

Symbol	Explanation
X, Y, Z	Uppercase characters represent arbitrary sets.
x, y, z	Lowercase characters represent elements of an arbitrary set.
$\mathbf{X}, \mathbf{Y}, \mathbf{Z}$	Bold, uppercase characters are used to represent point sets.
$\mathbf{x}, \mathbf{y}, \mathbf{z}$	Bold, lowercase characters are used to represent points, i.e., elements of point sets.
$\mathbb{N}$	The set $\mathbb{N} = \{0, 1, 2, 3, \ldots\}$.
$\mathbb{Z}, \mathbb{Z}^+, \mathbb{Z}^-$	The set of integers, positive integers, and negative integers, respectively.
$\mathbb{Z}_n$	The set $\mathbb{Z}_n = \{0, 1, \ldots, n-1\}$.
$\mathbb{Z}_n^+$	The set $\mathbb{Z}_n^+ = \{1, 2, \ldots, n\}$.
$\mathbb{Z}_{\pm n}$	The set $\mathbb{Z}_{\pm n} = \{-n+1, \ldots, -1, 0, 1, \ldots, n-1\}$.
$\mathbb{R}, \mathbb{R}^+, \mathbb{R}^-, \mathbb{R}^{\geq 0}$	The set of real numbers, positive real numbers, negative real numbers, and positive real numbers including 0, respectively.
$\mathbb{C}$	The set of complex numbers.
$\mathbb{F}$	An arbitrary set of values.
$\mathbb{F}_\infty$	The set $\mathbb{F}$ unioned with $\{\infty\}$.
$\mathbb{F}_{-\infty}$	The set $\mathbb{F}$ unioned with $\{-\infty\}$.
$\mathbb{F}_{\pm\infty}$	The set $\mathbb{F}$ unioned with $\{-\infty, \infty\}$.
$\varnothing$	The *empty set* (the set that has no elements).
2^X	The *power set* of X (the set of all subsets of X).
$\in$	"is an element of."
$\notin$	"is not an element of."
$\subset$	"is a subset of."

Symbol	Explanation
$X \bigcup Y$	*Union* $X \cup Y = \{z : z \in X \text{ or } z \in Y\}$.
$\displaystyle\bigcup_{\lambda \in \Lambda} X_\lambda$	Let $\{X_\lambda\}_{\lambda \in \Lambda}$ be a family of sets indexed by an indexing set Λ. $\displaystyle\bigcup_{\lambda \in \Lambda} X_\lambda = \{x : x \in X_\lambda \text{ for at least one } \lambda \in \Lambda\}$.
$\displaystyle\bigcup_{i=1}^{n} X_i$	$\displaystyle\bigcup_{i=1}^{n} X_i = X_1 \cup X_2 \cup \ldots \cup X_n$.
$\displaystyle\bigcup_{i=1}^{\infty} X_i$	$\displaystyle\bigcup_{i=1}^{\infty} X_i = \{x : x \in X_i \text{ for some } i \in \mathbb{Z}^+\}$.
$X \bigcap Y$	*Intersection* $X \cap Y = \{z : z \in X \text{ and } z \in Y\}$.
$\displaystyle\bigcap_{\lambda \in \Lambda} X_\lambda$	Let $\{X_\lambda\}_{\lambda \in \Lambda}$ be a family of sets indexed by an indexing set Λ. $\displaystyle\bigcap_{\lambda \in \Lambda} X_\lambda = \{x : x \in X_\lambda \text{ for all } \lambda \in \Lambda\}$.
$\displaystyle\bigcap_{i=1}^{n} X_i$	$\displaystyle\bigcap_{i=1}^{n} X_i = X_1 \cap X_2 \cap \ldots \cap X_n$.
$\displaystyle\bigcap_{i=1}^{\infty} X_i$	$\displaystyle\bigcap_{i=1}^{\infty} X_i = \{x : x \in X_i \text{ for all } i \in \mathbb{Z}^+\}$.
$X \times Y$	*Cartesian product* $X \times Y = \{(x, y) : x \in X, y \in Y\}$.
$\displaystyle\prod_{i=1}^{n} X_i$	$\displaystyle\prod_{i=1}^{n} X_i = \{(x_1, x_2, \ldots, x_n) : x_i \in X_i\}$.
$\displaystyle\prod_{i=1}^{\infty} X_i$	$\displaystyle\prod_{i=1}^{\infty} X_i = \{(x_1, x_2, x_3, \ldots) : x_i \in X_i\}$.
$\mathbb{F}^n$	The Cartesian product of n copies of $\mathbb{F}$, i.e., $\mathbb{F}^n = \displaystyle\prod_{i=1}^{n} \mathbb{F}$.
$X \setminus Y$	*Set difference* Let X and Y be subsets of some universal set U, $X \setminus Y = \{x \in X : x \notin Y\}$.
X'	*Complement* $X' = U \setminus X$, where U is the universal set that contains X.
$card(X)$	The *cardinality* of the set X.
$choice(X)$	A function that randomly selects an element from the set X.

Point and Point Set Operations

Symbol	Explanation
$\mathbf{x} + \mathbf{y}$	If $\mathbf{x}, \mathbf{y} \in \mathbb{R}^n$, then $\mathbf{x} + \mathbf{y} = (x_1 + y_1, \ldots, x_n + y_n)$.
$\mathbf{x} - \mathbf{y}$	If $\mathbf{x}, \mathbf{y} \in \mathbb{R}^n$, then $\mathbf{x} - \mathbf{y} = (x_1 - y_1, \ldots, x_n - y_n)$.

Symbol	Explanation
$\mathbf{x} \cdot \mathbf{y}$	If $\mathbf{x}, \mathbf{y} \in \mathbb{R}^n$, then $\mathbf{x} \cdot \mathbf{y} = (x_1 y_1, \ldots, x_n y_n)$.
$\mathbf{x}/\mathbf{y}$	If $\mathbf{x}, \mathbf{y} \in \mathbb{R}^n$, then $\mathbf{x}/\mathbf{y} = (x_1/y_1, \ldots, x_n/y_n)$.
$\mathbf{x} \vee \mathbf{y}$	If $\mathbf{x}, \mathbf{y} \in \mathbb{R}^n$, then $\mathbf{x} \vee \mathbf{y} = (x_1 \vee y_1, \ldots, x_n \vee y_n)$.
$\mathbf{x} \wedge \mathbf{y}$	If $\mathbf{x}, \mathbf{y} \in \mathbb{R}^n$, then $\mathbf{x} \wedge \mathbf{y} = (x_1 \wedge y_1, \ldots, x_n \wedge y_n)$.
$\mathbf{x}\gamma\mathbf{y}$	In general, if $\mathbf{x}, \mathbf{y} \in \mathbb{R}^n$, and $\gamma : \mathbb{R} \times \mathbb{R} \to \mathbb{R}$, then $\mathbf{x}\gamma\mathbf{y} = (x_1 \gamma y_1, \ldots, x_n \gamma y_n)$.
$k\gamma\mathbf{x}$	If $k \in \mathbb{R}$, $\mathbf{x} \in \mathbb{R}^n$, and $\gamma : \mathbb{R} \times \mathbb{R} \to \mathbb{R}$, then $k\gamma\mathbf{x} = (k\gamma x_1, \ldots, k\gamma x_n)$.
$\mathbf{x}{\bullet}\mathbf{y}$	If $\mathbf{x}, \mathbf{y} \in \mathbb{R}^n$, then $\mathbf{x}{\bullet}\mathbf{y} = x_1 y_1 + x_2 y_2 + \cdots + x_n y_n$.
$\mathbf{x} \times \mathbf{y}$	If $\mathbf{x}, \mathbf{y} \in \mathbb{R}^n$, then $\mathbf{x} \times \mathbf{y} = (x_2 y_3 - x_3 y_2, x_3 y_1 - x_1 y_3, x_1 y_2 - x_2 y_1)$.
$\hat{\mathbf{x}}\mathbf{y}$	If $\mathbf{x} \in \mathbb{R}^n$ and $\mathbf{y} \in \mathbb{R}^m$, then $\hat{\mathbf{x}}\mathbf{y} = (x_1, \ldots, x_n, y_1, \ldots, y_m)$.
$-\mathbf{x}$	If $\mathbf{x} \in \mathbb{R}^n$, then $-\mathbf{x} = (-x_1, \ldots, -x_n)$.
$\lceil \mathbf{x} \rceil$	If $\mathbf{x} \in \mathbb{R}^n$, then $\lceil \mathbf{x} \rceil = (\lceil x_1 \rceil, \ldots, \lceil x_n \rceil)$.
$\lfloor \mathbf{x} \rfloor$	If $\mathbf{x} \in \mathbb{R}^n$, then $\lfloor \mathbf{x} \rfloor = (\lfloor x_1 \rfloor, \ldots, \lfloor x_n \rfloor)$.
$[\mathbf{x}]$	If $\mathbf{x} \in \mathbb{R}^n$, then $[\mathbf{x}] = ([x_1], \ldots, [x_n])$.
$p_i(\mathbf{x})$	If $\mathbf{x} = (x_1, x_2, \ldots, x_n) \in \mathbb{R}^n$, then $p_i(\mathbf{x}) = x_i$.
$\Sigma\mathbf{x}$	If $\mathbf{x} \in \mathbb{R}^n$, then $\Sigma\mathbf{x} = x_1 + x_2 + \cdots + x_n$.
$\Pi\mathbf{x}$	If $\mathbf{x} \in \mathbb{R}^n$, then $\Pi\mathbf{x} = x_1 x_2 \cdots x_n$.
$\vee\mathbf{x}$	If $\mathbf{x} \in \mathbb{R}^n$, then $\vee\mathbf{x} = x_1 \vee x_2 \vee \cdots \vee x_n$.
$\wedge\mathbf{x}$	If $\mathbf{x} \in \mathbb{R}^n$, then $\wedge\mathbf{x} = x_1 \wedge x_2 \wedge \cdots \wedge x_n$.
$\|\mathbf{x}\|_2$	If $\mathbf{x} \in \mathbb{R}^n$, then $\|\mathbf{x}\|_2 = \sqrt{x_1^2 + \cdots + x_n^2}$.
$\|\mathbf{x}\|_1$	If $\mathbf{x} \in \mathbb{R}^n$, then $\|\mathbf{x}\|_1 = |x_1| + |x_2| + \cdots + |x_n|$.
$\|\mathbf{x}\|_\infty$	If $\mathbf{x} \in \mathbb{R}^n$, then $\|\mathbf{x}\|_\infty = |x_1| \vee |x_2| \vee \cdots \vee |x_n|$.
$dim(\mathbf{x})$	If $\mathbf{x} \in \mathbb{R}^n$, then $dim(\mathbf{x}) = n$.
$\mathbf{X} + \mathbf{Y}$	If $\mathbf{X}, \mathbf{Y} \subseteq \mathbb{R}^n$, then $\mathbf{X} + \mathbf{Y} = \{\mathbf{x} + \mathbf{y} : \mathbf{x} \in \mathbf{X} \text{ and } \mathbf{y} \in \mathbf{Y}\}$.
$\mathbf{X} - \mathbf{Y}$	If $\mathbf{X}, \mathbf{Y} \subseteq \mathbb{R}^n$, then $\mathbf{X} - \mathbf{Y} = \{\mathbf{x} - \mathbf{y} : \mathbf{x} \in \mathbf{X} \text{ and } \mathbf{y} \in \mathbf{Y}\}$.
$\mathbf{X} + \mathbf{p}$	If $\mathbf{X} \subseteq \mathbb{R}^n$ and $p \in \mathbb{R}^n$, then $\mathbf{X} + \mathbf{p} = \{\mathbf{x} + \mathbf{p} : \mathbf{x} \in \mathbf{X}\}$.

Symbol	Explanation
$\mathbf{X} - \mathbf{p}$	If $\mathbf{X} \subseteq \mathbb{R}^n$ and $p \in \mathbb{R}^n$, then $\mathbf{X} - \mathbf{p} = \{\mathbf{x} - \mathbf{p} : \mathbf{x} \in \mathbf{X}\}$.
$\mathbf{X} \cup \mathbf{Y}$	If $\mathbf{X}, \mathbf{Y} \subseteq \mathbb{R}^n$, then $\mathbf{X} \cup \mathbf{Y} = \{\mathbf{z} : \mathbf{z} \in \mathbf{X} \; or \; \mathbf{z} \in \mathbf{Y}\}$.
$\mathbf{X} \backslash \mathbf{Y}$	If $\mathbf{X}, \mathbf{Y} \subseteq \mathbb{R}^n$, then $\mathbf{X} \backslash \mathbf{Y} = \{\mathbf{z} : \mathbf{z} \in \mathbf{X} \; and \; \mathbf{z} \notin \mathbf{Y}\}$.
$\mathbf{X} \triangle \mathbf{Y}$	If $\mathbf{X}, \mathbf{Y} \subseteq \mathbb{R}^n$, then $\mathbf{X} \triangle \mathbf{Y} = \{\mathbf{z} : \mathbf{z} \in \mathbf{X} \cup \mathbf{Y} \; and \; \mathbf{z} \notin \mathbf{X} \cap \mathbf{Y}\}$.
$\mathbf{X} \times \mathbf{Y}$	If $\mathbf{X}, \mathbf{Y} \subseteq \mathbb{R}^n$, then $\mathbf{X} \times \mathbf{Y} = \{(\mathbf{x}, \mathbf{y}) : \mathbf{x} \in \mathbf{X} \; and \; \mathbf{y} \in \mathbf{Y}\}$.
$-\mathbf{X}$	If $\mathbf{X} \subseteq \mathbb{R}^n$, then $-\mathbf{X} = \{-\mathbf{x} : \mathbf{x} \in \mathbf{X}\}$.
$\tilde{\mathbf{X}}$	If $\mathbf{X} \subseteq \mathbb{R}^n$, then $\tilde{\mathbf{X}} = \{\mathbf{z} : \mathbf{z} \in \mathbb{R}^n \; and \; \mathbf{z} \notin \mathbf{X}\}$.
$sup(\mathbf{X})$	If $\mathbf{X} \subseteq \mathbb{R}^n$, then $sup(\mathbf{X}) =$ the supremum of $\mathbf{X}$. If $\mathbf{X} = \{\mathbf{x}_1, \mathbf{x}_2, \ldots, \mathbf{x}_n\}$, then $sup(\mathbf{X}) = \mathbf{x}_1 \vee \mathbf{x}_2 \vee \ldots \vee \mathbf{x}_n$.
$\bigvee \mathbf{X}$	For a point set $\mathbf{X}$ with total order $\prec$, $\mathbf{x}_0 = \bigvee \mathbf{X} \Leftrightarrow \mathbf{x} \prec \mathbf{x}_0, \forall \mathbf{x} \in \mathbf{X} \backslash \{\mathbf{x}_0\}$.
$inf(\mathbf{X})$	If $\mathbf{X} \subseteq \mathbb{R}^n$, then $inf(\mathbf{X}) =$ the infimum of $\mathbf{X}$. If $\mathbf{X} = \{\mathbf{x}_1, \mathbf{x}_2, \ldots, \mathbf{x}_n\}$, then $inf(\mathbf{X}) = \mathbf{x}_1 \wedge \mathbf{x}_2 \wedge \ldots \wedge \mathbf{x}_n$.
$\bigwedge \mathbf{X}$	For a point set $\mathbf{X}$ with total order $\prec$, $\mathbf{x}_0 = \bigwedge \mathbf{X} \Leftrightarrow \mathbf{x}_0 \prec x, \forall \mathbf{x} \in \mathbf{X} \backslash \{\mathbf{x}_0\}$.
$choice(\mathbf{X})$	If $\mathbf{X} \subseteq \mathbb{R}^n$, then $choice(\mathbf{X}) \in \mathbf{X}$ (*randomly chosen element*).
$card(\mathbf{X})$	If $\mathbf{X} \subseteq \mathbb{R}^n$, then $card(\mathbf{X}) = the\ cardinality\ of\ \mathbf{X}$. In particular, if $\mathbf{X} = \{\mathbf{x}_1, \cdots, \mathbf{x}_n\}$, then $card(\mathbf{X}) = n$.

Morphology

In the following table, $\mathbf{A}, \mathbf{B}, \mathbf{D}$, and $\mathbf{E}$ denote subsets of $\mathbb{R}^n$.

Symbol	Explanation
$\mathbf{A}^*$	The reflection of $\mathbf{A}$ across the origin $\mathbf{0} = (0, 0, \ldots, 0) \in \mathbb{R}^n$.
$\mathbf{A}'$	The complement of $\mathbf{A}$; i.e., $\mathbf{A}' = \{\mathbf{x} \in \mathbb{R}^n : \mathbf{x} \notin \mathbf{A}\}$.
$\mathbf{A}_b$	$\mathbf{A}_b = \{\mathbf{a} + \mathbf{b} : \mathbf{a} \in \mathbf{A}\}$.
$\mathbf{A} \times \mathbf{B}$	*Minkowski addition* is defined as $\mathbf{A} \times \mathbf{B} = \{\mathbf{a} + \mathbf{b} : \mathbf{a} \in \mathbf{A}, \mathbf{b} \in \mathbf{B}\}$ (Section 7.2).
$\mathbf{A}/\mathbf{B}$	*Minkowski subtraction* is defined as $\mathbf{A}/\mathbf{B} = (\mathbf{A}' \times \mathbf{B}^*)'$ (Section 7.2).

Symbol	Explanation
$\mathbf{A} \circ \mathbf{B}$	The *opening of* $\mathbf{A}$ *by* $\mathbf{B}$ is denoted $\mathbf{A} \circ \mathbf{B}$ and is defined by $\mathbf{A} \circ \mathbf{B} = (\mathbf{A}/\mathbf{B}) \times \mathbf{B}$ (Section 7.3).
$\mathbf{A} \bullet \mathbf{B}$	The *closing of* $\mathbf{A}$ *by* $\mathbf{B}$ is denoted $\mathbf{A} \bullet \mathbf{B}$ and is defined by $\mathbf{A} \bullet \mathbf{B} = (\mathbf{A} \times \mathbf{B})/\mathbf{B}$ (Section 7.3).
$\mathbf{A} \circledast \mathbf{C}$	Let $\mathbf{C} = (\mathbf{D}, \mathbf{E})$ be an ordered pair of structuring elements. The *hit-and-miss transform* of the set $\mathbf{A}$ is given by $\mathbf{A} \circledast \mathbf{C} = \{\mathbf{p} : \mathbf{D_p} \subset \mathbf{A} \text{ and } \mathbf{E_p} \subset \mathbf{A'}\}$ (Section 7.5).

Functions and Scalar Operations

Symbol	Explanation
$f : X \to Y$	f is a *function* from X into Y.
$domain(f)$	The *domain* of the function $f : X \to Y$ is the set X.
$range(f)$	The *range* of the function $f : X \to Y$ is the set $\{f(x) : x \in X\}$.
f^{-1}	The *inverse* of the function f.
Y^X	The set of all functions from X into Y, i.e., if $f \in Y^X$, then $f : X \to Y$.
$f|_A$	Given a function $f : X \to Y$ and a subset $A \subset X$, the *restriction of* f *to* A, $f|_A : A \to Y$, is defined by $f|_A(a) = f(a)$ for $a \in A$.
$f|^g$	Given $f : A \to Y$ and $g : B \to Y$, the *extension of* f *to* g is defined by $f|^g(x) = \begin{cases} f(x) & \text{if } x \in A \\ g(x) & \text{if } x \in B \setminus A. \end{cases}$
$g \circ f$	Given two functions $f : X \to Y$ and $g : Y \to Z$, the *composition* $g \circ f : X \to Z$ is defined by $(g \circ f)(x) = g(f(x))$, for every $x \in X$.
$f + g$	Let f and g be real or complex-valued functions, then $(f + g)(x) = f(x) + g(x)$.
$f \cdot g$	Let f and g be real or complex-valued functions, then $(f \cdot g)(x) = f(x) \cdot g(x)$.
$k \cdot f$	Let f be a real or complex-valued function, and k be a real or complex number, then $f \in \mathbb{F}^X$, $(k \cdot f)(x) = k \cdot (f(x))$.
$|f|$	$|f|(x) = |f(x)|$, where f is a real (or complex)-valued function, and $|f(x)|$ denotes the absolute value (or magnitude) of $f(x)$.

Symbol	Explanation
1_X	The *identity function* $1_X : X \to X$ is given by $1_X(x) = x$.
$p_j : \prod_{i=1}^{n} X_i \to X_j$	The *projection function* p_j *onto the jth coordinate* is defined by $p_j(x_1, \ldots, x_j, \ldots, x_n) = x_j$.
$card(X)$	The *cardinality* of the set X.
$choice(X)$	A function which randomly selects an element from the set X.
$x \vee y$	For $x, y \in \mathbb{R}$, $x \vee y$ is the maximum of x and y.
$x \wedge y$	For $x, y \in \mathbb{R}$, $x \wedge y$ is the minimun of x and y.
$\lceil x \rceil$	For $x \in \mathbb{R}$ the *ceiling function* $\lceil x \rceil$ returns the smallest integer that is greater than or equal to x.
$\lfloor x \rfloor$	For $x \in \mathbb{R}$ the *floor function* $\lfloor x \rfloor$ returns the largest integer that is less than or equal to x.
$[x]$	For $x \in \mathbb{R}$ the *round function* returns the nearest integer to x. If there are two such integers it yields the integer with greater magnitude.
$x \bmod y$	For $x, y \in \mathbb{N}$, $x \bmod y = r$ if there exists $k, r \in \mathbb{N}$ with $r < y$ such that $x = yk + r$.
$\chi_S(x)$	The *characteristic function* χ_S is defined by $$\chi_S(x) = \begin{cases} 1 & \text{if } x \in S \\ 0 & \text{otherwise.} \end{cases}$$

Images and Image Operations

Symbol	Explanation
$\mathbf{a}, \mathbf{b}, \mathbf{c}$	Bold, lowercase characters are used to represent *images*. Image variables will usually be chosen from the beginning of the alphabet.
$\mathbf{a} \in \mathbb{F}^X$	The image $\mathbf{a}$ is an $\mathbb{F}$-*valued image* on X. The set $\mathbb{F}$ is called the *value set* of $\mathbf{a}$ and X the *spatial domain* of $\mathbf{a}$.
$\mathbf{1} \in \mathbb{F}^X$	Let $\mathbb{F}$ be a set with unit 1. Then $\mathbf{1}$ denotes an image, all of whose pixel values are 1.
$\mathbf{0} \in \mathbb{F}^X$	Let $\mathbb{F}$ be a set with zero 0. Then $\mathbf{0}$ denotes an image, all of whose pixel values are 0.
$\mathbf{a}|_Z$	The *domain restriction* of $\mathbf{a} \in \mathbb{F}^X$ to a subset $\mathbf{Z}$ of $\mathbf{X}$ is defined by $\mathbf{a}|_Z = \mathbf{a} \cap (\mathbf{Z} \times \mathbb{F})$. Thus, $\mathbf{a}|_Z \in \mathbb{F}^Z$.

Symbol	Explanation
$\mathbf{a}\|_S$	The *range restriction* of $\mathbf{a} \in \mathsf{F}^{\mathbf{X}}$ to the subset $S \subset \mathsf{F}$ is defined by $\mathbf{a}\|_S = \mathbf{a} \cap (\mathbf{X} \times S)$. The double-bar notation is used to focus attention on the fact that the restriction is applied to the second coordinate of $\mathbf{a} \subset \mathbf{X} \times \mathsf{F}$. Thus if $\mathbf{W} = \{\mathbf{x} \in \mathbf{X} : \mathbf{a}(\mathbf{x}) \in S\}$, then $\mathbf{a}\|_S \in S^{\mathbf{W}}$.
$\mathbf{a}\|_{(\mathbf{Z},S)}$	If $\mathbf{a} \in \mathsf{F}^{\mathbf{X}}$, $\mathbf{Z} \subset \mathbf{X}$, and $S \subset \mathsf{F}$, then the *restriction of $\mathbf{a}$ to $\mathbf{Z}$ and S* is defined as $\mathbf{a}\|_{(\mathbf{Z},S)} = \mathbf{a} \cap (\mathbf{Z} \times S)$. Thus if $\mathbf{W} = \{\mathbf{z} \in \mathbf{Z} : \mathbf{a}(\mathbf{z}) \in S\}$, $\mathbf{a}\|_{(\mathbf{Z},S)} \in S^{\mathbf{W}}$.
$\mathbf{a}\|^{\mathbf{b}}$	Let $\mathbf{X}$ and $\mathbf{Y}$ be subsets of the same topological space. The *extension of $\mathbf{a} \in \mathsf{F}^{\mathbf{X}}$ to $\mathbf{b} \in \mathsf{F}^{\mathbf{Y}}$* is defined by $$\mathbf{a}\|^{\mathbf{b}}(\mathbf{x}) = \begin{cases} \mathbf{a}(\mathbf{x}) & \text{if } \mathbf{x} \in \mathbf{X} \\ \mathbf{b}(\mathbf{x}) & \text{if } \mathbf{x} \in \mathbf{Y} \setminus \mathbf{X}. \end{cases}$$
$(\mathbf{a}\|\mathbf{b})$, $(\mathbf{a}_1\|\mathbf{a}_2\|\cdots\|\mathbf{a}_n)$	Row concatenation of images $\mathbf{a}$ and $\mathbf{b}$, respectively the row concatenation of images $\mathbf{a}_1, \mathbf{a}_2, \ldots, \mathbf{a}_n$.
$\begin{pmatrix} \mathbf{a} \\ - \\ \mathbf{b} \end{pmatrix}$	Column concatenation of images $\mathbf{a}$ and $\mathbf{b}$.
$f(\mathbf{a})$	If $\mathbf{a} \in \mathsf{F}^{\mathbf{X}}$ and $f : \mathsf{F} \to \mathsf{G}$, then the image $f(\mathbf{a}) \in \mathsf{G}^{\mathbf{X}}$ is given by $f \circ \mathbf{a}$, i.e., $f(\mathbf{a}) = \{(\mathbf{x}, \mathbf{c}(\mathbf{x})) : \mathbf{c}(\mathbf{x}) = f(\mathbf{a}(\mathbf{x})), \mathbf{x} \in \mathbf{X}\}$.
$\mathbf{a} \circ f$	If $f : \mathbf{Y} \to \mathbf{X}$ and $\mathbf{a} \in \mathsf{F}^{\mathbf{X}}$, the induced image $\mathbf{a} \circ f \in \mathsf{F}^{\mathbf{Y}}$ is defined by $\mathbf{a} \circ f = \{(\mathbf{y}, \mathbf{a}(f(\mathbf{y}))) : \mathbf{y} \in \mathbf{Y}\}$.
$\mathbf{a} \gamma \mathbf{b}$	If γ is a binary operation on F, then an induced operation on $\mathsf{F}^{\mathbf{X}}$ can be defined. Let $\mathbf{a}, \mathbf{b} \in \mathsf{F}^{\mathbf{X}}$; the induced operation is given by $\mathbf{a} \gamma \mathbf{b} = \{(\mathbf{x}, \mathbf{c}(\mathbf{x})) : \mathbf{c}(x) = \mathbf{a}(\mathbf{x}) \gamma \mathbf{b}(\mathbf{x}), \mathbf{x} \in \mathbf{X}\}$.
$k \gamma \mathbf{a}$	Let $k \in \mathsf{F}$, $\mathbf{a} \in \mathsf{F}^{\mathbf{X}}$, and γ be a binary operation on F. An induced scalar operation on images is defined by $k \gamma \mathbf{a} = \{(\mathbf{x}, \mathbf{c}(\mathbf{x})) : \mathbf{c}(\mathbf{x}) = k \gamma \mathbf{a}(\mathbf{x}), \mathbf{x} \in \mathbf{X}\}$.
$\mathbf{a}^{\mathbf{b}}$	Let $\mathbf{a}, \mathbf{b} \in \mathbb{R}^{\mathbf{X}}$; $\mathbf{a}^{\mathbf{b}} = \left\{ (\mathbf{x}, \mathbf{c}(\mathbf{x})) : \mathbf{c}(\mathbf{x}) = \mathbf{a}(\mathbf{x})^{\mathbf{b}(\mathbf{x})}, \mathbf{x} \in \mathbf{X} \right\}$.
$log_{\mathbf{b}}\mathbf{a}$	Let $\mathbf{a}, \mathbf{b} \in (\mathbb{R}^{+})^{\mathbf{X}}$; $log_{\mathbf{b}}\mathbf{a} = \left\{ (\mathbf{x}, \mathbf{c}(\mathbf{x})) : \mathbf{c}(\mathbf{x}) = log_{\mathbf{b}(\mathbf{x})}\mathbf{a}(x), \mathbf{x} \in \mathbf{X} \right\}$.
$\mathbf{a}^{*}$	If $\mathbf{a} \in \mathsf{F}^{\mathbf{X}}$ and F has a conjugation operation $*$, then the pointwise conjugate of image $\mathbf{a}$, $\mathbf{a}^{*}(\mathbf{x}) = (\mathbf{a}(\mathbf{x}))^{*}$.
$\Gamma\mathbf{a}$	$\Gamma\mathbf{a}$ denotes reduction by a generic reduce operation $\Gamma : \mathsf{F}^{\mathbf{X}} \to \mathsf{F}$ (Section 1.4).

The following four items are specific examples of the global reduce operation. Each assumes $\mathbf{a} \in \mathbb{R}^{\mathbf{X}}$ and $\mathbf{X} = \{\mathbf{x}_1, \mathbf{x}_2, \ldots, \mathbf{x}_n\}$.

Symbol	Explanation
$\sum \mathbf{a}$	$\sum \mathbf{a} = \mathbf{a}(\mathbf{x}_1) + \mathbf{a}(\mathbf{x}_2) + \cdots + \mathbf{a}(\mathbf{x}_n)$.
$\prod \mathbf{a}$	$\prod \mathbf{a} = \mathbf{a}(\mathbf{x}_1) \cdot \mathbf{a}(\mathbf{x}_2) \cdot \cdots \cdot \mathbf{a}(\mathbf{x}_n)$.
$\bigvee \mathbf{a}$	$\bigvee \mathbf{a} = \mathbf{a}(\mathbf{x}_1) \vee \mathbf{a}(\mathbf{x}_2) \vee \cdots \vee \mathbf{a}(\mathbf{x}_n)$.
$\bigwedge \mathbf{a}$	$\bigwedge \mathbf{a} = \mathbf{a}(\mathbf{x}_1) \wedge \mathbf{a}(\mathbf{x}_2) \wedge \cdots \wedge \mathbf{a}(\mathbf{x}_n)$.
$\mathbf{a} \bullet \mathbf{b}$	Dot product, $\mathbf{a} \bullet \mathbf{b} = \Sigma(\mathbf{a} \cdot \mathbf{b}) = \sum\limits_{\mathbf{x} \in \mathbf{X}} (\mathbf{a}(\mathbf{x}) \cdot \mathbf{b}(\mathbf{x}))$.
$\tilde{\mathbf{a}}$	Complementation of a set-valued image $\mathbf{a}$.
$\mathbf{a}^c$	Complementation of a Boolean image $\mathbf{a}$.
$\mathbf{a}'$	Transpose of image $\mathbf{a}$.

Templates and Template Operations

Symbol	Explanation
$\mathbf{s}, \mathbf{t}, \mathbf{u}$	Bold, lowercase characters are used to represent *templates*. Usually characters from the middle of the alphabet are used as template variables.
$\mathbf{t} \in \left(\mathbb{F}^{\mathbf{X}}\right)^{\mathbf{Y}}$	A template is an image whose pixel values are images. In particular, an $\mathbb{F}$-*valued template from* $\mathbf{Y}$ *to* $\mathbf{X}$ is a function $\mathbf{t} : \mathbf{Y} \to \mathbb{F}^{\mathbf{X}}$. Thus, $\mathbf{t} \in \left(\mathbb{F}^{\mathbf{X}}\right)^{\mathbf{Y}}$ and $\mathbf{t}$ is an $\mathbb{F}^{\mathbf{X}}$-valued image on $\mathbf{Y}$.
$\mathbf{t}_{\mathbf{y}}$	Let $\mathbf{t} \in \left(\mathbb{F}^{\mathbf{X}}\right)^{\mathbf{Y}}$. For each $\mathbf{y} \in \mathbf{Y}$, $\mathbf{t}_{\mathbf{y}} = \mathbf{t}(\mathbf{y})$. The image $\mathbf{t}_{\mathbf{y}} \in \mathbb{F}^{\mathbf{X}}$ is given by $\mathbf{t}_{\mathbf{y}} = \{(\mathbf{x}, \mathbf{t}_{\mathbf{y}}(\mathbf{x})) : \mathbf{x} \in \mathbf{X}\}$.
$S(\mathbf{t}_{\mathbf{y}})$	If $\mathbb{F} \in \{\mathbb{R}, \mathbb{C}\}$ and $\mathbf{t} \in \left(\mathbb{F}^{\mathbf{X}}\right)^{\mathbf{Y}}$, then the *support* of $\mathbf{t}$ is denoted by $S(\mathbf{t}_{\mathbf{y}})$ and is defined by $S(\mathbf{t}_{\mathbf{y}}) = \{\mathbf{x} \in \mathbf{X} : \mathbf{t}_{\mathbf{y}}(\mathbf{x}) \neq 0\}$.
$S_{\infty}(\mathbf{t}_{\mathbf{y}})$	If $\mathbf{t} \in \left(\mathbb{R}_{\infty}^{\mathbf{X}}\right)^{\mathbf{Y}}$, then $S_{\infty}(\mathbf{t}_{\mathbf{y}}) = \{\mathbf{x} \in \mathbf{X} : \mathbf{t}_{\mathbf{y}}(\mathbf{x}) \neq \infty\}$.
$S_{-\infty}(\mathbf{t}_{\mathbf{y}})$	If $\mathbf{t} \in \left(\mathbb{R}_{-\infty}^{\mathbf{X}}\right)^{\mathbf{Y}}$, then $S_{-\infty}(\mathbf{t}_{\mathbf{y}}) = \{\mathbf{x} \in \mathbf{X} : \mathbf{t}_{\mathbf{y}}(\mathbf{x}) \neq -\infty\}$.
$S_{\pm\infty}(\mathbf{t}_{\mathbf{y}})$	If $\mathbf{t} \in \left(\mathbb{R}_{\pm\infty}^{\mathbf{X}}\right)^{\mathbf{Y}}$, then $S_{\pm\infty}(\mathbf{t}_{\mathbf{y}}) = \{\mathbf{x} \in \mathbf{X} : \mathbf{t}_{\mathbf{y}}(\mathbf{x}) \neq \pm\infty\}$.

Symbol	Explanation
$\mathbf{t}(p)$	A parameterized $\mathbb{F}$-*valued template from* $\mathbf{Y}$ *to* $\mathbf{X}$ *with parameters in* P is a function of the form $\mathbf{t} : P \to \left(\mathbb{F}^{\mathbf{X}}\right)^{\mathbf{Y}}$.
$\mathbf{t}'$	Let $\mathbf{t} \in \left(\mathbb{F}^{\mathbf{X}}\right)^{\mathbf{Y}}$. The *transpose* $\mathbf{t}' \in \left(\mathbb{F}^{\mathbf{Y}}\right)^{\mathbf{X}}$ is defined as $\mathbf{t}'_{\mathbf{x}}(\mathbf{y}) = \mathbf{t}_{\mathbf{y}}(\mathbf{x})$.

Image-Template Operations

In the table below, $\mathbf{X}$ is a finite subset of $\mathbb{R}^n$.

Symbol	Explanation
$\mathbf{a}\,⑦\,\mathbf{t}$	Let $(\mathbb{F}, \gamma, \bigcirc)$ be a semiring and $\mathbf{a} \in \mathbb{F}^{\mathbf{X}}$, $\mathbf{t} \in \left(\mathbb{F}^{\mathbf{X}}\right)^{\mathbf{Y}}$, then the *generic right convolution product of* $\mathbf{a}$ *with* $\mathbf{t}$ is defined as $\mathbf{a}\,⑦\,\mathbf{t} = \left\{ (\mathbf{y}, \mathbf{b}(\mathbf{y})) : \mathbf{y} \in \mathbf{Y}, \ \mathbf{b}(\mathbf{y}) = \underset{\mathbf{x}\in\mathbf{X}}{\Gamma}\, \mathbf{a}(\mathbf{x}) \bigcirc \mathbf{t}_{\mathbf{y}}(\mathbf{x}) \right\}$.
$\mathbf{t}\,⑦\,\mathbf{a}$	With the conditions above, except that now $\mathbf{t} \in \left(\mathbb{F}^{\mathbf{Y}}\right)^{\mathbf{X}}$, the *generic left convolution product of* $\mathbf{a}$ *with* $\mathbf{t}$ is defined as $\mathbf{t}\,⑦\,\mathbf{a} = \left\{ (\mathbf{y}, \mathbf{b}(\mathbf{y})) : \mathbf{y} \in \mathbf{Y}, \mathbf{b}(\mathbf{y}) = \underset{\mathbf{x}\in\mathbf{X}}{\Gamma}\, \mathbf{a}(\mathbf{x}) \bigcirc \mathbf{t}_{\mathbf{x}}(\mathbf{y}) \right\}$.
$\mathbf{a}\,\oplus\,\mathbf{t}$	Let $\mathbf{Y} \subset \mathbb{R}^m$, $\mathbf{a} \in \mathbb{F}^{\mathbf{X}}$, and $\mathbf{t} \in \left(\mathbb{F}^{\mathbf{X}}\right)^{\mathbf{Y}}$, where $\mathbb{F} \in \{\mathbb{C}, \mathbb{R}\}$. The *right linear convolution product* is defined as $\mathbf{a}\,\oplus\,\mathbf{t} = \left\{ (\mathbf{y}, \mathbf{b}(\mathbf{y})) : \mathbf{y} \in \mathbf{Y}, \mathbf{b}(\mathbf{y}) = \sum_{\mathbf{x}\in\mathbf{X}\cap S(\mathbf{t}_{\mathbf{y}})} \mathbf{a}(\mathbf{x}) \cdot \mathbf{t}_{\mathbf{y}}(\mathbf{x}) \right\}$.
$\mathbf{t}\,\oplus\,\mathbf{a}$	With the conditions above, except that $\mathbf{t} \in \left(\mathbb{F}^{\mathbf{Y}}\right)^{\mathbf{X}}$, the *left linear convolution product* is defined as $\mathbf{t}\,\oplus\,\mathbf{a} = \left\{ (\mathbf{y}, \mathbf{b}(\mathbf{y})) : \mathbf{y} \in \mathbf{Y}, \mathbf{b}(\mathbf{y}) = \sum_{\mathbf{x}\in\mathbf{X}\cap S(\mathbf{t}'_{\mathbf{y}})} \mathbf{a}(\mathbf{x}) \cdot \mathbf{t}_{\mathbf{x}}(\mathbf{y}) \right\}$.
$\mathbf{a}\,\boxtimes\,\mathbf{t}$	For $\mathbf{a} \in \mathbb{R}_{\pm\infty}^{\mathbf{X}}$ and $\mathbf{t} \in \left(\mathbb{R}_{\pm\infty}^{\mathbf{X}}\right)^{\mathbf{Y}}$, the *right morphological max convolution product* is defined by $\mathbf{a}\boxtimes\mathbf{t} = \left\{ (\mathbf{y}, \mathbf{b}(\mathbf{y})) : \mathbf{y} \in \mathbf{Y}, \mathbf{b}(\mathbf{y}) = \bigvee_{\mathbf{x}\in\mathbf{X}\cap S_{-\infty}(\mathbf{t}_{\mathbf{y}})} \mathbf{a}(\mathbf{x}) + \mathbf{t}_{\mathbf{y}}(\mathbf{x}) \right\}$.
$\mathbf{t}\,\boxtimes\,\mathbf{a}$	For $\mathbf{a} \in \mathbb{R}_{\pm\infty}^{\mathbf{X}}$ and $\mathbf{t} \in \left(\mathbb{R}_{\pm\infty}^{\mathbf{Y}}\right)^{\mathbf{X}}$, the *left morphological max convolution product* is defined by $\mathbf{t}\boxtimes\mathbf{a} = \left\{ (\mathbf{y}, \mathbf{b}(\mathbf{y})) : \mathbf{y} \in \mathbf{Y}, \mathbf{b}(\mathbf{y}) = \bigvee_{\mathbf{x}\in\mathbf{X}\cap S_{-\infty}(\mathbf{t}'_{\mathbf{y}})} \mathbf{a}(\mathbf{x}) + \mathbf{t}_{\mathbf{x}}(\mathbf{y}) \right\}$.
$\mathbf{a}\,\boxtimes\,\mathbf{t}$	For $\mathbf{a} \in \mathbb{R}_{\pm\infty}^{\mathbf{X}}$ and $\mathbf{t} \in \left(\mathbb{R}_{\pm\infty}^{\mathbf{X}}\right)^{\mathbf{Y}}$, the *right morphological min convolution product* is defined by $\mathbf{a}\boxtimes\mathbf{t} = \left\{ (\mathbf{y}, \mathbf{b}(\mathbf{y})) : \mathbf{y} \in \mathbf{Y}, \ \mathbf{b}(\mathbf{y}) = \bigwedge_{\mathbf{x}\in\mathbf{X}\cap S_{\infty}(\mathbf{t}_{\mathbf{y}})} \mathbf{a}(\mathbf{x}) +' \mathbf{t}_{\mathbf{y}}(\mathbf{x}) \right\}$.

Symbol	Explanation

$\mathbf{t} \boxed{\wedge} \mathbf{a}$

For $\mathbf{a} \in \mathbb{R}^{\mathbf{X}}_{\pm\infty}$ and $\mathbf{t} \in \left(\mathbb{R}^{\mathbf{Y}}_{\pm\infty}\right)^{\mathbf{X}}$, the *left morphological min convolution product* is defined by

$$\mathbf{t}\boxed{\wedge}\mathbf{a}=\left\{(\mathbf{y},\mathbf{b}(\mathbf{y})) : \mathbf{y}\in\mathbf{Y}, \mathbf{b}(\mathbf{y}) = \bigwedge_{\mathbf{x}\in\mathbf{X}\cap\mathsf{S}_\infty(\mathbf{t}'_\mathbf{y})} \mathbf{a}(\mathbf{x}) +' \mathbf{t}_\mathbf{x}(\mathbf{y})\right\}.$$

$\mathbf{a} \boxed{\vee} \mathbf{t}$

For $\mathbf{a} \in \left(\mathbb{R}^{\geq 0}_{\infty}\right)^{\mathbf{X}}$ and $\mathbf{t} \in \left(\left(\mathbb{R}^{\geq 0}_{\infty}\right)^{\mathbf{X}}\right)^{\mathbf{Y}}$, the *right multiplicative max convolution product* is defined by

$$\mathbf{a}\boxed{\vee}\mathbf{t}=\left\{(\mathbf{y},\mathbf{b}(\mathbf{y})) : \mathbf{y}\in\mathbf{Y}, \mathbf{b}(\mathbf{y}) = \bigvee_{\mathbf{x}\in\mathbf{X}\cap\mathsf{S}(\mathbf{t}_\mathbf{y})} \mathbf{a}(\mathbf{x}) \times \mathbf{t}_\mathbf{y}(\mathbf{x})\right\}.$$

$\mathbf{t} \boxed{\vee} \mathbf{a}$

For $\mathbf{a} \in \left(\mathbb{R}^{\geq 0}_{\infty}\right)^{\mathbf{X}}$ and $\mathbf{t} \in \left(\left(\mathbb{R}^{\geq 0}_{\infty}\right)^{\mathbf{Y}}\right)^{\mathbf{X}}$, the *left multiplicative max convolution product* is defined by

$$\mathbf{t}\boxed{\vee}\mathbf{a}=\left\{(\mathbf{y},\mathbf{b}(\mathbf{y})) : \mathbf{y}\in\mathbf{Y}, \mathbf{b}(\mathbf{y}) = \bigvee_{\mathbf{x}\in\mathbf{X}\cap\mathsf{S}(\mathbf{t}'_\mathbf{y})} \mathbf{a}(\mathbf{x}) \times \mathbf{t}_\mathbf{x}(\mathbf{y})\right\}.$$

$\mathbf{a} \boxed{\wedge} \mathbf{t}$

For $\mathbf{a} \in \left(\mathbb{R}^{\geq 0}_{\infty}\right)^{\mathbf{X}}$ and $\mathbf{t} \in \left(\left(\mathbb{R}^{\geq 0}_{\infty}\right)^{\mathbf{X}}\right)^{\mathbf{Y}}$, the *right multiplicative min convolution product* is defined by

$$\mathbf{a}\boxed{\wedge}\mathbf{t}=\left\{(\mathbf{y},\mathbf{b}(\mathbf{y})) : \mathbf{y}\in\mathbf{Y}, \mathbf{b}(\mathbf{y}) = \bigwedge_{\mathbf{x}\in\mathbf{X}\cap\mathsf{S}_\infty(\mathbf{t}_\mathbf{y})} \mathbf{a}(\mathbf{x}) \times' \mathbf{t}_\mathbf{y}(\mathbf{x})\right\}.$$

$\mathbf{t} \boxed{\wedge} \mathbf{a}$

For $\mathbf{a} \in \left(\mathbb{R}^{\geq 0}_{\infty}\right)^{\mathbf{X}}$ and $\mathbf{t} \in \left(\left(\mathbb{R}^{\geq 0}_{\infty}\right)^{\mathbf{Y}}\right)^{\mathbf{X}}$, the *left multiplicative min convolution product* is defined by

$$\mathbf{t}\boxed{\wedge}\mathbf{a}=\left\{(\mathbf{y},\mathbf{b}(\mathbf{y})) : \mathbf{y}\in\mathbf{Y}, \mathbf{b}(\mathbf{y}) = \bigwedge_{\mathbf{x}\in\mathbf{X}\cap\mathsf{S}_\infty(\mathbf{t}'_\mathbf{y})} \mathbf{a}(\mathbf{x}) \times' \mathbf{t}_\mathbf{x}(\mathbf{y})\right\}.$$

Neighborhoods and Neighborhood Operations

Symbol	Explanation

M, N

Italic uppercase characters are used to denote *neighborhoods*.

$N \in \left(2^{\mathbf{X}}\right)^{\mathbf{Y}}$

A neighborhood is an image whose pixel values are sets of points. In particular, a *neighborhood from* $\mathbf{Y}$ *to* $\mathbf{X}$ is a function $N : \mathbf{Y} \to 2^{\mathbf{X}}$.

$N(p)$

A parameterized neighborhood *from* $\mathbf{Y}$ *to* $\mathbf{X}$ *with parameters in* P is a function of the form $N : P \to \left(2^{\mathbf{X}}\right)^{\mathbf{Y}}$.

N'

Let $N \in \left(2^{\mathbf{X}}\right)^{\mathbf{Y}}$, the *transpose* $N' \in \left(2^{\mathbf{Y}}\right)^{\mathbf{X}}$ is defined as $N'(\mathbf{x}) = \{\mathbf{y} \in \mathbf{Y} : \mathbf{x} \in N(\mathbf{y})\}$, that is, $\mathbf{x} \in N(\mathbf{y})$ iff $\mathbf{y} \in N'(\mathbf{x})$.

$N_1 \oplus N_2$

The *dilation* of N_1 by N_2 is defined by
$$N(\mathbf{y}) = \bigcup_{\mathbf{p}\in N_2(\mathbf{y})} (N_1(\mathbf{y}) + (\mathbf{p} - \mathbf{y})).$$

Image-Neighborhood Operations

In the table below, $\mathbf{X}$ is a finite subset of $\mathbb{R}^n$.

Symbol	Explanation
$\mathbb{F}^{\mathbf{X}}|_N$	If $N \in \left(2^{\mathbf{X}}\right)^{\mathbf{Y}}$, then $\mathbb{F}^{\mathbf{X}}|_N = \left\{\mathbf{a}|_{N(\mathbf{y})} : \mathbf{a} \in \mathbb{F}^{\mathbf{X}}, \mathbf{y} \in \mathbf{Y}\right\}$.
$\mathbf{a} \circledr N$	Given $\mathbf{a} \in \mathbb{F}^{\mathbf{X}}$, $N \in \left(2^{\mathbf{X}}\right)^{\mathbf{Y}}$, and reduce operation $\Gamma : \mathbb{F}^{\mathbf{X}}|_N \to \mathbb{F}$, the *generic right reduction of* $\mathbf{a}$ *with* N is defined as $(\mathbf{a} \circledr N)(\mathbf{y}) = \Gamma \mathbf{a}|_{N(\mathbf{y})} = \Gamma_{\mathbf{x} \in N(\mathbf{y})} \mathbf{a}(\mathbf{x})$. Thus, $\mathbf{a} \circledr N = \left\{(\mathbf{y}, \mathbf{b}(\mathbf{y})) : \mathbf{y} \in \mathbf{Y}, \mathbf{b}(\mathbf{y}) = \Gamma \mathbf{a}|_{N(\mathbf{y})}\right\}$.
$N \circledl \mathbf{a}$	With the conditions above, except that now $N \in \left(2^{\mathbf{X}}\right)^{\mathbf{Y}}$, the *generic left reduction of* $\mathbf{a}$ *with* $\mathbf{t}$ is defined as $(N \circledl \mathbf{a}) = (\mathbf{a} \circledr N')$.
$\mathbf{a} \circleda N$	Given $\mathbf{a} \in \mathbb{R}^{\mathbf{Y}}$, and the *image average* function $a : \mathbb{R}^{\mathbf{Y}}|_N \to \mathbb{R}$, the right reduction of $\mathbf{a}$ with N yields the neighborhood averaging operation, $(\mathbf{a} \circleda N)(\mathbf{y}) = a\left(\mathbf{a}|_{N(\mathbf{y})}\right)$.
$\mathbf{a} \circledm N$	Given $\mathbf{a} \in \mathbb{R}^{\mathbf{Y}}$, and the *image median* function $m : \mathbb{R}^{\mathbf{Y}}|_N \to \mathbb{R}$, the right reduction of $\mathbf{a}$ with N yields the neighborhood median filtered image, $(\mathbf{a} \circledm N)(\mathbf{y}) = m\left(\mathbf{a}|_{N(\mathbf{y})}\right)$.

Matrix and Vector Operations

In the table below, A and B represent matrices.

Symbol	Explanation
A^*	The *conjugate* of matrix A.
A'	The *transpose* of matrix A.
$A \times B, \; AB$	The matrix *product* of matrices A and B.
$A \otimes B$	The *tensor product* of matrices A and B.
$A \oplus_p B$	The *p-product* of matrices A and B.
$A \oplus'_p B$	The dual *p-product* of matrices A and B, defined by $A \oplus'_p B = \left(B' \oplus_p A'\right)'$.

References

[1] G. Ritter, "Image algebra." Unpublished manuscript, available via anonymous ftp from `ftp://ftp.cise.ufl.edu/pub/src/ia/documents`, 1994.

To our brothers,
Friedrich Karl and
Scott Winfield

Contents

CHAPTER 1
IMAGE ALGEBRA

1.1. Introduction

Since the field of image algebra is a recent development it will be instructive to provide some background information. In the broad sense, image algebra is a mathematical theory concerned with the transformation and analysis of images. Although much emphasis is focused on the analysis and transformation of digital images, the main goal is the establishment of a comprehensive and unifying theory of image transformations, image analysis, and image understanding in the discrete as well as the continuous domain [1].

The idea of establishing a unifying theory for the various concepts and operations encountered in image and signal processing is not new. Over thirty years ago, Unger proposed that many algorithms for image processing and image analysis could be implemented in parallel using *cellular array* computers [2]. These cellular array computers were inspired by the work of von Neumann in the 1950s [3, 4]. Realization of von Neumann's cellular array machines was made possible with the advent of VLSI technology. NASA's massively parallel processor or MPP and the CLIP series of computers developed by Duff and his colleagues represent the classic embodiment of von Neumann's original automaton [5, 6, 7, 8, 9]. A more general class of cellular array computers are pyramids and Thinking Machines Corporation's Connection Machines [10, 11, 12]. In an abstract sense, the various versions of Connection Machines are universal cellular automatons with an additional mechanism added for nonlocal communication.

Many operations performed by these cellular array machines can be expressed in terms of simple elementary operations. These elementary operations create a mathematical basis for the theoretical formalism capable of expressing a large number of algorithms for image processing and analysis. In fact, a common thread among designers of parallel image processing architectures is the belief that large classes of image transformations can be described by a small set of standard rules that induce these architectures. This belief led to the creation of mathematical formalisms that were used to aid in the design of special-purpose parallel architectures. Matheron and Serra's Texture Analyzer [13], ERIM's (Environmental Research Institute of Michigan) Cytocomputer [14, 15, 16], Martin Marietta's GAPP [17, 18, 19], and Lockheed Martin's PAL processor [20] are examples of this approach.

The formalism associated with these cellular architectures is that of pixel neighborhood arithmetic and mathematical morphology. Mathematical morphology is the part of image processing concerned with image filtering and analysis by structuring elements. It grew out of the early work of Minkowski and Hadwiger [21, 22, 23], and entered the modern era through the work of Matheron and Serra of the Ecole des Mines in Fontainebleau, France [24, 25, 26, 27]. Matheron and Serra not only formulated the modern concepts of morphological image transformations, but also designed and built the Texture Analyzer System. Since those early days, morphological operations have been applied from low-level, to intermediate, to high-level vision problems. Among some recent research papers on morphological image processing are Crimmins and Brown [28], Haralick et al. [29, 30], Maragos and Schafer [31, 32, 33], Davidson [34, 35, 36], Dougherty [37, 38], Goutsias [39, 40], Koskinen and Astola [41], and Sivakumar and Goutsias [42].

Serra and Sternberg were the first to unify morphological concepts and methods into a coherent algebraic theory specifically designed for image processing and image analysis. Sternberg was also the first to use the term "image algebra" [43, 44]. In the mid 1980s, Maragos introduced a new theory unifying a large class of linear and nonlinear systems under the theory of mathematical morphology [45]. More recently, Davidson completed the mathematical foundation of mathematical morphology by formulating its embedding into the lattice algebra known as *Mini-Max algebra* [46, 47, 48]. However, despite these profound accomplishments, morphological methods have some well-known limitations. For example, such fairly common image processing techniques as feature extraction based on convolution, Fourier-like transformations, chain coding, histogram equalization transforms, image rotation, and image registration and rectification are — with the exception of a few simple cases — either extremely difficult or impossible to express in terms of morphological operations. The failure of a morphologically based image algebra to express a fairly straightforward U.S. government-furnished FLIR (forward-looking infrared) algorithm was demonstrated by Miller of Perkin-Elmer [49].

The failure of an image algebra based solely on morphological operations to provide a universal image processing algebra is due to its set-theoretic formulation, which rests on the Minkowski addition and subtraction of sets [23]. These operations ignore the linear domain, transformations between different domains (spaces of different sizes and dimensionality), and transformations between different value sets (algebraic structures), e.g., sets consisting of real-, complex-, or vector-valued numbers. The image algebra discussed in this text includes these concepts and extends the morphological operations [1].

The development of image algebra grew out of a need, by the U.S. Air Force Systems Command, for a common image-processing language. Defense contractors do not use a standardized, mathematically rigorous and efficient structure that is specifically designed for image manipulation. Documentation by contractors of algorithms for image processing and rationale underlying algorithm design is often accomplished via word description or analogies that are extremely cumbersome and often ambiguous. The result of these *ad hoc* approaches has been a proliferation of nonstandard notation and increased research and development cost. In response to this chaotic situation, the Air Force Armament Laboratory (AFATL — now known as Wright Laboratory MNGA) of the Air Force Systems Command, in conjunction with the Defense Advanced Research Project Agency (DARPA), supported the early development of image algebra with the intent that the fully developed structure would subsequently form the basis of a common image-processing language. The goal of AFATL was the development of a complete, unified algebraic structure that provides a common mathematical environment for image-processing algorithm development, optimization, comparison, coding, and performance evaluation. The development of this structure proved highly successful, capable of fulfilling the tasks set forth by the government, and is now commonly known as image algebra.

Because of the goals set by the government, the theory of image algebra provides for a language which, if properly implemented as a standard image processing environment, can greatly reduce research and development costs. Since the foundation of this language is purely mathematical and independent of any future computer architecture or language, the longevity of an image algebra standard is assured. Furthermore, savings due to commonality of language and increased productivity could dwarf any reasonable initial investment for adapting image algebra as a standard environment for image processing.

Although commonality of language and cost savings are two major reasons for considering image algebra as a standard language for image processing, there exists a multitude of other reasons for desiring the broad acceptance of image algebra as a component of all image processing development systems. Premier among these is the predictable influence of an image algebra standard on future image processing technology. In this, it can be compared to the influence on scientific reasoning and the advancement of science due to the replacement of the myriad of different number systems (e.g., Roman, Syrian, Hebrew, Egyptian, Chinese, etc.)   by the now common Indo-Arabic notation. Additional benefits provided by the use of image algebra are

- The elemental image algebra operations are small in number, translucent, simple, and provide a method of transforming images that is easily learned and used;

- Image algebra operations and operands provide the capability of expressing all image-to-image transformations;

- Theorems governing image algebra make computer programs based on image algebra notation amenable to both machine dependent and machine independent optimization techniques;

- The algebraic notation provides a deeper understanding of image manipulation operations due to conciseness and brevity of code and is capable of suggesting new techniques;

- The notational adaptability to programming languages allows the substitution of extremely short and concise image algebra expressions for equivalent blocks of code, and therefore increases programmer productivity;

- Image algebra provides a rich mathematical structure that can be exploited to relate image processing problems to other mathematical areas;

- Without image algebra, a programmer will never benefit from the bridge that exists between an image algebra programming language and the multitude of mathematical structures, theorems, and identities that are related to image algebra;

- There is no competing notation that adequately provides all these benefits.

The role of image algebra in computer vision and image processing tasks and theory should not be confused with the government's Ada programming language effort. The goal of the development of the Ada programming language was to provide a single high-order language in which to implement embedded systems. The special architectures being developed nowadays for image processing applications are not often capable of directly executing Ada language programs, often due to support of parallel processing models not accommodated by Ada's tasking mechanism. Hence, most applications designed for such processors are still written in special assembly or microcode languages. Image algebra, on the other hand, provides a level of specification, directly derived from the underlying mathematics on which image processing is based and that is compatible with both sequential and parallel architectures.

Enthusiasm for image algebra must be tempered by the knowledge that image algebra, like any other field of mathematics, will never be a finished product but remain a continuously evolving mathematical theory concerned with the unification of image processing and computer vision tasks. Much of the mathematics associated with image algebra and its implication to computer vision remains largely unchartered territory which awaits discovery. For example, very little work has been done in relating image algebra to computer vision techniques which employ tools from such diverse areas as knowledge representation, graph theory, and surface representation.

Several image algebra programming languages have been developed. These include image algebra Fortran (IAF) [50], an image algebra Ada (IAA) translator [51], image algebra Connection Machine *Lisp [52, 53], an image algebra language (IAL) implementation on transputers [54, 55], and an image algebra C++ class library (iac++) [56, 57]. Unfortunately, there is often a tendency among engineers to confuse or equate these languages with image algebra. An image algebra programming language is *not* image algebra, which is a mathematical theory. An image algebra-based programming language typically implements a particular subalgebra of the full image algebra. In addition, simplistic implementations can result in poor computational performance. Restrictions and limitations in implementation are usually due to a combination of factors, the most pertinent being development costs and hardware and software environment constraints. They are not limitations of image algebra, and they should not be confused with the capability of image algebra as a mathematical tool for image manipulation.

Image algebra is a *heterogeneous* or *many-valued algebra* in the sense of Birkhoff and Lipson [58, 1], with multiple sets of operands and operators. Manipulation of images for purposes of image enhancement, analysis, and understanding involves operations not only on images, but also on different types of values and quantities associated with these images. Thus, the basic operands of image algebra are images and the values and quantities associated with these images. Roughly speaking, an image consists of two things, a collection of *points* and a set of *values* associated with these points. Images are therefore endowed with two types of information, namely the spatial relationship of the points, and also some type of numeric or other descriptive information associated with these points. Consequently, the field of image algebra bridges two broad mathematical areas, the theory of point sets and the algebra of value sets, and investigates their interrelationship. In the sections that follow we discuss point and value sets as well as images, templates, and neighborhoods that characterize some of their interrelationships.

1.2.   Point Sets

A *point set* is simply a topological space. Thus, a point set consists of two things, a collection of objects called *points* and a topology which provides for such notions as *nearness* of two points, the *connectivity* of a subset of the point set, the *neighborhood* of a point, *boundary points*, and *curves* and *arcs*. Point sets are typically denoted by capital bold letters from the end of the alphabet, i.e., $\mathbf{W}$, $\mathbf{X}$, $\mathbf{Y}$, and $\mathbf{Z}$.

Points (elements of point sets) are typically denoted by lower case bold letters from the end of the alphabet, namely $\mathbf{x}$, $\mathbf{y}$, $\mathbf{z} \in \mathbf{X}$. Note also that if $\mathbf{x} \in \mathbb{R}^n$, then $\mathbf{x}$ is of form $\mathbf{x} = (x_1, x_2, \ldots, x_n)$, where for each $i = 1, 2, \ldots, n$, x_i denotes a real number called the *ith coordinate* of $\mathbf{x}$.

The most common point sets occurring in image processing are discrete subsets of n–dimensional Euclidean space $\mathbb{R}^n$ with $n = 1$, 2, or 3 together with the discrete topology. However, other topologies such as the *von Neumann topology* and the *odd-even product topology* are also commonly used topologies in computer vision [1].

There is no restriction on the *shape* of the discrete subsets of $\mathbb{R}^n$ used in applications of image algebra to solve vision problems. Point sets can assume arbitrary shapes. In particular, shapes can be rectangular, circular, or snake-like. Some of the more pertinent point sets are the set of integer points $\mathbb{Z}$ (here we view $\mathbb{Z} \subset \mathbb{R}^1$), the n–dimensional lattice $\mathbb{Z}^n \subset \mathbb{R}^n$ (i.e., $\mathbb{Z}^n = \mathbb{Z} \times \mathbb{Z} \times \cdots \times \mathbb{Z} = \{\mathbf{x} \in \mathbb{R}^\mathbf{n} : \mathbf{x} = (x_1, \ldots, x_n), x_i \in \mathbb{Z} \text{ for } i = 1, \ldots, n\}$) with $n = 2$ or $n = 3$, and rectangular subsets of $\mathbb{Z}^2$. Two of the most often encountered rectangular point sets are of form

$$\mathbf{X} = \mathbb{Z}_m \times \mathbb{Z}_n = \left\{ (x_1, x_2) \in \mathbb{Z}^2 \, : \, 0 \leq x_1 \leq m - 1, \, 0 \leq x_2 \leq n - 1 \right\},$$

or

$$\mathbf{X} = \mathbb{Z}_m^+ \times \mathbb{Z}_n^+ = \left\{ (x_1, x_2) \in \mathbb{Z}^2 \, : \, 1 \leq x_1 \leq m, \, 1 \leq x_2 \leq n \right\}.$$

We follow standard practice and represent these rectangular point sets by listing the points in matrix form. Figure 1.2.1 provides a graphical representation of the point set $\mathbf{X} = \mathbb{Z}_m^+ \times \mathbb{Z}_n^+$.

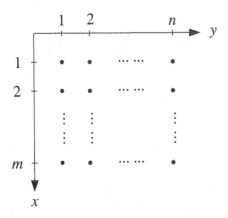

Figure 1.2.1. The rectangular point set $\mathbf{X} = \mathbb{Z}_m^+ \times \mathbb{Z}_n^+$.

Point Operations

As mentioned, some of the more pertinent point sets are discrete subsets of the vector space $\mathbb{R}^n$. These point sets inherit the usual elementary vector space operations. Thus, for example, if $\mathbf{X} \subset \mathbb{Z}^n$ (or $\mathbf{X} \subset \mathbb{R}^n$) and $\mathbf{x} = (x_1, \ldots, x_n)$, $\mathbf{y} = (y_1, \ldots, y_n) \in \mathbf{X}$, then the sum of the points $\mathbf{x}$ and $\mathbf{y}$ is defined as

$$\mathbf{x} + \mathbf{y} = (x_1 + y_1, \ldots, x_n + y_n),$$

while the multiplication and addition of a scalar $k \in \mathbb{Z}$ (or $k \in \mathbb{R}$) and a point $\mathbf{x}$ is given by

$$k \cdot \mathbf{x} = (k \cdot x_1, \ldots, k \cdot x_n)$$

and

$$k + \mathbf{x} = (k + x_1, \ldots, k + x_n),$$

respectively. Point subtraction is also defined in the usual way.

In addition to these standard vector space operations, image algebra also incorporates three basic types of point multiplication. These are the *Hadamard product*, the *cross product* (or *vector product*) for points in $\mathbb{Z}^3$ (or $\mathbb{R}^3$), and the *dot product* which are defined by

$$\mathbf{x} \cdot \mathbf{y} = (x_1 \cdot y_1, \ldots, x_n \cdot y_n),$$
$$\mathbf{x} \times \mathbf{y} = (x_2 \cdot y_3 - x_3 \cdot y_2, \, x_3 \cdot y_1 - x_1 \cdot y_3, \, x_1 \cdot y_2 - x_2 \cdot y_1),$$

and

$$\mathbf{x} \bullet \mathbf{y} = x_1 \cdot y_1 + x_2 \cdot y_2 + \cdots + x_n \cdot y_n,$$

respectively.

Note that the sum of two points, the Hadamard product, and the cross product are binary operations that take as input two points and produce another point. Therefore, these operations can be viewed as mappings $\mathbf{X} \times \mathbf{X} \to \mathbf{X}$ whenever $\mathbf{X}$ is closed under these operations. In contrast, the binary operation of dot product is a scalar and not another vector. This provides an example of a mapping $\mathbf{X} \times \mathbf{X} \to \mathbb{F}$, where $\mathbb{F}$ denotes the appropriate field of scalars. Another such mapping, associated with metric spaces, is the distance function $\mathbf{X} \times \mathbf{X} \to \mathbb{R}$ which assigns to each pair of points $\mathbf{x}$ and $\mathbf{y}$ the *distance* from $\mathbf{x}$ to $\mathbf{y}$. The most common distance functions occurring in image processing are the *Euclidean* distance, the *city block* or *diamond* distance, and the *chessboard* distance which are defined by

$$d(\mathbf{x}, \mathbf{y}) = \left[\sum_{k=1}^{n} (x_k - y_k)^2 \right]^{\frac{1}{2}} ,$$

$$\rho(\mathbf{x}, \mathbf{y}) = \sum_{k=1}^{n} |x_k - y_k| ,$$

and

$$\delta(\mathbf{x}, \mathbf{y}) = max\{|x_k - y_k| : 1 \le k \le n\} ,$$

respectively.

Distances can be conveniently computed in terms of the *norm* of a point. The three norms of interest here are derived from the standard L^p norms

$$\|\mathbf{x}\|_p = \left(\sum_{i=1}^{n} |x_i|^p \right)^{1/p} .$$

The L^∞ norm is given by

$$\|\mathbf{x}\|_\infty = \bigvee_{i=1}^{n} |x_i| ,$$

where $\bigvee_{i=1}^{n} |x_i| = max\{|x_1|, \ldots, |x_n|\}$. Specifically, the *Euclidean norm* is given by $\|\mathbf{x}\|_2 = \sqrt{x_1^2 + \cdots + x_n^2}$. Thus, $d(\mathbf{x}, \mathbf{y}) = \|\mathbf{x} - \mathbf{y}\|_2$. Similarly, the city block distance can be computed using the formulation $\rho(\mathbf{x}, \mathbf{y}) = \|\mathbf{x} - \mathbf{y}\|_1$ and the chessboard distance by using $\delta(\mathbf{x}, \mathbf{y}) = \|\mathbf{x} - \mathbf{y}\|_\infty$.

Note that the p-norm of a point $\mathbf{x}$ is a *unary* operation, namely a function $\| \ \|_p : \mathbf{X} \to \mathbb{R}$. Another assemblage of functions $\mathbf{X} \to \mathbb{R}$ which play a major role in various applications are the projection functions. Given $\mathbf{X} \subset \mathbb{R}^n$, then the *$i$th projection* on $\mathbf{X}$, where $i \in \{1, \ldots, n\}$, is denoted by p_i and defined by $p_i(\mathbf{x}) = x_i$, where x_i denotes the ith coordinate of $\mathbf{x}$.

Characteristic functions and neighborhood functions are two of the most frequently occurring unary operations in image processing. In order to define these operations, we need to recall the notion of a *power set* of a set. The power set of a set S is defined as the set of all subsets of S and is denoted by 2^S. Thus, if $\mathbf{Z}$ is a point set, then $2^{\mathbf{Z}} = \{\mathbf{X} : \mathbf{X} \subset \mathbf{Z}\}$.

Given $\mathbf{X} \in 2^{\mathbf{Z}}$ (i.e., $\mathbf{X} \subset \mathbf{Z}$), then the *characteristic function* associated with $\mathbf{X}$ is the function

$$\chi_{\mathbf{x}} : \mathbf{Z} \to \{0, 1\}$$

defined by

$$\chi_{\mathbf{x}}(\mathbf{z}) = \begin{cases} 1 & if \ \mathbf{z} \in \mathbf{X} \\ 0 & if \ \mathbf{z} \notin \mathbf{X} . \end{cases}$$

For a pair of point sets $\mathbf{X}$ and $\mathbf{Z}$, a *neighborhood system for* $\mathbf{X}$ *in* $\mathbf{Z}$, or equivalently, a *neighborhood function from* $\mathbf{X}$ *to* $\mathbf{Z}$, is a function

$$N : \mathbf{X} \rightarrow 2^{\mathbf{Z}} .$$

It follows that for each point $\mathbf{x} \in \mathbf{X}$, $N(\mathbf{x}) \subset \mathbf{Z}$. The set $N(\mathbf{x})$ is called a *neighborhood for* $\mathbf{x}$.

There are two neighborhood functions on subsets of $\mathbb{Z}^2$ which are of particular importance in image processing. These are the *von Neumann* neighborhood and the *Moore* neighborhood. The von Neumann neighborhood $N : \mathbf{X} \rightarrow 2^{\mathbb{Z}^2}$ is defined by

$$N(\mathbf{x}) = \{\mathbf{y} \ : \ \mathbf{y} = (x_1 \pm j, \ x_2) \ or \ \mathbf{y} = (x_1, \ x_2 \pm k), \ j, \ k \in \{0, 1\} \},$$

where $\mathbf{x} = (x_1, \ x_2) \in \mathbf{X} \subset \mathbb{Z}^2$, while the Moore neighborhood $M : \mathbf{X} \rightarrow 2^{\mathbb{Z}^2}$ is defined by

$$M(\mathbf{x}) = \{\mathbf{y} \ : \ \mathbf{y} = (x_1 \pm j, \ x_2 \pm k), \ j, \ k \in \{0, 1\} \}.$$

Figure 1.2.2 provides a pictorial representation of these two neighborhood functions; the hashed center area represents the point $\mathbf{x}$ and the adjacent cells represent the adjacent points. The von Neumann and Moore neighborhoods are also called the *four neighborhood* and *eight neighborhood*, respectively. They are *local neighborhoods* since they only include the directly adjacent points of a given point.

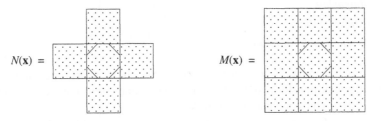

Figure 1.2.2. The von Neumann neighborhood $N(\mathbf{x})$
and the Moore neighborhood $M(\mathbf{x})$ of a point $\mathbf{x}$.

There are many other point operations that are useful in expressing computer vision algorithms in succinct algebraic form. For instance, in certain interpolation schemes it becomes necessary to switch from points with real-valued coordinates (floating point coordinates) to corresponding integer-valued coordinate points. One such method uses the induced *floor* operation $\lfloor \ \rfloor : \mathbb{R}^n \rightarrow \mathbb{Z}^n$ defined by $\lfloor \mathbf{x} \rfloor = (\lfloor x_1 \rfloor, \lfloor x_2 \rfloor, \ldots, \lfloor x_n \rfloor)$, where $\mathbf{x} = (x_1, x_2, \ldots, x_n) \in \mathbb{R}^n$ and $\lfloor x_i \rfloor \in \mathbb{Z}$ denotes the largest integer *less than or equal to* x_i (i.e., $\lfloor x_i \rfloor \leq x_i$ and if $k \in \mathbb{Z}$ with $k \leq x_i$, then $k \leq \lfloor x_i \rfloor$).

Summary of Point Operations

We summarize some of the more pertinent point operations. Some image algebra implementations such as iac++ provide many additional point operations [59].

Binary operations.   Let $\mathbf{x} = (x_1, x_2, \ldots, x_n)$, $\mathbf{y} = (y_1, y_2, \ldots, y_n) \in \mathbb{R}^n$, and $\mathbf{z} = (z_1, z_2, \ldots, z_m) \in \mathbb{R}^m$.

addition	$\mathbf{x} + \mathbf{y} = (x_1 + y_1, \ldots, x_n + y_n)$
subtraction	$\mathbf{x} - \mathbf{y} = (x_1 - y_1, \ldots, x_n - y_n)$
multiplication	$\mathbf{x} \cdot \mathbf{y} = (x_1 y_1, \ldots, x_n y_n)$
division	$\mathbf{x}/\mathbf{y} = (x_1/y_1, \ldots, x_n/y_n)$
supremum	$sup(\mathbf{x}, \mathbf{y}) = (x_1 \vee y_1, \ldots, x_n \vee y_n)$
infimum	$inf(\mathbf{x}, \mathbf{y}) = (x_1 \wedge y_1, \ldots, x_n \wedge y_n)$
dot product	$\mathbf{x}\bullet\mathbf{y} = x_1 y_1 + x_2 y_2 + \cdots + x_n y_n$
cross product $(n = 3)$	$\mathbf{x} \times \mathbf{y} = (x_2 y_3 - x_3 y_2, x_3 y_1 - x_1 y_3, x_1 y_2 - x_2 y_1)$
concatenation	$\mathbf{x}\hat{\,}\mathbf{z} = (x_1, \ldots, x_n, z_1, \ldots, z_m)$
scalar operations	$k\gamma\mathbf{x} = (k\gamma x_1, \ldots, k\gamma x_n)$,
	where $\gamma \in \{+, -, *, \vee, \wedge\}$

Unary operations.   In the following let $\mathbf{x} = (x_1, x_2, \ldots, x_n) \in \mathbb{R}^n$.

negation	$-\mathbf{x} = (-x_1, \ldots, -x_n)$
ceiling	$\lceil \mathbf{x} \rceil = (\lceil x_1 \rceil, \ldots, \lceil x_n \rceil)$
floor	$\lfloor \mathbf{x} \rfloor = (\lfloor x_1 \rfloor, \ldots, \lfloor x_n \rfloor)$
rounding	$[\mathbf{x}] = ([x_1], \ldots, [x_n])$
projection	$p_i(\mathbf{x}) = x_i$
sum	$\Sigma\mathbf{x} = x_1 + x_2 + \cdots + x_n$
product	$\Pi\mathbf{x} = x_1 x_2 \cdots x_n$
maximum	$\vee\mathbf{x} = x_1 \vee x_2 \vee \cdots \vee x_n$
minimum	$\wedge\mathbf{x} = x_1 \wedge x_2 \wedge \cdots \wedge x_n$
Euclidean norm	$\|\mathbf{x}\|_2 = \sqrt{x_1^2 + \cdots + x_n^2}$
L^1 *norm*	$\|\mathbf{x}\|_1 = |x_1| + |x_2| + \cdots + |x_n|$
L^∞ *norm*	$\|\mathbf{x}\|_\infty = |x_1| \vee |x_2| \vee \cdots \vee |x_n|$
dimension	$dim(\mathbf{x}) = n$
neighborhood	$N(\mathbf{x}) \subset \mathbb{R}^n$
characteristic function	$\chi_{\mathbf{x}}(\mathbf{z}) = \begin{cases} 1 & if \ \mathbf{z} \in \mathbf{X} \\ 0 & if \ \mathbf{z} \notin \mathbf{X} \end{cases}$

It is important to note that several of the above unary operations are special instances of spatial transformations $\mathbf{X} \rightarrow \mathbf{Y}$. Spatial transforms play a vital role in many image processing and computer vision tasks.

In the above summary we only considered points with real- or integer-valued coordinates. Points of other spaces have their own induced operations. For example, typical operations on points of $\mathbf{X} = (\mathbb{Z}_2)^n$ (i.e., Boolean-valued points) are the usual logical operations of AND, OR, XOR, and complementation.

Point Set Operations

Point arithmetic leads in a natural way to the notion of set arithmetic. Given a vector space $\mathbf{Z}$, then for $\mathbf{X}, \mathbf{Y} \in 2^{\mathbf{Z}}$ (i.e., $\mathbf{X}, \mathbf{Y} \subset \mathbf{Z}$) and an arbitrary point $\mathbf{p} \in \mathbf{Z}$ we define the following arithmetic operations:

addition	$\mathbf{X} + \mathbf{Y} = \{\mathbf{x} + \mathbf{y} : \mathbf{x} \in \mathbf{X} \ and \ \mathbf{y} \in \mathbf{Y}\}$
subtraction	$\mathbf{X} - \mathbf{Y} = \{\mathbf{x} - \mathbf{y} : \mathbf{x} \in \mathbf{X} \ and \ \mathbf{y} \in \mathbf{Y}\}$
point addition	$\mathbf{X} + \mathbf{p} = \{\mathbf{x} + \mathbf{p} : \mathbf{x} \in \mathbf{X}\}$
point subtraction	$\mathbf{X} - \mathbf{p} = \{\mathbf{x} - \mathbf{p} : \mathbf{x} \in \mathbf{X}\}$

Another set of operations on $2^{\mathbf{Z}}$ are the usual set operations of *union*, *intersection*, *set difference* (or *relative complement*), *symmetric difference*, and *Cartesian product* as defined below.

union	$\mathbf{X} \cup \mathbf{Y} = \{\mathbf{z} \, : \, \mathbf{z} \in \mathbf{X} \; or \; \mathbf{z} \in \mathbf{Y}\}$
intersection	$\mathbf{X} \cap \mathbf{Y} = \{\mathbf{z} \, : \, \mathbf{z} \in \mathbf{X} \; and \; \mathbf{z} \in \mathbf{Y}\}$
set difference	$\mathbf{X} \backslash \mathbf{Y} = \{\mathbf{z} \, : \, \mathbf{z} \in \mathbf{X} \; and \; \mathbf{z} \notin \mathbf{Y}\}$
symmetric difference	$\mathbf{X} \triangle \mathbf{Y} = \{\mathbf{z} \, : \, \mathbf{z} \in \mathbf{X} \cup \mathbf{Y} \; and \; \mathbf{z} \notin \mathbf{X} \cap \mathbf{Y}\}$
Cartesian product	$\mathbf{X} \times \mathbf{Y} = \{(\mathbf{x}, \mathbf{y}) \, : \, \mathbf{x} \in \mathbf{X} \; and \; \mathbf{y} \in \mathbf{Y}\}$

Note that with the exception of the Cartesian product, the set obtained for each of the above operations is again an element of $2^{\mathbf{Z}}$.

Another common set theoretic operation is *set complementation*. For $\mathbf{X} \in 2^{\mathbf{Z}}$, the *complement of* $\mathbf{X}$ is denoted by $\tilde{\mathbf{X}}$, and defined as $\tilde{\mathbf{X}} = \{\mathbf{z} \, : \, \mathbf{z} \in \mathbf{Z} \; and \; \mathbf{z} \notin \mathbf{X}\}$. In contrast to the binary set operations defined above, set complementation is a unary operation. However, complementation can be computed in terms of the binary operation of set difference by observing that $\tilde{\mathbf{X}} = \mathbf{Z} \backslash \mathbf{X}$.

In addition to complementation there are various other common unary operations which play a major role in algorithm development using image algebra. Among these is the *cardinality* of a set which, when applied to a finite point set, yields the number of elements in the set, and the *choice* function which, when applied to a set, selects a randomly chosen point from the set. The cardinality of a set $\mathbf{X}$ will be denoted by $card(\mathbf{X})$. Note that

$$card : 2^{\mathbf{Z}} \to \mathbb{N} \; \left(\text{for all finite elements of } 2^{\mathbf{Z}}\right),$$

while

$$choice : 2^{\mathbf{Z}} \to \mathbf{Z} \, .$$

That is, $card(\mathbf{X}) \in \mathbb{N}$ and $choice(\mathbf{X}) = \mathbf{x}$, where $\mathbf{x}$ is some randomly chosen element of $\mathbf{X}$.

As was the case for operations on points, algebraic operations on point sets are too numerous to discuss at length in a short treatise as this. Therefore, we again only summarize some of the more frequently occurring unary operations.

Summary of Unary Point Set Operations

In the following $\mathbf{X} \subset \mathbb{R}^n$.

negation	$-\mathbf{X} = \{-\mathbf{x} \, : \, \mathbf{x} \in \mathbf{X}\}$
complementation	$\tilde{\mathbf{X}} = \{\mathbf{z} \, : \, \mathbf{z} \in \mathbb{R}^n \; and \; \mathbf{z} \notin \mathbf{X}\}$
supremum	$sup(\mathbf{X}) \; (for \; finite \; point \; set \; \mathbf{X})$
infimum	$inf(\mathbf{X}) \; (for \; finite \; point \; set \; \mathbf{X})$
choice function	$choice(\mathbf{X}) \in \mathbf{X} \; (randomly \; chosen \; element)$
cardinality	$card(\mathbf{X}) = the \; cardinality \; of \; \mathbf{X}$

The interpretation of $sup(\mathbf{X})$ is as follows. Suppose $\mathbf{X}$ is finite, say $\mathbf{X} = \{\mathbf{x}_1, \mathbf{x}_2, \ldots, \mathbf{x}_k\}$. Then $sup(\mathbf{X}) = sup(\ldots sup(sup(sup(\mathbf{x}_1, \mathbf{x}_2), \mathbf{x}_3), \mathbf{x}_4), \ldots, \mathbf{x}_k)$, where $sup(\mathbf{x}_i, \mathbf{x}_j)$ denotes the binary operation of the supremum of two points defined earlier. For example, if $\mathbf{x}_i = (x_i, y_i)$ for $i = 1, \ldots, k$, then $sup(\mathbf{X}) = (x_1 \vee x_2 \vee \cdots \vee x_k, y_1 \vee y_2 \vee \cdots \vee y_k)$. More generally, $sup(\mathbf{X})$ is defined to be the *least upper bound* of $\mathbf{X}$ (if it exists). The infimum of $\mathbf{X}$ is interpreted in a similar fashion.

If $\mathbf{X}$ is finite and has a total order, then we also define the *maximum* and *minimum* of $\mathbf{X}$, denoted by $\bigvee \mathbf{X}$ and $\bigwedge \mathbf{X}$, respectively, as follows. Suppose $\mathbf{X} = \{\mathbf{x}_1, \mathbf{x}_2, \ldots, \mathbf{x}_k\}$ and $\mathbf{x}_1 \prec \mathbf{x}_2 \prec \cdots \prec \mathbf{x}_k$, where the symbol $\prec$ denotes the particular total order on $\mathbf{X}$.

Then $\bigvee \mathbf{X} = \mathbf{x}_k$ and $\bigwedge \mathbf{X} = \mathbf{x}_1$. The most commonly used order for a subset $\mathbf{X}$ of $\mathbb{Z}^2$ is the row scanning order. Note also that in contrast to the supremum or infimum, the maximum and minimum of a (finite totally ordered) set is always a member of the set.

1.3. Value Sets

A *heterogeneous algebra* is a collection of nonempty sets of possibly different types of elements together with a set of finitary operations which provide the rules of combining various elements in order to form a new element. For a precise definition of a heterogeneous algebra we refer the reader to Ritter [1]. Note that the collection of point sets, points, and scalars together with the operations described in the previous section form a heterogeneous algebra.

A *homogeneous algebra* is a heterogeneous algebra with only one set of operands. In other words, a homogeneous algebra is simply a set together with a finite number of operations. Homogeneous algebras will be referred to as *value sets* and will be denoted by capital blackboard font letters, e.g., $\mathbb{E}$, $\mathbb{F}$, and $\mathbb{G}$. There are several value sets that occur more often than others in digital image processing. These are the set of integers, real numbers (floating point numbers), the complex numbers, binary numbers of fixed length k, the extended real numbers (which include the symbols $+\infty$ and/or $-\infty$), and the extended non–negative real numbers. We denote these sets by $\mathbb{Z}$, $\mathbb{R}$, $\mathbb{C}$, $\mathbb{Z}_{2^k}$, $\mathbb{R}_{+\infty} = \mathbb{R} \cup \{+\infty\}$, $\mathbb{R}_{-\infty} = \mathbb{R} \cup \{-\infty\}$, $\mathbb{R}_{\pm\infty} = \mathbb{R} \cup \{+\infty, -\infty\}$, and $\mathbb{R}_{\infty}^{\geq 0} = \mathbb{R}^+ \cup \{0, +\infty\}$, respectively, where the symbol $\mathbb{R}^+$ denotes the set of positive real numbers.

Operations on Value Sets

The operations on and between elements of a given value set $\mathbb{F}$ are the usual elementary operations associated with $\mathbb{F}$. Thus, if $\mathbb{F} \in \{\mathbb{Z}, \mathbb{R}, \mathbb{Z}_{2^k}\}$, then the binary operations are the usual arithmetic and logic operations of addition, multiplication, and maximum, and the complementary operations of subtraction, division, and minimum. If $\mathbb{F} = \mathbb{C}$, then the binary operations are addition, subtraction, multiplication, and division. Similarly, we allow the usual elementary unary operations associated with these sets such as the absolute value, conjugation, as well as trigonometric, logarithmic and exponential functions as these are available in all higher-level scientific programming languages.

For the set $\mathbb{R}_{\pm\infty}$ we need to extend the arithmetic and logic operations of $\mathbb{R}$ as follows:

$$a + (-\infty) = (-\infty) + a = -\infty \qquad\qquad a \in \mathbb{R}_{-\infty}$$

$$a + \infty = \infty + a = \infty \qquad\qquad a \in \mathbb{R}_{\infty}$$

$$(-\infty) + \infty = \infty + (-\infty) = -\infty$$

$$a \vee (-\infty) = (-\infty) \vee a = a \qquad\qquad a \in \mathbb{R}_{\pm\infty}$$

Note that the element $-\infty$ acts as a null element in the system $(\mathbb{R}_{\pm\infty}, \vee, +)$ if we view the operation $+$ as multiplication and the operation $\vee$ as addition. The same cannot be said about the element ∞ in the system $(\mathbb{R}_{\pm\infty}, \wedge, +)$ since $(-\infty) + \infty = \infty + (-\infty) = -\infty$. In order to remedy this situation we define the *dual* structure $(\mathbb{R}_{\pm\infty}, \wedge, +')$ of $(\mathbb{R}_{\pm\infty}, \vee, +)$ as follows:

$$a +' b = a + b \qquad\qquad a, b \in \mathbb{R}$$

$$a +' (-\infty) = (-\infty) +' a = -\infty \qquad\qquad a \in \mathbb{R}_{-\infty}$$

$$a +' \infty = \infty +' a = \infty \qquad\qquad a \in \mathbb{R}_{\infty}$$

$$(-\infty) +' \infty = \infty +' (-\infty) = \infty$$

$$a \wedge \infty = \infty \wedge a = a \qquad\qquad a \in \mathbb{R}_{\pm\infty}$$

Now the element $+\infty$ acts as a null element in the system $(\mathbb{R}_{+\infty}, \wedge, +')$. Observe, however, that the dual additions $+$ and $+'$ introduce an asymmetry between $-\infty$ and $+\infty$. The resultant structure $(\mathbb{R}_{\pm\infty}, \vee, \wedge, +, +')$ is known as a *bounded lattice ordered group* [1].

Dual structures provide for the notion of dual elements. For each $r \in \mathbb{R}_{\pm\infty}$ we define its *dual* or *conjugate* r^* by $r^* = -r$, where $-(-\infty) = \infty$. The following duality laws are a direct consequence of this definition:

(1) $\qquad\qquad (r^*)^* = r$

(2) $\qquad\qquad (r \wedge t)^* = r^* \vee t^*$ and $(r \vee t)^* = r^* \wedge t^*$.

Closely related to the *additive* bounded lattice ordered group described above is the *multiplicative* bounded lattice ordered group $(\mathbb{R}_\infty^{\geq 0}, \vee, \wedge, \times, \times')$. Here the dual $\times'$ of ordinary multiplication is defined as

$$a \times' b = a \times b \ \forall \, a, b \in \mathbb{R}^{\geq 0} = \mathbb{R}^+ \cup \{0\}$$

with both multiplicative operations extended as follows:

$$a \times \infty = \infty \times a = \infty \qquad\qquad a \in \mathbb{R}_\infty^+$$
$$a \times' \infty = \infty \times' a = \infty \qquad\qquad a \in \mathbb{R}_\infty^+$$
$$0 \times \infty = \infty \times 0 = 0$$
$$0 \times' \infty = \infty \times' 0 = \infty$$

Hence, the element 0 acts as a null element in the system $(\mathbb{R}_\infty^{\geq 0}, \vee, \times)$ and the element $+\infty$ acts as a null element in the system $(\mathbb{R}_\infty^{\geq 0}, \wedge, \times')$. The conjugate r^* of an element $r \in \mathbb{R}_\infty^{\geq 0}$ of this value set is defined by

$$r^* \equiv \begin{cases} r^{-1} & \text{if } r \in \mathbb{R}^+ \\ 0 & \text{if } r = +\infty \\ +\infty & \text{if } r = 0 \end{cases}.$$

Another algebraic structure with duality which is of interest in image algebra is the value set $\left(\mathbb{Z}_2^*, \vee, \wedge, \tilde{+}, \tilde{+}'\right)$, where $\mathbb{Z}_2^* = (\mathbb{Z}_2)_{\pm\infty} = \mathbb{Z}_2 \cup \{\infty, -\infty\} = \{0, 1, -\infty, \infty\}$. The logical operations $\vee$ and $\wedge$ are the usual binary operations of max (or) and min (and), respectively, while the dual additive operations $\tilde{+}$ and $\tilde{+}'$ are defined by the tables shown in Figure 1.3.1.

$\tilde{+}$	0	1	$-\infty$	∞
0	1	0	∞	$-\infty$
1	0	1	∞	$-\infty$
$-\infty$	∞	∞	∞	∞
∞	$-\infty$	$-\infty$	∞	$-\infty$

$\tilde{+}'$	0	1	$-\infty$	∞
0	1	0	∞	$-\infty$
1	0	1	∞	$-\infty$
$-\infty$	∞	∞	∞	$-\infty$
∞	$-\infty$	$-\infty$	$-\infty$	∞

Figure 1.3.1. The dual additive operations $\tilde{+}$ and $\tilde{+}'$.

Note that the addition $\tilde{+}$ (as well as $\tilde{+}'$) restricted to $\mathbb{Z}_2 = \{0, 1\}$ is the complement of the exclusive-or operation, *xor*, and computes the values for the truth table of the biconditional statement $p \leftrightarrow q$ (i.e., p if and only if q).

The operations on the value set $\mathbb{Z}_2^*$ can be easily generalized to its k-fold Cartesian product $\mathbb{Z}_{2^k}^* = \mathbb{Z}_2^* \times \mathbb{Z}_2^* \times \cdots \times \mathbb{Z}_2^*$. Specifically, if $m = (m_1, \ldots, m_k) \in \mathbb{Z}_{2^k}^*$ and $n = (n_1, \ldots, n_k) \in \mathbb{Z}_{2^k}^*$, where $m_i, n_i \in \mathbb{Z}_2^*$ for $i = 1, \ldots, k$, then $m \tilde{+} n = (m_1 \tilde{+} n_1, \ldots, m_k \tilde{+} n_k)$.

The addition $\tilde{+}$ should not be confused with the usual addition $mod2^k$ on $\mathbb{Z}_{2^k}$. In fact, for $m, n \in \mathbb{Z}_{2^k}$ $m \tilde{+} n = ((m_1 + n_1)', \ldots, (m_k + n_k)')$, where

$$(m_i + n_i)' = \begin{cases} 0 & if \ (m_i + n_i)mod2 = 1 \\ 1 & if \ (m_i + n_i)mod2 = 0 \, . \end{cases}$$

Many point sets are also value sets. For example, the point set $\mathbf{X} = \mathbb{R}^n$ is a metric space as well as a vector space with the usual operation of vector addition. Thus, $(\mathbb{R}^n, +)$, where the symbol "+" denotes vector addition, will at various times be used both as a point set and as a value set. Confusion as to usage will not arise as usage should be clear from the discussion.

Summary of Pertinent Numeric Value Sets

In order to focus attention on the value sets most often used in this treatise we provide a listing of their algebraic structures:

(a) $(\mathbb{R}, \vee, \wedge, +, \cdot)$

(b) $(\mathbb{C}, +, \cdot)$

(c) $(\mathbb{Z}, \vee, \wedge, +, \cdot)$

(d) $(\mathbb{Z}_{2^k}, \vee, \wedge, +, \cdot)$

(e) $(\mathbb{R}_{\pm\infty}, \vee, \wedge, +, +')$

(f) $(\mathbb{R}_\infty^{\geq 0}, \vee, \wedge, \times, \times')$

(g) $\left(\mathbb{Z}_2^*, \vee, \wedge, \tilde{+}, \tilde{+}'\right)$

In contrast to structure c, the addition and multiplication in structure d is addition and multiplication $mod2^k$.

These listed structures represent the pertinent global structures. In various applications only certain subalgebras of these algebras are used. For example, the subalgebras $(\mathbb{R}_{-\infty}, \vee, +)$ and $(\mathbb{R}_{+\infty}, \wedge, +')$ of $(\mathbb{R}_{\pm\infty}, \vee, \wedge, +, +')$ play special roles in morphological processing. Similarly, the subalgebra $(\mathbb{N}, \vee, \wedge, +)$ of $(\mathbb{Z}, \vee, \wedge, +, \cdot)$, where $\mathbb{N} = \{0, 1, 2, \ldots, n, \ldots\}$, is the only pertinent applicable algebra in certain cases.

The complementary binary operations, whenever they exist, are assumed to be part of the structures. Thus, for example, subtraction and division which can be defined in terms of addition and multiplication, respectively, are assumed to be part of $(\mathbb{R}, \vee, \wedge, +, \cdot)$.

Value Set Operators

As for point sets, given a value set $\mathbb{F}$, the operations on $2^\mathbb{F}$ are again the usual operations of union, intersection, set difference, etc. If, in addition, $\mathbb{F}$ is a lattice, then the operations of infimum and supremum are also included. A brief summary of value set operators is given below.

For the following operations assume that $A, B \in 2^{\mathsf{F}}$ for some value set F.

union	$A \cup B = \{c : c \in A \ or \ c \in B\}$
intersection	$A \cap B = \{c : c \in A \ and \ c \in B\}$
set difference	$A \backslash B = \{c : c \in A \ and \ c \notin B\}$
symmetric difference	$A \triangle B = \{c : c \in A \cup B \ and \ c \notin A \cap B\}$
Cartesian product	$A \times B = \{(a,b) : a \in A \ and \ b \in B\}$
choice function	$choice(A) \in A$
cardinality	$card(A) = cardinality \ of \ A$
supremum	$sup(A) = supremum \ of \ A$
infimum	$inf(A) = infimum \ of \ A$

1.4. Images

The primary operands in image algebra are images, templates, and neighborhoods. Of these three classes of operands, images are the most fundamental since templates and neighborhoods can be viewed as special cases of the general concept of an image. In order to provide a mathematically rigorous definition of an image that covers the plethora of objects called an "image" in signal processing and image understanding, we define an image in general terms, with a minimum of specification. In the following we use the notation A^B to denote the set of all functions $B \rightarrow A$ (i.e., $A^B = \{f : f \ is \ a \ function \ from \ B \ to \ A\}$).

Definition: Let F be a value set and $\mathbf{X}$ a point set. An F-*valued image on* $\mathbf{X}$ is any element of $\mathsf{F}^{\mathbf{X}}$. Given an F–valued image $\mathbf{a} \in \mathsf{F}^{\mathbf{X}}$ (i.e., $\mathbf{a} : \mathbf{X} \rightarrow \mathsf{F}$), then F is called the *set of possible range values* of $\mathbf{a}$ and $\mathbf{X}$ the *spatial domain* of $\mathbf{a}$.

It is often convenient to let the graph of an image $\mathbf{a} \in \mathsf{F}^{\mathbf{X}}$ represent $\mathbf{a}$. The graph of an image is also referred to as the *data structure representation* of the image. Given the data structure representation $\mathbf{a} = \{(\mathbf{x}, \mathbf{a}(\mathbf{x})) : \mathbf{x} \in \mathbf{X}\}$, then an element $(\mathbf{x}, \mathbf{a}(\mathbf{x}))$ of the data structure is called a *picture element* or *pixel*. The first coordinate $\mathbf{x}$ of a pixel is called the *pixel location* or *image point*, and the second coordinate $\mathbf{a}(\mathbf{x})$ is called the *pixel value* of $\mathbf{a}$ at location $\mathbf{x}$.

The above definition of an image covers all mathematical images on topological spaces with range in an algebraic system. Requiring $\mathbf{X}$ to be a topological space provides us with the notion of nearness of pixels. Since $\mathbf{X}$ is not directly specified we may substitute any space required for the analysis of an image or imposed by a particular sensor and scene. For example, $\mathbf{X}$ could be a subset of $\mathbb{Z}^3$ or $\mathbb{R}^3$ with $\mathbf{x} \in \mathbf{X}$ of form $\mathbf{x} = (x, y, t)$, where the first coordinates (x, y) denote spatial location and t a time variable.

Similarly, replacing the unspecified value set F with $\mathbb{Z}_{2^k}$ or $\mathsf{F} = (\mathbb{Z}_{2^k}, \mathbb{Z}_{2^m}, \mathbb{Z}_{2^n})$ provides us with digital integer-valued and digital vector-valued images, respectively. An implication of these observations is that our image definition also characterizes any type of discrete or continuous *physical image*.

Induced Operations on Images

Operations on and between F-valued images are the natural induced operations of the algebraic system F. For example, if γ is a binary operation on F, then γ induces a binary operation — again denoted by γ — on $\mathsf{F}^{\mathbf{X}}$ defined as follows:

Let $\mathbf{a}, \mathbf{b} \in \mathbb{F}^{\mathbf{X}}$. Then

$$\mathbf{a}\gamma\mathbf{b} = \{(\mathbf{x}, \mathbf{c}(\mathbf{x})) \, : \, \mathbf{c}(\mathbf{x}) = \mathbf{a}(\mathbf{x})\gamma\mathbf{b}(\mathbf{x}), \ \mathbf{x} \in \mathbf{X}\}.$$

For example, suppose $\mathbf{a}, \mathbf{b} \in \mathbb{R}^{\mathbf{X}}$ and our value set is the algebraic structure of the real numbers $(\mathbb{R}, +, \cdot, \vee, \wedge)$. Replacing γ by the binary operations $+$, $\cdot$, $\vee$, and $\wedge$ we obtain the *basic* binary operations

$$\mathbf{a} + \mathbf{b} = \{(\mathbf{x}, \mathbf{c}(\mathbf{x})) \, : \, \mathbf{c}(\mathbf{x}) = \mathbf{a}(\mathbf{x}) + \mathbf{b}(\mathbf{x}), \ \mathbf{x} \in \mathbf{X}\},$$
$$\mathbf{a} \cdot \mathbf{b} = \{(\mathbf{x}, \mathbf{c}(\mathbf{x})) \, : \, \mathbf{c}(\mathbf{x}) = \mathbf{a}(\mathbf{x}) \cdot \mathbf{b}(\mathbf{x}), \ \mathbf{x} \in \mathbf{X}\},$$
$$\mathbf{a} \vee \mathbf{b} = \{(\mathbf{x}, \mathbf{c}(\mathbf{x})) \, : \, \mathbf{c}(\mathbf{x}) = \mathbf{a}(\mathbf{x}) \vee \mathbf{b}(\mathbf{x}), \ \mathbf{x} \in \mathbf{X}\},$$

and

$$\mathbf{a} \wedge \mathbf{b} = \{(\mathbf{x}, \mathbf{c}(\mathbf{x})) \, : \, \mathbf{c}(\mathbf{x}) = \mathbf{a}(\mathbf{x}) \wedge \mathbf{b}(\mathbf{x}), \ \mathbf{x} \in \mathbf{X}\}$$

on real-valued images. Obviously, all four operations are commutative and associative.

In addition to the binary operation between images, the binary operation γ on $\mathbb{F}$ also induces the following scalar operations on images:

For $k \in \mathbb{F}$ and $\mathbf{a} \in \mathbb{F}^{\mathbf{X}}$,

$$k\gamma\mathbf{a} = \{(\mathbf{x}, \mathbf{c}(\mathbf{x})) \, : \, \mathbf{c}(\mathbf{x}) = k\gamma\mathbf{a}(\mathbf{x}), \ \mathbf{x} \in \mathbf{X}\}$$

and

$$\mathbf{a}\gamma k = \{(\mathbf{x}, \mathbf{c}(\mathbf{x})) \, : \, \mathbf{c}(\mathbf{x}) = \mathbf{a}(\mathbf{x})\gamma k , \ \mathbf{x} \in \mathbf{X}\}.$$

Thus, for $k \in \mathbb{R}$, we obtain the following scalar multiplication and addition of real-valued images:

$$k \cdot \mathbf{a} = \{(\mathbf{x}, \mathbf{c}(\mathbf{x})) \, : \, \mathbf{c}(\mathbf{x}) = k \cdot \mathbf{a}(\mathbf{x}), \ \mathbf{x} \in \mathbf{X}\}$$

and

$$k + \mathbf{a} = \{(\mathbf{x}, \mathbf{c}(\mathbf{x})) \, : \, \mathbf{c}(\mathbf{x}) = k + \mathbf{a}(\mathbf{x}), \ \mathbf{x} \in \mathbf{X}\}.$$

It follows from the commutativity of real numbers that,

$$k \cdot \mathbf{a} = \mathbf{a} \cdot k \text{ and } k + \mathbf{a} = \mathbf{a} + k.$$

Although much of image processing is accomplished using real-, integer-, binary-, or complex-valued images, many higher-level vision tasks require manipulation of vector- and set-valued images. A set-valued image is of form $\mathbf{a} : \mathbf{X} \to 2^{\mathbb{F}}$. Here the underlying value set is $(2^{\mathbb{F}}, \cup, \cap, \tilde{\ })$, where the tilde symbol denotes complementation. Hence, the operations on set-valued images are those induced by the Boolean algebra of the value set. For example, if $\mathbf{a}, \mathbf{b} \in (2^{\mathbb{F}})^{\mathbf{X}}$, then

$$\mathbf{a} \cup \mathbf{b} = \{(\mathbf{x}, \mathbf{c}(\mathbf{x})) \, : \, \mathbf{c}(\mathbf{x}) = \mathbf{a}(\mathbf{x}) \cup \mathbf{b}(\mathbf{x}), \ \mathbf{x} \in \mathbf{X}\},$$
$$\mathbf{a} \cap \mathbf{b} = \{(\mathbf{x}, \mathbf{c}(\mathbf{x})) \, : \, \mathbf{c}(\mathbf{x}) = \mathbf{a}(\mathbf{x}) \cap \mathbf{b}(\mathbf{x}), \ \mathbf{x} \in \mathbf{X}\},$$

and

$$\tilde{\mathbf{a}} = \left\{(\mathbf{x}, \mathbf{c}(\mathbf{x})) \, : \, \mathbf{c}(\mathbf{x}) = \widetilde{\mathbf{a}(\mathbf{x})}, \ \mathbf{x} \in \mathbf{X}\right\},$$

where $\widetilde{\mathbf{a}(\mathbf{x})} = \mathbb{F}\backslash\mathbf{a}(\mathbf{x})$.

The operation of complementation is, of course, a unary operation. A particularly useful unary operation on images which is induced by a binary operation on a value set is known as the *global reduce operation*. More precisely, if γ is an associative and commutative binary operation on $\mathbb{F}$ and $\mathbf{X}$ is finite, say $\mathbf{X} = \{\mathbf{x}_1, \mathbf{x}_2, \ldots, \mathbf{x}_n\}$, then γ induces a unary operation

$$\Gamma : \mathbb{F}^{\mathbf{X}} \rightarrow \mathbb{F}$$

called the *global reduce operation induced by* γ, which is defined as

$$\Gamma \mathbf{a} = \underset{\mathbf{x} \in \mathbf{X}}{\Gamma} \, \mathbf{a}(\mathbf{x}) = \overset{n}{\underset{k=1}{\Gamma}} \mathbf{a}(\mathbf{x}_k) = \mathbf{a}(\mathbf{x}_1)\gamma\mathbf{a}(\mathbf{x}_2)\gamma \cdots \gamma\mathbf{a}(\mathbf{x}_n).$$

Thus, for example, if $\mathbb{F} = \mathbb{R}$ and γ is the operation of addition ($\gamma = +$), then $\Gamma = \Sigma$ and

$$\sum \mathbf{a} = \sum_{\mathbf{x} \in \mathbf{X}} \mathbf{a}(\mathbf{x}) = \mathbf{a}(\mathbf{x}_1) + \mathbf{a}(\mathbf{x}_2) + \cdots + \mathbf{a}(\mathbf{x}_n).$$

In all, the value set $(\mathbb{R}, +, \cdot, \vee, \wedge)$ provides for four basic global reduce operations, namely $\sum \mathbf{a}$, $\prod \mathbf{a}$, $\bigvee \mathbf{a}$, and $\bigwedge \mathbf{a}$.

Induced Unary Operations and Functional Composition

In the previous section we discussed unary operations on elements of $\mathbb{F}^{\mathbf{X}}$ induced by a binary operation γ on $\mathbb{F}$. Typically, however, unary image operations are induced directly by unary operations on $\mathbb{F}$. Given a unary operation $f : \mathbb{F} \rightarrow \mathbb{F}$, then the induced unary operation $\mathbb{F}^{\mathbf{X}} \rightarrow \mathbb{F}^{\mathbf{X}}$ is again denoted by f and is defined by

$$f(\mathbf{a}) = \{(\mathbf{x}, \mathbf{c}(\mathbf{x})) : \mathbf{c}(\mathbf{x}) = f(\mathbf{a}(\mathbf{x})), \mathbf{x} \in \mathbf{X}\}.$$

Note that in this definition we view the composition $f \circ \mathbf{a}$ as a unary operation on $\mathbb{F}^{\mathbf{X}}$ with operand $\mathbf{a}$. This subtle distinction has the important consequence that f is viewed as a unary operation — namely a function from $\mathbb{F}^{\mathbf{X}}$ to $\mathbb{F}^{\mathbf{X}}$ — and $\mathbf{a}$ as an *argument* of f. For example, substituting $\mathbb{R}$ for $\mathbb{F}$ and the sine function $sin : \mathbb{R} \rightarrow \mathbb{R}$ for f, we obtain the induced operation $sin : \mathbb{R}^{\mathbf{X}} \rightarrow \mathbb{R}^{\mathbf{X}}$, where

$$sin(\mathbf{a}) = \{(\mathbf{x}, \mathbf{c}(\mathbf{x})) : \mathbf{c}(\mathbf{x}) = sin(\mathbf{a}(\mathbf{x})), \mathbf{x} \in \mathbf{X}\}.$$

As another example, consider the characteristic function

$$\chi_{\geq k}(r) = \begin{cases} 1 & \text{if } r \geq k \\ 0 & \text{otherwise.} \end{cases}$$

Then for any $\mathbf{a} \in \mathbb{R}^{\mathbf{X}}$, $\chi_{\geq k}(\mathbf{a})$ is the Boolean (two-valued) image on $\mathbf{X}$ with value 1 at location $\mathbf{x}$ if $\mathbf{a}(\mathbf{x}) \geq k$ and value 0 if $\mathbf{a}(\mathbf{x}) < k$. An obvious application of this operation is the thresholding of an image. Given a floating point image $\mathbf{a}$ and using the characteristic function

$$\chi_{[j,k]}(r) = \begin{cases} 1 & \text{if } j \leq r \leq k \\ 0 & \text{otherwise,} \end{cases}$$

then the image $\mathbf{b}$ in the image algebra expression

$$\mathbf{b} := \mathbf{a} \cdot \chi_{[j,k]}(\mathbf{a})$$

is given by

$$\mathbf{b} = \{(\mathbf{x}, \mathbf{b}(\mathbf{x})) \, : \, \mathbf{b}(\mathbf{x}) = \mathbf{a}(\mathbf{x}) \text{ if } j \leq \mathbf{a}(\mathbf{x}) \leq k, \text{ otherwise } \mathbf{b}(\mathbf{x}) = 0\}.$$

The unary operations on an image $\mathbf{a} \in \mathsf{F}^{\mathbf{X}}$ discussed thus far have resulted either in a scalar (an element of F) by use of the global reduction operation, or another F-valued image by use of the composition $f \circ \mathbf{a} = f(\mathbf{a})$. More generally, given a function $f : \mathsf{F} \to \mathsf{G}$, then the composition $f \circ \mathbf{a}$ provides for a unary operation which changes an F-valued image into a G-valued image $f(\mathbf{a})$. Taking the same viewpoint, but using a function f between spatial domains instead, provides a scheme for realizing naturally induced operations for spatial manipulation of image data. In particular, if $f : \mathbf{Y} \to \mathbf{X}$ and $\mathbf{a} \in \mathsf{F}^{\mathbf{X}}$, then we define the induced image $\mathbf{a} \circ f \in \mathsf{F}^{\mathbf{Y}}$ by

$$\mathbf{a} \circ f = \{(\mathbf{y}, \mathbf{a}(f(\mathbf{y}))) \, : \, \mathbf{y} \in \mathbf{Y}\}.$$

Thus, the operation defined by the above equation transforms an F-valued image defined over the space $\mathbf{X}$ into an F-valued image defined over the space $\mathbf{Y}$.

Examples of spatial based image transformations are affine and perspective transforms. For instance, suppose $\mathbf{a} \in \mathbb{R}^{\mathbf{X}}$, where $\mathbf{X} \subset \mathbb{Z}^2$ is a rectangular $m \times n$ array. If $1 \leq k \leq \frac{m}{2}$ and $f : \mathbf{X} \to \mathbf{X}$ is defined as

$$f(x, y) = \begin{cases} (x, y) & \text{if } k \leq x \\ (2k - x, y) & \text{if } x < k \end{cases},$$

then $\mathbf{a} \circ f$ is a one sided reflection of $\mathbf{a}$ across the line $x = k$. Further examples are provided by several of the algorithms presented in this text.

Simple shifts of an image can be achieved by using either a spatial transformation or *point addition*. In particular, given $\mathbf{a} \in \mathsf{F}^{\mathbf{X}}$, $\mathbf{X} \subset \mathbb{Z}^2$, and $\mathbf{y} \in \mathbb{Z}^2$, we define a shift of $\mathbf{a}$ by $\mathbf{y}$ as

$$\mathbf{a} + \mathbf{y} = \{(\mathbf{z}, \mathbf{b}(\mathbf{z})) \, : \, \mathbf{b}(\mathbf{z}) = \mathbf{a}(\mathbf{z} - \mathbf{y}), \; \mathbf{z} - \mathbf{y} \in \mathbf{X}\}.$$

Note that $\mathbf{a} + \mathbf{y}$ is an image on $\mathbf{X} + \mathbf{y}$ since $\mathbf{z} - \mathbf{y} \in \mathbf{X} \Leftrightarrow \mathbf{z} \in \mathbf{X} + \mathbf{y}$, which provides for the equivalent formulation

$$\mathbf{a} + \mathbf{y} = \{(\mathbf{z}, \mathbf{b}(\mathbf{z})) \, : \, \mathbf{b}(\mathbf{z}) = \mathbf{a}(\mathbf{z} - \mathbf{y}), \; \mathbf{z} \in \mathbf{X} + \mathbf{y}\}.$$

Of course, one could just as well define a spatial transformation $f : \mathbf{X} + \mathbf{y} \to \mathbf{X}$ by $f(\mathbf{z}) = \mathbf{z} - \mathbf{y}$ in order to obtain the identical shifted image $\mathbf{a} + \mathbf{y} = \mathbf{a} \circ f$.

Another simple unary image operation that can be defined in terms of a spatial map is *image transposition*. Given an image $\mathbf{a} \in \mathsf{F}^{\mathbb{Z}_m \times \mathbb{Z}_n}$, then the *transpose* of $\mathbf{a}$, denoted by $\mathbf{a}'$, is defined as $\mathbf{a}' \equiv \mathbf{a} \circ f$, where $f : \mathbb{Z}_n \times \mathbb{Z}_m \to \mathbb{Z}_m \times \mathbb{Z}_n$ is given by $f(x, y) = (y, x)$.

Binary Operations Induced by Unary Operations

Various unary operations image operations induced by functions $f : \mathsf{F} \to \mathsf{F}$ can be generalized to binary operations on $\mathsf{F}^{\mathbf{X}}$. As a simple illustration, consider the exponentiation function $f : \mathbb{R}^{\geq 0} \to \mathbb{R}$ defined by $f(r) = r^k$, where k denotes some non-negative real number. Then f induces the exponentiation operation

$$\mathbf{a}^k = \left\{ (\mathbf{x}, \mathbf{b}(\mathbf{x})) \, : \, \mathbf{b}(\mathbf{x}) = [\mathbf{a}(\mathbf{x})]^k, \; \mathbf{x} \in \mathbf{X} \right\},$$

where $\mathbf{a}$ is a non-negative real-valued image on $\mathbf{X}$. We may extend this operation to a binary image operation as follows: if $\mathbf{a}, \mathbf{b} \in (\mathbb{R}^{\geq 0})^{\mathbf{X}}$, then

$$\mathbf{a}^{\mathbf{b}} = \left\{ (\mathbf{x}, \mathbf{c}(\mathbf{x})) \; : \; \mathbf{c}(\mathbf{x}) = \mathbf{a}(\mathbf{x})^{\mathbf{b}(\mathbf{x})}, \; \mathbf{x} \in \mathbf{X} \right\}.$$

The notion of exponentiation can be extended to negative valued images as long as we follow the rules of arithmetic and restrict this binary operation to those pairs of real-valued images for which $\mathbf{a}(\mathbf{x})^{\mathbf{b}(\mathbf{x})} \in \mathbb{R} \; \forall \mathbf{x} \in \mathbf{X}$. This avoids creation of complex, undefined, and indeterminate pixel values such as $(-1)^{\frac{1}{2}}$, $\frac{1}{0^2}$, and 0^0, respectively. However, there is one exception to these rules of standard arithmetic. The algebra of images provides for the existence of *pseudo inverses*. For $\mathbf{a} \in \mathbb{R}^{\mathbf{X}}$, the *pseudo inverse* of $\mathbf{a}$, which for reason of simplicity is denoted by $\mathbf{a}^{-1}$ is defined as

$$\mathbf{a}^{-1} = \left\{ (\mathbf{x}, \mathbf{b}(\mathbf{x})) \; : \; \mathbf{b}(\mathbf{x}) = \frac{1}{\mathbf{a}(\mathbf{x})} \text{ if } \mathbf{a}(\mathbf{x}) \neq 0 \text{ otherwise } \mathbf{b}(\mathbf{x}) = 0 \right\}.$$

Note that if some pixel values of $\mathbf{a}$ are zero, then $\mathbf{a} \cdot \mathbf{a}^{-1} \neq \mathbf{1}$, where $\mathbf{1}$ denotes unit image all of whose pixel values are 1. However, the equality $\mathbf{a} \cdot \mathbf{a}^{-1} \cdot \mathbf{a} = \mathbf{a}$ always holds. Hence the name "pseudo inverse."

The inverse of exponentiation is defined in the usual way by taking logarithms. Specifically,

$$log_{\mathbf{b}}\mathbf{a} = \left\{ (\mathbf{x}, \mathbf{c}(\mathbf{x})) \; : \; \mathbf{c}(\mathbf{x}) = log_{\mathbf{b}(\mathbf{x})}\mathbf{a}(\mathbf{x}), \; \mathbf{x} \in \mathbf{X} \right\}.$$

As for real numbers, $log_{\mathbf{b}}\mathbf{a}$ is defined only for positive images; i.e., $\mathbf{a}, \mathbf{b} \in (\mathbb{R}^{+})^{\mathbf{X}}$.

Another set of examples of binary operations induced by unary operations are the characteristic functions for comparing two images. For $\mathbf{a}, \mathbf{b} \in \mathbb{R}^{\mathbf{X}}$ we define

$$\chi_{\leq \mathbf{b}}(\mathbf{a}) = \{(\mathbf{x}, \mathbf{c}(\mathbf{x})) : \; \mathbf{c}(\mathbf{x}) = 1 \text{ if } \mathbf{a}(\mathbf{x}) \leq \mathbf{b}(\mathbf{x}), \text{ otherwise } \mathbf{c}(\mathbf{x}) = 0\}$$
$$\chi_{< \mathbf{b}}(\mathbf{a}) = \{(\mathbf{x}, \mathbf{c}(\mathbf{x})) : \; \mathbf{c}(\mathbf{x}) = 1 \text{ if } \mathbf{a}(\mathbf{x}) < \mathbf{b}(\mathbf{x}), \text{ otherwise } \mathbf{c}(\mathbf{x}) = 0\}$$
$$\chi_{= \mathbf{b}}(\mathbf{a}) = \{(\mathbf{x}, \mathbf{c}(\mathbf{x})) : \; \mathbf{c}(\mathbf{x}) = 1 \text{ if } \mathbf{a}(\mathbf{x}) = \mathbf{b}(\mathbf{x}), \text{ otherwise } \mathbf{c}(\mathbf{x}) = 0\}$$
$$\chi_{\geq \mathbf{b}}(\mathbf{a}) = \{(\mathbf{x}, \mathbf{c}(\mathbf{x})) : \; \mathbf{c}(\mathbf{x}) = 1 \text{ if } \mathbf{a}(\mathbf{x}) \geq \mathbf{b}(\mathbf{x}), \text{ otherwise } \mathbf{c}(\mathbf{x}) = 0\}$$
$$\chi_{> \mathbf{b}}(\mathbf{a}) = \{(\mathbf{x}, \mathbf{c}(\mathbf{x})) : \; \mathbf{c}(\mathbf{x}) = 1 \text{ if } \mathbf{a}(\mathbf{x}) > \mathbf{b}(\mathbf{x}), \text{ otherwise } \mathbf{c}(\mathbf{x}) = 0\}$$
$$\chi_{\neq \mathbf{b}}(\mathbf{a}) = \{(\mathbf{x}, \mathbf{c}(\mathbf{x})) : \; \mathbf{c}(\mathbf{x}) = 1 \text{ if } \mathbf{a}(\mathbf{x}) \neq \mathbf{b}(\mathbf{x}), \text{ otherwise } \mathbf{c}(\mathbf{x}) = 0\}.$$

Functional Specification of Image Operations

The basic concepts of elementary function theory provide the underlying foundation of a functional specification of image processing techniques. This is a direct consequence of viewing images as functions. The most elementary concepts of function theory are the notions of domain, range, restriction, and extension of a function.

Image restrictions and extensions are used to restrict images to regions of particular interest and to embed images into larger images, respectively. Employing standard mathematical notation, the *restriction* of $\mathbf{a} \in \mathbb{F}^{\mathbf{X}}$ to a subset $\mathbf{Z}$ of $\mathbf{X}$ is denoted by $\mathbf{a}|_{\mathbf{Z}}$, and defined by

$$\mathbf{a}|_{\mathbf{Z}} \equiv \mathbf{a} \cap (\mathbf{Z} \times \mathbb{F}) = \{(\mathbf{x}, \mathbf{a}(\mathbf{x})) : \; \mathbf{x} \in \mathbf{Z}\}.$$

Thus, $\mathbf{a}|_{\mathbf{Z}} \in \mathbb{F}^{\mathbf{Z}}$. In practice, the user may specify $\mathbf{Z}$ explicitly by providing bounds for the coordinates of the points of $\mathbf{Z}$.

There is nothing magical about restricting $\mathbf{a}$ to a subset $\mathbf{Z}$ of its domain $\mathbf{X}$. We can just as well define restrictions of images to subsets of the range values. Specifically, if $S \subset \mathbb{F}$ and $\mathbf{a} \in \mathbb{F}^{\mathbf{X}}$, then the *restriction* of $\mathbf{a}$ to S is denoted by $\mathbf{a}\|_S$ and defined as

$$\mathbf{a}\|_S \equiv \mathbf{a} \cap (\mathbf{X} \times S).$$

In terms of the pixel representation of $\mathbf{a}\|_S$ we have $\mathbf{a}\|_S = \{(\mathbf{x}, \mathbf{a}(\mathbf{x})) : \mathbf{a}(\mathbf{x}) \in S\}$. The double-bar notation is used to focus attention on the fact that the restriction is applied to the second coordinate of $\mathbf{a} \subset \mathbf{X} \times \mathbb{F}$.

Image restrictions in terms of subsets of the value set $\mathbb{F}$ is an extremely useful concept in computer vision as many image processing tasks are restricted to image domains over which the image values satisfy certain properties. Of course, one can always write this type of restriction in terms of a first coordinate restriction by setting $\mathbf{Z} = \{\mathbf{x} \in \mathbf{X} : \mathbf{a}(\mathbf{x}) \in S\}$ so that $\mathbf{a}\|_S = \mathbf{a}|_{\mathbf{Z}}$. However, writing a program statement such as $\mathbf{b} := \mathbf{a}|_{\mathbf{Z}}$ is of little value since $\mathbf{Z}$ is implicitly specified in terms of S; i.e., $\mathbf{Z}$ must be determined in terms of the property "$\mathbf{a}(\mathbf{x}) \in S$." Thus, $\mathbf{Z}$ would have to be precomputed, adding to the computational overhead as well as increased code. In contrast, direct restriction of the second coordinate values to an explicitly specified set S avoids these problems and provides for easier implementation.

As mentioned, restrictions to the range set provide a useful tool for expressing various algorithmic procedures. For instance, if $\mathbf{a} \in \mathbb{R}^{\mathbf{X}}$ and S is the interval $(k, \infty) \subset \mathbb{R}$, where k denotes some given threshold value, then $\mathbf{a}\|_{(k,\infty)}$ denotes the image $\mathbf{a}$ restricted to all those points of $\mathbf{X}$ where $\mathbf{a}(\mathbf{x})$ exceeds the value k. In order to reduce notation, we define $\mathbf{a}\|_{>k} \equiv \mathbf{a}\|_{(k,\infty)}$. Similarly,

$$\mathbf{a}\|_{\geq k} \equiv \mathbf{a}\|_{[k,\infty)}, \quad \mathbf{a}\|_{<k} \equiv \mathbf{a}\|_{(-\infty,k)}, \quad \mathbf{a}\|_k \equiv \mathbf{a}\|_{\{k\}}, \quad \text{and} \quad \mathbf{a}\|_{\leq k} \equiv \mathbf{a}\|_{(-\infty,k]}.$$

As in the case of characteristic functions, a more general form of range restriction is given when S corresponds to a set-valued image $S \in \left(2^{\mathbb{F}}\right)^{\mathbf{X}}$; i.e., $S(\mathbf{x}) \subset \mathbb{F} \; \forall \mathbf{x} \in \mathbf{X}$. In this case we define

$$\mathbf{a}\|_S = \{(\mathbf{x}, \mathbf{a}(\mathbf{x})) : \mathbf{a}(\mathbf{x}) \in S(\mathbf{x})\}.$$

For example, for $\mathbf{a}, \mathbf{b} \in \mathbb{R}^{\mathbf{X}}$ we define

$$\mathbf{a}\|_{\leq \mathbf{b}} \equiv \{(\mathbf{x}, \mathbf{a}(\mathbf{x})) : \mathbf{a}(\mathbf{x}) \leq \mathbf{b}(\mathbf{x})\}, \quad \mathbf{a}\|_{<\mathbf{b}} \equiv \{(\mathbf{x}, \mathbf{a}(\mathbf{x})) : \mathbf{a}(\mathbf{x}) < \mathbf{b}(\mathbf{x})\},$$
$$\mathbf{a}\|_{\geq \mathbf{b}} \equiv \{(\mathbf{x}, \mathbf{a}(\mathbf{x})) : \mathbf{a}(\mathbf{x}) \geq \mathbf{b}(\mathbf{x})\}, \quad \mathbf{a}\|_{>\mathbf{b}} \equiv \{(\mathbf{x}, \mathbf{a}(\mathbf{x})) : \mathbf{a}(\mathbf{x}) > \mathbf{b}(\mathbf{x})\},$$
$$\mathbf{a}\|_{\mathbf{b}} \equiv \{(\mathbf{x}, \mathbf{a}(\mathbf{x})) : \mathbf{a}(\mathbf{x}) = \mathbf{b}(\mathbf{x})\}, \quad \mathbf{a}\|_{\neq \mathbf{b}} \equiv \{(\mathbf{x}, \mathbf{a}(\mathbf{x})) : \mathbf{a}(\mathbf{x}) \neq \mathbf{b}(\mathbf{x})\}.$$

Combining the concepts of first and second coordinate (domain and range) restrictions provides the general definition of an image restriction. If $\mathbf{a} \in \mathbb{F}^{\mathbf{X}}$, $\mathbf{Z} \subset \mathbf{X}$, and $S \subset \mathbb{F}$, then the *restriction of* $\mathbf{a}$ *to* $\mathbf{Z}$ *and* S is defined as

$$\mathbf{a}|_{(\mathbf{Z}, S)} = \mathbf{a} \cap (\mathbf{Z} \times S).$$

It follows that $\mathbf{a}|_{(\mathbf{Z}, S)} = \{(\mathbf{x}, \mathbf{a}(\mathbf{x})) : \mathbf{x} \in \mathbf{Z} \text{ and } \mathbf{a}(\mathbf{x}) \in S\}$, $\mathbf{a}|_{(\mathbf{X}, S)} = \mathbf{a}\|_S$, and $\mathbf{a}|_{(\mathbf{Z}, \mathbb{F})} = \mathbf{a}|_{\mathbf{Z}}$.

The *extension* of $\mathbf{a} \in \mathbb{F}^{\mathbf{X}}$ to $\mathbf{b} \in \mathbb{F}^{\mathbf{Y}}$ on $\mathbf{Y}$, where $\mathbf{X}$ and $\mathbf{Y}$ are subsets of the same topological space, is denoted by $\mathbf{a}|^{\mathbf{b}}$ and defined by

$$\mathbf{a}|^{\mathbf{b}}(\mathbf{x}) = \begin{cases} \mathbf{a}(\mathbf{x}) & \text{if } \mathbf{x} \in \mathbf{X} \\ \mathbf{b}(\mathbf{x}) & \text{if } \mathbf{x} \in \mathbf{Y}\backslash\mathbf{X}. \end{cases}$$

In actual practice, the user will have to specify the function $\mathbf{b}$.

Two of the most important concepts associated with a function are its domain and range. In the field of image understanding, it is convenient to view these concepts as functions that map images to sets associated with certain image properties. Specifically, we view the concept of range as a function

$$range : \mathsf{F}^{\mathbf{X}} \rightarrow 2^{\mathsf{F}}$$

defined by $range(\mathbf{a}) = \{r \in \mathsf{F} : r = \mathbf{a}(\mathbf{x}) \; for \; some \; \mathbf{x} \in \mathbf{X}\}$.

Similarly, the concept of domain is viewed as the function

$$domain : \mathsf{F}^{\mathbf{X}}|_{(2^{\mathbf{X}} \times 2^{\mathsf{F}})} \rightarrow 2^{\mathbf{X}},$$

where

$$\mathsf{F}^{\mathbf{X}}|_{(2^{\mathbf{X}} \times 2^{\mathsf{F}})} = \{\mathbf{b} : \mathbf{b} = \mathbf{a}|_{(\mathbf{Z},S)}, \; \mathbf{a} \in \mathsf{F}^{\mathbf{X}}, \; \mathbf{Z} \in 2^{\mathbf{X}}, \; S \in 2^{\mathsf{F}}\}$$

and *domain* is defined by

$$domain(\mathbf{b}) = \{\mathbf{x} \in \mathbf{X} : \mathbf{a}|_{(\mathbf{Z},S)}(\mathbf{x}) = \mathbf{b}(\mathbf{x}) = r \; for \; some \; r \in \mathsf{F}\}.$$

These mappings can be used to extract point sets and value sets from regions of images of particular interest. For example, the statement

$$s := domain(\mathbf{a}\|_{>k})$$

yields the *set* of all points (pixel locations) where $\mathbf{a}(\mathbf{x})$ exceeds k, namely $s = \{\mathbf{x} \in \mathbf{X} : \mathbf{a}(\mathbf{x}) > k\}$. The statement

$$s := range(\mathbf{a}\|_{>k})$$

on the other hand, results in a subset of $\mathbb{R}$ instead of $\mathbf{X}$.

Closely related to spatial transformations and functional composition is the notion of *image concatenation*. Concatenation serves as a tool for simplifying algorithm code, adding translucency to code, and to provide a link to the usual block notion used in matrix algebra. Given $\mathbf{a} \in \mathsf{F}^{\mathbb{Z}_m \times \mathbb{Z}_k}$ and $\mathbf{b} \in \mathsf{F}^{\mathbb{Z}_m \times \mathbb{Z}_n}$, then the *row-order concatenation of* $\mathbf{a}$ *with* $\mathbf{b}$ is denoted by $(\mathbf{a} \mid \mathbf{b})$ and is defined as

$$(\mathbf{a} \mid \mathbf{b}) \equiv \mathbf{a}|^{\mathbf{b} + (0,k)}.$$

Note that $(\mathbf{a} \mid \mathbf{b}) \in \mathsf{F}^{\mathbb{Z}_m \times \mathbb{Z}_{n+k}}$.

Assuming the correct dimensionality in the first coordinate, concatenation of any number of images is defined inductively using the formula $(\mathbf{a} \mid \mathbf{b} \mid \mathbf{c}) = ((\mathbf{a} \mid \mathbf{b})\mid \mathbf{c})$ so that in general we have

$$(\mathbf{a}_1|\mathbf{a}_2| \cdots |\mathbf{a}_l) = ((\mathbf{a}_1|\mathbf{a}_2| \cdots |\mathbf{a}_{l-1})|\mathbf{a}_l).$$

Column-order concatenation can be defined in a similar manner or by simple transposition; i.e.,

$$\begin{pmatrix} \mathbf{a}_1 \\ - \\ \mathbf{a}_2 \\ - \\ \vdots \\ - \\ \mathbf{a}_l \end{pmatrix} = (\mathbf{a}_1|\mathbf{a}_2| \cdots |\mathbf{a}_l)'.$$

Multi-Valued Image Operations

Although general image operations described in the previous sections apply to both single and multi-valued images as long as there is no specific value type associated with the generic value set $\mathbb{F}$, there exist a large number of multi-valued image operations that are quite distinct from single-valued image operations. As the general theory of multi-valued image operations is beyond the scope of this treatise, we shall restrict our attention to some specific operations on vector-valued images while referring the reader interested in more intricate details to Ritter [1]. However, it is important to realize that vector-valued images are a special case of multi-valued images.

If $\mathbb{F} = \mathbb{R}^n$ and $\mathbf{a} \in \mathbb{F}^X$, then $\mathbf{a}(\mathbf{x})$ is a vector of form $\mathbf{a}(\mathbf{x}) = (\mathbf{a}_1(\mathbf{x}), \ldots, \mathbf{a}_n(\mathbf{x}))$ where for each $i = 1, \ldots, n$, $\mathbf{a}_i(\mathbf{x}) \in \mathbb{R}$. Thus, an image $\mathbf{a} \in (\mathbb{R}^n)^X$ is of form $\mathbf{a} = (\mathbf{a}_1, \ldots, \mathbf{a}_n)$ and with each *vector value* $\mathbf{a}(\mathbf{x})$ there are associated n real values $\mathbf{a}_i(\mathbf{x})$.

Real-valued image operations generalize to the usual vector operations on $(\mathbb{R}^n)^X$. In particular, if $\mathbf{a}, \mathbf{b} \in (\mathbb{R}^n)^X$, then

$$\mathbf{a} + \mathbf{b} = (\mathbf{a}_1 + \mathbf{b}_1, \ldots, \mathbf{a}_n + \mathbf{b}_n)$$
$$\mathbf{a} \cdot \mathbf{b} = (\mathbf{a}_1 \cdot \mathbf{b}_1, \ldots, \mathbf{a}_n \cdot \mathbf{b}_n)$$
$$\mathbf{a} \vee \mathbf{b} = (\mathbf{a}_1 \vee \mathbf{b}_1, \ldots, \mathbf{a}_n \vee \mathbf{b}_n)$$
$$\mathbf{a} \wedge \mathbf{b} = (\mathbf{a}_1 \wedge \mathbf{b}_1, \ldots, \mathbf{a}_n \wedge \mathbf{b}_n).$$

If $\mathbf{r} = (r_1, \ldots, r_n) \in \mathbb{R}^n$, then we also have

$$\mathbf{r} + \mathbf{a} = (r_1 + \mathbf{a}_1, \ldots, r_n + \mathbf{a}_n),$$
$$\mathbf{r} \cdot \mathbf{a} = (r_1 \cdot \mathbf{a}_1, \ldots, r_n \cdot \mathbf{a}_n),$$

etc. In the special case where $\mathbf{r} = (r, r, \ldots, r)$, we simply use the scalar $r \in \mathbb{R}$ and define $r + \mathbf{a} \equiv \mathbf{r} + \mathbf{a}$, $r \cdot \mathbf{a} \equiv \mathbf{r} \cdot \mathbf{a}$, and so on.

As before, binary operations on multi-valued images are induced by the corresponding binary operation $\gamma : \mathbb{R}^n \times \mathbb{R}^n \to \mathbb{R}^n$ on the value set $\mathbb{R}^n$. It turns out to be useful to generalize this concept by replacing the binary operation γ by a sequence of binary operations $\gamma_j : \mathbb{R}^n \times \mathbb{R}^n \to \mathbb{R}$, where $j = 1, \ldots, n$, and defining

$$\mathbf{a}\gamma\mathbf{b} \equiv (\mathbf{a}\gamma_1\mathbf{b}, \mathbf{a}\gamma_2\mathbf{b}, \ldots, \mathbf{a}\gamma_n\mathbf{b}).$$

For example, if $\gamma_j : \mathbb{R}^n \times \mathbb{R}^n \to \mathbb{R}$ is defined by

$$(x_1, \ldots, x_n)\gamma_j(y_1, \ldots, y_n) = max\{x_i \vee y_j : 1 \leq i \leq j\},$$

then for $\mathbf{a}, \mathbf{b} \in (\mathbb{R}^n)^X$ and $\mathbf{c} = \mathbf{a}\gamma\mathbf{b}$, the components of $\mathbf{c}(\mathbf{x}) = (\mathbf{c}_1(\mathbf{x}), \ldots, \mathbf{c}_n(\mathbf{x}))$ have values

$$\mathbf{c}_j(\mathbf{x}) = \mathbf{a}(\mathbf{x})\gamma_j\mathbf{b}(\mathbf{x}) = max\{\mathbf{a}_i(\mathbf{x}) \vee \mathbf{a}_j(\mathbf{x}) : 1 \leq i \leq j\}$$

for $j = 1, \ldots, n$.

As another example, suppose γ_1 and γ_2 are two binary operations $\mathbb{R}^2 \times \mathbb{R}^2 \to \mathbb{R}$ defined by

$$(x_1, x_2)\gamma_1(y_1, y_2) = x_1y_1 - x_2y_2$$

and

$$(x_1, x_2)\gamma_2(y_1, y_2) = x_1y_2 + x_2y_1,$$

respectively. Now if $\mathbf{a}, \mathbf{b} \in \left(\mathbb{R}^2\right)^{\mathbf{X}}$ represent two complex-valued images, then the product $\mathbf{c} = \mathbf{a}\gamma\mathbf{b}$ represents pointwise complex multiplication, namely

$$\mathbf{c}(\mathbf{x}) = (\mathbf{a}_1(\mathbf{x})\mathbf{b}_1(\mathbf{x}) - \mathbf{a}_2(\mathbf{x})\mathbf{b}_2(\mathbf{x}), \ \mathbf{a}_1(\mathbf{x})\mathbf{b}_2(\mathbf{x}) + \mathbf{a}_2(\mathbf{x})\mathbf{b}_1(\mathbf{x})).$$

Basic operations on single and multi-valued images can be combined to form image processing operations of arbitrary complexity. Two such operations that have proven to be extremely useful in processing real vector-valued images are the *winner take all jth-coordinate maximum* and minimum of two images. Specifically, if $\mathbf{a}, \mathbf{b} \in \left(\mathbb{R}^n\right)^{\mathbf{X}}$, then the *jth-coordinate maximum* of $\mathbf{a}$ and $\mathbf{b}$ is defined as

$$\mathbf{a} \vee |_j \mathbf{b} = \{(\mathbf{x}, \mathbf{c}(\mathbf{x})) \ : \ \mathbf{c}(\mathbf{x}) = \mathbf{a}(\mathbf{x}) \text{ if } \mathbf{a}_j(\mathbf{x}) \geq \mathbf{b}_j(\mathbf{x}), \text{ otherwise } \mathbf{c}(\mathbf{x}) = \mathbf{b}(\mathbf{x})\},$$

while the *jth-coordinate minimum* is defined as

$$\mathbf{a} \wedge |_j \mathbf{b} = \{(\mathbf{x}, \mathbf{c}(\mathbf{x})) \ : \ \mathbf{c}(\mathbf{x}) = \mathbf{a}(\mathbf{x}) \text{ if } \mathbf{a}_j(\mathbf{x}) \leq \mathbf{b}_j(\mathbf{x}), \text{ otherwise } \mathbf{c}(\mathbf{x}) = \mathbf{b}(\mathbf{x})\}.$$

Unary operations on vector-valued images are defined in a similar componentwise fashion. Given a function $f : \mathbb{R} \to \mathbb{R}$, then f induces a function $\mathbb{R}^n \to \mathbb{R}^n$, again denoted by f, which is defined by

$$f(x_1, x_2, \ldots, x_n) \equiv (f(x_1), f(x_2), \ldots, f(x_n)).$$

These functions provide for one type of unary operations on vector-valued images. In particular, if $\mathbf{a} = (\mathbf{a}_1, \mathbf{a}_2, \ldots, \mathbf{a}_n) \in \left(\mathbb{R}^n\right)^{\mathbf{X}}$, then

$$f(\mathbf{a}) \equiv f \circ \mathbf{a} = (f(\mathbf{a}_1), f(\mathbf{a}_2), \ldots, f(\mathbf{a}_n)).$$

Thus, if $f = sin : \mathbb{R} \to \mathbb{R}$, then

$$sin(\mathbf{a}) = (sin(\mathbf{a}_1), \ldots, sin(\mathbf{a}_n)).$$

Similarly, if $f = \chi_{\geq k}$, then

$$\chi_{\geq k}(\mathbf{a}) = \left(\chi_{\geq k}(\mathbf{a}_1), \ldots, \chi_{\geq k}(\mathbf{a}_n)\right).$$

Any function $f : \mathbb{R}^n \to \mathbb{R}^n$ gives rise to a sequence of functions $f_j = p_j \circ f : \mathbb{R}^n \to \mathbb{R}$, where $j = 1, \ldots, n$. Conversely, given a sequence of functions $f_j : \mathbb{R}^n \to \mathbb{R}$, where $j = 1, \ldots, n$, then we can define a function $f : \mathbb{R}^n \to \mathbb{R}^n$ by

$$f(\mathbf{x}) \equiv (f_1(\mathbf{x}), f_2(\mathbf{x}), \ldots, f_n(\mathbf{x})),$$

where $\mathbf{x} = (x_1, \ldots, x_n) \in \mathbb{R}^n$. Such functions provide for a more complex type of unary image operations since by definition

$$f(\mathbf{a}) = (f_1(\mathbf{a}), \ldots, f_m(\mathbf{a})) = \{(\mathbf{x}, \mathbf{b}(\mathbf{x})) \ : \ \mathbf{b}(\mathbf{x}) = (f_1(\mathbf{a}(\mathbf{x})), \ldots, f_m(\mathbf{a}(\mathbf{x})))\},$$

which means that the construction of each new coordinate depends on *all* the original coordinates. To provide a specific example, define $f_1 : \mathbb{R}^2 \to \mathbb{R}$ by $f_1(x, y) = sin(x) + cosh(y)$ and $f_2 : \mathbb{R}^2 \to \mathbb{R}$ by $f_2(x, y) = cos(x) + sinh(y)$. Then the induced function

$f : \left(\mathbb{R}^2\right)^{\mathbf{X}} \to \left(\mathbb{R}^2\right)^{\mathbf{X}}$ given by $f = (f_1, f_2)$. Applying f to an image $\mathbf{a} \in \left(\mathbb{R}^2\right)^{\mathbf{X}}$ results in the image

$$f(\mathbf{a}) = \{(\mathbf{x}, \mathbf{b}(\mathbf{x})) : \ \mathbf{b}(\mathbf{x}) = (sin(\mathbf{a}_1(\mathbf{x})) + cosh(\mathbf{a}_2(\mathbf{x})), \ cos(\mathbf{a}_1(\mathbf{x}))$$
$$+ \ sinh(\mathbf{a}_2(\mathbf{x}))), \ \mathbf{x} \in \mathbf{X} \}.$$

Thus, if we represent complex numbers as points in $\mathbb{R}^2$ and $\mathbf{a}$ denotes a complex-valued image, then $f(\mathbf{a})$ is a pointwise application of the complex sine function.

Global reduce operations are also applied componentwise. For example, if $\mathbf{a} \in (\mathbb{R}^n)^{\mathbf{X}}$, and $k = card(\mathbf{X})$, then

$$\Sigma\mathbf{a} = (\Sigma\mathbf{a}_1, \dots, \Sigma\mathbf{a}_n)$$

$$= \left(\sum_{j=1}^{k} \mathbf{a}_1(\mathbf{x}_j), \dots, \sum_{j=1}^{k} \mathbf{a}_n(\mathbf{x}_j) \right) \in \mathbb{R}^n.$$

In contrast, the summation $\sum\limits_{i=1}^{n} \mathbf{a}_i = \sum\limits_{i=1}^{n} p_i(\mathbf{a}) \in \mathbb{R}^{\mathbf{X}}$ since each $\mathbf{a}_i \in \mathbb{R}^{\mathbf{X}}$. Note that the projection function p_i is a unary operation $(\mathbb{R}^n)^{\mathbf{X}} \to \mathbb{R}^{\mathbf{X}}$.

Similarly,

$$\vee\,\mathbf{a} = (\vee\mathbf{a}_1, \dots, \vee\mathbf{a}_n),$$
$$\wedge\,\mathbf{a} = (\wedge\mathbf{a}_1, \dots, \wedge\mathbf{a}_n),$$

and

$$\Pi\mathbf{a} = (\Pi\mathbf{a}_1, \dots, \Pi\mathbf{a}_n).$$

Summary of Image Operations

The lists below summarize some of the more significant image operations.

Binary image operations.

It is assumed that only appropriately valued images are employed for the operations listed below. Thus, the operations of maximum and minimum apply to real- or integer-valued images but not complex-valued images. Similarly, union and intersection apply only to set-valued images.

generic	$\mathbf{a}\gamma\mathbf{b} = \{(\mathbf{x}, \mathbf{c}(\mathbf{x})) : \ \mathbf{c}(\mathbf{x}) = \mathbf{a}(\mathbf{x})\gamma\mathbf{b}(\mathbf{x}), \ \mathbf{x} \in \mathbf{X}\}$
addition	$\mathbf{a} + \mathbf{b} = \{(\mathbf{x}, \mathbf{c}(\mathbf{x})) : \ \mathbf{c}(\mathbf{x}) = \mathbf{a}(\mathbf{x}) + \mathbf{b}(\mathbf{x}), \ \mathbf{x} \in \mathbf{X}\}$
multiplication	$\mathbf{a} \cdot \mathbf{b} = \{(\mathbf{x}, \mathbf{c}(\mathbf{x})) : \ \mathbf{c}(\mathbf{x}) = \mathbf{a}(\mathbf{x}) \cdot \mathbf{b}(\mathbf{x}), \ \mathbf{x} \in \mathbf{X}\}$
maximum	$\mathbf{a} \vee \mathbf{b} = \{(\mathbf{x}, \mathbf{c}(\mathbf{x})) : \ \mathbf{c}(\mathbf{x}) = \mathbf{a}(\mathbf{x}) \vee \mathbf{b}(\mathbf{x}), \ \mathbf{x} \in \mathbf{X}\}$
minimum	$\mathbf{a} \wedge \mathbf{b} = \{(\mathbf{x}, \mathbf{c}(\mathbf{x})) : \ \mathbf{c}(\mathbf{x}) = \mathbf{a}(\mathbf{x}) \wedge \mathbf{b}(\mathbf{x}), \ \mathbf{x} \in \mathbf{X}\}$
scalar addition	$k + \mathbf{a} = \{(\mathbf{x}, \mathbf{c}(\mathbf{x})) : \ \mathbf{c}(\mathbf{x}) = k + \mathbf{a}(\mathbf{x}), \ \mathbf{x} \in \mathbf{X}\}$
scalar multiplication	$k \cdot \mathbf{a} = \{(\mathbf{x}, \mathbf{c}(\mathbf{x})) : \ \mathbf{c}(\mathbf{x}) = k \cdot \mathbf{a}(\mathbf{x}), \ \mathbf{x} \in \mathbf{X}\}$
point addition	$\mathbf{a} + \mathbf{y} = \{(\mathbf{z}, \mathbf{b}(\mathbf{z})) : \ \mathbf{b}(\mathbf{z}) = \mathbf{a}(\mathbf{z} - \mathbf{y}), \ \mathbf{z} \in \mathbf{X} + \mathbf{y}\}$
union	$\mathbf{a} \cup \mathbf{b} = \{(\mathbf{x}, \mathbf{c}(\mathbf{x})) : \ \mathbf{c}(\mathbf{x}) = \mathbf{a}(\mathbf{x}) \cup \mathbf{b}(\mathbf{x}), \ \mathbf{x} \in \mathbf{X}\}$
intersection	$\mathbf{a} \cap \mathbf{b} = \{(\mathbf{x}, \mathbf{c}(\mathbf{x})) : \ \mathbf{c}(\mathbf{x}) = \mathbf{a}(\mathbf{x}) \cap \mathbf{b}(\mathbf{x}), \ \mathbf{x} \in \mathbf{X}\}$
exponentiation	$\mathbf{a}^{\mathbf{b}} = \left\{(\mathbf{x}, \mathbf{c}(\mathbf{x})) : \ \mathbf{c}(\mathbf{x}) = \mathbf{a}(\mathbf{x})^{\mathbf{b}(\mathbf{x})}, \ \mathbf{x} \in \mathbf{X}\right\}$
logarithm	$log_{\mathbf{b}}\mathbf{a} = \{(\mathbf{x}, \mathbf{c}(\mathbf{x})) : \ \mathbf{c}(\mathbf{x}) = log_{\mathbf{b}(\mathbf{x})}\mathbf{a}(\mathbf{x}), \ \mathbf{x} \in \mathbf{X}\}$

concatenation $(\mathbf{a}|\mathbf{b}) = \mathbf{a}|^{\mathbf{b}+(0,k)}, \ \mathbf{a} \in \mathbb{F}^{Z_m \times Z_k}, \ \mathbf{b} \in \mathbb{F}^{Z_m \times Z_n}$

concatenation $\begin{pmatrix} \mathbf{a} \\ - \\ \mathbf{b} \end{pmatrix} = (\mathbf{a}|\mathbf{b})'$

characteristics

$$\chi_{\leq \mathbf{b}}(\mathbf{a}) = \{(\mathbf{x}, \mathbf{c}(\mathbf{x})) : \ \mathbf{c}(\mathbf{x}) = 1 \text{ if } \mathbf{a}(\mathbf{x}) \leq \mathbf{b}(\mathbf{x}), \text{ otherwise } \mathbf{c}(\mathbf{x}) = 0\}$$
$$\chi_{< \mathbf{b}}(\mathbf{a}) = \{(\mathbf{x}, \mathbf{c}(\mathbf{x})) : \ \mathbf{c}(\mathbf{x}) = 1 \text{ if } \mathbf{a}(\mathbf{x}) < \mathbf{b}(\mathbf{x}), \text{ otherwise } \mathbf{c}(\mathbf{x}) = 0\}$$
$$\chi_{= \mathbf{b}}(\mathbf{a}) = \{(\mathbf{x}, \mathbf{c}(\mathbf{x})) : \ \mathbf{c}(\mathbf{x}) = 1 \text{ if } \mathbf{a}(\mathbf{x}) = \mathbf{b}(\mathbf{x}), \text{ otherwise } \mathbf{c}(\mathbf{x}) = 0\}$$
$$\chi_{\geq \mathbf{b}}(\mathbf{a}) = \{(\mathbf{x}, \mathbf{c}(\mathbf{x})) : \ \mathbf{c}(\mathbf{x}) = 1 \text{ if } \mathbf{a}(\mathbf{x}) \geq \mathbf{b}(\mathbf{x}), \text{ otherwise } \mathbf{c}(\mathbf{x}) = 0\}$$
$$\chi_{> \mathbf{b}}(\mathbf{a}) = \{(\mathbf{x}, \mathbf{c}(\mathbf{x})) : \ \mathbf{c}(\mathbf{x}) = 1 \text{ if } \mathbf{a}(\mathbf{x}) > \mathbf{b}(\mathbf{x}), \text{ otherwise } \mathbf{c}(\mathbf{x}) = 0\}$$
$$\chi_{\neq \mathbf{b}}(\mathbf{a}) = \{(\mathbf{x}, \mathbf{c}(\mathbf{x})) : \ \mathbf{c}(\mathbf{x}) = 1 \text{ if } \mathbf{a}(\mathbf{x}) \neq \mathbf{b}(\mathbf{x}), \text{ otherwise } \mathbf{c}(\mathbf{x}) = 0\}$$

Whenever $\mathbf{b}$ is a constant image, say $\mathbf{b} = k$ (i.e., $\mathbf{b}(\mathbf{x}) = k \ \forall \mathbf{x} \in \mathbf{X}$), then we simply write $\mathbf{a}^k$ for $\mathbf{a}^\mathbf{b}$ and $log_k \mathbf{a}$ for $log_\mathbf{b} \mathbf{a}$. Similarly, we have $k + \mathbf{a}$, $\chi_{\leq k}(\mathbf{a})$, $\chi_{< k}(\mathbf{a})$, etc.

Unary image operations.

As in the case of binary operations, we again assume that only appropriately valued images are employed for the operations listed below.

value transform $f \circ \mathbf{a} = f(\mathbf{a}) = \{(\mathbf{x}, \mathbf{c}(\mathbf{x})) : \ \mathbf{c}(\mathbf{x}) = f(\mathbf{a}(\mathbf{x})), \ \mathbf{x} \in \mathbf{X}\}$

spatial transform $\mathbf{a} \circ f = \{(\mathbf{y}, \mathbf{a}(f(\mathbf{y}))) : \ \mathbf{y} \in \mathbf{Y}\}$

domain restriction $\mathbf{a}|_Z = \{(\mathbf{x}, \mathbf{a}(\mathbf{x})) : \ \mathbf{x} \in \mathbf{Z}\}$

range restriction $\mathbf{a}\|_S = \{(\mathbf{x}, \mathbf{a}(\mathbf{x})) : \ \mathbf{a}(\mathbf{x}) \in S\}$

extension $\mathbf{a}|^\mathbf{b} = \left\{(\mathbf{x}, \mathbf{c}(\mathbf{x})) : \ \mathbf{c}(\mathbf{x}) = \begin{cases} \mathbf{a}(\mathbf{x}) & \text{if } \mathbf{x} \in \mathbf{X} \\ \mathbf{b}(\mathbf{x}) & \text{if } \mathbf{x} \in \mathbf{Y} \backslash \mathbf{X} \end{cases}\right\}$

domain $domain(\mathbf{a}) = \{\mathbf{x} \in \mathbf{X} : \ \exists r \in \mathbb{F} \ s.t. \ \mathbf{a}(\mathbf{x}) = r\}$

range $range(\mathbf{a}) = \{r \in \mathbb{F} : \ \exists \mathbf{x} \in \mathbf{X} \ s.t. \ r = \mathbf{a}(\mathbf{x})\}$

generic reduction $\Gamma \, \mathbf{a} = \mathbf{a}(\mathbf{x}_1) \gamma \mathbf{a}(\mathbf{x}_2) \gamma \cdots \gamma \mathbf{a}(\mathbf{x}_n)$

image sum $\sum \mathbf{a} = \sum_{\mathbf{x} \in \mathbf{X}} \mathbf{a}(\mathbf{x}) = \mathbf{a}(\mathbf{x}_1) + \mathbf{a}(\mathbf{x}_2) + \cdots + \mathbf{a}(\mathbf{x}_n)$

image product $\prod \mathbf{a} = \prod_{\mathbf{x} \in \mathbf{X}} \mathbf{a}(\mathbf{x}) = \mathbf{a}(\mathbf{x}_1) \cdot \mathbf{a}(\mathbf{x}_2) \cdot \cdots \cdot \mathbf{a}(\mathbf{x}_n)$

image maximum $\bigvee \mathbf{a} = \bigvee_{\mathbf{x} \in \mathbf{X}} \mathbf{a}(\mathbf{x}) = \mathbf{a}(\mathbf{x}_1) \vee \mathbf{a}(\mathbf{x}_2) \vee \cdots \vee \mathbf{a}(\mathbf{x}_n)$

image minimum $\bigwedge \mathbf{a} = \bigwedge_{\mathbf{x} \in \mathbf{X}} \mathbf{a}(\mathbf{x}) = \mathbf{a}(\mathbf{x}_1) \wedge \mathbf{a}(\mathbf{x}_2) \wedge \cdots \wedge \mathbf{a}(\mathbf{x}_n)$

image complement $\tilde{\mathbf{a}} = \left\{(\mathbf{x}, \mathbf{c}(\mathbf{x})) : \ \mathbf{c}(\mathbf{x}) = \widetilde{\mathbf{a}(\mathbf{x})}, \ \mathbf{x} \in \mathbf{X}\right\}$

pseudo inverse $\mathbf{a}^{-1} = \left\{(\mathbf{x}, \mathbf{b}(\mathbf{x})) : \ \mathbf{b}(\mathbf{x}) = \begin{cases} \frac{1}{\mathbf{a}(\mathbf{x})} & \text{if } \mathbf{a}(\mathbf{x}) \neq 0 \\ 0 & \text{otherwise} \end{cases}\right\}$

image transpose $\mathbf{a}' = \{((x, y), \mathbf{a}'(x, y)) : \ \mathbf{a}'(x, y) = \mathbf{a}(y, x), \ (y, x) \in \mathbf{X}\}$

1.5.   Templates

Templates are images whose *values* are images. The notion of a template, as used in image algebra, unifies and generalizes the usual concepts of templates, masks, windows, and neighborhood functions into one general mathematical entity. In addition, templates generalize the notion of structuring elements as used in mathematical morphology [27, 60].

Definition. A *template* is an image whose pixel values are images (functions). In particular, an $\mathbb{F}$-*valued template from* $\mathbf{Y}$ *to* $\mathbf{X}$ is a function $\mathbf{t} : \mathbf{Y} \to \mathbb{F}^{\mathbf{X}}$. Thus, $\mathbf{t} \in \left(\mathbb{F}^{\mathbf{X}}\right)^{\mathbf{Y}}$ and $\mathbf{t}$ is an $\mathbb{F}^{\mathbf{X}}$-valued image on $\mathbf{Y}$.

For notational convenience we define $\mathbf{t_y} \equiv \mathbf{t}(\mathbf{y})$ $\forall \mathbf{y} \in \mathbf{Y}$. The image $\mathbf{t_y}$ has representation

$$\mathbf{t_y} = \{(\mathbf{x}, \mathbf{t_y}(\mathbf{x})) : \mathbf{x} \in \mathbf{X}\}.$$

The pixel values $\mathbf{t_y}(\mathbf{x})$ of this image are called the *weights* of the template at point $\mathbf{y}$.

If $\mathbf{t}$ is a real- or complex-valued template from $\mathbf{Y}$ to $\mathbf{X}$, then the *support* of $\mathbf{t_y}$ is denoted by $S(\mathbf{t_y})$ and is defined as

$$S(\mathbf{t_y}) = \{\mathbf{x} \in \mathbf{X} : \mathbf{t_y}(\mathbf{x}) \neq 0\}.$$

More generally, if $\mathbf{t} \in \left(\mathbb{F}^{\mathbf{X}}\right)^{\mathbf{Y}}$ and $\mathbb{F}$ is an algebraic structure with a zero element 0, then the support of $\mathbf{t_y}$ will be defined as $S(\mathbf{t_y}) = \{\mathbf{x} \in \mathbf{X} : \mathbf{t_y}(\mathbf{x}) \neq 0\}$.

For extended real-valued templates we also define the following supports *at infinity*:

$$S_\infty(\mathbf{t_y}) = \{\mathbf{x} \in \mathbf{X} : \mathbf{t_y}(\mathbf{x}) \neq \infty\}$$

and

$$S_{-\infty}(\mathbf{t_y}) = \{\mathbf{x} \in \mathbf{X} : \mathbf{t_y}(\mathbf{x}) \neq -\infty\}.$$

If $\mathbf{X}$ is a space with an operation $+$ such that $(\mathbf{X}, +)$ is a group, then a template $\mathbf{t} \in \left(\mathbb{F}^{\mathbf{X}}\right)^{\mathbf{X}}$ is said to be *translation invariant* (with respect to the operation $+$) if and only if for each triple $\mathbf{x}, \mathbf{y}, \mathbf{z} \in \mathbf{X}$ we have that $\mathbf{t_y}(\mathbf{x}) = \mathbf{t_{y+z}}(\mathbf{x} + \mathbf{z})$. Templates that are not translation invariant are called *translation variant* or, simply, *variant* templates. A large class of translation invariant templates with finite support have the nice property that they can be defined pictorially. For example, let $\mathbf{X} = \mathbb{Z}^2$ and $\mathbf{y} = (x,y)$ be an arbitrary point of $\mathbf{X}$. Set $\mathbf{x_1} = (x, y-1)$, $\mathbf{x_2} = (x+1, y)$, and $\mathbf{x_3} = (x+1, y-1)$. Define $\mathbf{t} \in \left(\mathbb{R}^{\mathbf{X}}\right)^{\mathbf{X}}$ by defining the weights $\mathbf{t_y}(\mathbf{y}) = 1$, $\mathbf{t_y}(\mathbf{x_1}) = 3$, $\mathbf{t_y}(\mathbf{x_2}) = 2$, $\mathbf{t_y}(\mathbf{x_3}) = 4$, and $\mathbf{t_y}(\mathbf{x}) = 0$ whenever $\mathbf{x}$ is not an element of $\{\mathbf{y}, \mathbf{x_1}, \mathbf{x_2}, \mathbf{x_3}\}$. Note that it follows from the definition of $\mathbf{t}$ that $S(\mathbf{t_y}) = \{\mathbf{y}, \mathbf{x_1}, \mathbf{x_2}, \mathbf{x_3}\}$. Thus, at any arbitrary point $\mathbf{y}$, the configuration of the support and weights of $\mathbf{t_y}$ is as shown in Figure 1.5.1. The shaded cell in the pictorial representation of $\mathbf{t_y}$ indicates the location of the point $\mathbf{y}$.

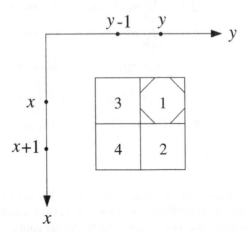

Figure 1.5.1. Pictorial representation of a translation invariant template.

There are certain collections of templates that can be defined explicitly in terms of parameters. These parameterized templates are of great practical importance.

Definition. A *parameterized* $\mathbb{F}$-*valued template from* **Y** *to* **X** *with parameters in* P is a function of form $\mathbf{t} : P \to \left(\mathbb{F}^{\mathbf{X}}\right)^{\mathbf{Y}}$. The set P is called the *set of parameters* and each $p \in P$ is called a *parameter* of **t**.

Thus, a parameterized $\mathbb{F}$-valued template from **Y** to **X** gives rise to a family of regular $\mathbb{F}$-valued templates from **Y** to **X**, namely $\left\{\mathbf{t}(p) \in \left(\mathbb{F}^{\mathbf{X}}\right)^{\mathbf{Y}} : p \in P\right\}$.

Image-Template Products

The definition of an image-template product provides the rules for combining images with templates and templates with templates. The definition of this product includes the usual correlation and convolution products used in digital image processing. Suppose $\mathbb{F}$ is a value set with two binary operations $\bigcirc$ and γ, where $\bigcirc$ distributes over γ, and γ is associative and commutative. If $\mathbf{t} \in \left(\mathbb{F}^{\mathbf{X}}\right)^{\mathbf{Y}}$, then for each $\mathbf{y} \in \mathbf{Y}$, $\mathbf{t_y} \in \mathbb{F}^{\mathbf{X}}$. Thus, if $\mathbf{a} \in \mathbb{F}^{\mathbf{X}}$, where **X** is finite, then $\mathbf{a} \bigcirc \mathbf{t_y} \in \mathbb{F}^{\mathbf{X}}$ and $\Gamma\left(\mathbf{a} \bigcirc \mathbf{t_y}\right) \in \mathbb{F}$. It follows that the binary operations $\bigcirc$ and γ induce a binary operation

$$\textcircled{γ} : \mathbb{F}^{\mathbf{X}} \times \left(\mathbb{F}^{\mathbf{X}}\right)^{\mathbf{Y}} \to \mathbb{F}^{\mathbf{Y}} ,$$

where

$$\mathbf{b} = \mathbf{a} \,\textcircled{γ}\, \mathbf{t} \in \mathbb{F}^{\mathbf{Y}}$$

is defined by

$$\mathbf{b}(\mathbf{y}) = \Gamma(\mathbf{a} \bigcirc \mathbf{t_y}) = \underset{\mathbf{x} \in \mathbf{X}}{\Gamma} \left(\mathbf{a}(\mathbf{x}) \bigcirc \mathbf{t_y}(\mathbf{x})\right) .$$

Therefore, if $\mathbf{X} = \{\mathbf{x}_1, \mathbf{x}_2, \dots, \mathbf{x}_n\}$, then

$$\mathbf{b}(\mathbf{y}) = (\mathbf{a}(\mathbf{x}_1) \bigcirc \mathbf{t_y}(\mathbf{x}_1)) \gamma (\mathbf{a}(\mathbf{x}_2) \bigcirc \mathbf{t_y}(\mathbf{x}_2)) \gamma \cdots \gamma (\mathbf{a}(\mathbf{x}_n) \bigcirc \mathbf{t_y}(\mathbf{x}_n)) .$$

The expression $\mathbf{a} \,\textcircled{γ}\, \mathbf{t}$ is called the *right convolution product of* **a** *with* **t**. Note that while **a** is an image on **X**, the product $\mathbf{a} \,\textcircled{γ}\, \mathbf{t}$ is an image on **Y**. Thus, templates allow for the transformation of an image from one type of domain to an entirely different domain type.

Replacing $(\mathbb{F}, \gamma, \bigcirc)$ by $(\mathbb{R}, +, \cdot)$ changes $\mathbf{b} = \mathbf{a} \,\textcircled{γ}\, \mathbf{t}$ into

$$\mathbf{b} = \mathbf{a} \oplus \mathbf{t} ,$$

the *linear image-template product* or simply the *convolution* of **a** with **t**, where

$$\mathbf{b}(\mathbf{y}) = \sum_{\mathbf{x} \in \mathbf{X}} (\mathbf{a}(\mathbf{x}) \cdot \mathbf{t_y}(\mathbf{x})) ,$$

$\mathbf{a} \in \mathbb{R}^{\mathbf{X}}$, and $\mathbf{t} \in \left(\mathbb{R}^{\mathbf{X}}\right)^{\mathbf{Y}}$.

Every template $\mathbf{s} \in \left(\mathbb{F}^{\mathbf{Y}}\right)^{\mathbf{X}}$ has a transpose $\mathbf{s}' \in \left(\mathbb{F}^{\mathbf{X}}\right)^{\mathbf{Y}}$ which is defined $\mathbf{s}'_{\mathbf{y}}(\mathbf{x}) = \mathbf{s_x}(\mathbf{y})$. Obviously, $(\mathbf{s}')' = \mathbf{s}$ and $\mathbf{s}'$ reverses the mapping order from $\mathbf{X} \to \mathbb{F}^{\mathbf{Y}}$

to $\mathbf{Y} \to \mathbb{F}^{\mathbf{X}}$. By definition, $s'_{\mathbf{y}} \bigcirc a \in \mathbb{F}^{\mathbf{X}}$ and $\Gamma\left(s'_{\mathbf{y}} \bigcirc a\right) \in \mathbb{F}$, whenever $a \in \mathbb{F}^{\mathbf{X}}$ and $s \in \left(\mathbb{F}^{\mathbf{Y}}\right)^{\mathbf{X}}$. Hence the binary operations $\bigcirc$ and γ induce another product operation

$$\textcircled{\gamma} : \left(\mathbb{F}^{\mathbf{Y}}\right)^{\mathbf{X}} \times \mathbb{F}^{\mathbf{X}} \to \mathbb{F}^{\mathbf{Y}},$$

where

$$\mathbf{b} = s\,\textcircled{\gamma}\,a \in \mathbb{F}^{\mathbf{Y}}$$

is defined by

$$\mathbf{b}(\mathbf{y}) = \Gamma\left(s'_{\mathbf{y}} \bigcirc a\right) = \underset{\mathbf{x} \in \mathbf{X}}{\Gamma}\left(s'_{\mathbf{y}}(\mathbf{x}) \bigcirc a(\mathbf{x})\right).$$

The expression $s\,\textcircled{\gamma}\,a$ is called the *left convolution product of* a *with* s.

When computing $s\,\textcircled{\gamma}\,a$, it is not necessary to use the transpose s' since

$$\underset{\mathbf{x} \in \mathbf{X}}{\Gamma}\left(s'_{\mathbf{y}}(\mathbf{x}) \bigcirc a(\mathbf{x})\right) = \underset{\mathbf{x} \in \mathbf{X}}{\Gamma}\left(s_{\mathbf{x}}(\mathbf{y}) \bigcirc a(\mathbf{x})\right).$$

This allows us to redefine the transformation $\mathbf{b} = s\,\textcircled{\gamma}\,a$ as

$$\mathbf{b}(\mathbf{y}) = \underset{\mathbf{x} \in \mathbf{X}}{\Gamma}\left(s_{\mathbf{x}}(\mathbf{y}) \bigcirc a(\mathbf{x})\right).$$

For the remainder of this section, we assume that $(\mathbb{F}, \gamma)$ is a monoid and let 0 denote the zero of $\mathbb{F}$ under the operation γ. Suppose $a \in \mathbb{F}^{\mathbf{X}}$ and $t \in \left(\mathbb{F}^{\mathbf{Z}}\right)^{\mathbf{Y}}$, where $\mathbf{X}$ and $\mathbf{Z}$ are subsets of the same space. Since $\mathbb{F}$ is a monoid, the operator $\textcircled{\gamma}$ can be extended to a mapping

$$\textcircled{\gamma} : \mathbb{F}^{\mathbf{X}} \times \left(\mathbb{F}^{\mathbf{Z}}\right)^{\mathbf{Y}} \to \mathbb{F}^{\mathbf{Y}},$$

where $\mathbf{b} = a\,\textcircled{\gamma}\,t$ is defined by

$$\mathbf{b}(\mathbf{y}) = \begin{cases} \underset{\mathbf{x} \in \mathbf{X} \cap \mathbf{Z}}{\Gamma}\left(a(\mathbf{x}) \bigcirc t_{\mathbf{y}}(\mathbf{x})\right) & \text{if } \mathbf{X} \cap \mathbf{Z} \neq \varnothing \\ 0 & \text{if } \mathbf{X} \cap \mathbf{Z} = \varnothing. \end{cases}$$

The left convolution product $s\,\textcircled{\gamma}\,a$ is defined in a similar fashion. Subsequent examples will demonstrate that the ability of replacing $\mathbf{X}$ with $\mathbf{Z}$ greatly simplifies the issue of template implementation and the use of templates in algorithm development.

Significant reduction in the number of computations involving the image-template product can be achieved if $(\mathbb{F}, \gamma, \bigcirc)$ is a commutative semiring. Recall that if $t \in \left(\mathbb{F}^{\mathbf{Z}}\right)^{\mathbf{Y}}$, then the support of t at a point $\mathbf{y} \in \mathbf{Y}$ with respect to the operation γ is defined as $S(t_{\mathbf{y}}) = \{\mathbf{x} \in \mathbf{Z} : t_{\mathbf{y}}(\mathbf{x}) \neq 0\}$. Since $t_{\mathbf{y}}(\mathbf{x}) = 0$ whenever $\mathbf{x} \notin S(t_{\mathbf{y}})$, we have that $a(\mathbf{x}) \bigcirc t_{\mathbf{y}}(\mathbf{x}) = 0$ whenever $\mathbf{x} \notin S(t_{\mathbf{y}})$ and, therefore,

$$\underset{\mathbf{x} \in \mathbf{X} \cap \mathbf{Z}}{\Gamma}\left(a(\mathbf{x}) \bigcirc t_{\mathbf{y}}(\mathbf{x})\right) = \underset{\mathbf{x} \in \mathbf{X} \cap S(t_{\mathbf{y}})}{\Gamma}\left(a(\mathbf{x}) \bigcirc t_{\mathbf{y}}(\mathbf{x})\right).$$

It follows that the computation of the new pixel value $\mathbf{b}(\mathbf{y})$ does not depend on the size of $\mathbf{X}$, but on the size of $S(t_{\mathbf{y}})$. Therefore, if $k = card(\mathbf{X} \cap S(t_{\mathbf{y}}))$, then the computation of $\mathbf{b}(\mathbf{y})$ requires a total of $2k - 1$ operations of type γ and $\bigcirc$.

As pointed out earlier, substitution of different value sets and specific binary operations for γ and $\bigcirc$ results in a wide variety of different image transforms. Our prime examples are the ring $(\mathbb{R}, +, \cdot)$ and the value sets $(\mathbb{R}_{\pm\infty}, \vee, \wedge, +, +')$ and $(\mathbb{R}_{\infty}^{\geq 0}, \vee, \wedge, \times, \times')$. The structure $(\mathbb{R}_{\pm\infty}, \vee, \wedge, +, +')$ provides for two lattice products:

$$\mathbf{b} = a \boxed{\vee} t,$$

where

$$\mathbf{b(y)} = \bigvee_{\mathbf{x} \in \mathbf{X} \cap S_{-\infty}(\mathbf{t_y})} [\mathbf{a(x)} + \mathbf{t_y(x)}] \,,$$

and

$$\mathbf{b} = \mathbf{a} \boxminus \mathbf{t} \,,$$

where

$$\mathbf{b(y)} = \bigwedge_{\mathbf{x} \in \mathbf{X} \cap S_{\infty}(\mathbf{t_y})} [\mathbf{a(x)} +' \mathbf{t_y(x)}] \,.$$

In order to distinguish between these two types of lattice transforms, we call the operator $\boxvee$ the *morphological max convolution operator* and $\boxminus$ the *morphological min convolution operator*. It follows from our earlier discussion that if $\mathbf{X} \cap S_{-\infty}(\mathbf{t_y}) = \varnothing$, then the value of $\mathbf{b(y)}$ is $-\infty$, the *zero* of $\mathbb{R}_{\pm\infty}$ under the operation of $\vee$. Similarly, if $\mathbf{X} \cap S_{\infty}(\mathbf{t_y}) = \varnothing$, then $\mathbf{b(y)} = \infty$.

The left morphological max and min operations are defined by

$$\mathbf{t} \boxvee \mathbf{a} = \left\{ (\mathbf{y}, \mathbf{b(y)}) \,:\, \mathbf{b(y)} = \bigvee_{\mathbf{x} \in \mathbf{X} \cap S_{-\infty}(\mathbf{t_y})} [\mathbf{t_x(y)} + \mathbf{a(x)}] \,,\, \mathbf{y} \in \mathbf{Y} \right\}$$

and

$$\mathbf{t} \boxminus \mathbf{a} = \left\{ (\mathbf{y}, \mathbf{b(y)}) \,:\, \mathbf{b(y)} = \bigwedge_{\mathbf{x} \in \mathbf{X} \cap S_{\infty}(\mathbf{t_y})} [\mathbf{t_x(y)} +' \mathbf{a(x)}] \,,\, \mathbf{y} \in \mathbf{Y} \right\} \,,$$

respectively. The relationship between the morphological max and min is given in terms of lattice duality by

$$\mathbf{a} \boxminus \mathbf{t} = (\mathbf{t}^* \boxvee \mathbf{a}^*)^* \,,$$

where the image $\mathbf{a}^*$ is defined by $\mathbf{a}^*(\mathbf{x}) = [\mathbf{a(x)}]^*$, and the conjugate (or dual) of $\mathbf{t} \in \left(\mathbb{R}_{\pm\infty}^{\mathbf{X}}\right)^{\mathbf{Y}}$ is the template $\mathbf{t}^* \in \left(\mathbb{R}_{\pm\infty}^{\mathbf{Y}}\right)^{\mathbf{X}}$ defined by $\mathbf{t_x^*(y)} = [\mathbf{t_y(x)}]^*$. It follows that $\mathbf{t_x^*(y)} = -\mathbf{t_y'(x)}$.

The value set $\left(\mathbb{R}_{\infty}^{\geq 0}, \vee, \wedge, \times, \times'\right)$ also provides for two lattice products. Specifically, we have

$$\mathbf{b} = \mathbf{a} \varovee \mathbf{t} \,,$$

where

$$\mathbf{b(y)} = \bigvee_{\mathbf{x} \in \mathbf{X} \cap S(\mathbf{t_y})} [\mathbf{a(x)} \times \mathbf{t_y(x)}] \,,$$

and

$$\mathbf{b} = \mathbf{a} \varowedge \mathbf{t} \,,$$

where

$$\mathbf{b(y)} = \bigwedge_{\mathbf{x} \in \mathbf{X} \cap S_{\infty}(\mathbf{t_y})} [\mathbf{a(x)} \times' \mathbf{t_y(x)}] \,.$$

Here 0 is the zero of $\mathbb{R}_{\infty}^{\geq 0}$ under the operation of $\vee$, so that $\mathbf{b(y)} = 0$ whenever $\mathbf{X} \cap S(\mathbf{t_y}) = \varnothing$. Similarly, $\mathbf{b(y)} = \infty$ whenever $\mathbf{X} \cap S_{\infty}(\mathbf{t_y}) = \varnothing$.

The lattice products $\bigcirc\!\!\!\!\vee$ and $\bigcirc\!\!\!\!\wedge$ are called the *multiplicative maximum* and *multiplicative minimum*, respectively. The *left multiplicative max* and *left multiplicative min* are defined as

$$t \,\bigcirc\!\!\!\!\vee\, a = \left\{ (\mathbf{y}, \mathbf{b}(\mathbf{y})) \,:\, \mathbf{b}(\mathbf{y}) = \bigvee_{\mathbf{x} \in \mathbf{X} \cap S_\infty(t'_\mathbf{y})} [\mathbf{t_x}(\mathbf{y}) \times \mathbf{a}(\mathbf{x})] \,,\, \mathbf{y} \in \mathbf{Y} \right\}$$

and

$$t \,\bigcirc\!\!\!\!\wedge\, a = \left\{ (\mathbf{y}, \mathbf{b}(\mathbf{y})) \,:\, \mathbf{b}(\mathbf{y}) = \bigwedge_{\mathbf{x} \in \mathbf{X} \cap S_\infty(t'_\mathbf{y})} [\mathbf{t_x}(\mathbf{y}) \times' \mathbf{a}(\mathbf{x})] \,,\, \mathbf{y} \in \mathbf{Y} \right\},$$

respectively. The duality relation between the multiplicative max and min is given by

$$\mathbf{a} \,\bigcirc\!\!\!\!\wedge\, t = (t^* \,\bigcirc\!\!\!\!\vee\, \mathbf{a}^*)^*,$$

where $\mathbf{a}^*(\mathbf{x}) = (\mathbf{a}(\mathbf{x}))^*$ and $\mathbf{t}_\mathbf{x}^*(\mathbf{y}) = [\mathbf{t_y}(\mathbf{x})]^*$. Here r^* denotes the conjugate of r in $\mathbb{R}_\infty^{\geq 0}$.

Summary of Image-Template Products

In the following list of pertinent image-template products $\mathbf{a} \in \mathbb{F}^\mathbf{X}$ and $t \in (\mathbb{F}^\mathbf{X})^\mathbf{Y}$. Again, for each operation we assume the appropriate value set $\mathbb{F}$.

right generic convolution product

$$\mathbf{a} \,\circledcirc\, t = \left\{ (\mathbf{y}, \mathbf{b}(\mathbf{y})) \,:\, \mathbf{b}(\mathbf{y}) = \overset{\Gamma}{\underset{\mathbf{x} \in \mathbf{X}}{}} (\mathbf{a}(\mathbf{x}) \bigcirc \mathbf{t_y}(\mathbf{x})) \,,\, \mathbf{y} \in \mathbf{Y} \right\}$$

right linear convolution product

$$\mathbf{a} \,\oplus\, t = \left\{ (\mathbf{y}, \mathbf{b}(\mathbf{y})) \,:\, \mathbf{b}(\mathbf{y}) = \sum_{\mathbf{x} \in \mathbf{X}} (\mathbf{a}(\mathbf{x}) \cdot \mathbf{t_y}(\mathbf{x})) \,,\, \mathbf{y} \in \mathbf{Y} \right\}$$

right morphological max convolution product

$$\mathbf{a} \,\boxplus\, t = \left\{ (\mathbf{y}, \mathbf{b}(\mathbf{y})) \,:\, \mathbf{b}(\mathbf{y}) = \bigvee_{\mathbf{x} \in \mathbf{X}} [\mathbf{a}(\mathbf{x}) + \mathbf{t_y}(\mathbf{x})] \,,\, \mathbf{y} \in \mathbf{Y} \right\}$$

right morphological min convolution product

$$\mathbf{a} \,\boxplus'\, t = \left\{ (\mathbf{y}, \mathbf{b}(\mathbf{y})) \,:\, \mathbf{b}(\mathbf{y}) = \bigwedge_{\mathbf{x} \in \mathbf{X}} [\mathbf{a}(\mathbf{x}) +' \mathbf{t_y}(\mathbf{x})] \,,\, \mathbf{y} \in \mathbf{Y} \right\}$$

right multiplicative max convolution product

$$\mathbf{a} \,\circledcirc\!\!\!\!\vee\, t = \left\{ (\mathbf{y}, \mathbf{b}(\mathbf{y})) \,:\, \mathbf{b}(\mathbf{y}) = \bigvee_{\mathbf{x} \in \mathbf{X}} [\mathbf{a}(\mathbf{x}) \times \mathbf{t_y}(\mathbf{x})] \,,\, \mathbf{y} \in \mathbf{Y} \right\}$$

right multiplicative min convolution product

$$\mathbf{a} \,\circledcirc\!\!\!\!\wedge\, t = \left\{ (\mathbf{y}, \mathbf{b}(\mathbf{y})) \,:\, \mathbf{b}(\mathbf{y}) = \bigwedge_{\mathbf{x} \in \mathbf{X}} [\mathbf{a}(\mathbf{x}) \times' \mathbf{t_y}(\mathbf{x})] \,,\, \mathbf{y} \in \mathbf{Y} \right\}$$

right xor max convolution product

$$\mathbf{a} \,\tilde{\boxplus}\, t = \left\{ (\mathbf{y}, \mathbf{b}(\mathbf{y})) \,:\, \mathbf{b}(\mathbf{y}) = \bigvee_{\mathbf{x} \in \mathbf{X}} [\mathbf{a}(\mathbf{x}) \tilde{+} \mathbf{t_y}(\mathbf{x})] \,,\, \mathbf{y} \in \mathbf{Y} \right\}$$

right xor min convolution product

$$\mathbf{a} \,\tilde{\boxplus}'\, t = \left\{ (\mathbf{y}, \mathbf{b}(\mathbf{y})) \,:\, \mathbf{b}(\mathbf{y}) = \bigwedge_{\mathbf{x} \in \mathbf{X}} [\mathbf{a}(\mathbf{x}) \tilde{+}' \mathbf{t_y}(\mathbf{x})] \,,\, \mathbf{y} \in \mathbf{Y} \right\}$$

In the next set of operations, $\mathbf{t} \in \left(F^Y \right)^X$.

left generic convolution product

$$\mathbf{t} \, \bigcirc\!\!\!\!\gamma \, \mathbf{a} = \left\{ (\mathbf{y}, \mathbf{b}(\mathbf{y})) \, : \, \mathbf{b}(\mathbf{y}) = \underset{\mathbf{x} \in X}{\Gamma} \, (\mathbf{t_x}(\mathbf{y}) \, \bigcirc \, \mathbf{a}(\mathbf{x})), \, \mathbf{y} \in \mathbf{Y} \right\}$$

left linear convolution product

$$\mathbf{t} \, \oplus \mathbf{a} = \left\{ (\mathbf{y}, \mathbf{b}(\mathbf{y})) \, : \, \mathbf{b}(\mathbf{y}) = \sum_{\mathbf{x} \in X} (\mathbf{t_x}(\mathbf{y}) \cdot \mathbf{a}(\mathbf{x})), \, \mathbf{y} \in \mathbf{Y} \right\}$$

left morphological max convolution product

$$\mathbf{t} \, \boxtimes \, \mathbf{a} = \left\{ (\mathbf{y}, \mathbf{b}(\mathbf{y})) \, : \, \mathbf{b}(\mathbf{y}) = \bigvee_{\mathbf{x} \in X} \, [\mathbf{t_x}(\mathbf{y}) + \mathbf{a}(\mathbf{x})], \, \mathbf{y} \in \mathbf{Y} \right\}$$

left morphological min convolution product

$$\mathbf{t} \, \boxtimes \, \mathbf{a} = \left\{ (\mathbf{y}, \mathbf{b}(\mathbf{y})) \, : \, \mathbf{b}(\mathbf{y}) = \bigwedge_{\mathbf{x} \in X} \, [\mathbf{t_x}(\mathbf{y}) +' \mathbf{a}(\mathbf{x})], \, \mathbf{y} \in \mathbf{Y} \right\}$$

left multiplicative max convolution product

$$\mathbf{t} \, \bigcirc\!\!\!\!\vee \mathbf{a} = \left\{ (\mathbf{y}, \mathbf{b}(\mathbf{y})) \, : \, \mathbf{b}(\mathbf{y}) = \bigvee_{\mathbf{x} \in X} \, [\mathbf{t_x}(\mathbf{y}) \times \mathbf{a}(\mathbf{x})], \, \mathbf{y} \in \mathbf{Y} \right\}$$

left multiplicative min convolution product

$$\mathbf{t} \, \bigcirc\!\!\!\!\wedge \mathbf{a} = \left\{ (\mathbf{y}, \mathbf{b}(\mathbf{y})) \, : \, \mathbf{b}(\mathbf{y}) = \bigwedge_{\mathbf{x} \in X} \, [\mathbf{t_x}(\mathbf{y}) \times' \mathbf{a}(\mathbf{x})], \, \mathbf{y} \in \mathbf{Y} \right\}$$

left xor max convolution product

$$\mathbf{t} \, \tilde{\boxtimes} \, \mathbf{a} = \left\{ (\mathbf{y}, \mathbf{b}(\mathbf{y})) \, : \, \mathbf{b}(\mathbf{y}) = \bigvee_{\mathbf{x} \in X} \, \left[\mathbf{t_x}(\mathbf{y}) \tilde{+} \mathbf{a}(\mathbf{x}) \right], \, \mathbf{y} \in \mathbf{Y} \right\}$$

left xor min convolution product

$$\mathbf{t} \, \tilde{\boxtimes} \, \mathbf{a} = \left\{ (\mathbf{y}, \mathbf{b}(\mathbf{y})) \, : \, \mathbf{b}(\mathbf{y}) = \bigwedge_{\mathbf{x} \in X} \, \left[\mathbf{t_x}(\mathbf{y}) \tilde{+}' \mathbf{a}(\mathbf{x}) \right], \, \mathbf{y} \in \mathbf{Y} \right\}$$

Binary and Unary Template Operations

Since templates are images, all unary and binary image operations discussed earlier apply to templates as well. Any binary operation γ on $\mathbb{F}$ induces a binary operation (again denoted by γ) on $(\mathbb{F}^{\mathbf{X}})^{\mathbf{Y}}$ as follows: for each pair $\mathbf{s}, \mathbf{t} \in (\mathbb{F}^{\mathbf{X}})^{\mathbf{Y}}$ the induced operation $\mathbf{s}\gamma\mathbf{t}$ is defined in terms of the induced binary image operation on $\mathbb{F}^{\mathbf{X}}$, namely $(\mathbf{s}\gamma\mathbf{t})_{\mathbf{y}} \equiv \mathbf{s}_{\mathbf{y}}\gamma\mathbf{t}_{\mathbf{y}} \quad \forall \mathbf{y} \in \mathbf{Y}$. Thus, if $\mathbb{F} = \mathbb{R}$, $\mathbf{s}, \mathbf{t} \in (\mathbb{R}^{\mathbf{X}})^{\mathbf{Y}}$, and $\gamma = +$, then $(\mathbf{s} + \mathbf{t})_{\mathbf{y}} = \mathbf{s}_{\mathbf{y}} + \mathbf{t}_{\mathbf{y}}$, where $\mathbf{s}_{\mathbf{y}} + \mathbf{t}_{\mathbf{y}}$ denotes the pointwise sum of the two images $\mathbf{s}_{\mathbf{y}} \in \mathbb{R}^{\mathbf{X}}$ and $\mathbf{t}_{\mathbf{y}} \in \mathbb{R}^{\mathbf{X}}$.

The unary template operations of prime importance are the global reduce operations. Suppose $\mathbf{Y}$ is a finite point set, say $\mathbf{Y} = \{\mathbf{y}_1, \mathbf{y}_2, \dots, \mathbf{y}_n\}$, and $\mathbf{t} \in (\mathbb{F}^{\mathbf{X}})^{\mathbf{Y}}$. Any binary semigroup operation γ on $\mathbb{F}$ induces a global reduce operation

$$\Gamma : (\mathbb{F}^{\mathbf{X}})^{\mathbf{Y}} \to \mathbb{F}^{\mathbf{X}}$$

which is defined by

$$\Gamma\mathbf{t} = \mathop{\Gamma}_{\mathbf{y} \in \mathbf{Y}} \mathbf{t}_{\mathbf{y}} = \mathop{\Gamma}_{k=1}^{n} \mathbf{t}_{\mathbf{y}_k} = \mathbf{t}_{\mathbf{y}_1}\gamma\mathbf{t}_{\mathbf{y}_2}\gamma \cdots \gamma\mathbf{t}_{\mathbf{y}_n} .$$

Thus, for example, if $\mathbb{F} = \mathbb{R}$ and γ is the operation of addition ($\gamma = +$), then $\Gamma = \Sigma$ and

$$\sum\mathbf{t} = \sum_{\mathbf{y} \in \mathbf{Y}} \mathbf{t}_{\mathbf{y}} = \mathbf{t}_{\mathbf{y}_1} + \mathbf{t}_{\mathbf{y}_2} + \cdots + \mathbf{t}_{\mathbf{y}_n} .$$

Therefore, $\sum\mathbf{t}$ is an image, namely the sum of a finite number of images.

In all, the value set $(\mathbb{R}, +, \cdot, \vee, \wedge)$ provides for four basic global reduce operations, namely $\sum\mathbf{t}$, $\prod\mathbf{t}$, $\bigvee\mathbf{t}$, and $\bigwedge\mathbf{t}$.

If the value set $\mathbb{F}$ has two binary operations γ and $\bigcirc$ so that $(\mathbb{F}, \gamma, \bigcirc)$ is a ring (or semiring), then under the induced operations $\left((\mathbb{F}^{\mathbf{X}})^{\mathbf{Y}}, \gamma, \bigcirc\right)$ is also a ring (or semiring). Analogous to the image-template product, the binary operations $\bigcirc$ and γ induce a template convolution product

$$\textcircled{\gamma} : (\mathbb{F}^{\mathbf{Z}})^{\mathbf{X}} \times (\mathbb{F}^{\mathbf{X}})^{\mathbf{Y}} \to (\mathbb{F}^{\mathbf{Z}})^{\mathbf{Y}}$$

defined as follows. Suppose $\mathbf{s} \in (\mathbb{F}^{\mathbf{Z}})^{\mathbf{X}}$, $\mathbf{t} \in (\mathbb{F}^{\mathbf{X}})^{\mathbf{Y}}$, and $\mathbf{X}$ a finite point set. Then the template product $\mathbf{r} = \mathbf{s}\,\textcircled{\gamma}\,\mathbf{t}$, where $\mathbf{r} \in (\mathbb{F}^{\mathbf{Z}})^{\mathbf{Y}}$, is defined as

$$\mathbf{r}_{\mathbf{y}}(\mathbf{z}) = \mathop{\Gamma}_{\mathbf{x} \in \mathbf{X}} (\mathbf{s}_{\mathbf{x}}(\mathbf{z}) \bigcirc \mathbf{t}_{\mathbf{y}}(\mathbf{x})) \quad \forall \mathbf{y} \in \mathbf{Y} \text{ and } \forall \mathbf{z} \in \mathbf{Z} .$$

Thus, if $\mathbf{s} \in (\mathbb{R}^{\mathbf{Z}})^{\mathbf{X}}$ and $\mathbf{t} \in (\mathbb{R}^{\mathbf{X}})^{\mathbf{Y}}$, then $\mathbf{r} = \mathbf{s} \oplus \mathbf{t}$ is given by the formula

$$\mathbf{r}_{\mathbf{y}}(\mathbf{z}) = \sum_{\mathbf{x} \in \mathbf{X}} \mathbf{s}_{\mathbf{x}}(\mathbf{z}) \cdot \mathbf{t}_{\mathbf{y}}(\mathbf{x}) .$$

The lattice product $\mathbf{r} = \mathbf{s} \boxtimes \mathbf{t}$ is defined in a similar manner. For $\mathbf{s} \in (\mathbb{R}_{\pm\infty}^{\mathbf{Z}})^{\mathbf{X}}$ and $\mathbf{t} \in (\mathbb{R}_{\pm\infty}^{\mathbf{X}})^{\mathbf{Y}}$, the product template $\mathbf{r}$ is given by

$$\mathbf{r}_{\mathbf{y}}(\mathbf{z}) = \bigvee_{\mathbf{x} \in \mathbf{X}} [\mathbf{s}_{\mathbf{x}}(\mathbf{z}) + \mathbf{t}_{\mathbf{y}}(\mathbf{x})] .$$

The following example provides a specific instance of the above product formulation.

Example: Suppose $s, t \in \left(\mathbb{R}^{\mathbb{Z}^2}\right)^{\mathbb{Z}^2}$ are the following translation invariant templates:

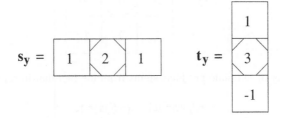

$$s_y = \boxed{\begin{array}{ccc} 1 & 2 & 1 \end{array}}$$

$$t_y = \begin{array}{c} 1 \\ 3 \\ -1 \end{array}$$

Then the template product $r = s \oplus t$ is the template defined by

$$r_y = \begin{array}{|c|c|c|} \hline 1 & 2 & 1 \\ \hline 3 & 6 & 3 \\ \hline -1 & -2 & -1 \\ \hline \end{array}$$

If $s, t \in \left(\mathbb{R}_{\pm\infty}^{\mathbb{Z}^2}\right)^{\mathbb{Z}^2}$ are defined as above with values $-\infty$ outside the support, then the template product $r = s \boxtimes t$ is the template defined by

$$r_y = \begin{array}{|c|c|c|} \hline 2 & 3 & 2 \\ \hline 4 & 5 & 4 \\ \hline 0 & 1 & 0 \\ \hline \end{array}$$

The template t is not an $\mathbb{R}_\infty^{\geq 0}$-valued template. To provide an example of the template product $s \ovee t$, we redefine t as

$$t_y = \begin{array}{c} 1 \\ 3 \\ 1 \end{array}$$

Then $\mathbf{r} = \mathbf{s} \, \boxed{\vee} \, \mathbf{t}$ is given by

$$
\mathbf{r_y} =
\begin{array}{|c|c|c|}
\hline
1 & 2 & 1 \\
\hline
3 & 6 & 3 \\
\hline
1 & 2 & 1 \\
\hline
\end{array}
$$

The utility of template products stems from the fact that in semirings the equation

$$\mathbf{a} \, \boxed{\gamma} \, (\mathbf{s} \, \boxed{\gamma} \, \mathbf{t}) = (\mathbf{a} \, \boxed{\gamma} \, \mathbf{s}) \, \boxed{\gamma} \, \mathbf{t}$$

holds [1]. This equation can be utilized in order to reduce the computational burden associated with typical convolution problems. For example, if $\mathbf{r} \in \left(\mathbb{R}^{\mathbb{Z}^2} \right)^{\mathbb{Z}^2}$ is defined by $\forall \mathbf{y} \in \mathbb{Z}^2$, then

$$
\mathbf{r_y} =
\begin{array}{|c|c|c|}
\hline
4 & 6 & -4 \\
\hline
6 & 9 & -6 \\
\hline
-4 & -6 & 4 \\
\hline
\end{array}
$$

$$\mathbf{a} \, \boxed{\oplus} \, \mathbf{r} = \mathbf{a} \, \boxed{\oplus} \, (\mathbf{s} \, \boxed{\oplus} \, \mathbf{t}) = (\mathbf{a} \, \boxed{\oplus} \, \mathbf{s}) \, \boxed{\oplus} \, \mathbf{t} \, ,$$

where

$$
\mathbf{s_y} =
\begin{array}{|c|c|c|}
\hline
2 & 3 & -2 \\
\hline
\end{array}
\qquad
\mathbf{t_y} =
\begin{array}{|c|}
\hline
2 \\
\hline
3 \\
\hline
-2 \\
\hline
\end{array}
$$

The construction of the new image $\mathbf{b} := \mathbf{a} \, \boxed{\oplus} \, \mathbf{r}$ requires nine multiplications and eight additions per pixel (if we ignore boundary pixels). In contrast, the computation of the image $\mathbf{b} := (\mathbf{a} \, \boxed{\oplus} \, \mathbf{s}) \, \boxed{\oplus} \, \mathbf{t}$ requires only six multiplications and four additions per pixel. For large images (e.g., size 1024×1024) this amounts to significant savings in computation.

Summary of Unary and Binary Template Operations

In the following $\mathbf{s}, \mathbf{t} \in \left(\mathbb{F}^X \right)^Y$ and $\mathbb{F}$ denotes the appropriate value set.

generic binary operation	$\mathbf{s} \gamma \mathbf{t}$:	$(\mathbf{s} \gamma \mathbf{t})_\mathbf{y} \equiv \mathbf{s_y} \gamma \mathbf{t_y}$
template sum	$\mathbf{s} + \mathbf{t}$:	$(\mathbf{s} + \mathbf{t})_\mathbf{y} \equiv \mathbf{s_y} + \mathbf{t_y}$
max of two templates	$\mathbf{s} \vee \mathbf{t}$:	$(\mathbf{s} \vee \mathbf{t})_\mathbf{y} \equiv \mathbf{s_y} \vee \mathbf{t_y}$

min of two templates	$s \wedge t : \quad (s \wedge t)_{\mathbf{y}} \equiv s_{\mathbf{y}} \wedge t_{\mathbf{y}}$
generic reduce operation	$\Gamma t \equiv \Gamma_{\mathbf{y} \in \mathbf{Y}} t_{\mathbf{y}} = \overset{n}{\underset{k=1}{\Gamma}} t_{\mathbf{y}_k} = t_{\mathbf{y}_1} \gamma t_{\mathbf{y}_2} \gamma \cdots \gamma t_{\mathbf{y}_n}$
sum reduce	$\sum t \equiv \sum_{\mathbf{y} \in \mathbf{Y}} t_{\mathbf{y}} = t_{\mathbf{y}_1} + t_{\mathbf{y}_2} + \cdots + t_{\mathbf{y}_n}$
product reduce	$\prod t \equiv \prod_{\mathbf{y} \in \mathbf{Y}} t_{\mathbf{y}} = t_{\mathbf{y}_1} \cdot t_{\mathbf{y}_2} \cdots \cdot t_{\mathbf{y}_n}$
max reduce	$\bigvee t \equiv \bigvee_{\mathbf{y} \in \mathbf{Y}} t_{\mathbf{y}} = t_{\mathbf{y}_1} \vee t_{\mathbf{y}_2} \vee \cdots \vee t_{\mathbf{y}_n}$
min reduce	$\bigwedge t \equiv \bigwedge_{\mathbf{y} \in \mathbf{Y}} t_{\mathbf{y}} = t_{\mathbf{y}_1} \wedge t_{\mathbf{y}_2} \wedge \cdots \wedge t_{\mathbf{y}_n}$

In the next list, $s \in \left(\mathbb{F}^{\mathbf{Z}} \right)^{\mathbf{X}}$, $t \in \left(\mathbb{F}^{\mathbf{X}} \right)^{\mathbf{Y}}$, $\mathbf{X}$ is a finite point set, and $\mathbb{F}$ denotes the appropriate value set.

generic template product	$r = s \textcircled{\gamma} t : \quad r_{\mathbf{y}}(z) = \Gamma_{\mathbf{x} \in \mathbf{X}} (s_{\mathbf{x}}(z) \bigcirc t_{\mathbf{y}}(\mathbf{x}))$
linear template product	$r = s \oplus t : \quad r_{\mathbf{y}}(z) = \sum_{\mathbf{x} \in \mathbf{X}} s_{\mathbf{x}}(z) \cdot t_{\mathbf{y}}(\mathbf{x})$
additive max product	$r = s \boxtimes t : \quad r_{\mathbf{y}}(z) = \bigvee_{\mathbf{x} \in \mathbf{X}} s_{\mathbf{x}}(z) + t_{\mathbf{y}}(\mathbf{x})$
additive min product	$r = s \boxtimes t : \quad r_{\mathbf{y}}(z) = \bigwedge_{\mathbf{x} \in \mathbf{X}} s_{\mathbf{x}}(z) + t_{\mathbf{y}}(\mathbf{x})$
multiplicative max product	$r = s \oslash t : \quad r_{\mathbf{y}}(z) = \bigvee_{\mathbf{x} \in \mathbf{X}} s_{\mathbf{x}}(z) \cdot t_{\mathbf{y}}(\mathbf{x})$
multiplicative min product	$r = s \oslash t : \quad r_{\mathbf{y}}(z) = \bigwedge_{\mathbf{x} \in \mathbf{X}} s_{\mathbf{x}}(z) \cdot t_{\mathbf{y}}(\mathbf{x})$

1.6. Recursive Templates

In this section we introduce the notions of recursive templates and recursive template operations, which are direct extensions of the notions of templates and the corresponding template operations discussed in the preceding section.

A recursive template is defined in terms of a regular template from some point set $\mathbf{X}$ to another point set $\mathbf{Y}$ with some partial order imposed on $\mathbf{Y}$.

> **Definition.** A *partially ordered set* $(P, \prec)$ (or *poset*) is a set P together with a binary relation $\prec$, satisfying the following three axioms for arbitrary $x, y, z \in P$:
>
> (i) $x \prec x$ (reflexive)
>
> (ii) $x \prec y$ and $y \prec x \Rightarrow x = y$ (antisymmetric)
>
> (iii) $x \prec y$ and $y \prec z \Rightarrow x \prec z$ (transitive)

Now suppose that $\mathbf{X}$ is a point set, $\mathbf{Y}$ is a partially ordered point set with partial order $\prec$, and $\mathbb{F}$ a monoid. An $\mathbb{F}$-*valued recursive template* $\mathbf{t}$ *from* $\mathbf{Y}$ *to* $\mathbf{X}$ is a function $\mathbf{t} = (\mathbf{t}_{\not\prec}, \mathbf{t}_{\prec}) : \mathbf{Y} \to \left(\mathbb{F}^{\mathbf{X}}, \mathbb{F}^{\mathbf{Y}} \right)$, where $\mathbf{t}_{\not\prec} : \mathbf{Y} \to \mathbb{F}^{\mathbf{X}}$ and $\mathbf{t}_{\prec} : \mathbf{Y} \to \mathbb{F}^{\mathbf{Y}}$, such that

1. $\mathbf{y} \notin S(\mathbf{t}_{\prec}(\mathbf{y}))$ and
2. for each $\mathbf{z} \in S(\mathbf{t}_{\prec}(\mathbf{y}))$, $\mathbf{z} \prec \mathbf{y}$.

Thus, for each $\mathbf{y} \in \mathbf{Y}$, $\mathbf{t}_{\nprec}(\mathbf{y})$ is an $\mathbb{F}$-valued image on $\mathbf{X}$ and $\mathbf{t}_{\prec}(\mathbf{y})$ is an $\mathbb{F}$-valued image on $\mathbf{Y}$.

In most applications, the relation $\mathbf{X} \subset \mathbf{Y}$ or $\mathbf{X} = \mathbf{Y}$ usually holds. Also, for consistency of notation and for notational convenience, we define $\mathbf{t}_{\nprec \mathbf{y}} \equiv \mathbf{t}_{\nprec}(\mathbf{y})$ and $\mathbf{t}_{\prec \mathbf{y}} \equiv \mathbf{t}_{\prec}(\mathbf{y})$ so that $\mathbf{t}_{\mathbf{y}} = (\mathbf{t}_{\nprec \mathbf{y}}, \mathbf{t}_{\prec \mathbf{y}})$. The support of $\mathbf{t}$ at a point $\mathbf{y}$ is defined as $S(\mathbf{t}_{\mathbf{y}}) = (S(\mathbf{t}_{\nprec \mathbf{y}}), S(\mathbf{t}_{\prec \mathbf{y}}))$. The set of all $\mathbb{F}$-valued recursive templates from $\mathbf{Y}$ to $\mathbf{X}$ will be denoted by $(\mathbb{F}^{\mathbf{X}}, \mathbb{F}^{\mathbf{Y}})^{(\mathbf{Y}, \prec)}$.

In analogy to our previous definition of translation invariant templates, if $\mathbf{X}$ is closed under the operation $+$, then a recursive template $\mathbf{t} \in (\mathbb{F}^{\mathbf{X}}, \mathbb{F}^{\mathbf{X}})^{(\mathbf{X}, \prec)}$ is called *translation invariant* if for each triple $\mathbf{x}, \mathbf{y}, \mathbf{z} \in \mathbf{X}$, we have $\mathbf{t}_{\mathbf{y}}(\mathbf{x}) = \mathbf{t}_{\mathbf{y}+\mathbf{z}}(\mathbf{x} + \mathbf{z})$, or equivalently, $\mathbf{t}_{\nprec \mathbf{y}}(\mathbf{x}) = \mathbf{t}_{\nprec \mathbf{y}+\mathbf{z}}(\mathbf{x} + \mathbf{z})$ and $\mathbf{t}_{\prec \mathbf{y}}(\mathbf{x}) = \mathbf{t}_{\prec \mathbf{y}+\mathbf{z}}(\mathbf{x} + \mathbf{z})$. An example of an invariant recursive template is shown in Figure 1.6.1.

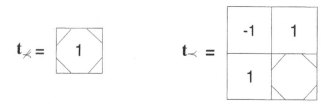

Figure 1.6.1. An example of an integer-valued invariant recursive template from $\mathbb{Z}^2$ to $\mathbb{Z}^2$.

If $\mathbf{t}$ is an invariant recursive template and has only one pixel defined on the target point of its nonrecursive support $S(\mathbf{t}_{\nprec \mathbf{y}})$, then $\mathbf{t}$ is called a *simplified* recursive template. Pictorially, a simplified recursive template can be drawn the same way as a nonrecursive template since the recursive part and the nonrecursive part do not overlap. In particular, the recursive template shown in Figure 1.6.1 can be redrawn as illustrated in Figure 1.6.2

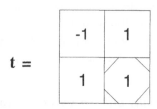

Figure 1.6.2. An example of an integer-valued simplified recursive template.

The notions of transpose and dual of a recursive template are defined in terms of those for nonrecursive templates. In particular, the *transpose* $\mathbf{t}'$ of a recursive template $\mathbf{t}$ is defined as $\mathbf{t}' = (\mathbf{t}'_{\nprec}, \mathbf{t}'_{\prec})$. Similarly, if $\mathbf{t} \in (\mathbb{R}^{\mathbf{X}}_{\pm\infty}, \mathbb{R}^{\mathbf{X}}_{\pm\infty})^{(\mathbf{X}, \prec)}$, then the *additive dual* of $\mathbf{t}$ is defined by $\mathbf{t}^* = (\mathbf{t}^*_{\nprec}, \mathbf{t}^*_{\prec})$. The *multiplicative dual* for recursive $\mathbb{R}^{\geq 0}_{\infty}$-valued templates is defined in a likewise fashion.

Operations between Images and Recursive Templates

In order to facilitate the discussion on recursive templates operations, we begin by extending the notions of the linear convolution product $\oplus$, the morphological max $\boxdot$, and the multiplicative max $\varovee$ to the corresponding recursive operations $\oplus_{\prec}$, $\boxdot_{\prec}$, and $\varovee_{\prec}$, respectively.

Let $\mathbf{X}$ and $\mathbf{Y}$ be finite subsets of $\mathbb{R}^n$ with $\mathbf{Y}$ partially ordered by $\prec$. If $\mathbf{a} \in \mathbb{R}^{\mathbf{X}}$ and $\mathbf{t} \in \left(\mathbb{R}^{\mathbf{X}}, \mathbb{R}^{\mathbf{Y}}\right)^{(\mathbf{Y}, \prec)}$, then the *recursive linear convolution product* $\mathbf{a} \oplus_{\prec} \mathbf{t}$ is defined by

$$\mathbf{a} \oplus_{\prec} \mathbf{t} = \left\{ (\mathbf{y}, \mathbf{b}(\mathbf{y})) : \mathbf{y} \in \mathbf{Y}, \ \mathbf{b}(\mathbf{y}) = \sum_{\mathbf{x} \in S(\mathbf{t}_{\not\prec \mathbf{y}})} (\mathbf{a}(\mathbf{x}) \cdot \mathbf{t}_{\not\prec \mathbf{y}}(\mathbf{x})) + \sum_{\mathbf{z} \in S(\mathbf{t}_{\prec \mathbf{y}})} (\mathbf{b}(\mathbf{z}) \cdot \mathbf{t}_{\prec \mathbf{y}}(\mathbf{z})) \right\}.$$

The recursive template operation $\oplus_{\prec}$ computes a new pixel value $\mathbf{b}(\mathbf{y})$ based on both the pixel values $\mathbf{a}(\mathbf{x})$ of the source image and some *previously* calculated new pixel values $\mathbf{b}(\mathbf{z})$ which are determined by the partial order $\prec$ and the region of support of the participating template. By definition of a recursive template, $\mathbf{z} \prec \mathbf{y}$ for every $\mathbf{z} \in S(\mathbf{t}_{\prec \mathbf{y}})$ and $\mathbf{y} \notin S(\mathbf{t}_{\prec \mathbf{y}})$. Therefore, $\mathbf{b}(\mathbf{y})$ is always recursively computable. Some partial orders that are commonly used in two-dimensional recursive transforms are forward and backward raster scanning and serpentine scanning.

It follows from the definition of $\oplus_{\prec}$ that the computation of a new pixel $\mathbf{b}(\mathbf{y})$ can be done only after all its predecessors (ordered by $\prec$) have been computed. Thus, in contrast to nonrecursive template operations, recursive template operations are not computed in a globally parallel fashion.

Note that if the recursive template $\mathbf{t}$ is defined such that $S(\mathbf{t}_{\prec \mathbf{y}}) = \varnothing$ for all $\mathbf{y} \in \mathbf{Y}$, then one obtains the usual nonrecursive template operation

$$\mathbf{a} \oplus_{\prec} \mathbf{t} = \left\{ (\mathbf{y}, \mathbf{b}(\mathbf{y})) : \mathbf{b}(\mathbf{y}) = \sum_{\mathbf{x} \in S(\mathbf{t}_{\not\prec \mathbf{y}})} (\mathbf{a}(\mathbf{x}) \cdot \mathbf{t}_{\not\prec \mathbf{y}}(\mathbf{x})), \ \mathbf{y} \in \mathbf{Y} \right\}.$$

Hence, recursive template operations are natural extensions of nonrecursive template operations.

Recursive morphological max and multiplicative max are defined in a similar fashion. Specifically, if $\mathbf{a} \in \mathbb{R}_{\pm\infty}^{\mathbf{X}}$ and $\mathbf{t} \in \left(\mathbb{R}_{\pm\infty}^{\mathbf{X}}, \mathbb{R}_{\pm\infty}^{\mathbf{X}}\right)^{(\mathbf{X}, \prec)}$, then

$$\mathbf{b} = \mathbf{a} \boxed{\vee}_{\prec} \mathbf{t}$$

is defined by

$$\mathbf{b}(\mathbf{y}) = \bigvee_{\mathbf{x} \in S_{-\infty}(\mathbf{t}_{\not\prec \mathbf{y}})} [\mathbf{a}(\mathbf{x}) + \mathbf{t}_{\not\prec \mathbf{y}}(\mathbf{x})] \vee \bigvee_{\mathbf{z} \in S_{-\infty}(\mathbf{t}_{\prec \mathbf{y}})} [\mathbf{b}(\mathbf{z}) + \mathbf{t}_{\prec \mathbf{y}}(\mathbf{z})].$$

For $\mathbf{a} \in \left(\mathbb{R}_{\infty}^{\geq 0}\right)^{\mathbf{X}}$ and $\mathbf{t} \in \left(\left(\mathbb{R}_{\infty}^{\geq 0}\right)^{\mathbf{X}}, \left(\mathbb{R}_{\infty}^{\geq 0}\right)^{\mathbf{Y}}\right)^{(\mathbf{Y}, \prec)}$,

$$\mathbf{b} = \mathbf{a} \boxed{\vee\kern-0.6em\vee}_{\prec} \mathbf{t}$$

is defined by

$$\mathbf{b}(\mathbf{y}) = \bigvee_{\mathbf{x} \in S(\mathbf{t}_{\not\prec \mathbf{y}})} [\mathbf{a}(\mathbf{x}) \times \mathbf{t}_{\not\prec \mathbf{y}}(\mathbf{x})] \vee \bigvee_{\mathbf{z} \in S(\mathbf{t}_{\prec \mathbf{y}})} [\mathbf{b}(\mathbf{z}) \times \mathbf{t}_{\prec \mathbf{y}}(\mathbf{z})].$$

The operations of the recursive morphological min and multiplicative min ($\boxed{\wedge}_{\prec}$ and $\boxed{\wedge\kern-0.6em\wedge}_{\prec}$) are defined in the same straightforward fashion.

Recursive morphological max and min as well as recursive multiplicative max and min are nonlinear operations. However, the recursive linear convolution remains a linear operation.

The basic recursive template operations described above can be easily generalized to the generic recursive image-template convolution product by simple substitution of the specific operations, such as multiplication and addition, by the generic operations $\bigcirc$ and γ. More precisely, given a semiring $(\mathbb{F}, \gamma, \bigcirc)$ with identity, then one can define the generic recursive product

$$\textcircled{?}_{\prec} : \mathbb{F}^{\mathbf{X}} \times (\mathbb{F}^{\mathbf{X}}, \mathbb{F}^{\mathbf{Y}})^{(\mathbf{Y},\prec)} \to \mathbb{F}^{\mathbf{Y}}$$

by defining $\mathbf{b} = \mathbf{a}\,\textcircled{?}_{\prec}\mathbf{t}$ by

$$\mathbf{b}(\mathbf{y}) = \underset{\mathbf{z} \in S\left(\mathbf{t}_{\not\prec \mathbf{y}}\right)}{\Gamma} [\mathbf{a}(\mathbf{x}) \bigcirc \mathbf{t}_{\not\prec \mathbf{y}}(\mathbf{x})]\,\gamma\, \underset{\mathbf{z} \in S\left(\mathbf{t}_{\prec \mathbf{y}}\right)}{\Gamma} [\mathbf{b}(\mathbf{z}) \bigcirc \mathbf{t}_{\prec \mathbf{y}}(\mathbf{z})] \, .$$

Again, in addition to the basic recursive template operations discussed earlier, a wide variety of recursive template operations can be derived from the generalized recursive rule by substituting different binary operations for $\bigcirc$ and γ. Additionally, parameterized recursive templates are defined in the same manner as parameterized nonrecursive templates; namely as functions

$$\mathbf{t} : P \to (\mathbb{F}^{\mathbf{X}}, \mathbb{F}^{\mathbf{Y}})^{(\mathbf{Y},\prec)},$$

where P denotes the set of parameters, and $\mathbf{t}(p) = \left(\mathbf{t}(p)_{\not\prec}, \mathbf{t}(p)_{\prec}\right)$ with $\mathbf{t}(p)_{\not\prec} \in \left(\mathbb{F}^{\mathbf{X}}\right)^{\mathbf{Y}}$ and $\mathbf{t}(p)_{\prec} \in \left(\mathbb{F}^{\mathbf{Y}}\right)^{(\mathbf{Y},\prec)}$.

Summary of Recursive Template Operations

In the following list of pertinent recursive image-template products $\mathbf{a} \in \mathbb{F}^{\mathbf{X}}$ and $\mathbf{t} \in (\mathbb{F}^{\mathbf{X}}, \mathbb{F}^{\mathbf{Y}})^{(\mathbf{Y},\prec)}$. As before, for each operation we assume the appropriate value set $\mathbb{F}$.

recursive generic convolution product

$$\mathbf{a}\,\textcircled{?}_{\prec}\mathbf{t} = \left\{ (\mathbf{y}, \mathbf{b}(\mathbf{y})) \; : \; \mathbf{b}(\mathbf{y}) = \mathbf{y} \in \mathbf{Y}, \underset{\mathbf{z} \in S\left(\mathbf{t}_{\not\prec \mathbf{y}}\right)}{\Gamma} [\mathbf{a}(\mathbf{x}) \bigcirc \mathbf{t}_{\not\prec \mathbf{y}}(\mathbf{x})]\gamma \underset{\mathbf{z} \in S\left(\mathbf{t}_{\prec \mathbf{y}}\right)}{\Gamma} [\mathbf{b}(\mathbf{z}) \bigcirc \mathbf{t}_{\prec \mathbf{y}}(\mathbf{z})] \right\}$$

recursive linear convolution product

$$\mathbf{a} \oplus_{\prec}\mathbf{t} = \left\{ (\mathbf{y}, \mathbf{b}(\mathbf{y})) \; : \; \mathbf{b}(\mathbf{y}) = \mathbf{y} \in \mathbf{Y}, \underset{\mathbf{x} \in S(\mathbf{t}_{\not\prec \mathbf{y}})}{\sum} (\mathbf{a}(\mathbf{x}) \cdot \mathbf{t}_{\not\prec \mathbf{y}}(\mathbf{x})) + \underset{\mathbf{z} \in S(\mathbf{t}_{\prec \mathbf{y}})}{\sum} (\mathbf{b}(\mathbf{z}) \cdot \mathbf{t}_{\prec \mathbf{y}}(\mathbf{z})) \right\}$$

recursive morphological max convolution product

$$\mathbf{a}\,\boxed{\vee}_{\prec}\mathbf{t} = \left\{ (\mathbf{y}, \mathbf{b}(\mathbf{y})) \; : \; \mathbf{b}(\mathbf{y}) = \mathbf{y} \in \mathbf{Y}, \underset{\mathbf{x} \in S_{-\infty}(\mathbf{t}_{\not\prec \mathbf{y}})}{\bigvee} [\mathbf{a}(\mathbf{x}) + \mathbf{t}_{\not\prec \mathbf{y}}(\mathbf{x})] \vee \underset{\mathbf{z} \in S_{-\infty}(\mathbf{t}_{\prec \mathbf{y}})}{\bigvee} [\mathbf{b}(\mathbf{z}) + \mathbf{t}_{\prec \mathbf{y}}(\mathbf{z})] \right\}$$

recursive morphological min convolution product

$$
\mathbf{a} \, \boxed{\triangle}_{\prec} \, \mathbf{t} =
\left\{
\begin{array}{l}
(\mathbf{y}, \mathbf{b}(\mathbf{y})) \, : \, \mathbf{b}(\mathbf{y}) = \mathbf{y} \in \mathbf{Y}, \;\; \bigwedge_{\mathbf{x} \in S_\infty(\mathbf{t}_{\not\prec \mathbf{y}})} [\mathbf{a}(\mathbf{x}) +' \mathbf{t}_{\not\prec \mathbf{y}}(\mathbf{x})] \wedge \\[2ex]
\hspace{5cm} \bigwedge_{\mathbf{z} \in S_\infty(\mathbf{t}_{\prec \mathbf{y}})} [\mathbf{b}(\mathbf{z}) +' \mathbf{t}_{\prec \mathbf{y}}(\mathbf{z})]
\end{array}
\right\}
$$

recursive multiplicative max convolution product

$$
\mathbf{a} \, \boxed{\vee}_{\prec} \, \mathbf{t} =
\left\{
\begin{array}{l}
(\mathbf{y}, \mathbf{b}(\mathbf{y})) \, : \, \mathbf{b}(\mathbf{y}) = \mathbf{y} \in \mathbf{Y}, \;\; \bigvee_{\mathbf{x} \in S(\mathbf{t}_{\not\prec \mathbf{y}})} [\mathbf{a}(\mathbf{x}) \times \mathbf{t}_{\not\prec \mathbf{y}}(\mathbf{x})] \vee \\[2ex]
\hspace{5cm} \bigvee_{\mathbf{z} \in S(\mathbf{t}_{\prec \mathbf{y}})} [\mathbf{b}(\mathbf{z}) \times \mathbf{t}_{\prec \mathbf{y}}(\mathbf{z})]
\end{array}
\right\}
$$

right multiplicative min convolution product

$$
\mathbf{a} \, \boxed{\wedge}_{\prec} \, \mathbf{t} =
\left\{
\begin{array}{l}
(\mathbf{y}, \mathbf{b}(\mathbf{y})) \, : \, \mathbf{b}(\mathbf{y}) = \mathbf{y} \in \mathbf{Y}, \;\; \bigwedge_{\mathbf{x} \in S_\infty(\mathbf{t}_{\not\prec \mathbf{y}})} [\mathbf{a}(\mathbf{x}) \times' \mathbf{t}_{\not\prec \mathbf{y}}(\mathbf{x})] \wedge \\[2ex]
\hspace{5cm} \bigwedge_{\mathbf{z} \in S_\infty(\mathbf{t}_{\prec \mathbf{y}})} [\mathbf{b}(\mathbf{z}) \times' \mathbf{t}_{\prec \mathbf{y}}(\mathbf{z})]
\end{array}
\right\}
$$

The definition of the left recursive convolution product $\mathbf{t} \, \boxed{\gamma}_{\prec} \mathbf{a}$ is also straightforward. However, for sake of brevity and since the different left products are not required for the remainder of this text, we dispense with their formulation. Additional facts about recursive convolution products, their properties and applications can be found in [1, 61, 62].

1.7. Neighborhoods

There are several types of template operations that are more easily implemented in terms of neighborhood operations. Typically, neighborhood operations replace template operations whenever the values in the support of a template consist only of the unit elements of the value set associated with the template. A template $\mathbf{t} \in (\mathbb{F}^{\mathbf{X}})^{\mathbf{Y}}$ with the property that for each $\mathbf{y} \in \mathbf{Y}$, the values in the support of $\mathbf{t_y}$ consist only of the unit of $\mathbb{F}$ is called a *unit template*.

For example, the invariant template $\mathbf{t} \in \left(\mathbb{R}^{\mathbb{Z}^2} \right)^{\mathbb{Z}^2}$ shown in Figure 1.7.1 is a unit template with respect to the value set $(\mathbb{R}, +, \cdot)$ since the value 1 is the unit with respect to multiplication.

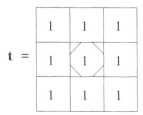

Figure 1.7.1. The unit Moore template for the value set $(\mathbb{R}, +, \cdot)$.

Similarly, the template $\mathbf{r} \in \left(\mathbb{R}_{-\infty}^{\mathbb{Z}^2} \right)^{\mathbb{Z}^2}$ shown in Figure 1.7.2 is a unit template with respect to the value set $(\mathbb{R}_{-\infty}, \vee, +)$ since the value 0 is the unit with respect to the operation $+$.

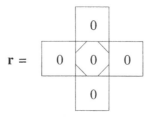

$$\mathbf{r} =$$

Figure 1.7.2. The unit von Neumann template for the value set $(\mathbb{R}_{-\infty}, \vee, +)$.

If $\mathbf{X} \subset \mathbb{Z}^2$ is an $m \times n$ array of points, $\mathbf{a} \in \mathbb{R}^{\mathbf{X}}$, and $\mathbf{t} \in \left(\mathbb{R}^{\mathbb{Z}^2} \right)^{\mathbb{Z}^2}$ is the 3×3 unit Moore template, then the values of the $m \times n$ image $\mathbf{b}$ obtained from the statement $\mathbf{b} := \mathbf{a} \oplus \mathbf{t}$ are computed by using the equation

$$\mathbf{b}(\mathbf{y}) = \sum_{\mathbf{x} \in \mathbf{X} \cap S(\mathbf{t_y})} \mathbf{a}(\mathbf{x}) \cdot \mathbf{t_y}(\mathbf{x}) = \sum_{\mathbf{x} \in \mathbf{X} \cap S(\mathbf{t_y})} \mathbf{a}(\mathbf{x}) \cdot 1 .$$

We need to point out that the difference between the mathematical equality $\mathbf{b} = \mathbf{a} \oplus \mathbf{t}$ and the pseudocode statement $\mathbf{b} := \mathbf{a} \oplus \mathbf{t}$ is that in the latter the new image is computed only for those points $\mathbf{y}$ for which $\mathbf{X} \cap S(\mathbf{t_y}) \neq \varnothing$. Observe that since $\mathbf{a}(\mathbf{x}) \cdot 1 = \mathbf{a}(\mathbf{x})$ and $M(\mathbf{y}) = S(\mathbf{t_y})$, where $M(\mathbf{y})$ denotes the Moore neighborhood of $\mathbf{y}$ (see Figure 1.2.2), it follows that

$$\mathbf{b}(\mathbf{y}) = \sum_{\mathbf{x} \in \mathbf{X} \cap M(\mathbf{y})} \mathbf{a}(\mathbf{x}) .$$

This observation leads to the notion of *neighborhood reduction*. In implementation, neighborhood reduction avoids unnecessary multiplication by the unit element and, as we shall shortly demonstrate, neighborhood reduction also avoids some standard boundary problems associated with image-template products.

To precisely define the notion of neighborhood reduction we need a more general notion of the reduce operation $\Gamma : \mathbb{F}^{\mathbf{X}} \to \mathbb{F}$, which was defined in terms of a binary operation γ on $\mathbb{F}$. The more general form of Γ is a function

$$\Gamma : \mathbb{F}^{\mathbf{X}}|_N \to \mathbb{F} ,$$

where $N \in \left(2^{\mathbf{X}} \right)^{\mathbf{Y}}$ and $\mathbb{F}^{\mathbf{X}}|_N = \left\{ \mathbf{a}|_{N(\mathbf{y})} : \mathbf{a} \in \mathbb{F}^{\mathbf{X}}, \mathbf{y} \in \mathbf{Y} \right\}$.

For example, if $\mathbb{F}^{\mathbf{X}} = \mathbb{R}^{\mathbf{X}}$, where $\mathbf{X} \subset \mathbb{Z}^2$ is an $m \times n$ array of points and $N \in \left(2^{\mathbf{X}} \right)^{\mathbb{Z}^2}$, then one such function could be defined as

$$\Sigma : \mathbb{R}^{\mathbf{X}}|_N \to \mathbb{R} ,$$

where $\sum \left(\mathbf{a}|_{N(\mathbf{y})} \right) = \sum_{\mathbf{x} \in N(\mathbf{y})} \mathbf{a}(\mathbf{x})$. Another example would be to define

$$\Gamma : \mathbb{R}^{\mathbf{X}}|_N \to \mathbb{R}$$

as $\Gamma\left(\mathbf{a}|_{N(\mathbf{y})}\right) = \frac{1}{card(N(\mathbf{y}))} \sum_{\mathbf{x} \in N(\mathbf{y})} \mathbf{a}(\mathbf{x})$, then Γ implements the averaging function, which we shall denote by *average*. Similarly, for integer-valued images, the median reduction

$$median : \mathbb{N}^{\mathbf{X}}|_N \rightharpoonup \mathbb{N}$$

is defined as $median\left(\mathbf{a}|_{N(\mathbf{y})}\right) = median\{\mathbf{a}(\mathbf{x}_{i_1}), \mathbf{a}(\mathbf{x}_{i_2}), \dots, \mathbf{a}(\mathbf{x}_{i_k})\}$, where $N(\mathbf{y}) = \{\mathbf{x}_{i_1}, \mathbf{x}_{i_2}, \dots, \mathbf{x}_{i_k}\}$.

Now suppose $\mathbf{X} \subset \mathbf{Z}$, $\mathbf{t} \in \left(\mathbb{F}^{\mathbf{Z}}\right)^{\mathbf{Y}}$ is a unit template with respect to the operation $\bigcirc$ of the semiring $(\mathbb{F}, \gamma, \bigcirc)$, $N : \mathbf{Y} \rightarrow 2^{\mathbf{Z}}$ is a neighborhood system defined by $N(\mathbf{y}) = S(\mathbf{t}_{\mathbf{y}})$, and $\mathbf{a} \in \mathbb{F}^{\mathbf{X}}$. It then follows that $\mathbf{b} := \mathbf{a} \, \textcircled{γ} \, \mathbf{t}$ is given by

$$\mathbf{b}(\mathbf{y}) = \underset{\mathbf{x} \in \mathbf{X} \cap S(\mathbf{t}_{\mathbf{y}})}{\Gamma} \left(\mathbf{a}(\mathbf{x}) \bigcirc \mathbf{t}_{\mathbf{y}}(\mathbf{x})\right) = \underset{\mathbf{x} \in \mathbf{X} \cap N(\mathbf{y})}{\Gamma} \mathbf{a}(\mathbf{x}).$$

This observation leads to the following definition of an *image-neighborhood convolution product*. Given $\mathbf{X} \subset \mathbf{Z}$, $\mathbf{a} \in \mathbb{F}^{\mathbf{X}}$, a neighborhood system $N : \mathbf{Y} \rightarrow 2^{\mathbf{Z}}$ (i.e., $N \in \left(2^{\mathbf{Z}}\right)^{\mathbf{Y}}$), and a reduction function $\Gamma : \mathbb{F}^{\mathbf{X}}|_N \rightarrow \mathbb{F}$, then the *image-neighborhood convolution product* $\mathbf{b} := \mathbf{a} \, \textcircled{Γ} \, N$ is defined by

$$\mathbf{b}(\mathbf{y}) = \Gamma\left(\mathbf{a}|_{\mathbf{X} \cap N(\mathbf{y})}\right)$$

for each $\mathbf{y} \in \mathbf{Y}$. Note that the product $\textcircled{$\Gamma$}$ is similar to the image template product $\textcircled{$\gamma$}$ in that $\textcircled{$\Gamma$}$ is a function

$$\textcircled{Γ} : \mathbb{F}^{\mathbf{X}} \times \left(2^{\mathbf{Z}}\right)^{\mathbf{Y}} \rightarrow \mathbb{F}^{\mathbf{Y}}.$$

In particular, if $\mathbf{a} \in \mathbb{R}^{\mathbf{X}}$, $M : \mathbb{Z}^2 \rightarrow 2^{\mathbb{Z}^2}$ is the Moore neighborhood, and $\mathbf{t} \in \left(\mathbb{R}^{\mathbb{Z}^2}\right)^{\mathbb{Z}^2}$ is the 3×3 unit Moore template defined earlier, then $\mathbf{a} \oplus \mathbf{t} = \mathbf{a} \oplus M$. Likewise, $\mathbf{a} \boxtimes \mathbf{r} = \mathbf{a} \boxtimes N$, where $\mathbf{r} \in \left(\mathbb{R}_{-\infty}^{\mathbb{Z}^2}\right)^{\mathbb{Z}^2}$ denotes the von Neumann unit template (Figure 1.7.2) and N denotes the von Neumann neighborhood (1.2.2). The latter equality stems from the fact that if $\mathbf{b} := \mathbf{a} \boxtimes \mathbf{r}$ and $\mathbf{c} := \mathbf{a} \boxtimes N$, then since $\mathbf{r}_{\mathbf{y}}(\mathbf{x}) = 0$ for all $\mathbf{x} \in \mathbf{X} \cap S_{-\infty}(\mathbf{r}_{\mathbf{y}})$ and $S_{-\infty}(\mathbf{r}_{\mathbf{y}}) = N(\mathbf{y})$ for all points $\mathbf{y} \in \mathbb{Z}^2$, we have that

$$\mathbf{b}(\mathbf{y}) = \bigvee_{\mathbf{x} \in \mathbf{X} \cap S_{-\infty}(\mathbf{r}_{\mathbf{y}})} \mathbf{a}(\mathbf{x}) + \mathbf{r}_{\mathbf{y}}(\mathbf{x}) = \bigvee_{\mathbf{x} \in \mathbf{X} \cap N(\mathbf{y})} \mathbf{a}(\mathbf{x}) = \mathbf{c}(\mathbf{y}).$$

Unit templates act like *characteristic* functions in that they do not weigh a pixel, but simply note which pixels are in their support and which are not. When employed in the image-template operations of their semiring, they only serve to collect a number of values that need to be reduced by the gamma operation. For this reason, unit templates are also referred to as *characteristic templates*. Now suppose that we wish to describe a translation invariant unit template with a specific support such as the 3×3 support of the Moore template $\mathbf{t}$ shown in Figure 1.7.1. Suppose further that we would like this template to be used with a variety of reduction operations, for instance, summation and maximum. In fact, we *cannot* describe such an operand without regard of the image-template operation $\textcircled{$\gamma$}$ by which it will be used. For us to derive the expected results, the template must map all points in its support to the unitary value with respect to the combining operation $\bigcirc$. Thus, for the reduce operation of summation $\sum$, the unit values in the support must be 1, while for the maximum reduce operation $\bigvee$, the values in the support must all be 0. Therefore, we cannot define a single template operand to characterize a neighborhood for reduction without regard to the image-template operation to be used to reduce the values within the neighborhood. However, we can capture exactly the information of interest in unit templates with the simple notion of neighborhood function. Thus, for example, the Moore neighborhood M can be used to add the values in every 3×3 neighborhood as well as to find the maximum or minimum in such a neighborhood by using the statements $\mathbf{a} \oplus M$, $\mathbf{a} \boxtimes M$, and $\mathbf{a} \boxtimes M$, respectively. This is one advantage for replacing unit templates with neighborhoods.

Another advantage of using neighborhoods instead of templates can be seen by considering the simple example of image smoothing by local averaging. Suppose $\mathbf{a} \in \mathbb{R}^{\mathbf{X}}$, where $\mathbf{X} \subset \mathbb{Z}^2$ is an $m \times n$ array of points, and $\mathbf{t} \in \left(\mathbb{R}^{\mathbb{Z}^2} \right)^{\mathbb{Z}^2}$ is the 3×3 unit Moore template with unit values 1. The image $\mathbf{b}$ obtained from the statement $\mathbf{b} := \frac{1}{9}(\mathbf{a} \oplus \mathbf{t})$ represents the image obtained from $\mathbf{a}$ by local averaging since the new pixel value $\mathbf{b}(\mathbf{y})$ is given by

$$\mathbf{b}(\mathbf{y}) = \frac{1}{9} \sum_{\mathbf{x} \in \mathbf{X} \cap S(\mathbf{t_y})} \mathbf{a}(\mathbf{x}) \cdot \mathbf{t_y}(\mathbf{x}) = \frac{1}{9} \sum_{\mathbf{x} \in \mathbf{X} \cap S(\mathbf{t_y})} \mathbf{a}(\mathbf{x}).$$

Of course, there will be a boundary effect. In particular, if $\mathbf{X} = \{(i,j) : 1 \le i \le m, \, 1 \le j \le n\}$, then

$$\mathbf{b}(1,1) = \frac{1}{9}(\mathbf{a}(1,1) + \mathbf{a}(1,2) + \mathbf{a}(2,1) + \mathbf{a}(2,2)),$$

which is not the average of four points. One may either ignore this boundary effect (the most common choice), or one may use one of several schemes to prevent it [1]. However, each of these schemes adds to the computational burden. A simpler and more elegant way is to use the Moore neighborhood function M combined with the averaging reduction $a \equiv average$. The simple statement $\mathbf{b} := \mathbf{a} \, @ \, M$ provides for the desired locally averaged image without boundary effect.

Neighborhood composition plays an important role in algorithm optimization and simplification of algebraic expressions. Given two neighborhood functions $N_1, N_2 : \mathbb{R}^n \to 2^{\mathbb{R}^n}$, then the *dilation* of N_1 by N_2, denoted by $N_1 \oplus N_2$, is a neighborhood function $N : \mathbb{R}^n \to 2^{\mathbb{R}^n}$ which is defined as

$$N(\mathbf{y}) = \bigcup_{\mathbf{p} \in N_2(\mathbf{y})} (N_1(\mathbf{y}) + (\mathbf{p} - \mathbf{y})),$$

where $N(\mathbf{y}) + \mathbf{q} \equiv \{\mathbf{x} + \mathbf{q} : \mathbf{x} \in N(\mathbf{y})\}$. Just as for template composition, algorithm optimization can be achieved by use of the equation $\mathbf{a} \oplus (N_1 \oplus N_2) = (\mathbf{a} \oplus N_1) \oplus N_2$ for appropriate neighborhood functions and neighborhood reduction functions Γ. For $k \in \mathbb{N}$, the kth *iterate* of a neighborhood $N : \mathbb{R}^n \to 2^{\mathbb{R}^n}$ is defined inductively as $N^k = N^{k-1} \oplus N$, where $N^0(\mathbf{y}) = \{\mathbf{y}\} \; \forall \mathbf{y} \in \mathbb{R}^n$.

Most neighborhood functions used in image processing are translation invariant subsets of $\mathbb{R}^n$ (in particular, subsets of $\mathbb{Z}^2 \subset \mathbb{R}^2$). A neighborhood function $N : \mathbb{R}^n \to 2^{\mathbb{R}^n}$ is said to be translation invariant if $N(\mathbf{y} + \mathbf{p}) = N(\mathbf{y}) + \mathbf{p}$ for every point $\mathbf{p} \in \mathbb{R}^n$. Given a translation invariant neighborhood N, we define its *reflection* or *conjugate* N^* by $N^*(\mathbf{y}) = N^*(\mathbf{0}) + \mathbf{y}$, where $N^*(\mathbf{0}) = \{-\mathbf{x} : \mathbf{x} \in N(\mathbf{0})\}$ and $\mathbf{0} = (0, 0, \dots, 0) \in \mathbb{R}^n$ denotes the origin. Conjugate neighborhoods play an important role in morphological image processing.

Note also that for a translation invariant neighborhood N, the kth iterate of N can be expressed in terms of the *sum* of sets

$$N^k(\mathbf{y}) = N^{k-1}(\mathbf{y}) + N(\mathbf{0}).$$

Furthermore,   since $N^{k-1}(\mathbf{y}) + N(\mathbf{0}) = \bigcup_{\mathbf{q}\in N(\mathbf{0})} \left(N^{k-1}(\mathbf{y})+\mathbf{q}\right)$ and

$\bigcup_{\mathbf{q}\in N(\mathbf{0})} \left(N^{k-1}(\mathbf{y})+\mathbf{q}\right) = \bigcup_{\mathbf{p}\in N(\mathbf{y})} \left(N^{k-1}(\mathbf{0})+\mathbf{p}\right)$,   we   have   the   symmetric   relation

$N^{k}(\mathbf{y}) = N^{k-1}(\mathbf{0}) + N(\mathbf{y})$.

Summary of Image-Neighborhood Products

In the following list of pertinent image-neigborhood products $\mathbf{a} \in \mathbb{F}^{\mathbf{X}}$, $\mathbf{X} \subset \mathbf{Z}$, and $N \in \left(2^{\mathbf{Z}}\right)^{\mathbf{Y}}$. Again, for each operation we assume the appropriate value set $\mathbb{F}$.

generic neighborhood reduction

$$\mathbf{a}\,\textcircled{r}\,N = \left\{(\mathbf{y}, \mathbf{b}(\mathbf{y})) : \mathbf{b}(\mathbf{y}) = \Gamma\left(\mathbf{a}|_{N(\mathbf{y})}\right), \mathbf{y} \in \mathbf{Y}\right\}$$

neigborhood sum

$$\mathbf{a}\,\oplus\,N = \left\{(\mathbf{y}, \mathbf{b}(\mathbf{y})) : \mathbf{b}(\mathbf{y}) = \sum_{\mathbf{x}\in \mathbf{X}\cap N(\mathbf{y})} \mathbf{a}(\mathbf{x}), \mathbf{y} \in \mathbf{Y}\right\}$$

neighborhood maximum

$$\mathbf{a}\,\boxdot\,N = \left\{(\mathbf{y}, \mathbf{b}(\mathbf{y})) : \mathbf{b}(\mathbf{y}) = \bigvee_{\mathbf{x}\in \mathbf{X}\cap N(\mathbf{y})} a(\mathbf{x}), \mathbf{y} \in \mathbf{Y}\right\}$$

neighborhood minimum

$$\mathbf{a}\,\boxbslash\,N = \left\{(\mathbf{y}, \mathbf{b}(\mathbf{y})) : \mathbf{b}(\mathbf{y}) = \bigwedge_{\mathbf{x}\in \mathbf{X}\cap N(\mathbf{y})} \mathbf{a}(\mathbf{x}), \mathbf{y} \in \mathbf{Y}\right\}$$

Note that

$$\mathbf{a}\,\circledvee\,N = \left\{(\mathbf{y}, \mathbf{b}(\mathbf{y})) : \mathbf{b}(\mathbf{y}) = \bigvee_{\mathbf{x}\in \mathbf{X}\cap N(\mathbf{y})} \mathbf{a}(\mathbf{x}), \mathbf{y} \in \mathbf{Y}\right\}$$

and, therefore, $\mathbf{a}\,\circledvee\,N = \mathbf{a}\,\boxdot\,N$. Similarly, $\mathbf{a}\,\circledwedge\,N = \mathbf{a}\,\boxbslash\,N$.

Although we did not address the issues of parameterized neighborhoods and recursive neighborhood operations, it should be clear that these are defined in the usual way by simple substitution of the appropriate neighborhood function for the corresponding Boolean template. For example, a parameterized neighborhood with parameters in the set P is a function $N : P \to \left(2^{\mathbf{Z}}\right)^{\mathbf{X}}$. Thus, for each parameter $p \in P$, $N(p)$ is a neighborhood system for $\mathbf{X}$ in $\mathbf{Z}$ since $N(p) : \mathbf{X} \to 2^{\mathbf{Z}}$. Similarly, a recursive neighborhood system for a partially ordered set $(\mathbf{X}, \prec)$ is a function $N = (N_{\nprec}, N_{\prec}) : \mathbf{X} \to \left(2^{\mathbf{Z}}, 2^{\mathbf{X}}\right)$ satisfying the conditions that for each $\mathbf{x} \in \mathbf{X}$, $\mathbf{x} \notin N_{\prec}(\mathbf{x})$, and for each $\mathbf{z} \in N_{\prec}(\mathbf{x})$, $\mathbf{z} \prec \mathbf{x}$.

1.8.   The p-Product

It is well known that in the linear domain template convolution products and image-template convolution products are equivalent to matrix products and vector-matrix products, respectively [63, 1]. The notion of a generalized matrix product was developed in order to provide a general matrix theory approach to image-template products and template convolution products in both the linear and nonlinear domains. This generalized matrix or p-product was first defined in Ritter [64]. This new matrix operation includes the matrix and vector products of linear algebra, the matrix product of minimax algebra [65], as well as generalized convolutions as special cases [64]. It provides for a transformation that combines the same or different types of values (or objects) into values of a possibly different type from those initially used in the combining operation. It has been shown that the p-product can be applied to express various image processing transforms in computing form [66, 67, 68]. In this document, however, we consider only products between matrices having the same type of values. In the subsequent discussion, $\mathbb{F} \in \{\mathbb{R}, \mathbb{C}\}$ and the set of all $m \times n$ matrices with entries from $\mathbb{F}$ will be denoted by $\mathbb{F}_{m \times n}$. We will follow the usual convention of setting $\mathbb{F}^n = \mathbb{F}_{1 \times n}$ and view $\mathbb{F}^n$ as the set of all n-dimensional row vectors with entries from $\mathbb{F}$. Similarly, the set of all m-dimensional column vectors with entries from $\mathbb{F}$ is given by $(\mathbb{F}^m)' = [\mathbb{F}_{1 \times m}]' = \mathbb{F}_{m \times 1}$.

Let m, n, and p be positive integers with p dividing both m and n. Define the following correspondences:

$$c_p : \mathbb{Z}_p^+ \times \mathbb{Z}_{n/p}^+ \to \mathbb{Z}_n^+$$
$$\text{by } c_p(k, j) = (k - 1)\frac{n}{p} + j,$$
$$\text{where } 1 \leq j \leq \frac{n}{p}, \text{and } 1 \leq k \leq p,$$

and

$$r_p : \mathbb{Z}_{m/p}^+ \times \mathbb{Z}_p^+ \to \mathbb{Z}_m^+$$
$$\text{by } r_p(i, k) = (i - 1)p + k,$$
$$\text{where } 1 \leq k \leq p, \text{and } 1 \leq i \leq \frac{m}{p}.$$

Since $r_p(i, k) < r_p(i', k') \Leftrightarrow i < i'$ or $i = i'$ and $k < k'$, r_p linearizes the array $\mathbb{Z}_{m/p}^+ \times \mathbb{Z}_p^+$ using the row scanning order as shown:

$$
\begin{bmatrix}
\begin{smallmatrix}1\\(1,1)\end{smallmatrix} & \begin{smallmatrix}2\\(1,2)\end{smallmatrix} & \cdots & \begin{smallmatrix}k\\(1,k)\end{smallmatrix} & \cdots & \begin{smallmatrix}p\\(1,p)\end{smallmatrix} \\
\begin{smallmatrix}p+1\\(2,1)\end{smallmatrix} & \begin{smallmatrix}p+2\\(2,2)\end{smallmatrix} & \cdots & \begin{smallmatrix}p+k\\(2,k)\end{smallmatrix} & & \begin{smallmatrix}2p\\(2,p)\end{smallmatrix} \\
\vdots & \vdots & & \vdots & & \vdots \\
\begin{smallmatrix}(i-1)p+1\\(i,1)\end{smallmatrix} & \begin{smallmatrix}\\(i,2)\end{smallmatrix} & \cdots & \begin{smallmatrix}(i-1)p+k\\(i,k)\end{smallmatrix} & \cdots & \begin{smallmatrix}ip\\(i,p)\end{smallmatrix} \\
\vdots & \vdots & & \vdots & & \vdots \\
\begin{smallmatrix}((m/p)-1)p+1\\(m/p,1)\end{smallmatrix} & \begin{smallmatrix}\\(m/p,2)\end{smallmatrix} & \cdots & \begin{smallmatrix}\\(m/p,k)\end{smallmatrix} & \cdots & \begin{smallmatrix}(m/p)p=m\\(m/p,p)\end{smallmatrix}
\end{bmatrix}
$$

It follows that the *row-scanning order* on $\mathbb{Z}_{m/p}^+ \times \mathbb{Z}_p^+$ is given by

$$(i, k) \leq (i', k') \iff r_p(i, k) \leq r_p(i', k')$$

or, equivalently, by

$$(i, k) \leq (i', k') \iff i < i' \text{ or } i = i' \text{ and } k \leq k'.$$

We define the one-to-one correspondence

$$f_p : \mathbb{Z}_l^+ \times \mathbb{Z}_{m/p}^+ \times \mathbb{Z}_p^+ \rightarrow \mathbb{Z}_l^+ \times \mathbb{Z}_m^+$$
$$\text{by } f_p : (x, y, z) \mapsto (x, r_p(y, z)).$$

The one-to-one correspondence allows us to re-index the entries of a matrix $A = (a_{s,t}) \in \mathbb{F}_{l \times m}$ in terms of a triple index $a_{s,(i,k)}$ by using the convention

$$a_{s,(i,k)} = a_{s,t} \iff r_p(i, k) = t,$$
$$\text{where } 1 \leq i \leq m/p \text{ and } 1 \leq k \leq p.$$

Example: Suppose $l = 2$, $m = 6$ and $p = 2$. Then $m/p = 3$, $1 \leq k \leq p = 2$, and $1 \leq i \leq m/p = 3$. Hence for $A = (a_{s,t}) \in \mathbb{F}_{2 \times 6}$, we have

$$A = \begin{pmatrix} a_{11} & a_{12} & a_{13} & a_{14} & a_{15} & a_{16} \\ a_{21} & a_{22} & a_{23} & a_{24} & a_{25} & a_{26} \end{pmatrix}$$

$$= \begin{pmatrix} a_{1,(1,1)} & a_{1,(1,2)} & a_{1,(2,1)} & a_{1,(2,2)} & a_{1,(3,1)} & a_{1,(3,2)} \\ a_{2,(1,1)} & a_{2,(1,2)} & a_{2,(2,1)} & a_{2,(2,2)} & a_{2,(3,1)} & a_{2,(3.2)} \end{pmatrix}.$$

The factor $\mathbb{Z}_n^+$ of the Cartesian product $\mathbb{Z}_n^+ \times \mathbb{Z}_q^+$ is decomposed in a similar fashion. Here the row-scanning map is given by

$$c_p : \mathbb{Z}_p^+ \times \mathbb{Z}_{n/p}^+ \rightarrow \mathbb{Z}_n^+$$
$$\text{where } c_p(k, j) = (k - 1)(n/p) + j,$$
$$1 \leq j \leq n/p, \text{ and } 1 \leq k \leq p.$$

This allows us to re-index the entries of a matrix $B = (b_{s,t}) \in M_{n \times q}(\mathbb{F})$ in terms of a triple index $b_{(k,j),t}$ by using the convention

$$b_{(k,j),t} = b_{s,t} \iff c_p(k, j) = s,$$
$$\text{where } 1 \leq k \leq p \text{ and } 1 \leq j \leq n/p.$$

Example: Suppose $n = 4$, $q = 3$ and $p = 2$. Then $n/p = 2$, $1 \le k \le p = 2$, and $1 \le j \le n/p = 2$. Hence for $B = (b_{s,t}) \in \mathbb{F}_{n \times q}$, we have

$$B = \begin{pmatrix} b_{11} & b_{12} & b_{13} \\ b_{21} & b_{22} & b_{23} \\ b_{31} & b_{32} & b_{33} \\ b_{41} & b_{42} & b_{43} \end{pmatrix} = \begin{pmatrix} b_{(1,1),1} & b_{(1,1),2} & b_{(1,1),3} \\ b_{(1,2),1} & b_{(1,2),2} & b_{(1,2),3} \\ b_{(2,1),1} & b_{(2,1),2} & b_{(2,1),3} \\ b_{(2,2),1} & b_{(2,2),2} & b_{(2,2),3} \end{pmatrix}.$$

Now let $A = (a_{sj'}) \in \mathbb{F}_{l \times m}$ and $B = (b_{i',t}) \in \mathbb{F}_{n \times q}$. Using the maps r_p and c_p, A and B can be rewritten as

$$A = \left(a_{s,(i,k)}\right)_{l \times m}, \text{ where } 1 \le s \le l,\, 1 \le r_p(i,k) = j' \le m, \text{ and}$$
$$B = \left(b_{(k,j),t}\right)_{n \times q}, \text{ where } 1 \le c_p(k,j) = i' \le n \text{ and } 1 \le t \le q.$$

The *p-product* or *generalized matrix product* of A and B is denoted by $A \oplus_p B$, and is the matrix

$$C = A \oplus_p B \in \mathbb{F}_{l(n/p) \times (m/p)q}$$

defined by

$$c_{(s,j)(i,t)} = \sum_{k=1}^{p} \left(a_{s,(i,k)} b_{(k,j),t}\right) = \left(a_{s,(i,1)} b_{(1,j),t}\right) + \ldots + \left(a_{s,(i,p)} b_{(p,j),t}\right),$$

where $c_{(s,j)(i,t)}$ denotes the (s,j)th row and (i,t)th column entry of C. Here we use the lexicographical order $(s,j) < (s',j') \Leftrightarrow s < s'$ or if $s = s', j < j'$. Thus, the matrix C has the following form:

$$\begin{bmatrix}
c_{(1,1)(1,1)} & \cdots & c_{(1,1)(1,q)} & c_{(1,1)(2,1)} & \cdots & c_{(1,1)(2,q)} & \cdots & c_{(1,1)(i,t)} & \cdots & c_{(1,1)\left(\frac{m}{p},q\right)} \\
c_{(1,2)(1,1)} & \cdots & c_{(1,2)(1,q)} & c_{(1,2)(2,1)} & \cdots & c_{(1,2)(2,q)} & \cdots & c_{(1,2)(i,t)} & \cdots & c_{(1,2)\left(\frac{m}{p},q\right)} \\
\vdots & & \vdots & \vdots & & \vdots & & \vdots & & \vdots \\
c_{\left(1,\frac{n}{p}\right)(1,1)} & \cdots & c_{\left(1,\frac{n}{p}\right)(1,q)} & c_{\left(1,\frac{n}{p}\right)(2,1)} & \cdots & c_{\left(1,\frac{n}{p}\right)(2,q)} & \cdots & c_{\left(1,\frac{n}{p}\right)(i,t)} & \cdots & c_{\left(1,\frac{n}{p}\right)\left(\frac{m}{p},q\right)} \\
c_{(2,1)(1,1)} & \cdots & c_{(2,1)(1,q)} & c_{(2,1)(2,1)} & \cdots & c_{(2,1)(2,q)} & \cdots & c_{(2,1)(i,t)} & \cdots & c_{(2,1)\left(\frac{m}{p},q\right)} \\
\vdots & & \vdots & \vdots & & \vdots & & \vdots & & \vdots \\
c_{\left(2,\frac{n}{p}\right)(1,1)} & \cdots & c_{\left(2,\frac{n}{p}\right)(1,q)} & c_{\left(2,\frac{n}{p}\right)(2,1)} & \cdots & c_{\left(2,\frac{n}{p}\right)(2,q)} & \cdots & c_{\left(2,\frac{n}{p}\right)(i,t)} & \cdots & c_{\left(2,\frac{n}{p}\right)\left(\frac{m}{p},q\right)} \\
\vdots & & \vdots & \vdots & & \vdots & & \vdots & & \vdots \\
c_{(s,j)(1,1)} & \cdots & c_{(s,j)(1,q)} & c_{(s,j)(2,1)} & \cdots & c_{(s,j)(2,q)} & \cdots & \underline{c_{(s,j)(i,t)}} & \cdots & c_{(s,j)\left(\frac{m}{p},q\right)} \\
\vdots & & \vdots & \vdots & & \vdots & & \vdots & & \vdots \\
c_{(l,1)(1,1)} & \cdots & c_{(l,1)(1,q)} & c_{(l,1)(2,1)} & \cdots & c_{(l,1)(2,q)} & \cdots & c_{(l,1)(i,t)} & \cdots & c_{(l,1)\left(\frac{m}{p},q\right)} \\
\vdots & & \vdots & \vdots & & \vdots & & \vdots & & \vdots \\
c_{\left(l,\frac{n}{p}\right)(1,1)} & \cdots & c_{\left(l,\frac{n}{p}\right)(1,q)} & c_{\left(l,\frac{n}{p}\right)(2,1)} & \cdots & c_{\left(l,\frac{n}{p}\right)(2,q)} & \cdots & c_{\left(l,\frac{n}{p}\right)(i,t)} & \cdots & c_{\left(l,\frac{n}{p}\right)\left(\frac{m}{p},q\right)}
\end{bmatrix}$$

The entry $c_{(s,j)(i,t)}$ in the (s,j)-row and (i,t)-column is underlined for emphasis.

To provide an example, suppose that $l = 2$, $m = 6$, $n = 4$, and $q = 3$. Then for $p = 2$, one obtains $m/p = 3$, $n/p = 2$ and $1 \leq k \leq 2$. Now let

$$A = \begin{pmatrix} a_{11} & a_{12} & a_{13} & a_{14} & a_{15} & a_{16} \\ a_{21} & a_{22} & a_{23} & a_{24} & a_{25} & a_{26} \end{pmatrix} \in M_{2\times 6}(\mathbb{R})$$

and

$$B = \begin{pmatrix} b_{11} & b_{12} & b_{13} \\ b_{21} & b_{22} & b_{23} \\ b_{31} & b_{32} & b_{33} \\ b_{41} & b_{42} & b_{43} \end{pmatrix} \in \mathbb{R}_{4\times 3}.$$

Then the (2,1)-row and (2,3)-column element $c_{(2,1)(2,3)}$ of the matrix

$$C = A \oplus_2 B \in \mathbb{R}_{l(n/p)\times(m/p)q} = \mathbb{R}_{4\times 9}$$

is given by

$$c_{(2,1)(2,3)} = \sum_{k=1}^{2} a_{2,r_2(2,k)} \cdot b_{c_2(k,1),3}$$

$$= a_{2,r_2(2,1)} \cdot b_{c_2(1,1),3} + a_{2,r_2(2,2)} \cdot b_{c_2(2,1),3}$$

$$= a_{23} \cdot b_{13} + a_{24} \cdot b_{33}.$$

Thus, in order to compute $c_{(2,1)(2,3)}$, the two underlined elements of A are combined with the two underlined elements of B as illustrated:

$$\begin{pmatrix} a_{11} & a_{12} & a_{13} & a_{14} & a_{15} & a_{16} \\ a_{21} & a_{22} & \underline{a_{23}} & \underline{a_{24}} & a_{25} & a_{26} \end{pmatrix} \oplus_2 \begin{pmatrix} b_{11} & b_{12} & \underline{b_{13}} \\ b_{21} & b_{22} & b_{23} \\ b_{31} & b_{32} & \underline{b_{33}} \\ b_{41} & b_{42} & b_{43} \end{pmatrix}$$

$$= \begin{pmatrix} a_{1,r_2(1,1)} & a_{1,r_2(1,2)} & a_{1,r_2(2,1)} & a_{1,r_2(2,2)} & a_{1,r_2(3,1)} & a_{1,r_2(3,2)} \\ a_{2,r_2(1,1)} & a_{2,r_2(1,2)} & \underline{a_{2,r_2(2,1)}} & \underline{a_{2,r_2(2,2)}} & a_{2,r_2(3,1)} & a_{2,r_2(3.2)} \end{pmatrix} \oplus_2$$

$$\begin{pmatrix} b_{c_2(1,1),1} & b_{c_2(1,1),2} & \underline{b_{c_2(1,1),3}} \\ b_{c_2(1,2),1} & b_{c_2(1,2),2} & \overline{b}_{c_2(1,2),3} \\ b_{c_2(2,1),1} & b_{c_2(2,1),2} & \underline{b_{c_2(2,1),3}} \\ b_{c_2(2,2),1} & b_{c_2(2,2),2} & \overline{b}_{c_2(2,2),3} \end{pmatrix}$$

$$= \begin{pmatrix} c_{(1,1)(1,1)} & c_{(1,1)(1,2)} & \cdots & c_{(1,1)(2,3)} & \cdots & c_{(1,1)(3,3)} \\ c_{(1,2)(1,1)} & c_{(1,2)(1,2)} & \cdots & c_{(1,2)(2,3)} & \cdots & c_{(1,2)(3,3)} \\ c_{(2,1)(1,1)} & c_{(2,1)(1,2)} & \cdots & \underline{c_{(2,1)(2,3)}} & \cdots & c_{(2,1)(3,3)} \\ c_{(2,2)(1,1)} & c_{(2,2)(1,2)} & \cdots & c_{(2,2)(2,3)} & \cdots & c_{(2,2)(3,3)} \end{pmatrix}$$

$$= \begin{pmatrix} c_{11} & c_{12} & \cdots & c_{16} & \cdots & c_{19} \\ c_{21} & c_{22} & \cdots & c_{26} & \cdots & c_{29} \\ c_{31} & c_{32} & \cdots & \underline{c_{36}} & \cdots & c_{39} \\ c_{41} & c_{42} & \cdots & c_{46} & \cdots & c_{49} \end{pmatrix}.$$

In particular,

$$\begin{pmatrix} 1 & 2 & 0 & 5 & 4 & 3 \\ 2 & 3 & 4 & 1 & 0 & 6 \end{pmatrix} \oplus_2 \begin{pmatrix} 2 & 6 & 1 \\ 1 & 3 & 2 \\ 2 & 2 & 5 \\ 3 & 0 & 4 \end{pmatrix} = \begin{pmatrix} 6 & 10 & 11 & 10 & 10 & 25 & 14 & 30 & 19 \\ 7 & 3 & 10 & 15 & 0 & 20 & 13 & 12 & 20 \\ 10 & 18 & 17 & 10 & 26 & 9 & 12 & 12 & 30 \\ 11 & 6 & 16 & 7 & 12 & 12 & 18 & 0 & 24 \end{pmatrix}.$$

If

$$A = \begin{pmatrix} 1 & 0 \\ -1 & 1 \end{pmatrix} \text{ and } B = \begin{pmatrix} 4 \\ 2 \\ 6 \\ 3 \end{pmatrix},$$

then

$$(A \oplus_2 B)' = (4, 2, 2, 1) \neq (4, -2, 6, -3) = B' \oplus_2 A'.$$

This shows that the transpose property, which holds for the regular matrix product, is generally false for the p-product. The reason is that the p-product is not a dual operation in the transpose domain. In order to make the transpose property hold we define the *dual* operation $\oplus_p'$ of $\oplus_p$ by

$$A \oplus_p' B \equiv (B' \oplus_p A')'.$$

It follows that

$$A \oplus_p B = (B' \oplus_p' A')'$$

and the p-product is the dual operation of $\oplus_p'$. In particular, we now have the transpose property $(A \oplus_p B)' = B' \oplus_p' A'$.

Since the operation $\oplus_p'$ is defined in terms of matrix transposition, labeling of matrix indices are reversed. Specifically, if $A = (a_{st})$ is an $l \times m$ matrix, then A gets reindexed as $A = (a_{s,(kj)})$, using the convention

$$a_{s,(k,j)} = a_{s,t} \iff c_p(k, j) = t,$$
$$\text{where } 1 \leq j \leq m/p \text{ and } 1 \leq k \leq p.$$

Similarly, if $B = (b_{st})$ is an $n \times q$ matrix, then the entries of B are relabeled as $b_{(i,k),t}$, using the convention

$$b_{(i,k),t} = b_{s,t} \iff r_p(i, k) = s,$$
$$\text{where } 1 \leq k \leq p \text{ and } 1 \leq i \leq n/p.$$

The product $A \oplus_p' B = C$ is then defined by the equation

$$c_{(i,s)(t,j)} = \sum_{k=1}^{p} \left(a_{s,(k,j)} b_{(i,k),t} \right) = \left(a_{s,(1,j)} b_{(i,1),t} \right) + \ldots + \left(a_{s,(p,j)} b_{(i,p),t} \right).$$

Note that the dimension of C is $l \cdot \frac{n}{p} \times \frac{m}{p} \cdot q$.

To provide a specific example of the dual operation $\oplus_p'$, suppose that

$$A = \begin{pmatrix} a_{11} & a_{12} & a_{13} & a_{14} \\ a_{21} & a_{22} & a_{23} & a_{24} \\ a_{31} & a_{32} & a_{33} & a_{34} \end{pmatrix} \text{ and } B = \begin{pmatrix} b_{11} & b_{12} \\ b_{21} & b_{22} \\ b_{31} & b_{32} \\ b_{41} & b_{42} \\ b_{51} & b_{52} \\ b_{61} & b_{62} \end{pmatrix}.$$

In this case we have $l = 3$, $m = 4$, $n = 6$, and $q = 2$. Thus, for $p = 2$ and using the scheme described above, the reindexed matrices have form

$$A = \begin{pmatrix} a_{1,(1,1)} & a_{1,(1,2)} & a_{1,(2,1)} & a_{1,(2,2)} \\ a_{2,(1,1)} & a_{2,(1,2)} & a_{2,(2,1)} & a_{2,(2,2)} \\ a_{3,(1,1)} & a_{3,(1,2)} & a_{3,(2,1)} & a_{3,(2,2)} \end{pmatrix} \quad \text{and} \quad B = \begin{pmatrix} b_{(1,1),1} & b_{(1,1),2} \\ b_{(1,2),1} & b_{(1,2),2} \\ b_{(2,1),1} & b_{(2,1),2} \\ b_{(2,2),1} & b_{(2,2),2} \\ b_{(3,1),1} & b_{(3,1),2} \\ b_{(3,2),1} & b_{(3,2),2} \end{pmatrix}.$$

According to the dual product definition, the matrix $A \oplus_2' B = C$ is a 9×4 matrix given by

$$C = \begin{pmatrix} c_{11} & c_{12} & c_{13} & c_{14} \\ c_{21} & c_{22} & c_{23} & c_{24} \\ \vdots & \vdots & \vdots & \vdots \\ c_{61} & c_{62} & \underline{c_{63}} & c_{64} \\ \vdots & \vdots & \vdots & \vdots \\ c_{91} & c_{92} & c_{93} & c_{94} \end{pmatrix} = \begin{pmatrix} c_{(1,1)(1,1)} & c_{(1,1)(1,2)} & c_{(1,1)(2,1)} & c_{(1,1)(2,2)} \\ c_{(1,2)(1,1)} & c_{(1,2)(1,2)} & c_{(1,2)(2,1)} & c_{(1,2)(2,2)} \\ \vdots & \vdots & \vdots & \vdots \\ c_{(2,3)(1,1)} & c_{(2,3)(1,2)} & \underline{c_{(2,3)(2,1)}} & c_{(2,3)(2,2)} \\ \vdots & \vdots & \vdots & \vdots \\ c_{(3,3)(1,1)} & c_{(3,3)(1,2)} & c_{(3,3)(2,1)} & c_{(3,3)(2,2)} \end{pmatrix}.$$

The underlined element c_{63} is obtained by using the formula:

$$c_{63} = c_{(2,3)(2,1)} = \sum_{k=1}^{2} a_{3,(k,1)} b_{(2,k),2} = a_{3,(1,1)} b_{(2,1),2} + a_{3,(2,1)} b_{(2,2),2} .$$

Thus, in order to compute c_{63}, the two underlined elements of A are combined with the two underlined elements of B as illustrated:

$$\begin{pmatrix} a_{11} & a_{12} & a_{13} & a_{14} \\ a_{21} & a_{22} & a_{23} & a_{24} \\ \underline{a_{31}} & a_{32} & \underline{a_{33}} & a_{34} \end{pmatrix} \oplus_2' \begin{pmatrix} b_{11} & b_{12} \\ b_{21} & b_{22} \\ b_{31} & \underline{b_{32}} \\ b_{41} & \underline{b_{42}} \\ b_{51} & b_{52} \\ b_{61} & b_{62} \end{pmatrix}.$$

As a final observation, note that the matrices A, B, and C in this example have the form of the transposes of the matrices B, A, and C, respectively, of the previous example.

1.9. Exercises

1. a. Show that the addition of a scalar $k \in \mathbb{R}$ and a point $\mathbf{x} \in \mathbb{R}^n$ is a special case of the addition of two points $\mathbf{x}, \mathbf{y} \in \mathbb{R}^n$.
    b. Show that scalar multiplication of a point $\mathbf{x} \in \mathbb{R}^n$ is a special case of Hadamard multiplication of two points in $\mathbb{R}^n$.

2. Show that the operations of point addition and scalar multiplication satisfy:

    a. $k \cdot (\mathbf{x} + \mathbf{y}) = k \cdot \mathbf{x} + k \cdot \mathbf{y}$, where $k \in \mathbb{R}$ and $\mathbf{x}, \mathbf{y} \in \mathbb{R}^n$,
    b. $(k + h) \cdot \mathbf{x} = k \cdot \mathbf{x} + h \cdot \mathbf{x}$, where $k, h \in \mathbb{R}$ and $\mathbf{x} \in \mathbb{R}^n$.

3. Let d be a distance function on $\mathbb{R}^2$ and $\mathbf{x}$ a point in $\mathbb{R}^2$. The *neighborhood* (or sphere or ball) of radius r of $\mathbf{x}$ is denoted by $N_{r,d}(\mathbf{x})$ and defined as

$$N_{r,d}(\mathbf{x}) = \{\mathbf{y} \in \mathbb{R}^2 : d(\mathbf{x}, \mathbf{y}) \leq r\}.$$

Give a graphical representation of the neighborhood of the origin $\mathbf{x} = (0, 0)$ of radius 1 if

    a. d is the Euclidean distance,
    b. d is the chessboard distance,
    c. d is the city block distance.

4. Let $\mathbf{x}, \mathbf{y} \in \mathbb{R}^n$. For $p = 2$, show that:

    a. $\|\mathbf{x}\|_p \geq 0$,
    b. $\|\mathbf{x}\|_p = 0$ if and only if $\mathbf{x} = (0, 0)$,
    c. $\|\mathbf{x} - \mathbf{y}\|_p = \|\mathbf{y} - \mathbf{x}\|_p$,
    d. $\left| \sum_{k=1}^{n} x_k y_k \right| \leq \|\mathbf{x}\|_p \|\mathbf{y}\|_p$,
    e. $\|\mathbf{x} + \mathbf{y}\|_p \leq \|\mathbf{x}\|_p \|\mathbf{y}\|_p$.

5. Let $\mathbf{X}$, $\mathbf{Y}$, and $\mathbf{Z}$ be any three sets in $\mathbb{R}^n$. Show that

    a. $\mathbf{X} \cup (\mathbf{Y} \cap \mathbf{Z}) = (\mathbf{X} \cup \mathbf{Y}) \cap (\mathbf{X} \cup \mathbf{Z})$,
    b. $\mathbf{X} \cap (\mathbf{Y} \cup \mathbf{Z}) = (\mathbf{X} \cap \mathbf{Y}) \cup (\mathbf{Y} \cap \mathbf{Z})$,
    c. $\overline{(\mathbf{X} \cup \mathbf{Y})} = \overline{\mathbf{X}} \cap \overline{\mathbf{Y}}$,
    d. $\overline{(\mathbf{X} \cap \mathbf{Y})} = \overline{\mathbf{X}} \cup \overline{\mathbf{Y}}$,
    e. $\mathbf{X} \times (\mathbf{Y} \cup \mathbf{Z}) = (\mathbf{X} \times \mathbf{Y}) \cup (\mathbf{X} \times \mathbf{Z})$,
    f. $\mathbf{X} \times (\mathbf{Y} \cap \mathbf{Z}) = (\mathbf{X} \times \mathbf{Y}) \cap (\mathbf{X} \times \mathbf{Z})$.

6. Let $\mathbf{a} \in \mathbb{R}^{\mathbf{X}}$. Show that $\mathbf{a} \cdot \mathbf{a}^{-1} \cdot \mathbf{a} = \mathbf{a}$ and that $\mathbf{a}^{-1} \cdot \mathbf{a} \cdot \mathbf{a}^{-1} = \mathbf{a}^{-1}$.

7. Suppose $\mathbf{a}, \mathbf{b} \in \mathbb{R}^{\mathbf{X}}$ and $\mathbf{c} = (\mathbf{a} - \mathbf{b}) \vee \mathbf{0}$. Show that

$$\chi_{>b}(\mathbf{a}) = \mathbf{c}^{-1} \cdot \mathbf{c}.$$

8. Define the Boolean complement of a real-valued image $\mathbf{a} \in \mathbb{R}^{\mathbf{X}}$ by $\mathbf{a}^c$, where $\mathbf{a}^c = 1 - \mathbf{a}^{-1} \cdot \mathbf{a}$. If $\mathbf{b} \in \mathbb{R}^{\mathbf{X}}$ show that

    a. $\chi_{\leq b}(\mathbf{a}) = [\chi_{>b}(\mathbf{a})]^c$,
    b. $\chi_{\neq b}(\mathbf{a}) = [\chi_b(\mathbf{a})]^c$,
    c. $\mathbf{a}^c \cdot \mathbf{a} = 0$.

9. Suppose $f : \mathbb{R} \to \mathbb{R}$ denotes the absolute value function $f(r) = |r|$, $r \in \mathbb{R}$. Show that

$$f \circ \mathbf{a} = \mathbf{a} \vee (-\mathbf{a})$$

whenever $\mathbf{a} \in \mathbb{R}^{\mathbf{X}}$.

10. Suppose $\mathbf{s}, \mathbf{t} \in \left(\mathbb{R}^{\mathbb{Z}^2} \right)^{\mathbb{Z}^2}$ are the following translation invariant templates:

    a.   Find $(\mathbf{s} + \mathbf{t})_{\mathbf{y}}$, $(\mathbf{s} \cdot \mathbf{t})_y$, and $(\mathbf{s} \vee \mathbf{t})_{\mathbf{y}}$.
    b.   Let $\mathbf{r} = \mathbf{s} \, \boxtimes \, \mathbf{t}$. Give a graphical representation of $\mathbf{r}_{\mathbf{y}}$.

11. Suppose $\mathbf{t} \in \left(\mathbb{R}_{\pm\infty}^{X}\right)^{Y}$. The *dual* of $\mathbf{t}$ is denoted by $\mathbf{t}^{*}$ and is the template $\mathbf{t}^{*} \in \left(\mathbb{R}_{\pm\infty}^{Y}\right)^{X}$ defined by $\mathbf{t}_{\mathbf{x}}^{*}(\mathbf{y}) = \left[\mathbf{t}_{\mathbf{y}}(\mathbf{x})\right]^{*}$. Show that

$$\mathbf{s} \,\boxslash\, \mathbf{t} = \left(\mathbf{t}^{*} \,\boxminus\, \mathbf{s}^{*}\right)^{*}$$

where $\mathbf{s}, \mathbf{t} \in \left(\mathbb{R}_{\pm\infty}^{Y}\right)^{X}$.

12. Suppose

$$\mathbf{s_y} = \quad\begin{array}{|c|}\hline 3 \\\hline\end{array} \qquad \mathbf{t_y} = \begin{array}{|c|c|c|}\hline 1 & 1 & 1 \\\hline 1 & 2 & 1 \\\hline 1 & 1 & 1 \\\hline\end{array}$$

are two translation invariant real-valued templates on $\mathbb{Z}^{2}$. Provide a graphical representation of $\mathbf{r_y}$ where

    a. $\mathbf{r} = \mathbf{s} \oplus \mathbf{t}$, and

    b. $\mathbf{r} = \mathbf{s} \,\boxminus\, \mathbf{t}$.

13. A real- dual-valued image $\mathbf{b}$ is an element of $(\mathbb{R} \times \mathbb{R})^{X}$ or $\left(\mathbb{R}^{2}\right)^{X}$. Construct three real dual-valued 10×10 rectangular images $\mathbf{a}_{1}, \mathbf{a}_{2}, \mathbf{a}_{3}$. Compute the image

$$\mathbf{a} = \left(\vee|_{1}\right)_{i=0}^{3} \mathbf{a}_{i} = \left(\mathbf{a}_{1} \vee_{1} \mathbf{a}_{2}\right) \vee_{1} \mathbf{a}_{3}.$$

14. Let $\mathbf{a}_{1}, \mathbf{a}_{2}, \mathbf{a}_{3}$ be as in Exercise 13. Find

    a. $\mathbf{b} = \mathbf{a}_{1} + \mathbf{a}_{2} + \mathbf{a}_{3}$, and

    b. $\sum \mathbf{b}$.

15. Let N_{1} and N_{2} be two neighborhoods defined on $\mathbb{Z}^{2}$, $\mathbf{a}$ a rectangular $m \times n$ real-valued image and $\Gamma = \vee$. Show that

$$\mathbf{a} \oplus \left(N_{1} \oplus N_{2}\right) = \left(\mathbf{a} \oplus N_{1}\right) \oplus N_{2}.$$

16. Let

$$A = \begin{pmatrix} 1 & 2 & 0 & 5 & 4 & 3 \\ 2 & 3 & 4 & 1 & 0 & 6 \end{pmatrix} \quad \text{and}$$

$$B = \begin{pmatrix} 2 & 6 & 1 \\ 1 & 3 & 2 \\ 2 & 2 & 5 \\ 3 & 0 & 4 \end{pmatrix}.$$

Compute each of the following:

    a. $\left(A \oplus_{2} B\right)'$,

    b. $\left(B' \oplus_{2} A'\right)$,

    c. $B' \oplus_{2}' A'$.

17. a.   Suppose A and B are $l \times m$ matrices and C is a $u \times v$ matrix. Show that $(A + B) \oplus_{p} C = \left(A \oplus_{p} C\right) + \left(B \oplus_{p} C\right)$.

    b.   Suppose A is an $l \times m$ matrix and B and C are of dimension $u \times v$. Show that $A \oplus_{p} (B + C) = \left(A \oplus_{p} C\right) + \left(B \oplus_{p} C\right)$.

    c.   Let the dimension of A, B, and C be $u \times v$, $w \times l$, and $m \times n$, respectively. Suppose p divides both v and w, and q divides both l and m. Show that

$$A \oplus_{p} \left(B \oplus_{q} C\right) = \left(A \oplus_{p} B\right) \oplus_{q} C.$$

This is the associative law for the generalized matrix product.

1.10.   References

[1] G. Ritter, "Image algebra." Unpublished manuscript, available via anonymous ftp from `ftp://ftp.cise.ufl.edu/pub/src/ia/documents`, 1994.

[2] S. Unger, "A computer oriented toward spatial problems," *Proceedings of the IRE*, vol. 46, pp. 1144–1750, 1958.

[3] J. von Neumann, "The general logical theory of automata," in *Cerebral Mechanism in Behavior: The Hixon Symposium*, New York, NY: John Wiley & Sons, 1951.

[4] J. von Neumann, *Theory of Self-Reproducing Automata*. Urbana, IL: University of Illinois Press, 1966.

[5] K. Batcher, "Design of a massively parallel processor," *IEEE Transactions on Computers*, vol. 29, no. 9, pp. 836–840, 1980.

[6] M. Duff, D. Watson, T. Fountain, and G. Shaw, "A cellular logic array for image processing," *Pattern Recognition*, vol. 5, pp. 229–247, June 1973.

[7] M. J. B. Duff, "Review of CLIP image processing system," in *Proceedings of the National Computing Conference*, pp. 1055–1060, AFIPS, 1978.

[8] M. Duff, "Clip4," in *Special Computer Architectures for Pattern Processing* (K. Fu and T. Ichikawa, eds.), ch. 4, pp. 65–86, Boca Raton, FL: CRC Press, 1982.

[9] T. Fountain, K. Matthews, and M. Duff, "The CLIP7A image processor," *IEEE Pattern Analysis and Machine Intelligence*, vol. 10, no. 3, pp. 310–319, 1988.

[10] L. Uhr, "Pyramid multi-computer structures, and augmented pyramids," in *Computing Structures for Image Processing* (M. Duff, ed.), pp. 95–112, London: Academic Press, 1983.

[11] L. Uhr, *Algorithm-Structured Computer Arrays and Networks*. New York, NY: Academic Press, 1984.

[12] W. Hillis, *The Connection Machine*. Cambridge, MA: The MIT Press, 1985.

[13] J. Klein and J. Serra, "The texture analyzer," *Journal of Microscopy*, vol. 95, pp. 349–356, 1972.

[14] R. Lougheed and D. McCubbrey, "The cytocomputer: A practical pipelined image processor," in *Proceedings of the Seventh International Symposium on Computer Architecture*, pp. 411–418, 1980.

[15] S. Sternberg, "Biomedical image processing," *Computer*, vol. 16, pp. 22–34, Jan. 1983.

[16] R. Lougheed, "A high speed recirculating neighborhood processing architecture," in *Architectures and Algorithms for Digital Image Processing II*, vol. 534 of *Proceedings of SPIE*, pp. 22–33, 1985.

[17] E. Cloud and W. Holsztynski, "Higher efficiency for parallel processors," in *Proceedings IEEE Southcon 84*, (Orlando, FL), pp. 416–422, Mar. 1984.

[18] E. Cloud, "The geometric arithmetic parallel processor," in *Proceedings Frontiers of Massively Parallel Processing*, George Mason University, Oct. 1988.

[19] E. Cloud, "Geometric arithmetic parallel processor: Architecture and implementation," in *Parallel Architectures and Algorithms for Image Understanding* (V. Prasanna, ed.), (Boston, MA.), Academic Press, Inc., 1991.

[20] J. Wilson and E. Riedy, "Efficient SIMD evaluation of image processing programs," in *Parallel and Distributed Methods for Image Processing*, vol. 3166 of *Proceedings of SPIE*, (San Diego, CA), pp. 199–211, July 1997.

[21] H. Minkowski, "Volumen und oberflache," *Mathematische Annalen*, vol. 57, pp. 447–495, 1903.

[22] H. Minkowski, *Gesammelte Abhandlungen*. Leipzig-Berlin: Teubner Verlag, 1911.

[23] H. Hadwiger, *Vorlesungen Über Inhalt, Oberflæche und Isoperimetrie*. Berlin: Springer-Verlag, 1957.

[24] G. Matheron, *Random Sets and Integral Geometry*. New York: Wiley, 1975.

[25] J. Serra, "Introduction a la morphologie mathematique," booklet no. 3, Cahiers du Centre de Morphologie Mathematique, Fontainebleau, France, 1969.

[26] J. Serra, "Morphologie pour les fonctions "a peu pres en tout ou rien"," technical report, Cahiers du Centre de Morphologie Mathematique, Fontainebleau, France, 1975.

[27] J. Serra, *Image Analysis and Mathematical Morphology*. London: Academic Press, 1982.

[28] T. Crimmins and W. Brown, "Image algebra and automatic shape recognition," *IEEE Transactions on Aerospace and Electronic Systems*, vol. AES-21, pp. 60–69, Jan. 1985.

[29] R. Haralick, S. Sternberg, and X. Zhuang, "Image analysis using mathematical morphology: Part I," *IEEE Transactions on Pattern Analysis and Machine Intelligence*, vol. 9, pp. 532–550, July 1987.

[30] R. Haralick, L. Shapiro, and J. Lee, "Morphological edge detection," *IEEE Journal of Robotics and Automation*, vol. RA-3, pp. 142–157, Apr. 1987.

[31] P. Maragos and R. Schafer, "Morphological skeleton representation and coding of binary images," *IEEE Transactions on Acoustics, Speech, and Signal Processing*, vol. ASSP-34, pp. 1228–1244, Oct. 1986.

[32] P. Maragos and R. Schafer, "Morphological filters Part II : Their relations to median, order-statistic, and stack filters," *IEEE Transactions on Acoustics, Speech, and Signal Processing*, vol. ASSP-35, pp. 1170–1184, Aug. 1987.

[33] P. Maragos and R. Schafer, "Morphological filters Part I: Their set-theoretic analysis and relations to linear shift-invariant filters," *IEEE Transactions on Acoustics, Speech, and Signal Processing*, vol. ASSP-35, pp. 1153–1169, Aug. 1987.

[34] J. Davidson, "Simulated annealing and morphological neural networks," in *Image Algebra and Morphological Image Processing III*, vol. 1769 of *Proceedings of SPIE*, (San Diego, CA), pp. 119–127, July 1992.

[35] J. Davidson and A. Talukder, "Template identification using simulated annealing in morphology neural networks," in *Second Annual Midwest Electro-Technology Conference*, (Ames, IA), pp. 64–67, IEEE Central Iowa Section, Apr. 1993.

[36] J. Davidson and F. Hummer, "Morphology neural networks: An introduction with applications," *IEEE Systems Signal Processing*, vol. 12, no. 2, pp. 177–210, 1993.

[37] E. Dougherty, "Unification of nonlinear filtering in the context of binary logical calculus, part ii: Gray-scale filters," *Journal of Mathematical Imaging and Vision*, vol. 2, pp. 185–192, Nov. 1992.

[38] E. Dougherty, "Optimal binary morphological bandpass filters induced by granulometric spectral representation," *Journal of Mathematical Imaging and Vision*, vol. 7, pp. 175–191, Mar. 1997.

[39] D. Schonfeld and J. Goutsias, "Optimal morphological pattern restoration from noisy binary images," *IEEE Transactions on Pattern Analysis and Machine Intelligence*, vol. 13, pp. 14–29, Jan. 1991.

[40] J. Goutsias, "On the morphological analysis of discrete random shapes," *Journal of Mathematical Imaging and Vision*, vol. 2, pp. 193–216, Nov. 1992.

[41] L. Koskinen and J. Astola, "Asymptotic behaviour of morphological filters," *Journal of Mathematical Imaging and Vision*, vol. 2, pp. 117–136, Nov. 1992.

[42] K. Sivakumar and J. Goutsias, "Morphologically constrained grfs: Applications to texture synthesis and analysis," *IEEE Transactions on Pattern Analysis and Machine Intelligence*, vol. 21, pp. 99–131, Feb. 1999.

[43] S. R. Sternberg, "Language and architecture for parallel image processing," in *Proceedings of the Conference on Pattern Recognition in Practice*, (Amsterdam), May 1980.

[44] S. Sternberg, "Overview of image algebra and related issues," in *Integrated Technology for Parallel Image Processing* (S. Levialdi, ed.), London: Academic Press, 1985.

[45] P. Maragos, *A Unified Theory of Translation-Invariant Systems with Applications to Morphological Analysis and Coding of Images*. Ph.D. dissertation, Georgia Institute of Technology, Atlanta, 1985.

[46] J. Davidson, *Lattice Structures in the Image Algebra and Applications to Image Processing*. Ph.D. thesis, University of Florida, Gainesville, FL, 1989.

[47] J. Davidson, "Foundation and applications of lattice transforms in image processing," in *Advances in Electronics and Electron Physics* (P. Hawkes, ed.), vol. 84, pp. 61–130, New York, NY: Academic Press, 1992.

[48] J. Davidson, "Classification of lattice transformations in image processing," *Computer Vision, Graphics, and Image Processing: Image Understanding*, vol. 57, pp. 283–306, May 1993.

[49] P. Miller, "Development of a mathematical structure for image processing," optical division tech. rep., Perkin-Elmer, 1983.

[50] J. Wilson, D. Wilson, G. Ritter, and D. Langhorne, "Image algebra FORTRAN language, version 3.0," Tech. Rep. TR-89–03, University of Florida CIS Department, Gainesville, 1989.

[51] J. Wilson, "An introduction to image algebra Ada," in *Image Algebra and Morphological Image Processing II*, vol. 1568 of *Proceedings of SPIE*, (San Diego, CA), pp. 101–112, July 1991.

[52] J. Wilson, G. Fischer, and G. Ritter, "Implementation and use of an image processing algebra for programming massively parallel computers," in *Frontiers '88: The Second Symposium on the Frontiers of Massively Parallel Computation*, (Fairfax, VA), pp. 587–594, 1988.

[53] G. Fischer and M. Rowlee, "Computation of disparity in stereo images on the Connection Machine," in *Image Algebra and Morphological Image Processing*, vol. 1350 of *Proceedings of SPIE*, pp. 328–334, 1990.

[54] D. Crookes, P. Morrow, and P. McParland, "An algebra-based language for image processing on transputers," *IEE Third Annual Conference on Image Processing and its Applications*, vol. 307, pp. 457–461, July 1989.

[55] D. Crookes, P. Morrow, and P. McParland, "An implementation of image algebra on transputers," tech. rep., Department of Computer Science, Queen's University of Belfast, Northern Ireland, 1990.

[56] J. Wilson, "Supporting image algebra in the C++ language," in *Image Algebra and Morphological Image Processing IV*, vol. 2030 of *Proceedings of SPIE*, (San Diego, CA), pp. 315–326, July 1993.

[57] University of Florida Center for Computer Vision and Visualization, *Software User Manual for the* `iac++` *Class Library,* Version 1.0, 1994. Available via anonymous ftp from `ftp.cise.ufl.edu` in `/pub/ia/documents`.

[58] G. Birkhoff and J. Lipson, "Heterogeneous algebras," *Journal of Combinatorial Theory*, vol. 8, pp. 115–133, 1970.

[59] University of Florida Center for Computer Vision and Visualization, *Software User Manual for the* `iac++` *Class Library,* Version 2.0, 1995. Available via anonymous ftp from `ftp.cise.ufl.edu` in `/pub/ia/documents`.

[60] G. Ritter, J. Davidson, and J. Wilson, "Beyond mathematical morphology," in *Visual Communication and Image Processing II*, vol. 845 of *Proceedings of SPIE*, (Cambridge, MA), pp. 260–269, Oct. 1987.

[61] D. Li, *Recursive Operations in Image Algebra and Their Applications to Image Processing*. Ph.D. thesis, University of Florida, Gainesville, FL, 1990.

[62] D. Li and G. Ritter, "Recursive operations in image algebra," in *Image Algebra and Morphological Image Processing*, vol. 1350 of *Proceedings of SPIE*, (San Diego, CA), July 1990.

[63] R. Gonzalez and P. Wintz, *Digital Image Processing*. Reading, MA: Addison-Wesley, second ed., 1987.

[64] G. Ritter, "Heterogeneous matrix products," in *Image Algebra and Morphological Image Processing II*, vol. 1568 of *Proceedings of SPIE*, (San Diego, CA), pp. 92–100, July 1991.

[65] R. Cuninghame-Green, *Minimax Algebra: Lecture Notes in Economics and Mathematical Systems 166*. New York: Springer-Verlag, 1979.

[66] G. Ritter and H. Zhu, "The generalized matrix product and its applications," *Journal of Mathematical Imaging and Vision*, vol. 1(3), pp. 201–213, 1992.

[67] H. Zhu and G. X. Ritter, "The generalized matrix product and the wavelet transform," *Journal of Mathematical Imaging and Vision*, vol. 3, no. 1, pp. 95–104, 1993.

[68] H. Zhu and G. X. Ritter, "The *p*-product and its applications in signal processing," *SIAM Journal of Matrix Analysis and Applications*, vol. 16, no. 2, pp. 579–601, 1995.

CHAPTER 2
IMAGE ENHANCEMENT TECHNIQUES

2.1. Introduction

The purpose of image enhancement is to improve the visual appearance of an image, or to transform an image into a form that is better suited for human interpretation or machine analysis. Although there exists a multitude of image enhancement techniques, surprisingly, there does not exist a corresponding unifying theory of image enhancement. This is due to the absence of a general standard of image quality that could serve as a design criterion for image enhancement algorithms. In this chapter we consider several techniques that have proved useful in a wide variety of applications.

2.2. Averaging of Multiple Images

The purpose of averaging multiple images is to obtain an enhanced image by averaging the intensities of several images of the same scene. A detailed discussion concerning rationale and methodology can be found in Gonzalez and Wintz [1].

Image Algebra Formulation

For $i=1,\ldots,k$, let $\mathbf{a}_i \in \mathbb{R}^{\mathbf{X}}$ be a family of images of the same scene. The enhanced image, $\mathbf{a} \in \mathbb{R}^{\mathbf{X}}$, is given by

$$\mathbf{a} := \frac{1}{k} \sum_{i=1}^{k} \mathbf{a}_i.$$

For actual implementation the summation will probably involve the loop

$$\mathbf{a} := 0$$
$$\texttt{for } i \texttt{ in } 1..k \texttt{ loop}$$
$$\mathbf{a} := \mathbf{a} + \mathbf{a}_i$$
$$\texttt{end loop}$$
$$\mathbf{a} := \frac{1}{k}\mathbf{a}.$$

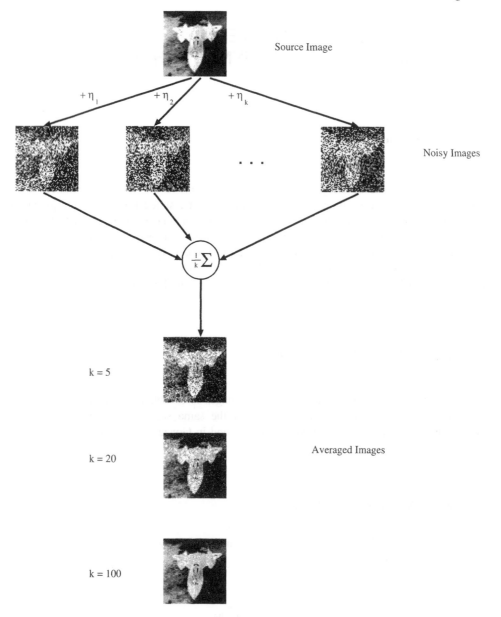

Figure 2.2.1. Averaging of multiple images for different values
of k. Additional explanations are given in the comments section.

Comments and Observations

Averaging multiple images is applicable when several noise degraded images,
$\mathbf{a}_1, \mathbf{a}_2, \ldots, \mathbf{a}_k$, of the same scene exist. Each $\mathbf{a}_i$ is assumed to have pixel values of the
form

$$\mathbf{a}_i(x) = \mathbf{a}_0(\mathbf{x}) + \eta_i(\mathbf{x}),$$

where $\mathbf{a}_0$ is the true (uncorrupted by noise) image and $\eta_i(\mathbf{x})$ is a random variable representing the introduction of noise (see Figure 2.2.1). The averaging multiple images technique
assumes that the noise is uncorrelated and has mean equal zero. Under these assumptions

the law of large numbers guarantees that as k increases, $\mathbf{a}(\mathbf{x}) = \frac{1}{k} \sum_{i=1}^{k} \mathbf{a}_i(\mathbf{x})$ approaches $\mathbf{a}_0(\mathbf{x})$. Thus, by averaging multiple images, it may be possible to assuage degradation due to noise. Clearly, it is necessary that the noisy images be registered so that corresponding pixels line up correctly.

2.3. Local Averaging

Local averaging smooths an image by locally reducing the variation in intensities. This is done by replacing the intensity level at a point by the average of the intensities in a neighborhood of the point.

Specifically, if $\mathbf{a}$ denotes the source image and $N(\mathbf{y})$ a neighborhood of $\mathbf{y}$ with $card(N(\mathbf{y})) = n \quad \forall \, \mathbf{y} \in \mathbb{Z}^2$, then the enhanced image $\mathbf{b}$ is given by

$$\mathbf{b}(\mathbf{y}) = \frac{1}{n} \sum_{\mathbf{x} \in N(\mathbf{y})} \mathbf{a}(\mathbf{x}).$$

Additional details about the effects of this simple technique can be found in Gonzalez and Wintz [1].

Image Algebra Formulation

Let $\mathbf{a} \in \mathbb{R}^{\mathbf{X}}$ be the source image, and $N(\mathbf{y}) \subset \mathbb{Z}^2$ a predefined neighborhood of $\mathbf{y} \in \mathbb{Z}^2$. Let $a : \mathbb{R}^{\mathbf{X}}|_N \to \mathbb{R}$ denote the averaging function (see Section 1.7). The result image $\mathbf{b} \in \mathbb{R}^{\mathbf{X}}$, derived by local averaging from $\mathbf{a} \in \mathbb{R}^{\mathbf{X}}$ is given by:

$$\mathbf{b} := \mathbf{a} \, \textcircled{a} \, N.$$

Comments and Observations

Local averaging traditionally imparts an artifact to the boundary of its result image. This is because the number of neighbors is smaller at the boundary of an image, so the average should be computed over fewer values. Simply dividing the sum of those neighbors by a fixed constant will not yield an accurate average. The image algebra specification does not yield such an artifact because the average of pixels is computed from the set of neighbors of each image pixel. No fixed divisor is specified.

2.4. Variable Local Averaging

Variable local averaging smooths an image by reducing the variation in intensities locally. This is done by replacing the intensity level at a point by the average of the intensities in a neighborhood of the point. In contrast to local averaging, this technique allows the size of the neighborhood configuration to vary. This is desirable for images that exhibit higher noise degradation toward the edges of the image [2, 3].

The actual mathematical formulation of this method is as follows. Suppose $\mathbf{a} \in \mathbb{R}^{\mathbf{X}}$ denotes the source image and $N : \mathbf{X} \to 2^{\mathbf{X}}$ a neighborhood function. If $n_{\mathbf{y}}$ denotes the number of points in $N(\mathbf{y}) \subset \mathbf{X}$, then the enhanced image $\mathbf{b}$ is given by

$$\mathbf{b}(\mathbf{y}) = \frac{1}{n_{\mathbf{y}}} \sum_{\mathbf{x} \in N(\mathbf{y})} \mathbf{a}(\mathbf{x}).$$

Image Algebra Formulation

Let $\mathbf{a} \in \mathbb{R}^{\mathbf{X}}$ denote the source image and $N : \mathbf{X} \to 2^{\mathbf{X}}$ the specific neighborhood configuration function. The enhanced image $\mathbf{b}$ is now given by

$$\mathbf{b} := \mathbf{a} \, \textcircled{a} \, N.$$

Comments and Observations

Although this technique is computationally more intense than local averaging, it may be more desirable if variations in noise degradation in different image regions can be determined beforehand by statistical or other methods. Note that if N is translation invariant, then the technique reduces to local averaging.

2.5.   Iterative Conditional Local Averaging

The goal of iterative conditional local averaging is to reduce additive noise in approximately piecewise constant images without blurring of edges. The method presented here is a simplified version of one of several methods proposed by Lev, Zucker, and Rosenfeld [4]. In this method, the value of the image $\mathbf{a}$ at location $\mathbf{y}$, $\mathbf{a}(\mathbf{y})$, is replaced by the average of the pixel values in a neighborhood of $\mathbf{y}$ whose values are approximately the same as $\mathbf{a}(\mathbf{y})$. The method is iterated (usually four to six times) until the image assumes the right visual fidelity as judged by a human observer.

For the precise formulation, let $\mathbf{a} \in \mathbb{R}^{\mathbf{X}}$ and for $\mathbf{y} \in \mathbf{X}$, let $N(\mathbf{y})$ denote the desired neighborhood of $\mathbf{y}$. Usually, $N(\mathbf{y})$ is a 3×3 Moore neighborhood. Define

$$S(\mathbf{y}) = \{\mathbf{x} \in N(\mathbf{y}) : |\mathbf{a}(\mathbf{y}) - \mathbf{a}(\mathbf{x})| < T\},$$

where T denotes a user-defined threshold, and set $n(\mathbf{y}) = card(S(\mathbf{y}))$.

The conditional local averaging operation has the form

$$\mathbf{a}_k(\mathbf{y}) = \frac{1}{n(\mathbf{y})} \sum_{\mathbf{x} \in S(\mathbf{y})} \mathbf{a}_{k-1}(\mathbf{x}),$$

where $\mathbf{a}_k(\mathbf{y})$ is the value at the kth iteration and $\mathbf{a}_0 = \mathbf{a}$.

Image Algebra Formulation

Let $\mathbf{a} \in \mathbb{R}^{\mathbf{X}}$ denote the source image and $N : \mathbf{X} \to 2^{\mathbf{X}}$ the desired neighborhood function. Select an appropriate threshold T and define the following parameterized neighborhood function:

$$[S(\mathbf{a})](\mathbf{y}) = \{\mathbf{x} \in N(\mathbf{y}) : |\mathbf{a}(\mathbf{y}) - \mathbf{a}(\mathbf{x})| < T\}.$$

The iterative conditional local averaging algorithm can now be written as

$$\mathbf{a}_k := \mathbf{a}_{k-1} \, \textcircled{a} \, S(\mathbf{a}_{k-1}),$$

where $\mathbf{a}_0 = \mathbf{a}$.

2.6. Gaussian Smoothing

Gaussian smoothing is employed in numerous image processing and computer vision algorithms. As pointed out by Marr and Hildreth [5], there are two considerations in choosing a smoothing filter. First, it must reduce the range of scales over which intensity changes occur, thus it should be smooth and relatively band-limited in the frequency domain. Second, it should represent an averaging of nearby points, thus it should be smooth and localized in the spatial domain as well. Simple spatial domain smoothing filters such as local averaging address the second consideration but fail to satisfy the first. In fact, these are in mutual conflict. Leipnik [6] has shown that the filter that provides localization best in both the spatial and frequency domains is the Gaussian, namely

$$G(x) = \frac{1}{\sigma\sqrt{2\pi}} e^{\frac{-x^2}{2\sigma^2}},$$

and in two dimensions, $G(x, y) = \frac{1}{2\pi\sigma^2} e^{\frac{-x^2-y^2}{2\sigma^2}}$.

The Gaussian smoothing **b** of an image **a** is achieved by convolving it with the Gaussian G:

$$\mathbf{b}(x, y) = \mathbf{a}(x, y) * G(x, y) = \int\limits_{-\infty}^{\infty} \int\limits_{-\infty}^{\infty} \mathbf{a}(\alpha, \beta) G(x - \alpha, y - \beta) d\alpha d\beta.$$

Image Algebra Formulation

Let $\mathbf{a} \in \mathbb{R}^{\mathbf{X}}$ for some 2–dimensional point set $\mathbf{X}$ be the source image. If we were to imagine the set $\mathbf{X}$ to have unbounded extent, then the Gaussian smoothing **b** of image **a** would given by

$$\mathbf{b} = \mathbf{a} \oplus \mathbf{s}$$

where $\mathbf{s} \in \left(\mathbb{R}^{\mathbf{X}}\right)^{\mathbf{X}}$ is defined by

$$\mathbf{s}_{(x,y)}(v, w) = \frac{1}{2\pi\sigma^2} e^{\frac{-(x-v)^2-(y-w)^2}{2\sigma^2}}.$$

In practice, we must smooth bounded images with a bounded template. In such cases, truncating the extent of the template will lead to introduction of a bias by the smoothing even at those points where the template support is entirely contained within the image point set. This is due to the integral of the truncated Gaussian being less than one. To achieve an unbiased result, the template must be constructed as follows:

$$\mathbf{s}_{(x,y)}(v, w) = \frac{1}{2k\pi\sigma^2} e^{\frac{-(x-v)^2-(y-w)^2}{2\sigma^2}}$$

where k is given by

$$k = \sum_{(x,y)\in S\left(\mathbf{s}_{(0,0)}\right)} \frac{1}{2\pi\sigma^2} e^{\frac{-x^2-y^2}{2\sigma^2}}.$$

Comments and Observations

The Gaussian is a separable function, that is, $G(x,y) = G(x)G(y)$. This means that convolutions with the 2–dimensional Gaussian can be computed with iterated convolutions of 1–dimensional Gaussians, namely,

$$\mathbf{a}(x,y) * G(x,y) = \mathbf{a}(x,y) * G(x) * G(y).$$

This can be computed with two image-template linear product operations:

$$\mathbf{b} = (\mathbf{a} \oplus \mathbf{s_1}) \oplus \mathbf{s_2}.$$

where

$$\mathbf{s}_{1(x,y)}(u,y) = \frac{1}{k_1 \sigma \sqrt{2\pi}} e^{\frac{-(x-u)^2}{2\sigma^2}},$$

$$\mathbf{s}_{2(x,y)}(x,v) = \frac{1}{k_2 \sigma \sqrt{2\pi}} e^{\frac{-(y-v)^2}{2\sigma^2}},$$

$$k_1 = \sum_{(x,y)\in S\left(\mathbf{s}_{1(0,0)}\right)} \frac{1}{\sigma\sqrt{2\pi}} e^{\frac{-x^2}{2\sigma^2}},$$

and

$$k_2 = \sum_{(x,y)\in S\left(\mathbf{s}_{2(0,0)}\right)} \frac{1}{\sigma\sqrt{2\pi}} e^{\frac{-y^2}{2\sigma^2}}.$$

Choice of the support of the template s must be made wisely, balancing the template's size against the properties of the smoothing filter. Too small an extent will yield a filter that no longer provides good locality in both space and frequency. Too large an extent will require extra processing time for no measurable benefit.

An appropriate choice of size should be driven by the standard deviation of the Gaussian, σ. More than 95% of the energy in the Gaussian is contained within a radius of 2σ units from the origin and more than 99% of the area is contained within a radius of 3σ units. Either of these represent reasonable choices for template radius for a variety of applications.

2.7. Max-Min Sharpening Transform

The max-min sharpening transform is an image enhancement technique which sharpens fuzzy boundaries and brings fuzzy gray level objects into focus. It also smoothens isolated peaks or valleys. It is an iterative technique that compares maximum and minimum values with respect to the central pixel value in a small neighborhood. The central pixel value is replaced by whichever of the extrema in its neighborhood is closest to its value.

The following specification of the max-min sharpening transform was formulated in Kramer and Bruchner [7]. Let $\mathbf{a} \in \mathbb{R}^{\mathbf{X}}$ be the source image and $N(\mathbf{y})$ denote a symmetric neighborhood of $\mathbf{y} \in \mathbf{X}$. Define

$$\mathbf{a}_M(\mathbf{y}) = \max\{\mathbf{a}(\mathbf{x}) : \mathbf{x} \in N(\mathbf{y})\}$$
$$\mathbf{a}_m(\mathbf{y}) = \min\{\mathbf{a}(\mathbf{x}) : \mathbf{x} \in N(\mathbf{y})\}.$$

The sharpening transform s is defined as

$$\mathbf{s}(\mathbf{a}(\mathbf{y})) = \begin{cases} \mathbf{a}_M(\mathbf{y}) & \text{if } \mathbf{a}_M(\mathbf{y}) - \mathbf{a}(\mathbf{y}) \le \mathbf{a}(\mathbf{y}) - \mathbf{a}_m(\mathbf{y}) \\ \mathbf{a}_m(\mathbf{y}) & \text{otherwise.} \end{cases}$$

The procedure can be iterated as

$$\mathbf{s}^{n+1}(\mathbf{a}(\mathbf{y})) = \mathbf{s}(\mathbf{s}^n(\mathbf{a}(\mathbf{y}))).$$

Image Algebra Formulation

Let $\mathbf{a} \in \mathbb{R}^{\mathbf{X}}$ be the source image, and $N(\mathbf{y})$ a desired neighborhood of $\mathbf{y} \in \mathbb{Z}^2$. The max-min sharpened image $\mathbf{s}$ is given by the following algorithm:

$$
\begin{aligned}
\mathbf{a}_M &:= \mathbf{a} \boxslash N \\
\mathbf{a}_m &:= \mathbf{a} \boxbslash N \\
\mathbf{b} &:= \mathbf{a}_M + \mathbf{a}_m - 2 \cdot \mathbf{a} \\
\mathbf{s} &:= \chi_{\leq 0}(\mathbf{b}) \cdot \mathbf{a}_M + \chi_{>0}(\mathbf{b}) \cdot \mathbf{a}_m \ .
\end{aligned}
$$

The algorithm is usually iterated until $\mathbf{s}$ stabilizes or objects in the image have assumed desirable fidelity (as judged by a human observer).

Comments and Observations

Figure 2.7.1 is a blurred image of four Chinese characters. Figure 2.7.2 show the results, after convergence, of applying the max-min sharpening algorithm to the blurred characters. Convergence required 128 iterations. The neighborhood N used in this example is the von Neumann neighborhood.

Figure 2.7.1. Blurred characters.

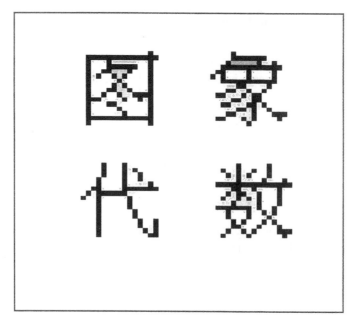

Figure 2.7.2. Result of applying max-min sharpening to blurred characters.

2.8.   Smoothing Binary Images by Association

The purpose of this smoothing method is to reduce the effects of noise in binary pictures. The basic idea is that the 1-elements due to noise are scattered uniformly while the 1-elements due to message information tend to be clustered together. The original image is partitioned into rectangular regions. If the number of 1's in each region exceeds a given threshold, then the region is not changed; otherwise, the 1's are set to zero. The regions are then treated as single cells, a cell being assigned a 1 if there is at least one 1 in the corresponding region and 0 otherwise. This new collection of cells can be viewed as a lower resolution image. The pixelwise minimum of the lower resolution image and the original image provides for the smoothed version of the original image. The smoothened version of the original image can again be partitioned by viewing the cells of the lower resolution image as pixels and partitioning these pixels into regions subject to the same threshold procedure. The precise specification of this algorithm is given by the image algebra formulation below.

Image Algebra Formulation

Let T denote a given threshold and $\mathbf{a} \in \{0, 1\}^{\mathbf{X}}$ be the source image with $\mathbf{X} \subset \mathbb{Z}^2$. For a fixed integer $k \geq 2$, define a neighborhood function $N(k) : \mathbf{X} \to 2^{\mathbf{X}}$ by

$$[N(k)](\mathbf{y}) = \left\{ \mathbf{x} \in \mathbf{X} : \left\lfloor \frac{\mathbf{x}}{k} \right\rfloor = \left\lfloor \frac{\mathbf{y}}{k} \right\rfloor \right\}.$$

Here $\left\lfloor \frac{\mathbf{x}}{k} \right\rfloor$ means that if $\mathbf{x} = (x, y)$, then $\left\lfloor \frac{\mathbf{x}}{k} \right\rfloor = \left(\left\lfloor \frac{x}{k} \right\rfloor, \left\lfloor \frac{y}{k} \right\rfloor \right)$.

The smoothed image $\mathbf{a}_1 \in \{0, 1\}^{\mathbf{X}}$ is computed by using the statement

$$\mathbf{a}_1 := \mathbf{a} \wedge \chi_{\geq T}(\mathbf{a} \oplus N(k)).$$

If recursion is desired, define

$$\mathbf{a}_i := \mathbf{a} \wedge \chi_{\geq T}(\mathbf{a}_{i-1} \oplus N(k + i - 1))$$

for $i > 0$, where $\mathbf{a}_0 = \mathbf{a}$.

The recursion algorithm may reintroduce pixels with values 1 that had been eliminated at a previous stage. The following alternative recursion formulation avoids this phenomenon:

$$\mathbf{a}_i := \mathbf{a}_{i-1} \wedge \chi_{\geq T}(\mathbf{a}_{i-1} \oplus N(k + i - 1)).$$

Comments and Observations

Figures 2.8.1 through 2.8.5 provide an example of this smoothing algorithm for $k = 2$ and $T = 2$. Note that $N(k)$ partitions the point set $\mathbf{X}$ into disjoint subsets since $[N(k)](\mathbf{y}) = [N(k)](\mathbf{z}) \Leftrightarrow \lfloor \frac{\mathbf{y}}{k} \rfloor = \lfloor \frac{\mathbf{z}}{k} \rfloor$. Obviously, the larger the number k, the larger the size of the cells $[N(k)](\mathbf{y})$. In the iteration, one views the cells $[N(k)](\mathbf{y})$ as pixels forming the next partition $[N(k + 1)](\mathbf{y})$. The effects of the two different iteration algorithms can be seen in Figures 2.8.4 and 2.8.5.

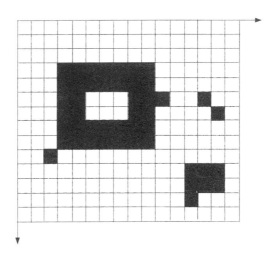

Figure 2.8.1. The binary source image $\mathbf{a}$.

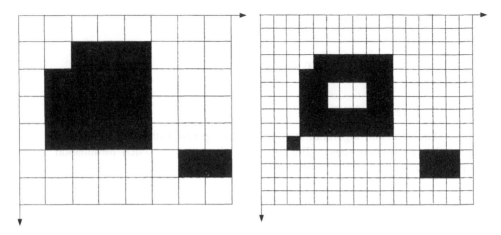

Figure 2.8.2. The lower-resolution image $\chi_{\geq 2}(\mathbf{a} \oplus N(2))$ is shown on the left and the smoothened version $\mathbf{a} \wedge \chi_{\geq 2}(\mathbf{a} \oplus N(2))$ of $\mathbf{a}$ on the right.

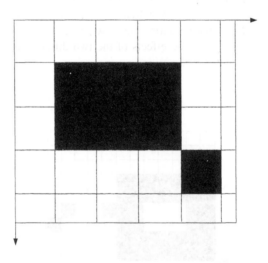

Figure 2.8.3. The lower-resolution image $\chi_{\geq 2}(\mathbf{a}_1 \oplus N(3))$ of the first iteration.

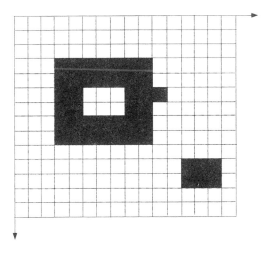

Figure 2.8.4. The smoothened version $\mathbf{a}_2 = \mathbf{a} \wedge \chi_{\geq 2}(\mathbf{a}_1 \oplus N(3))$ of $\mathbf{a}$.

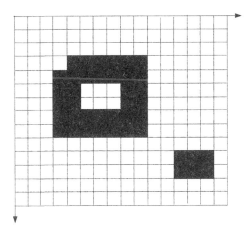

Figure 2.8.5. The image $\mathbf{a}_2 = \mathbf{a}_1 \wedge \chi_{\geq 2}(\mathbf{a}_1 \oplus N(3))$.

As can be ascertained from Figs. 2.8.1 through 2.8.5, several problems can arise when using this smoothing method. The technique as stated will not fill in holes caused by noise. It could be modified so that it fills in the rectangular regions if the number of 1's exceeds the threshold, but this would cause distortion of the objects in the scene. Objects that split across boundaries of adjacent regions may be eliminated by this algorithm. Also, if the image cannot be broken into rectangular regions of uniform size, other boundary-sensitive techniques may need to be employed to avoid inconsistent results near the image boundary.

Additionally, the neighborhood $N(k)$ is a translation variant neighborhood function that needs to be computed at each pixel location $\mathbf{y}$, resulting in possibly excessive computational overhead. For these reasons, morphological methods producing similar results may be preferable for image smoothing.

2.9.   Median Filter

The median filter is a smoothing technique that causes minimal edge blurring. However, it will remove isolated spikes and may destroy fine lines [1, 2, 8]. The technique involves replacing the pixel value at each point in an image by the median of the pixel values in a neighborhood about the point.

The choice of neighborhood and median selection method distinguish the various median filter algorithms. Neighborhood selection is dependent on the source image. The machine architecture will determine the best way to select the median from the neighborhood.

A sampling of two median filter algorithms is presented in this section. The first is for an arbitrary neighborhood. It shows how an image-template operation can be defined that finds the median value by sorting lists. The second formulation shows how the familiar bubble sort can be used to select the median over a 3×3 neighborhood.

Image Algebra Formulation

Let $a \in \mathbb{R}^{\mathbf{X}}$ be the source image, $N : \mathbf{X} \to 2^{\mathbf{X}}$ a neighborhood function. Let $m : \mathbb{R}^{\mathbf{X}}|_N \to \mathbb{R}$ denote the median function described in Section 1.7. The median filtered image $\mathbf{m}$ is given by

$$\mathbf{m} := \mathbf{a} \, \textcircled{m} \, N.$$

Alternate Image Algebra Formulation

The alternate formulation uses a bubble sort to compute the median value over a 3×3 neighborhood. Let $\mathbf{a} \in \mathbb{R}^{\mathbf{X}}$ be the source image. The images $\mathbf{a}_i$ are obtained by

$$\mathbf{a}_i := \mathbf{a} \circ f_i, \ 1 \leq i \leq 8,$$

where the functions f_i are defined as follows:

$$f_1(\mathbf{x}) = \mathbf{x} + (0, 1)$$
$$f_2(\mathbf{x}) = \mathbf{x} + (-1, 1)$$
$$f_3(\mathbf{x}) = \mathbf{x} + (-1, 0)$$
$$f_4(\mathbf{x}) = \mathbf{x} + (-1, -1)$$
$$f_5(\mathbf{x}) = \mathbf{x} + (0, -1)$$
$$f_6(\mathbf{x}) = \mathbf{x} + (1, -1)$$
$$f_7(\mathbf{x}) = \mathbf{x} + (1, 0)$$
$$f_8(\mathbf{x}) = \mathbf{x} + (1, 1).$$

The median image $\mathbf{m}$ is calculated with the following image algebra pseudocode:

$$\mathbf{a}_0 := \mathbf{a}$$

```
for i in 0..4 loop
   for j in i+1..8 loop
      b := a_i ∨ a_j
      a_j := a_i ∧ a_j
      a_i := b
   end loop
end loop
m := a_4.
```

Comments and Observations

The figures below offer a comparison between the results of applying averaging filter and a median filters. Figure 2.9.1 is the source image of a noisy jet. Figures 2.9.2 and 2.9.3 show the results of applying averaging and median filter, respectively, over 3×3 neighborhoods of the noisy jet image.

Figure 2.9.1. Noisy jet image.

Figure 2.9.2. Noisy jet image smoothed with averaging filter over a 3×3 neighborhood.

Figure 2.9.3. Result of applying median filter over
a 3 × 3 neighborhood to the noisy jet image.

Coffield [9] describes a stack comparator architecture which is particularly well
suited for image processing tasks that involve order statistic filtering. His methodology is
similar to the alternative image algebra formulation given above.

2.10.   Unsharp Masking

Unsharp masking involves blending an image's high-frequency components and
low-frequency components to produce an enhanced image [2, 10, 11, 12]. The blending
may sharpen or blur the source image depending on the proportion of each component in the
enhanced image. Enhancement takes place in the spatial domain. The precise formulation
of this procedure is given in the image algebra formulation below.

Image Algebra Formulation

Let $\mathbf{a} \in \mathbb{R}^{X}$ be the source image and let $\mathbf{b}$ be the image obtained from $\mathbf{a}$ by
applying an averaging mask (Section 2.3). The image $\mathbf{b}$ is the low-frequency component
of the source image and the high-frequency component is $\mathbf{a} - \mathbf{b}$. The enhanced image $\mathbf{c}$
produced by unsharp masking is given by

$$\mathbf{c} := \gamma \cdot (\mathbf{a} - \mathbf{b}) + \mathbf{b},$$

or, equivalently,

$$\mathbf{c} := \gamma \cdot \mathbf{a} + (1 - \gamma) \cdot \mathbf{b}.$$

A $\gamma \in \mathbb{R}$ between 0 and 1 results in a smoothing of the source image. A γ greater than
1 emphasizes the high-frequency components of the source image, which sharpens detail.
Figure 2.10.1 shows the result of applying the unsharp masking technique to a mammogram
for several values of γ. A 3 × 3 averaging neighborhood was used to produce the low-
frequency component image $\mathbf{b}$.

A more general formulation for unsharp masking is given by

$$\mathbf{c} := \alpha \cdot (\mathbf{a} - \mathbf{b}) + \beta \cdot \mathbf{b},$$

where $\alpha, \beta \in \mathbb{R}$. Here α is the weighting of the high-frequency component and β is the weighting of the low-frequency component.

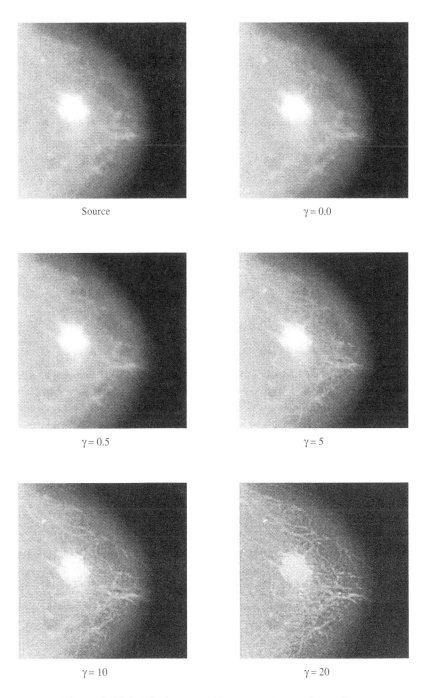

Figure 2.10.1. Unsharp masking at various values of γ.

Comments and Observations

Unsharp masking can be accomplished by simply convolving the source image with the appropriate template. For example, the unsharp masking can be done for the mammogram in Figure 2.10.1 using

$$c := a \oplus t,$$

where t is the template defined in Figure 2.10.2. The values of v and w are $\frac{1-\gamma}{9}$ and $\frac{8\gamma+1}{9}$, respectively.

$$t = $$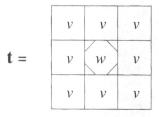

Figure 2.10.2. The Moore configuration template for unsharp masking using a simple convolution.

2.11. Local Area Contrast Enhancement

In this section we present two methods of *local contrast enhancement* from Harris [13] and Narendra and Fitch [14]. Each is a form of unsharp masking (Section 2.10) in which the weighting of the high-frequency component is a function of local standard deviation.

Image Algebra Formulation

Let $a \in \mathbb{R}^X$, and N be a neighborhood selected for local averaging (Section 2.3). The von Neumann or Moore neighborhoods are the most commonly used neighborhoods for this unsharp masking technique. The local mean image of a with respect to this neighborhood function is given by

$$m := a \,\textcircled{a}\, N.$$

The local standard deviation of a is given by the image

$$d := \beta \vee \left((a - m)^2 \,\textcircled{a}\, N \right)^{\frac{1}{2}}.$$

The image $\left((a - m)^2 \,\textcircled{a}\, N \right)^{\frac{1}{2}}$ actually represents the local standard deviation while $\beta > 0$ is a lower bound applied to d in order to avoid problems with division by zero in the next step of this technique.

The enhancement technique of [13] is a high-frequency emphasis scheme which uses local standard deviation to control gain. The enhanced image b is given by

$$b := \frac{1}{d} \cdot (a - m).$$

As seen in Section 2.10, $a - m$ represents a highpass filtering of a in the spatial domain. The local gain factor $1/d$ is inversely proportional to the local standard deviation. Thus, a larger gain will be applied to regions with low contrast. This enhancement technique is useful for images whose information of interest lies in the high-frequency component and whose contrast levels vary widely from region to region.

The local contrast enhancement technique of Narendra and Fitch [14] is similar to the one above, except for a slightly different gain factor and the addition of a low-frequency component. In this technique, let

$$m := \frac{1}{card(\mathbf{a})} \sum \mathbf{a}$$

denote the global mean of $\mathbf{a}$. The enhanced image $\mathbf{b}$ is given by

$$\mathbf{b} := \frac{\alpha \cdot m}{\mathbf{d}} \cdot (\mathbf{a} - \mathbf{m}) + \mathbf{m},$$

where $0 < \alpha < 1$. The addition of the low-frequency component $\mathbf{m}$ is used to restore the average intensity level within local regions.

We need to remark that in both techniques above it may be necessary to put limits on the gain factor to prevent excessive gain variations.

2.12. Histogram Equalization

Histogram equalization is a technique which rescales the range of an image's pixel values to produce an enhanced image whose pixel values are more uniformly distributed [15, 1, 16]. The enhanced image tends to have higher contrast.

The mathematical formulation is as follows. Let $\mathbf{a} \in \mathbb{Z}_l^{\mathbf{X}}$ denote the source image, $n = card(\mathbf{X})$, and n_j be the number of times the gray level j occurs in the image $\mathbf{a}$. Recall that $\mathbb{Z}_l = \{0, 1, \cdots, l-1\}$. The enhanced image $\mathbf{b}$ is given by

$$\mathbf{b}(\mathbf{x}) = l \sum_{j=0}^{\mathbf{a}(\mathbf{x})} \frac{n_j}{n} .$$

Image Algebra Formulation

Let $\mathbf{a} \in \mathbb{Z}_l^{\mathbf{X}}$ be the source image and let $\overline{c}$ denote the normalized cumulative histogram of $\mathbf{a}$ as defined in Section 10.11.

The enhanced image $\mathbf{b}$ is given by

$$\mathbf{b} := (l-1) \cdot (\overline{c} \circ \mathbf{a}).$$

Comments and Observations

Figure 2.12.1 is an illustration of the histogram equalization process. The original image and its histograms are on the left side of the figure. The original image appears dark (or underexposed). This darkness manifests itself in a bias toward the lower end of the gray scale in the original image's histogram. On the right side of Figure 2.12.1 is the equalized image and its histograms. The equalized histogram is distributed more uniformly. This more uniform distribution of pixel values has resulted in an image with better contrast.

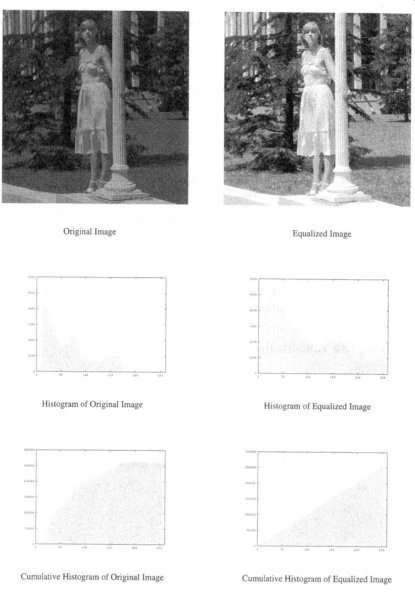

Figure 2.12.1. Left: Original image and
histograms. Right: Equalized image and histograms.

2.13.   Histogram Modification

The histogram equalization technique of Section 2.12 is a specific example of image enhancement using histogram modification. The goal of the histogram modification technique is to adjust the distribution of an image's pixel values so as to produce an enhanced image in which certain gray level ranges are highlighted. In the case of histogram equalization, the goal is to enhance the contrast of an image by producing an enhanced image in which the pixel values are evenly distributed, i.e., the histogram of the enhanced image is flat.

Histogram modification is accomplished via a transfer function

$$g = T(z),$$

where the variable z, $z_1 \le z \le z_n$, represents a gray level value of the source image and g, $g_1 \le g \le g_m$, is the variable that represents a gray level value of the enhanced image. The method of deriving a transfer function to produce an enhanced image with a prescribed distribution of pixel values can be found in either Gonzalez and Wintz [1] or Pratt [2].

Table 2.13.1 lists the transfer functions for several of the histogram modification examples featured in Pratt [2]. In the table $P(z)$ is the cumulative probability distribution of the input image and α is an empirically derived constant. The image algebra formulations for histogram modification based on the transfer functions in Table 2.13.1 are presented next.

Table 2.13.1 Histogram Transfer Functions

Name of modification	Transfer function $T(z)$
Uniform modification:	$g = [g_m - g_1] P(z) + g_1$
Exponential modification:	$g = g_1 - \frac{1}{\alpha} \ln [1 - P(z)]$
Rayleigh modification:	$g = g_1 + \left[2\alpha^2 \ln \left(\frac{1}{1-P(z)} \right) \right]^{1/2}$
Hyperbolic cube root modification:	$g = \left(\left[g_m^{1/3} - g_1^{1/3} \right] [P(z)] + g_1^{1/3} \right)^3$
Hyperbolic logarithmic modification:	$g = g_1 \left[\frac{g_m}{g_1} \right]^{P(z)}$

Image Algebra Formulation

Let $\mathbf{a} \in \mathbb{Z}_l^{\mathbf{X}}$ denote the source image and $\overline{c}$ the normalized cumulative histogram of $\mathbf{a}$ as defined in Section 10.11. Let g_1 and g_m denote the grey value bounds for the enhanced image. Table 2.13.2 below describes the image algebra formulation for obtaining the enhanced image $\mathbf{b}$ using the histogram transform functions defined in Table 2.13.1.

2.14. Lowpass Filtering

Lowpass filtering is an image enhancement process used to attenuate the high-frequency components of an image's Fourier transform [1]. Since high-frequency components are associated with sharp edges, lowpass filtering has the effect of smoothing the image.

Suppose $\mathbf{a} \in \mathbb{C}^{\mathbf{X}}$ where $\mathbf{X} = \mathbb{Z}_m \times \mathbb{Z}_n$. Let $\hat{\mathbf{a}}$ denote the Fourier transform of $\mathbf{a}$, and $\hat{\mathbf{h}}$ denote the lowpass filter transfer function. The enhanced image $\mathbf{g}$ is given by

$$\mathbf{g}(x, y) = \mathfrak{F}^{-1} \left\{ \hat{\mathbf{a}}(u, v) \cdot \hat{\mathbf{h}}(u, v) \right\},$$

where $\hat{\mathbf{h}}$ is the filter which attenuates high frequencies, and $\mathfrak{F}^{-1}$ denotes the inverse Fourier transform. Sections 8.2 and 8.4 present image algebra implementations of the Fourier transform and its inverse.

Table 2.13.2 Image Algebra Formulation of Histogram Transfer Functions

Name of modification	Image algebra formulation
Uniform modification:	$\mathbf{b} := [g_m - g_1](\overline{\mathbf{c}} \circ \mathbf{a}) + g_1$
Exponential modification:	$\mathbf{b} := g_1 - \frac{1}{\alpha} \ln[1 - (\overline{\mathbf{c}} \circ \mathbf{a})]$
Rayleigh modification:	$\mathbf{b} := g_1 + \left[2\alpha^2 \ln\left(\frac{1}{1-(\overline{\mathbf{c}}\circ\mathbf{a})}\right)\right]^{1/2}$
Hyperbolic cube root modification:	$\mathbf{b} := \left(\left[g_m^{1/3} - g_1^{1/3}\right](\overline{\mathbf{c}} \circ \mathbf{a}) + g_1^{1/3}\right)^3$
Hyperbolic logarithmic modification:	$\mathbf{b} := g_1\left[\frac{g_m}{g_1}\right]^{(\overline{\mathbf{c}}\circ\mathbf{a})}$

Let $\hat{\mathbf{d}}(u,v)$ be the distance from the point (u,v) to the origin of the frequency plane; that is,

$$\hat{\mathbf{d}}(u,v) = (u^2 + v^2)^{1/2}.$$

The transfer function of the *ideal lowpass filter* is given by

$$\hat{\mathbf{h}}(u,v) = \begin{cases} 1 & \text{if } \hat{\mathbf{d}}(u,v) \le d \\ 0 & \text{if } \hat{\mathbf{d}}(u,v) > d, \end{cases}$$

where d is a specified nonnegative quantity, which represents the *cutoff frequency*.

The transfer function of the *Butterworth lowpass filter* of order k is given by

$$\hat{\mathbf{h}}(u,v) = \frac{1}{1 + c\left[\hat{\mathbf{d}}(u,v)/d\right]^{2k}},$$

where c is a scaling constant. Typical values for c are 1 and $\sqrt{2} - 1$.

The transfer function of the *exponential lowpass filter* is given by

$$\hat{\mathbf{h}}(u,v) = exp\left[-a\left(\hat{\mathbf{d}}(u,v)/d\right)^k\right].$$

Typical values for a are 1 and $ln\left(\sqrt{2}\right)$.

The images that follow illustrate some of the properties of lowpass filtering. When filtering in the frequency domain, the origin of the image is assumed to be at the center of the display. The Fourier transform image has been shifted so that its origin appears at the center of the display (see Section 8.2).

Figure 2.14.1 is the original image of a noisy angiogram. The noise is in the form of a sinusoidal wave. Figure 2.14.2 represents the power spectrum image of the noisy angiogram. The noise component of the original image shows up as isolated spikes above and below the center of the frequency image. The noise spikes in the frequency domain are easily filtered out by an ideal lowpass filter whose cutoff frequency d falls within the distance from the center of the image to the spikes.

Figure 2.14.3 shows the result of applying an ideal lowpass filter to 2.14.2 and then mapping back to spatial coordinates via the inverse Fourier transform. Note how the washboard effect of the sinusoidal noise has been removed.

One artifact of lowpass filtering is the "ringing" which can be seen in Figure 2.14.2. Ringing is caused by the ideal filter's sharp cutoff between the low frequencies it lets pass and the high frequencies it suppresses. The Butterworth filter offers a smooth discrimination between frequencies, which results in less severe ringing.

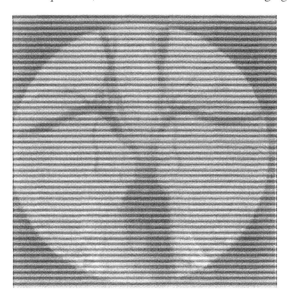

Figure 2.14.1. Noisy angiogram.

The lowpass filtering used above was successful in removing the noise from the angiogram; however, the filtering blurred the true image. The image of the angiogram before noise was added is seen in Figure 2.14.4. The blurring introduced by filtering is seen by comparing this image with Figure 2.14.3.

Lowpass filtering blurs an image because edges and other sharp transitions are associated with the high frequency content of an image's Fourier spectrum. The degree of blurring is related to the proportion of the spectrum's signal power that remains after filtering. The signal power P of $\mathbf{b} \in \mathbb{C}^{\mathbf{X}}$ is defined by

$$P(\mathbf{b}) = \sum_{\mathbf{x} \in \mathbf{X}} \left[Re^2\left(\mathbf{b}(\mathbf{x})\right) + Im^2\left(\mathbf{b}(\mathbf{x})\right) \right].$$

The percentage of power that remains after filtering $\mathbf{a}$ using an ideal filter $\hat{\mathbf{h}}_d$ with cutoff frequency d is

$$\frac{P\left(\hat{\mathbf{a}} \cdot \hat{\mathbf{h}}_d\right)}{P(\hat{\mathbf{a}})} \cdot 100.$$

Figure 2.14.2. Fourier transform of noisy angiogram.

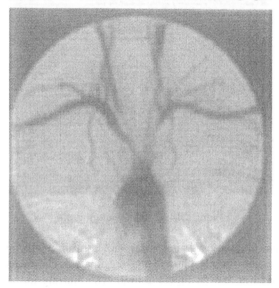

Figure 2.14.3. Filtered angiogram using ideal filter.

As d increases, more and higher frequencies are encompassed by the circular region that makes up the support of $\hat{\mathbf{h}}_d$. Thus, as d increases, the signal power of the filtered image increases and blurring decreases. The top two images of Figure 2.14.5 are those of an original image (peppers) and its power spectrum. The lower four images show the blurring that results from filtering with ideal lowpass filters whose cutoff frequencies preserve $90, 93, 97$, and 99% of the original image's signal power.

The blurring caused by lowpass filtering is not always undesirable. In fact, lowpass filtering may be used as a smoothing technique.

Image Algebra Formulation

The image algebra formulation of the lowpass filter is roughly that of the

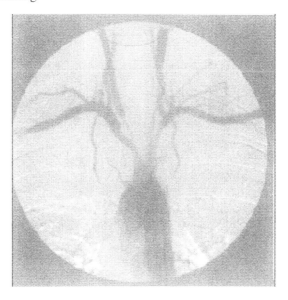

Figure 2.14.4. Angiogram before noise was introduced.

mathematical formulation presented earlier. However, there are a few subtleties that a programmer needs to be aware of when implementing lowpass filters.

The transfer function

$$\hat{\mathbf{h}}(u,v) = \begin{cases} 1 & \text{if } \left(u^2 + v^2\right)^{1/2} \leq d \\ 0 & \text{if } \left(u^2 + v^2\right)^{1/2} > d \end{cases}$$

represents a disk of radius d of unit height centered at the origin $(0,0)$. On the other hand, the Fourier transform of an image $\mathbf{a} \in \mathbb{R}^{\mathbf{X}}$, where $\mathbf{X} = \mathbb{Z}_m \times \mathbb{Z}_n$, results in a complex-valued image $\hat{\mathbf{a}} \in \mathbb{C}^{\mathbf{X}}$. In other words, the location of the origin with respect to $\hat{\mathbf{a}}$ is not the center of $\mathbf{X}$, which is the point $\left(\frac{m}{2}, \frac{n}{2}\right)$, but the upper left-hand corner of $\hat{\mathbf{a}}$. Therefore, the product $\hat{\mathbf{a}} \cdot \hat{\mathbf{h}}$ is undefined. Simply shifting the image $\hat{\mathbf{a}}$ so that the midpoint of $\mathbf{X}$ moves to the origin, or shifting the disk image so that the disk's center will be located at the midpoint of $\mathbf{X}$, will result in properly aligned images that can be multiplied, but will not result in a lowpass filter since the high frequencies, which are located at the corners of $\hat{\mathbf{a}}$, would be eliminated.

There are various options for the correct implementation of the lowpass filter. One such options is to center the spectrum of the Fourier transform at the midpoint of $\mathbf{X}$ (Section 8.3) and then multiplying the centered transform by the shifted disk image. The result of this process needs to be uncentered prior to applying the inverse Fourier transform. The exact specification of the ideal lowpass filter is given by the following algorithm.

Suppose $\hat{\mathbf{a}} \in \mathbb{C}^{\mathbf{X}}$ and $\mathbf{X} = \mathbb{Z}_m \times \mathbb{Z}_n$, where m and n are even integers. Define the point set

$$\mathbf{U} = \left\{ (x,y) : \left[\left(x - \frac{m}{2} \right)^2 + \left(y - \frac{n}{2} \right)^2 \right]^{1/2} \leq d \right\}$$

and set

$$\hat{\mathbf{h}} := \left(\mathbf{1}|_{\mathbf{U}} \right)|^{\mathbf{0}}.$$

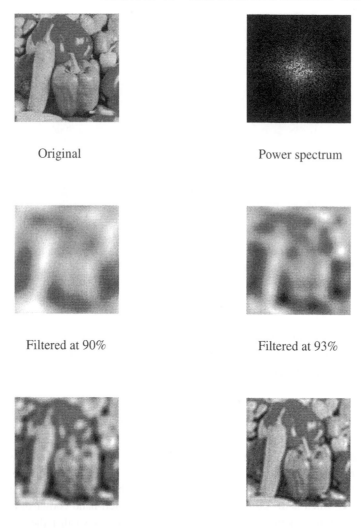

Original   Power spectrum

Filtered at 90%   Filtered at 93%

Filtered at 97%   Filtered at 99%

Figure 2.14.5. Lowpass filtering with ideal filter at various power levels.

The algorithm now becomes

$$\hat{\mathbf{a}} := \hat{\mathbf{a}} \circ center(\mathbf{X})$$
$$\mathbf{g} := \mathfrak{F}^{-1}\left[\left(\hat{\mathbf{a}} \cdot \hat{\mathbf{h}}\right) \circ center(\mathbf{X})\right].$$

In this algorithm, $\mathbf{1}$ and $\mathbf{0}$ denote the *complex*-valued unit and zero image, respectively.

If $\mathbf{U}$ is defined as $\mathbf{U} = \left\{(u,v) : \left(u^2 + v^2\right)^{1/2} \leq d\right\}$ instead, then it becomes necessary to define $\hat{\mathbf{h}}$ as $\hat{\mathbf{h}} := \left[\mathbf{1}|_{\mathbf{U}} + \left(\frac{m}{2}, \frac{n}{2}\right)\right]|^{\mathbf{0}}$, where the extension to $\mathbf{0}$ is on the array $\mathbf{X}$, prior to multiplying $\hat{\mathbf{a}}$ by $\hat{\mathbf{h}}$. Figure 2.14.6 provides a graphical interpretation of this algorithm.

Another method, which avoids centering the Fourier transform, is to specify the set $\mathbf{U} = \left\{(u,v) : \left(u^2 + v^2\right)^{1/2} \leq d\right\}$ and translate the image $\mathbf{1}|_{\mathbf{U}}$ directly to the corners of

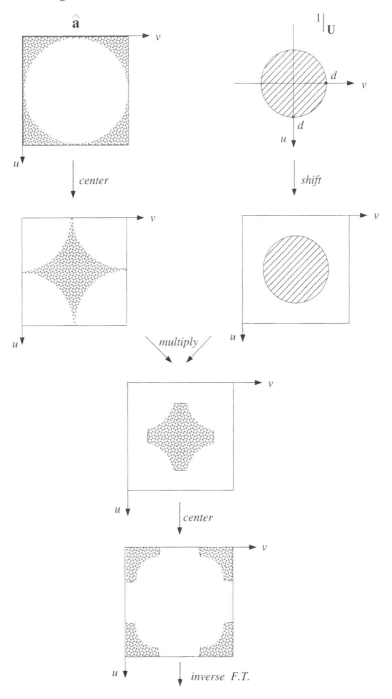

Figure 2.14.6. Illustration of the basic steps involved in the ideal lowpass filter.

X in order to obtain the desired multiplication result. Specifically, we obtain the following form of the lowpass filter algorithm:

$$\hat{\mathbf{h}} := [\mathbf{1}|_U + (\mathbf{1}|_U + (m,0)) + (\mathbf{1}|_U + (0,n)) + (\mathbf{1}|_U + (m,n))]|_X^0$$

$$\mathbf{g} := \mathfrak{F}^{-1}\left(\hat{\mathbf{a}} \cdot \hat{\mathbf{h}}\right).$$

Here $\mathbf{d}|_{\mathbf{X}}^{0}$ means restricting the image $\mathbf{d}$ to the point set $\mathbf{X}$ and then extending $\mathbf{d}$ on $\mathbf{X}$ to the zero image wherever it is not defined on $\mathbf{X}$. The basic steps of this algorithm are shown in Figure 2.14.7.

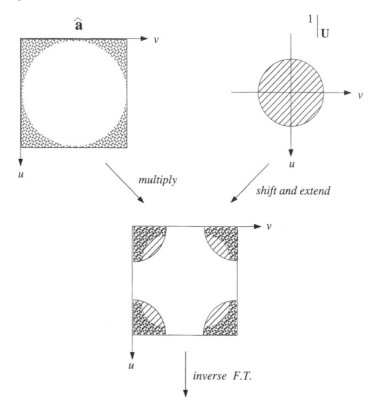

Figure 2.14.7. Illustration of the basic steps as specified by the alternate version of the lowpass filter.

As for the ideal lowpass filter, there are several way for implementing the kth-order Butterworth lowpass filter. The transfer function for this filter is defined over the point set $\mathbf{U} = \mathbb{Z}_{\pm \frac{m}{2}} \times \mathbb{Z}_{\pm \frac{n}{2}}$, and is defined by

$$\hat{\mathbf{h}}(u, v) = \frac{1}{1 + c \cdot \left[\frac{(u^2 + v^2)^{1/2}}{d} \right]^{2k}},$$

where c denotes the scaling constant and the value $\hat{\mathbf{h}}(u, v)$ needs to be converted to a complex value. For correct multiplication with the image $\hat{\mathbf{a}}$, the transfer function $\hat{\mathbf{h}}$ needs to be shifted to the corners of $\mathbf{X}$. The exact specification of the remainder of the algorithm is given by the following two lines of pseudocode:

$$\hat{\mathbf{h}} := \left[\hat{\mathbf{h}} + \left(\hat{\mathbf{h}} + (m, 0) \right) + \left(\hat{\mathbf{h}} + (0, n) \right) + \left(\hat{\mathbf{h}} + (m, n) \right) \right] |_{\mathbf{X}}$$

$$\mathbf{g} := \mathfrak{F}^{-1} \left(\hat{\mathbf{a}} \cdot \hat{\mathbf{h}} \right).$$

The transfer function for exponential lowpass filtering is given by

$$\hat{\mathbf{h}}(u, v) = exp \left[-a \left(\frac{\sqrt{u^2 + v^2}}{d} \right)^k \right],$$

where $(u, v) \in \mathbb{Z}_{\pm \frac{m}{2}} \times \mathbb{Z}_{\pm \frac{n}{2}}$ and $\hat{\mathbf{h}}(u, v)$ needs to be a complex valued. The remainder of the algorithm is identical to the Butterworth filter algorithm.

2.15. Highpass Filtering

Highpass filtering enhances the edge elements of an image based on the fact that edges and abrupt changes in gray levels are associated with the high-frequency components of an image's Fourier transform (Sections 8.2 and 8.4). As before, suppose $\hat{\mathbf{a}}$ denotes the Fourier transform of the source image $\mathbf{a}$. If $\hat{\mathbf{h}}$ denotes a transfer function which attenuates low frequencies and lets high frequencies pass, then the filtered enhancement of $\mathbf{a}$ is the inverse Fourier transform of the product of $\hat{\mathbf{a}}$ and $\hat{\mathbf{h}}$. That is, the enhanced image $\mathbf{g}$ is given by

$$\mathbf{g}(x, y) = \mathfrak{F}^{-1}\left\{ \hat{\mathbf{a}}(u, v) \cdot \hat{\mathbf{h}}(u, v) \right\},$$

where $\mathfrak{F}^{-1}$ denotes the inverse Fourier transform.

The formulation of the highpass transfer function is basically the complement of the lowpass transfer function. Specifically, the transfer function of the *ideal highpass filter* is given by

$$\hat{\mathbf{h}}(u, v) = \begin{cases} 0 & \text{if } \hat{\mathbf{d}}(u, v) \leq d \\ 1 & \text{if } \hat{\mathbf{d}}(u, v) > d, \end{cases}$$

where d is a specified nonnegative quantity which represents the *cutoff frequency* and $\hat{\mathbf{d}}(u, v) = \left(u^2 + v^2 \right)^{1/2}$.

The transfer function of the *Butterworth highpass filter* of order k is given by

$$\hat{\mathbf{h}}(u, v) = \begin{cases} \frac{1}{1 + c\left[d/\hat{\mathbf{d}}(u,v) \right]^{2k}} & \text{if } (u, v) \neq (0, 0) \\ 0 & \text{if } (u, v) = (0, 0), \end{cases}$$

where c is a scaling constant. Typical values for c are 1 and $\sqrt{2} - 1$.

The transfer function of the *exponential highpass filter* is given by

$$\hat{\mathbf{h}}(u, v) = \begin{cases} exp\left[-a\left(d/\hat{\mathbf{d}}(u, v) \right)^{k} \right] & \text{if } (u, v) \neq (0, 0) \\ 0 & \text{if } (u, v) = (0, 0). \end{cases}$$

Typical values for a are 1 and $ln\left(\sqrt{2} \right)$.

Image Algebra Formulation

Let $\hat{\mathbf{a}} \in \mathbb{C}^{\mathbf{X}}$ denote the source image, where $\mathbf{X} = \mathbb{Z}_m \times \mathbb{Z}_n$. Specify the point set $\mathbf{U} = \mathbb{Z}_{\pm \frac{m}{2}} \times \mathbb{Z}_{\pm \frac{n}{2}}$, and define $\hat{\mathbf{h}} \in \mathbb{C}^{\mathbf{U}}$ by

$$\hat{\mathbf{h}}(u, v) = \begin{cases} 0 & if \ \left(u^2 + v^2 \right)^{1/2} \leq d \\ 1 & if \ \left(u^2 + v^2 \right)^{1/2} > d. \end{cases}$$

Once the transfer function is specified, the remainder of the algorithm is analogous to the lowpass filter algorithm. Thus, one specification would be

$$\hat{\mathbf{h}} := \left[\hat{\mathbf{h}} + \left(\hat{\mathbf{h}} + (m, 0) \right) + \left(\hat{\mathbf{h}} + (0, n) \right) + \left(\hat{\mathbf{h}} + (m, n) \right) \right] |_{\mathbf{x}}$$

$$\mathbf{g} := \mathfrak{F}^{-1}\left(\hat{\mathbf{a}} \cdot \hat{\mathbf{h}} \right).$$

This pseudocode specification can also be used for the Butterworth and exponential highpass filter transfer functions which are described next.

The Butterworth highpass filter transfer function of order k with scaling constant c is given by

$$\hat{\mathbf{h}}(u, v) = \begin{cases} \frac{1}{1+c\left[d/(u^2+v^2)^{1/2}\right]^{2k}} & \text{if } (u, v) \neq (0,0) \\ 0 & \text{if } (u, v) = (0,0), \end{cases}$$

where $(u, v) \in \mathbf{U} = \mathbb{Z}_{\pm\frac{m}{2}} \times \mathbb{Z}_{\pm\frac{n}{2}}$.

The transfer function for exponential highpass filtering is given by

$$\hat{\mathbf{h}}(u, v) = \begin{cases} exp\left[-a\left(d/(u^2+v^2)^{1/2}\right)^k\right] & \text{if } (u, v) \neq (0,0) \\ 0 & \text{if } (u, v) = (0,0). \end{cases}$$

2.16.   Exercises

1.   Construct a synthetic image and add random perturbations to each pixel value. In your construction, choose a representation that clarifies the effects of local averaging when using the neighborhoods described below.

    a.   Implement the local averaging algorithm using the Moore neighborhood.
    b.   Repeat 1.a using the skew neighborhood

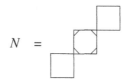

2.   Construct a synthetic image consisting of vertical stripes of different widths.

    a.   Implement the local averaging algorithms using the Moore neighborhood.
    b.   Repeat 2.a using the template

$$\mathbf{t_y} = \begin{array}{|c|c|c|} \hline 1 & 2 & 1 \\ \hline 2 & 4 & 2 \\ \hline 1 & 2 & 1 \\ \hline \end{array}$$

and multiply the resulting image by 1/16.
    c.   Explain the different effects of these two smoothing operations.

3.   Consider the following algorithm. At each pixel location, calculate the difference V between the pixel values of the two vertical neighbors above and below the pixel. Calculate the difference H between the pixel values of the two horizontal neighbors to the left and right of the pixel. If V exceeds H, then the value of the pixel is replaced by the weighted average of the pixel and its two horizontal neighbors. Otherwise, it is replaced by the average of the pixel and its two vertical neighbors.

    a.   Write an image algebra formulation of this algorithm.
    b.   Implement the algorithm.
    c.   The algorithm can be implemented in terms of templates with weights $+1$ and -1. Investigate the effects of this algorithm for different combinations of weights and also for repeated application.

4.   Prove that the Max-Min Sharpening Transform stabilizes, i.e., show that there exists a positive integer n such that $s^n(\mathbf{a}) = s^k(\mathbf{a})$ for all integers $k \geq n$.

5.   Show that while $(\mathbf{a} + \mathbf{b}) \, @ \, N = \mathbf{a} \, @ \, N + \mathbf{b} \, @ \, N$, it is not true in general that $(\mathbf{a} + \mathbf{b}) \, ⓜ N = \mathbf{a} \, ⓜ N + \mathbf{b} \, ⓜ N$. This shows that local averaging is a linear operation while median filtering is not.

6.   A generalization of unsharp masking that is often used for object enhancement and background suppressions is given by the operation

$$\mathbf{b}(i, j) = \left| \sum_{N_1} \mathbf{a}(l, k) - \sum_{N_2} \mathbf{a}(l, k) \right|$$

where $\mathbf{a}$ denotes the input image, N_1 is an $n \times n$ rectangular neighborhood centered at (i, j), N_2 is an $m \times m$ rectangular neighborhood centered at (i, j), and $n > m$.

    a.   Provide an image algebra formulation of this algorithm.
    b.   Implement this algorithm on a synthetic image where the object intensity is greater than the background and the object size can be covered by N_1.
    c.   Repeat 6.b except that the object intensity is less than the background.
    d.   Repeat 6.b except that the object size is greater than N_1 but less than N_2.

7.   Implement the Gaussian smoothing algorithm on at least two different images using different template support. Analyze your results.

8.*   Implement the five histogram modifications described in Table 2.13.2 and describe their effects and differences when applied to an image.

9.*   Develop a program to display histogram plots of digital images similar to those shown in Figure 2.12.1. Test your program on suitable images and provide an image algebra formulation of your program.

10.* It follows from Section 1.7 that if N denotes an $n \times n$ rectangular neighborhood and $\mathbf{t}$ a template with $N(\mathbf{y}) = S(\mathbf{t_y})$ for each $\mathbf{y}$, then the image $\mathbf{a} \oplus \mathbf{t}$ is equivalent to $\mathbf{a} \, @ \, N$ as long as one ignores boundary effects and $\mathbf{t_y}(\mathbf{x}) = 1/n^2$ whenever $\mathbf{x} \in S(\mathbf{t_y})$.

    a.   Show that the Fourier transform of $\mathbf{t_y}$ is given by $\hat{\mathbf{h}}(u, v) = \frac{1}{n} \sum_{k=0}^{n-1} \sum_{l=0}^{n-1} \frac{1}{n^2} e^{(-2\pi i u k)/n} \cdot e^{(-2\pi i v k)/n}$.
    b.   Using 10.a, establish a relationship between smoothing by averaging and lowpass filtering.

11.* Implement the three highpass filters described in Section 2.15.

12.* Consider the unsharp masking algorithm given by $\mathbf{b} := \mathbf{a} \oplus \mathbf{t}$, where $\mathbf{t}$ denotes an $n \times n$ rectangular template with $\mathbf{t_y}(\mathbf{y}) = 1$ and $\mathbf{t_y}(\mathbf{x}) = -\frac{1}{n-1}$ whenever $\mathbf{y} \neq \mathbf{x} \in S(\mathbf{t_y})$.

    a.   Find the Fourier transform of $\mathbf{t_y}$.
    b.   Using 12.a, establish a relationship between unsharp masking and highpass filtering.

2.17.   References

[1] R. Gonzalez and P. Wintz, *Digital Image Processing.* Reading, MA: Addison-Wesley, second ed., 1987.

[2] W. Pratt, *Digital Image Processing.* New York: John Wiley, 1978.

[3] G. Ritter, J. Wilson, and J. Davidson, "Image algebra: An overview," *Computer Vision, Graphics, and Image Processing*, vol. 49, pp. 297–331, Mar. 1990.

[4] A. Lev, S. Zucker, and A. Rosenfeld, "Iterative enhancement of noisy images," *IEEE Transactions on Systems, Man, and Cybernetics*, vol. SMC-7, pp. 435–442, June 1977.

[5] D. Marr and E. Hildreth, "Theory of edge detection," in *Proceedings of the Royal Society of London B*, vol. 207, pp. 187–217, 1980.

[6] R. Leipnik, "The extended entropy uncertainty principle," *Information and Control*, vol. 3, pp. 18–25, 1960.

[7] H. Kramer and J. Bruchner, "Iterations of a non-linear transformation for the enhancement of digital images," *Pattern Recognition*, vol. 7, pp. 53–58, 1975.

[8] R. Haralick and L. Shapiro, *Computer and Robot Vision*, vol. I. New York: Addison-Wesley, 1992.

[9] P. Coffield, "An architecture for processing image algebra operations," in *Image Algebra and Morphological Image Processing III*, vol. 1769 of *Proceedings of SPIE*, (San Diego, CA), pp. 178–189, July 1992.

[10] R. Gonzalez and R. Woods, *Digital Image Processing.* Reading, MA: Addison-Wesley, 1992.

[11] L. Loo, K. Doi, and C. Metz, "Investigation of basic imaging properties in digital radiography 4. Effect of unsharp masking on the detectability of simple patterns," *Medical Physics*, vol. 12, pp. 209–214, Mar. 1985.

[12] R. Cromartie and S. Pizer, "Edge-affected context for adaptive contrast enhancement," in *Ninth Conference on Information Processing in Medical Imaging*, 1991.

[13] J. Harris, "Constant variance enhancement: a digital processing technique," *Applied Optics*, vol. 16, pp. 1268–1271, May 1977.

[14] P. Narendra and R. Fitch, "Real-time adaptive contrast enhancement," *IEEE Transactions on Pattern Analysis and Machine Intelligence*, vol. PAMI-3, pp. 655–661, Nov. 1981.

[15] D. Ballard and C. Brown, *Computer Vision.* Englewood Cliffs, NJ: Prentice-Hall, 1982.

[16] A. Rosenfeld and A. Kak, *Digital Picture Processing.* New York, NY: Academic Press, 2nd ed., 1982.

CHAPTER 3
EDGE DETECTION AND
BOUNDARY FINDING TECHNIQUES

3.1. Introduction

Edge detection is an important operation in a large number of image processing applications such as image segmentation, character recognition, and scene analysis. An edge in an image is a contour across which the brightness of the image changes abruptly in magnitude or in the rate of change of magnitude.

Generally, it is difficult to specify *a priori* which edges correspond to relevant boundaries in an image. Image transforms which enhance and/or detect edges are usually task domain dependent. We, therefore, present a wide variety of commonly used transforms. Most of these transforms can be categorized as belonging to one of the following four classes: (1) simple windowing techniques to find the boundaries of Boolean objects, (2) transforms that approximate the gradient, (3) transforms that use multiple templates at different orientations, and (4) transforms that fit local intensity values with parametric edge models.

The goal of boundary finding is somewhat different from that of edge detection methods which are generally based on intensity information methods classified in the preceding paragraph. Gradient methods for edge detection, followed by thresholding, typically produce a number of undesired artifacts such as missing edge pixels and parallel edge pixels, resulting in thick edges. Edge thinning processes and thresholding may result in disconnected edge elements. Additional processing is usually required in order to group edge pixels into coherent boundary structures. The goal of boundary finding is to provide coherent *one-dimensional* boundary features from the individual local edge pixels.

3.2. Binary Image Boundaries

A boundary point of an object in a binary image is a point whose 4-neighborhood (or 8-neighborhood, depending on the boundary classification) intersects the object and its complement. Boundaries for binary images are classified by their connectivity and whether they lie within the object or its complement. The four classifications will be expanded upon in the discussion involving the mathematical formulation.

Binary image boundary transforms are thinning methods. They do not preserve the homotopy of the original image. Boundary transforms can be especially useful when used inside of other algorithms that require location of the boundary to perform their tasks. Many of the other thinning transforms in this chapter fall into this category.

The techniques outlined below work by using the appropriate neighborhood to either enlarge or reduce the region with the ⊻ or ⊼ operation, respectively. After the object has been enlarged or reduced, it is intersected with its original complement to produce the boundary image.

For $\mathbf{a} \in \{0,1\}^{\mathbf{X}}$, let $\mathbf{A}$ denote the support of $\mathbf{a}$. The boundary *image* $\mathbf{b} \in \{0,1\}^{\mathbf{X}}$ of $\mathbf{a}$ is classified by its connectivity and whether $\mathbf{B} \subset \mathbf{A}$ or $\mathbf{B} \subset \mathbf{A}'$, where $\mathbf{B}$ denotes the support of $\mathbf{b}$.

(a)   The image $\mathbf{b}$ is an *exterior 8-boundary image* if $\mathbf{B}$ is 8-connected, $\mathbf{B} \subset \mathbf{A}'$, and $\mathbf{B}$ is the set of points in $\mathbf{A}'$ whose 4-neighborhoods intersect $\mathbf{A}$. That is,

$$\mathbf{b}(\mathbf{x}) = \begin{cases} 1 & \text{if } N(\mathbf{x}) \cap \mathbf{A} \neq \varnothing \text{ and } \mathbf{x} \in \mathbf{A}' \\ 0 & \text{otherwise.} \end{cases}$$

(b)   The image $\mathbf{b}$ is an *interior 8-boundary image* if $\mathbf{B}$ is 8-connected, $\mathbf{B} \subset \mathbf{A}$, and $\mathbf{B}$ is the set of points in $\mathbf{A}$ whose 4-neighborhoods intersect $\mathbf{A}'$. The interior 8-boundary $\mathbf{b}$ can be expressed as

$$\mathbf{b}(\mathbf{x}) = \begin{cases} 1 & \text{if } N(\mathbf{x}) \cap \mathbf{A}' \neq \varnothing \text{ and } \mathbf{x} \in \mathbf{A} \\ 0 & \text{otherwise.} \end{cases}$$

(c)   The image $\mathbf{b}$ is an *exterior 4-boundary image* if $\mathbf{B}$ is 4-connected, $\mathbf{B} \subset \mathbf{A}'$, and $\mathbf{B}$ is the set of points in $\mathbf{A}'$ whose 8-neighborhoods intersect $\mathbf{A}$. That is, the image $\mathbf{b}$ is defined by

$$\mathbf{b}(\mathbf{x}) = \begin{cases} 1 & \text{if } M(\mathbf{x}) \cap \mathbf{A} \neq \varnothing \text{ and } \mathbf{x} \in \mathbf{A}' \\ 0 & \text{otherwise.} \end{cases}$$

(d)   The image $\mathbf{b}$ is an *interior 4-boundary* image if $\mathbf{B}$ is 4-connected, $\mathbf{B} \subset \mathbf{A}$, and $\mathbf{B}$ is the set of points in $\mathbf{A}$ whose 8-neighborhoods intersect $\mathbf{A}'$. Thus,

$$\mathbf{b}(\mathbf{x}) = \begin{cases} 1 & \text{if } M(\mathbf{x}) \cap \mathbf{A}' \neq \varnothing \text{ and } \mathbf{x} \in \mathbf{A} \\ 0 & \text{otherwise.} \end{cases}$$

Figure 3.2.1 below illustrates the boundaries just described. The center image is the original image. The 8-boundaries are to the left, and the 4-boundaries are to the right. Exterior boundaries are black. Interior boundaries are gray.

Figure 3.2.1. Interior and exterior 8-boundaries (left), original image (center), and interior and exterior 4-boundaries (right).

Image Algebra Formulation

The von Neumann neighborhood function N is used in the image algebra formulation for detecting 8-boundaries, while the Moore neighborhood function M is used for detection of 4-boundaries.

Let $\mathbf{a} \in \{0,1\}^{\mathbf{X}}$ be the source image. The boundary image will be denoted by $\mathbf{b}$.

(1) Exterior 8-boundary —

$$\mathbf{b} := (\mathbf{1} - \mathbf{a}) \cdot (\mathbf{a} \boxdot N)$$

(2) Interior 8-boundary —

$$\mathbf{b} := (\mathbf{1} - (\mathbf{a} \boxbackslash N)) \cdot \mathbf{a}$$

(3) Exterior 4-boundary —

$$\mathbf{b} := (\mathbf{1} - \mathbf{a}) \cdot (\mathbf{a} \boxdot M)$$

(4) Interior 4-boundary —

$$\mathbf{b} := (\mathbf{1} - (\mathbf{a} \boxbackslash M)) \cdot \mathbf{a}$$

Comments and Observations

These transforms are designed for binary images only. More sophisticated algorithms must be used for gray level images.

Noise around the boundary may adversely affect results of the algorithm. An algorithm such as the salt and pepper noise removal transform may be useful in cleaning up the boundary before the boundary transform is applied.

3.3. Edge Enhancement by Discrete Differencing

Discrete differencing is a local edge enhancement technique. It is used to sharpen edge elements in an image by discrete differencing in either the vertical or horizontal direction, or in a combined fashion [1, 2, 3, 4].

Let $\mathbf{a} \in \mathbb{R}^{\mathbf{X}}$ be the source image. The edge enhanced image $\mathbf{b} \in \mathbb{R}^{\mathbf{X}}$ can be obtained by one of the following difference methods:

(1) Horizontal differencing

$$\mathbf{b}(i, j) = \mathbf{a}(i, j) - \mathbf{a}(i, j + 1)$$

or

$$\mathbf{b}(i, j) = 2\mathbf{a}(i, j) - \mathbf{a}(i, j - 1) - \mathbf{a}(i, j + 1).$$

(2) Vertical differencing

$$\mathbf{b}(i, j) = \mathbf{a}(i, j) - \mathbf{a}(i + 1, j)$$

or

$$\mathbf{b}(i, j) = 2\mathbf{a}(i, j) - \mathbf{a}(i - 1, j) - \mathbf{a}(i + 1, j).$$

(3) Gradient approximation

$$\mathbf{b}(i, j) = |\mathbf{a}(i, j) - \mathbf{a}(i + 1, j)| + |\mathbf{a}(i, j) - \mathbf{a}(i, j + 1)|$$

or

$$\mathbf{b}(i, j) = |\mathbf{a}(i, j) - \mathbf{a}(i + 1, j + 1)| + |\mathbf{a}(i, j) - \mathbf{a}(i - 1, j + 1)|.$$

Image Algebra Formulation

Given the source image $\mathbf{a} \in \mathbb{R}^X$, the edge enhanced image $\mathbf{b} \in \mathbb{R}^X$ is obtained by the appropriate image-template convolution below.

(1) Horizontal differencing

$$\mathbf{b} := \mathbf{a} \oplus \mathbf{r},$$

where the invariant enhancement template $\mathbf{r}$ is defined by

$$\mathbf{r}_{(i,j)}(\mathbf{x}) = \begin{cases} 1 & \text{if } \mathbf{x} = (i,j) \\ -1 & \text{if } \mathbf{x} = (i, j+1) \\ 0 & \text{otherwise} \end{cases}$$

or

$$\mathbf{b} := \mathbf{a} \oplus \mathbf{s},$$

where

$$\mathbf{s}_{(i,j)}(\mathbf{x}) = \begin{cases} 2 & \text{if } \mathbf{x} = (i,j) \\ -1 & \text{if } \mathbf{x} \in \{(i, j-1), (i, j+1)\} \\ 0 & \text{otherwise.} \end{cases}$$

(2) Vertical differencing

$$\mathbf{b} := \mathbf{a} \oplus \mathbf{t}$$

where the template $\mathbf{t}$ is defined by

$$\mathbf{t}_{(i,j)}(\mathbf{x}) = \begin{cases} 1 & \text{if } \mathbf{x} = (i,j) \\ -1 & \text{if } \mathbf{x} = (i+1, j) \\ 0 & \text{otherwise} \end{cases}$$

or

$$\mathbf{b} := \mathbf{a} \oplus \mathbf{u},$$

where

$$\mathbf{u}_{(i,j)}(\mathbf{x}) = \begin{cases} 2 & \text{if } \mathbf{x} = (i,j) \\ -1 & \text{if } \mathbf{x} \in \{(i-1, j), (i+1, j)\} \\ 0 & \text{otherwise.} \end{cases}$$

(3) Gradient approximation

$$\mathbf{b} := |\, \mathbf{a} \oplus \mathbf{t}| + |\, \mathbf{a} \oplus \mathbf{r}\, |$$

or

$$\mathbf{b} := |\mathbf{a} \oplus \mathbf{v}| + |\mathbf{a} \oplus \mathbf{w}|,$$

where the templates $\mathbf{t}$ and $\mathbf{r}$ are defined as above and $\mathbf{v}$, $\mathbf{w}$ are defined by

$$\mathbf{v}_{(i,j)}(\mathbf{x}) = \begin{cases} 1 & \text{if } \mathbf{x} = (i,j) \\ -1 & \text{if } \mathbf{x} = (i+1, j+1) \\ 0 & \text{otherwise} \end{cases}$$

and

$$\mathbf{w}_{(i,j)}(\mathbf{x}) = \begin{cases} 1 & \text{if } \mathbf{x} = (i,j) \\ -1 & \text{if } \mathbf{x} = (i-1, j+1) \\ 0 & \text{otherwise.} \end{cases}$$

The templates can be described pictorially as

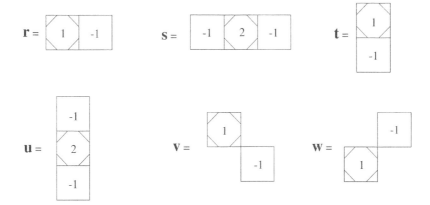

Comments and Observations

Figures 3.3.2 through 3.3.4 below are the edge enhanced images of the motorcycle image in Figure 3.3.1.

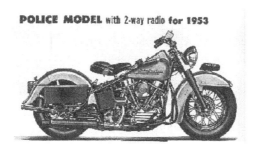

Figure 3.3.1. Original image.

Figure 3.3.2. Horizontal differencing: left $|\mathbf{a} \oplus \mathbf{r}|$, right $|\mathbf{a} \oplus \mathbf{s}|$.

Figure 3.3.3. Vertical differencing: left $|\mathbf{a} \oplus \mathbf{t}|$, right $|\mathbf{a} \oplus \mathbf{u}|$.

Figure 3.3.4. Gradient approximation: left $|\mathbf{a} \oplus \mathbf{t}| + |\mathbf{a} \oplus \mathbf{r}|$, right $|\mathbf{a} \oplus \mathbf{v}| + |\mathbf{a} \oplus \mathbf{w}|$.

3.4.   Roberts Edge Detector

The Roberts edge detector is another example of an edge enhancement technique that uses discrete differencing [5, 1]. Let $\mathbf{a} \in \mathbb{R}^{\mathbf{X}}$ be the source image. The edge enhanced image $\mathbf{b} \in \mathbb{R}^{\mathbf{X}}$ that the Roberts technique produces is defined by

$$\mathbf{b}(i,j) = \left((\mathbf{a}(i,j) - \mathbf{a}(i+1,j+1))^2 + (\mathbf{a}(i,j+1) - \mathbf{a}(i+1,j))^2\right)^{1/2}.$$

Image Algebra Formulation

Given the source image $\mathbf{a} \in \mathbb{R}^{\mathbf{X}}$, the edge enhanced image $\mathbf{b} \in \mathbb{R}^{\mathbf{X}}$ is given by

$$\mathbf{b} := \left((\mathbf{a} \oplus \mathbf{s})^2 + (\mathbf{a} \oplus \mathbf{t})^2\right)^{1/2},$$

where the templates $\mathbf{s}$ and $\mathbf{t}$ are defined by

$$\mathbf{s}_{(i,j)}(\mathbf{x}) = \begin{cases} 1 & \text{if } \mathbf{x} = (i,j) \\ -1 & \text{if } \mathbf{x} = (i+1,j+1) \\ 0 & \text{otherwise}; \end{cases}$$

$$\mathbf{t}_{(i,j)}(\mathbf{x}) = \begin{cases} 1 & \text{if } \mathbf{x} = (i,j) \\ -1 & \text{if } \mathbf{x} = (i+1,j-1) \\ 0 & \text{otherwise}. \end{cases}$$

The templates $\mathbf{s}$ and $\mathbf{t}$ can be represented pictorially as

$$\mathbf{s} = \boxed{\begin{array}{|c|c|} \hline 1 & \\ \hline & -1 \\ \hline \end{array}} \qquad \mathbf{t} = \boxed{\begin{array}{|c|c|} \hline & 1 \\ \hline -1 & \\ \hline \end{array}}$$

Comments and Observations

Figure 3.4.1 shows the result of applying the Roberts edge detector to the image of a motorcycle.

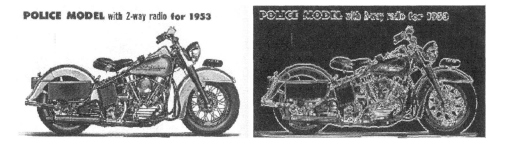

Figure 3.4.1. Motorcycle and its Roberts edge enhanced image.

3.5. Prewitt Edge Detector

The Prewitt edge detector calculates an edge gradient vector at each point of the source image. An edge enhanced image is produced from the magnitude of the gradient vector. An edge angle, which is equal to the angle the gradient makes to the horizontal axis, can also be assigned to each point of the source image [6, 7, 1, 8, 4].

Two masks are convolved with the source image to approximate the gradient vector. One mask represents the partial derivative with respect to x and the other the partial derivative with respect to y.

Let $\mathbf{a} \in \mathbb{R}^{\mathbf{X}}$ be the source image, and $a_0, a_1, \ldots, a_7$ denote the pixel values of the eight neighbors of (i, j) enumerated in the counterclockwise direction as follows:

a_3	a_2	a_1
a_4	(i,j)	a_0
a_5	a_6	a_7

Let $u = (a_5 + a_6 + a_7) - (a_1 + a_2 + a_3)$ and $v = (a_0 + a_1 + a_7) - (a_3 + a_4 + a_5)$. The edge enhanced image $\mathbf{b} \in \mathbb{R}^{\mathbf{X}}$ is given by

$$\mathbf{b}(i, j) = \left(u^2 + v^2\right)^{1/2}$$

and the edge direction image $\mathbf{d} \in \mathbb{R}^{\mathbf{X}}$ is given by

$$\mathbf{d}(i, j) = arctan\left(\frac{v}{u}\right).$$

Image Algebra Formulation

Let $\mathbf{a} \in \mathbb{R}^{\mathbf{X}}$ be the source image and let $\mathbf{s}$ and $\mathbf{t}$ be defined as follows:

$$\mathbf{s}_{(i,j)}(\mathbf{x}) = \begin{cases} -1 & \text{if } \mathbf{x} = (i-1, j-1),\ (i-1,j),\ \text{or } (i-1, j+1) \\ 1 & \text{if } \mathbf{x} = (i+1, j-1),\ (i+1,j),\ \text{or } (i+1, j+1) \\ 0 & \text{otherwise;} \end{cases}$$

$$\mathbf{t}_{(i,j)}(\mathbf{x}) = \begin{cases} -1 & \text{if } \mathbf{x} = (i-1, j-1),\ (i, j-1),\ \text{or } (i+1, j-1) \\ 1 & \text{if } \mathbf{x} = (i-1, j+1),\ (i, j+1),\ \text{or } (i+1, j+1) \\ 0 & \text{otherwise.} \end{cases}$$

Pictorially we have:

$$\mathbf{S} = \quad \begin{array}{|c|c|c|} \hline -1 & -1 & -1 \\ \hline & & \\ \hline 1 & 1 & 1 \\ \hline \end{array} \qquad\qquad \mathbf{t} = \quad \begin{array}{|c|c|c|} \hline -1 & & 1 \\ \hline -1 & & 1 \\ \hline -1 & & 1 \\ \hline \end{array}$$

The edge enhanced image $\mathbf{b} \in \mathbb{R}^{\mathbf{Y}}$ is given by

$$\mathbf{b} := \left(\left[(\mathbf{a} \oplus \mathbf{s})^2 + (\mathbf{a} \oplus \mathbf{t})^2 \right]^{1/2} \right).$$

The edge direction image $\mathbf{d} \in \mathbb{R}^{\mathbf{Y}}$ is given by

$$\mathbf{d} := arctan2\left((\mathbf{a} \oplus \mathbf{t})|_{domain(\mathbf{b}\|>0)},\ (\mathbf{a} \oplus \mathbf{s})|_{domain(\mathbf{b}\|>0)} \right).$$

Here we use the common programming language convention for the arctangent of two variables which defines $arctan2(u,v) \equiv arctan\left(\frac{u}{v}\right)$.

Comments and Observations

A variety of masks may be used to approximate the partial derivatives.

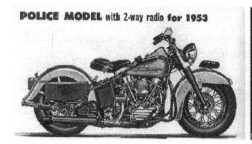 

Figure 3.5.1. Motorcycle and the image that represents
the magnitude component of the Prewitt edge detector.

3.6. Sobel Edge Detector

The Sobel edge detector is a nonlinear edge enhancement technique. It is another simple variation of the discrete differencing scheme for enhancing edges [9, 10, 1, 8, 4].

Let $\mathbf{a} \in \mathbb{R}^X$ be the source image, and $a_0, a_1, \ldots, a_7$ denote the pixel values of the eight neighbors of (i, j) enumerated in the counterclockwise direction as follows:

a_3	a_2	a_1
a_4	(i,j)	a_0
a_5	a_6	a_7

The Sobel edge magnitude image $\mathbf{m} \in \mathbb{R}^X$ is given by

$$\mathbf{m}(i, j) = \left(u^2 + v^2\right)^{1/2},$$

where

$$u = (a_5 + 2a_6 + a_7) - (a_1 + 2a_2 + a_3)$$

and

$$v = (2a_0 + a_1 + a_7) - (a_3 + 2a_4 + a_5).$$

The gradient direction image $\mathbf{d}$ is given by

$$\mathbf{d}(i, j) = arctan\left(\frac{u}{v}\right).$$

Image Algebra Formulation

Let $\mathbf{a} \in \mathbb{R}^X$ be the source image. The gradient magnitude image $\mathbf{m} \in \mathbb{R}^X$ is given by

$$\mathbf{m} := \left[(\mathbf{a} \oplus \mathbf{s})^2 + (\mathbf{a} \oplus \mathbf{t})^2\right]^{1/2},$$

where the templates $\mathbf{s}$ and $\mathbf{t}$ are defined as follows:

$$\mathbf{s}_{(i,j)}(\mathbf{x}) = \begin{cases} -1 & \text{if } \mathbf{x} = (i{-}1, j{-}1) \text{ or } (i{-}1, j{+}1) \\ 1 & \text{if } \mathbf{x} = (i{+}1, j{-}1) \text{ or } (i{+}1, j{+}1) \\ -2 & \text{if } \mathbf{x} = (i-1, j) \\ 2 & \text{if } \mathbf{x} = (i{+}1, j) \\ 0 & \text{otherwise.} \end{cases}$$

$$\mathbf{t}_{(i,j)}(\mathbf{x}) = \begin{cases} -1 & \text{if } \mathbf{x} = (i{-}1, j{-}1) \text{ or } (i{+}1, j{-}1) \\ 1 & \text{if } \mathbf{x} = (i{-}1, j{+}1) \text{ or } (i{+}1, j{+}1) \\ -2 & \text{if } \mathbf{x} = (i, j{-}1) \\ 2 & \text{if } \mathbf{x} = (i, j{+}1) \\ 0 & \text{otherwise.} \end{cases}$$

$\mathbf{s} =$ 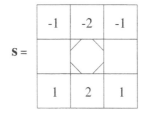 $\mathbf{t} =$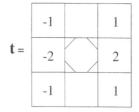

The gradient direction image is given by

$$\mathbf{d} := arctan2((\mathbf{a} \oplus \mathbf{s})|_{\mathbf{m}>0}, (\mathbf{a} \oplus \mathbf{t})|_{\mathbf{m}>0}),$$

where $arctan2(u, v) \equiv arctan\left(\frac{u}{v}\right)$.

Comments and Observations

The Sobel edge detection emphasizes horizontal and vertical edges over skewed edges; it is relatively insensitive to off-axis edges. Figure 3.6.1 shows the Sobel edge image of the motorcycle test image.

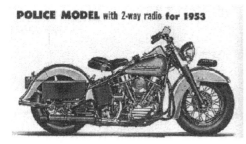 

Figure 3.6.1. Motorcycle and its Sobel edge enhanced image.

3.7.   Wallis Logarithmic Edge Detection

Under the Wallis edge detection scheme a pixel is an edge element if the logarithm of its value exceeds the average of the logarithms of its 4-neighbors by a fixed threshold [2, 1]. Suppose $\mathbf{a} \in (\mathbb{R}^+)^{\mathbf{X}}$ denotes a source image containing only positive values, and a_0, a_1, a_2, a_3 denote the values of the 4-neighbors of (i, j) enumerated as follows:

	a_1	
a_2	(i,j)	a_0
	a_3	

The edge enhanced image $\mathbf{b} \in \mathbb{R}^{\mathbf{X}}$ is given by

$$\mathbf{b}(i, j) = log_b(\mathbf{a}(i, j)) - \frac{1}{4}(log_b(a_0) + log_b(a_1) + log_b(a_2) + log_b(a_3))$$

or

$$\mathbf{b}(i,j) = \frac{1}{4}log_b\left(\frac{(\mathbf{a}(i,j))^4}{a_0 a_1 a_2 a_3}\right).$$

Image Algebra Formulation

Let $\mathbf{a} \in (\mathbb{R}^+)^\mathbf{X}$ be the source image. The edge enhanced image $\mathbf{b} \in \mathbb{R}^\mathbf{X}$ is given by

$$\mathbf{b} := \frac{1}{4}log_b(\mathbf{a}) \oplus \mathbf{t},$$

where the invariant enhancement template $\mathbf{t}\colon \mathbf{X} \to \mathbb{R}^\mathbf{X}$ is defined as follows:

$$\mathbf{t}_{(i,j)}(\mathbf{x}) = \begin{cases} 4 & \text{if } \mathbf{x} = (i,j) \\ -1 & \text{if } \mathbf{x} \in \{(i+1,j),(i,j-1),(i-1,j),(i,j+1)\} \\ 0 & \text{otherwise.} \end{cases}$$

Pictorially, $\mathbf{t}$ is given by

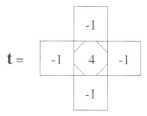

Comments and Observations

The Wallis edge detector is insensitive to a global multiplicative change in image values. The edge image of $\mathbf{a}$ will be the same as that of $n \cdot \mathbf{a}$ for any $n \in \mathbb{R}^+$.

Note that if the edge image is to be thresholded, it is not necessary to compute the logarithm of $\mathbf{a}$. That is, if

$$\mathbf{b} := \frac{1}{4}log_b(\mathbf{a}) \oplus \mathbf{t}$$

and

$$\mathbf{c} := \mathbf{a} \oplus \mathbf{t},$$

then

$$\mathbf{b}|_{>\tau} = \mathbf{c}|_{>b^{4\tau}}.$$

Figure 3.7.1 shows the result of applying the Wallis edge detector to the motorcycle image. The logarithm used for the edge image is base 2 ($b = 2$).

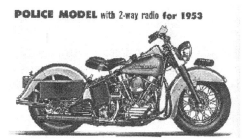

Figure 3.7.1. Motorcycle and its Wallis edge enhancement.

3.8. Frei-Chen Edge and Line Detection

A 3×3 subimage $\mathbf{b}$ of an image $\mathbf{a}$ may be thought of as a vector in $\mathbb{R}^9$. For example, the subimage shown in Figure 3.8.1 has vector representation

$$\mathbf{b} = \begin{bmatrix} b_0 \\ b_1 \\ \vdots \\ b_8 \end{bmatrix}.$$

The mathematical structure of the vector space $\mathbb{R}^9$ carries over to the vector space of 3×3 subimages in the obvious way.

$$\mathbf{b} = \begin{array}{|c|c|c|} \hline b_4 & b_3 & b_2 \\ \hline b_5 & b_0 & b_1 \\ \hline b_6 & b_7 & b_8 \\ \hline \end{array}$$

Figure 3.8.1. A 3×3 subimage.

Let V denote the vector space of 3×3 subimages. An orthogonal basis for V, $\mathcal{B}_V$, that is used for the *Frei-Chen* method is the one shown in Figure 3.8.2. The subspace E of V that is spanned by the subimages $\mathbf{v}_1, \mathbf{v}_2, \mathbf{v}_3$, and $\mathbf{v}_4$ is called the *edge subspace* of V. The Frei-Chen edge detection method bases its determination of edge points on the size of the angle between the subimage $\mathbf{b}$ and its projection on the edge subspace [11, 12, 13].

The angle θ_E between $\mathbf{b}$ and its projection on the edge subspace is given by

$$\theta_E = cos^{-1} \left(\frac{\displaystyle\sum_{i=1}^{4} (\mathbf{v}_i \bullet \mathbf{b})^2}{\displaystyle\sum_{i=1}^{9} (\mathbf{v}_i \bullet \mathbf{b})^2} \right)^{\frac{1}{2}}.$$

The $\bullet$ operator in the formula above is the familiar dot (or scalar) product defined for vector spaces. The dot product of $\mathbf{b}, \mathbf{c} \in V$ is given by

$$\mathbf{b} \bullet \mathbf{c} = \sum_{i=0}^{8} b_i c_i.$$

Small values of θ_E imply a better fit between $\mathbf{b}$ and the edge subspace.

For each point (x, y) in the source image $\mathbf{a}$, the Frei-Chen algorithm for edge detection calculates the angle θ_E between the projection of the 3×3 subimage centered at (x, y) and the edge subspace of V. The smaller the value of θ_E at (x, y) is, the better edge point (x, y) is deemed to be. After θ_E is calculated for each image point, a threshold τ is applied to select points for which $\theta_E \leq \tau$.

Figure 3.8.3 shows the results of applying the Frei-Chen edge detection algorithm to a source image of variety of peppers. Thresholds were applied for angles θ_E of 18, 19, 20, 21, and 22°.

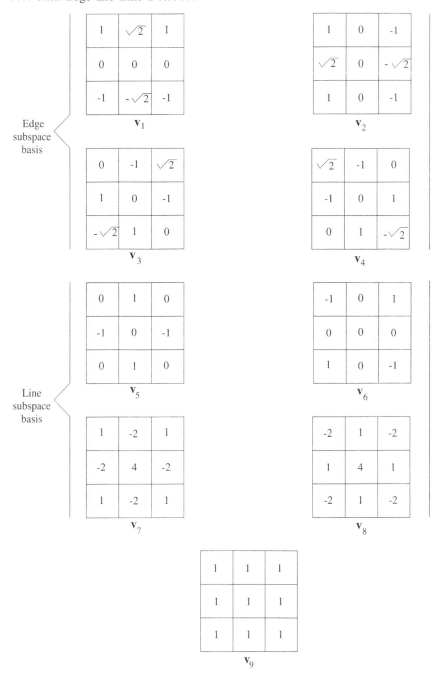

Figure 3.8.2. The basis $\mathcal{B}_V$ used for Frei-Chen feature detection.

Equivalent results can be obtained by thresholding based on the statistic

$$\frac{\sum_{i=1}^{4}\left(\mathbf{v}(i)\bullet\mathbf{b}\right)^2}{\mathbf{b}\bullet\mathbf{b}},$$

which is easier to compute. In this case, a larger value indicates a stronger edge point.

The Frei-Chen method can also be used to detect lines. Subimages $\mathbf{v}_5, \mathbf{v}_6, \mathbf{v}_7$, and $\mathbf{v}_8$ form the basis of the *line subspace* of V. The Frei-Chen edge detection method

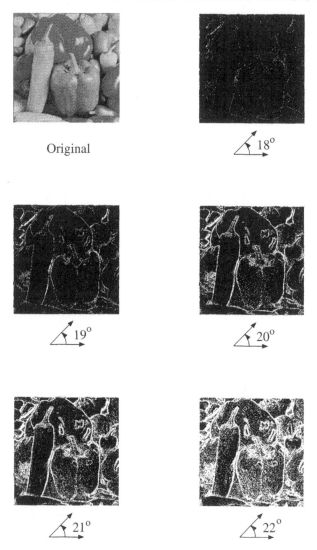

Figure 3.8.3. Edge detection using the Frei-Chen method.

bases its determination of lines on the size of the angle between the subimage **b** and its projection on the line subspace. Thresholding for line detection is done using the statistic

$$\frac{\sum_{i=5}^{8} \left(\mathbf{v}(i) \bullet \mathbf{b} \right)^2}{\mathbf{b} \bullet \mathbf{b}}.$$

Larger values indicate stronger line points.

Image Algebra Formulation

Let $\mathbf{a} \in \mathbb{R}^{\mathbf{X}}$ be the source image and let $\mathbf{v}(i)_{\mathbf{y}}$ denote the parameterized template whose values are defined by the image $\mathbf{v}_i$ of Figure 3.8.2. The center cell of the image $\mathbf{v}_i$ is taken to be the location of $\mathbf{y}$ for the template. For a given threshold level τ, the

Frei-Chen edge image **e** is given by

$$\mathbf{e} := \chi_{\geq\tau}\left(\frac{\sum\limits_{i=1}^{4}\left(\mathbf{a}\oplus\mathbf{v}(i)\right)^2}{\mathbf{a}^2\oplus\mathbf{v}(9)}\right).$$

The Frei-Chen line image **l** is given by

$$\mathbf{l} := \chi_{\geq\tau}\left(\frac{\sum\limits_{i=5}^{8}\left(\mathbf{a}\oplus\mathbf{v}(i)\right)^2}{\mathbf{a}^2\oplus\mathbf{v}(9)}\right).$$

3.9. Kirsch Edge Detector

The Kirsch edge detector [14] applies eight masks at each point of an image in order to determine an edge gradient direction and magnitude. Let $\mathbf{a}\in\mathbb{R}^{\mathbf{X}}$ be the source image. For each (i,j) we denote $a_0, a_1, \ldots, a_7$ as pixel values of the eight neighbors of (i,j) enumerated in the counterclockwise direction as follows:

a_3	a_2	a_1
a_4	(i,j)	a_0
a_5	a_6	a_7

The image $\mathbf{m}\in\mathbb{R}^{\mathbf{X}}$ that represents the magnitude of the edge gradient is given by

$$\mathbf{m}(i,j) = max\{1, max\{|5s_k - 3t_k| \ : \ k = 0, \ldots, 7\}\},$$

where $s_k = a_k + a_{k+1} + a_{k+2}$ and $t_k = a_{k+3} + a_{k+4} + \cdots + a_{k+7}$. The subscripts are evaluated *modulo* 8. Further details about this method of directional edge detection are given in [11, 8, 1, 14, 4].

The image $\mathbf{d}\in\{0,1,\ldots,7\}^{\mathbf{X}}$ that represents the gradient direction is defined by

$$\mathbf{d}(i,j) := d, \text{ where } |5s_d - 3t_d| = \max_{0\leq k\leq 7}\{|5s_k - 3t_k|\}.$$

The image $\mathbf{d}\cdot 45°$ can be used if it is desirable to express gradient angle in degree measure. Figure 3.9.1 shows the magnitude image that results from applying the Kirsch edge detector to the motorcycle image.

Figure 3.9.1. Kirsch edge detector applied to the motorcycle image.

Image Algebra Formulation

Let $M = \{-1, 0, 1\} \times \{-1, 0, 1\}$ denote the Moore neighborhood about the origin, and $M' = M \setminus \{(0,0)\}$ denote the deleted Moore neighborhood.

Let f be the function that specifies the counterclockwise enumeration of M' as shown below. Note that $f \in \mathbb{Z}_8^{M'}$ and for each 8-neighbor $\mathbf{x}$ of a point (i, j), the point $\mathbf{x}'$ defined by $\mathbf{x}' = \mathbf{x} - (i, j)$ is a member of M'.

$$f = \begin{array}{|c|c|c|} \hline 3 & 2 & 1 \\ \hline 4 & & 0 \\ \hline 5 & 6 & 7 \\ \hline \end{array}$$

For each $k \in \{0, 1, \ldots, 7\}$, we define $\mathbf{t}(k)$ as follows:

$$\mathbf{t}(k)_{(i,j)}(\mathbf{x}) = \begin{cases} 5 & \text{if } \mathbf{x}' \in M' \text{ and } f(\mathbf{x}') + k \bmod 8 \leq 2 \\ -3 & \text{if } \mathbf{x}' \in M' \text{and } f(\mathbf{x}') + k \bmod 8 > 2 \\ 0 & \text{otherwise.} \end{cases}$$

Thus, $\mathbf{t}$ can be pictorially represented as follows:

$$\mathbf{t}(0) = \begin{array}{|c|c|c|} \hline -3 & 5 & 5 \\ \hline -3 & & 5 \\ \hline -3 & -3 & -3 \\ \hline \end{array} \quad \mathbf{t}(1) = \begin{array}{|c|c|c|} \hline -3 & -3 & 5 \\ \hline -3 & & 5 \\ \hline -3 & -3 & 5 \\ \hline \end{array} \quad \mathbf{t}(2) = \begin{array}{|c|c|c|} \hline -3 & -3 & -3 \\ \hline -3 & & 5 \\ \hline -3 & 5 & 5 \\ \hline \end{array} \quad \mathbf{t}(3) = \begin{array}{|c|c|c|} \hline -3 & -3 & -3 \\ \hline -3 & & -3 \\ \hline 5 & 5 & 5 \\ \hline \end{array}$$

$$\mathbf{t}(4) = \begin{array}{|c|c|c|} \hline -3 & -3 & -3 \\ \hline 5 & & -3 \\ \hline 5 & 5 & -3 \\ \hline \end{array} \quad \mathbf{t}(5) = \begin{array}{|c|c|c|} \hline 5 & -3 & -3 \\ \hline 5 & & -3 \\ \hline 5 & -3 & -3 \\ \hline \end{array} \quad \mathbf{t}(6) = \begin{array}{|c|c|c|} \hline 5 & 5 & -3 \\ \hline 5 & & -3 \\ \hline -3 & -3 & -3 \\ \hline \end{array} \quad \mathbf{t}(7) = \begin{array}{|c|c|c|} \hline 5 & 5 & 5 \\ \hline -3 & & -3 \\ \hline -3 & -3 & -3 \\ \hline \end{array}$$

For the source image $\mathbf{a} \in \mathbb{R}^X$ let $\mathbf{b}_k := |\mathbf{a} \oplus \mathbf{t}(k)|$. The Kirsch edge magnitude image $\mathbf{m}$ is given by

$$\mathbf{m} := 1 \bigvee \left(\bigvee_{k=0}^{7} \mathbf{b}_k \right).$$

The gradient direction image $\mathbf{d}$ is given by

$$\mathbf{d} := \left(\bigwedge_{k=0}^{7} (k+1) \cdot \chi_{\mathbf{b}_k}(\mathbf{m}) \right) - 1.$$

In actual implementation, the algorithm for computing the image $\mathbf{m}$ will probably involve the loop

$$\mathbf{m} := 1$$
$$\textbf{for } k \textbf{ in } 0..7 \textbf{ loop}$$
$$\mathbf{m} := \mathbf{m} \vee |\mathbf{a} \oplus \mathbf{t}(k)|$$
$$\textbf{end loop.}$$

A similar loop argument can be used to construct the direction image $\mathbf{d}$.

3.10.   Directional Edge Detection

The directional edge detector is an edge detection technique based on the use of directional derivatives [15, 16]. It identifies pixels as possible edge elements and assigns one of n directions to them. This is accomplished by convolving the image with a set of $\frac{n}{2}$ edge masks. Each mask has two directions associated with it. A pixel is assigned the direction associated with the largest magnitude obtained by the convolutions. Suppose that the convolution with the ith mask yielded the largest magnitude. If the result of the convolution is positive, then the direction θ_i is assigned; otherwise, the direction $\theta_i + 180$ (*modulo* 360) is assigned.

For a given source image $\mathbf{a} \in \mathbb{R}^\mathbf{X}$, let $\mathbf{a}_i = \mathbf{a} * \mathbf{m}_i$, where $\mathbf{m}_0$, $\mathbf{m}_1$, ... , $\mathbf{m}_{(n/2)-1}$ denote the edge masks and $*$ denotes convolution. The edge magnitude image $\mathbf{m} \in \mathbb{R}^\mathbf{X}$ is given by

$$\mathbf{m}(\mathbf{x}) = \max_{0 \le i \le \frac{n}{2} - 1} |\mathbf{a}_i(\mathbf{x})|.$$

Let θ_i be the direction associated with the mask $\mathbf{m}_i$. Suppose that $\mathbf{m}(\mathbf{x}) = |\mathbf{a}_k(\mathbf{x})|$. The direction image $\mathbf{d} \in \mathbb{R}^\mathbf{X}$ is given by

$$\mathbf{d}(\mathbf{x}) = (\theta_k + \chi_{<0}(\mathbf{a}_k(\mathbf{x})) \cdot 180) \bmod 360.$$

As an illustration, the directional edge technique is applied to the infrared image of a causeway across a bay as shown in Figure 3.10.1. The six masks of Figure 3.10.2 are used to determine one of twelve directions, $0°, 30°, 60°, \ldots,$ or $330°$, to be assigned to each point of the source image. The result of applying the directional edge detection algorithm is seen in Figure 3.10.3. In the figure, directions are presented as tangent vectors to the edge rather than gradient vectors (the tangent to the edge is orthogonal to the gradient). With this representation, the bridge and causeway can be characterized by two bands of vectors. The vectors in one band run along a line in one direction, while those in the other band run along a line in the opposite direction.

The edge image representation is different from the source image representation in that it is a plot of a vector field. The edge magnitude image was thresholded (see Section 4.2), and vectors were attached to the points that survived the thresholding process. The vector at each point has length proportional to the edge magnitude at the point, and points in the direction of the edge.

Figure 3.10.1. Source image of causeway with bridge.

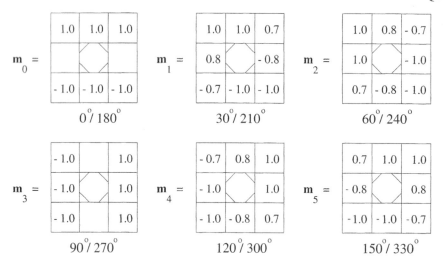

Figure 3.10.2. Edge mask with their associated directions.

Image Algebra Formulation

Let $a \in \mathbb{R}^{\mathbf{X}}$ be the source image. For $i \in \left\{0, 1, \cdots, \frac{n}{2} - 1\right\}$, let

$$\mathbf{a}_i := (0, \mathrm{i}) + f(\mathbf{a} \oplus \mathbf{m}_i),$$

where $f(r)$ is defined as

$$f(r) = \left(|r|, \frac{n}{2} \cdot \chi_{<0}(r)\right).$$

The result image $\mathbf{b}$ is given by

$$\mathbf{b} := (\vee |_1)_{i=0}^{\frac{n}{2}-1} \mathbf{a}_i.$$

Note that the result $\mathbf{b}$ contains at each point both the maximal edge magnitude and the direction associated with that magnitude. As formulated above, integers 0 through $n - 1$ to represent the n directions.

Comments and Observations

Trying to include the edge direction information in Figure 3.10.3 on one page has resulted in a rather "busy" display. However, edge direction information is often very useful. For example, the adjacent parallel edges in Figure 3.10.3 that run in opposite directions are characteristic of bridge-like structures.

The major problem with the display of Figure 3.10.3 is that, even after thresholding, thick (more than one pixel wide) edges remain. The thinning edge direction technique of Section 5.7 is one remedy to this problem. It uses edge magnitude and edge direction information to reduce edges to one pixel wide thickness.

One other comment on directional edge detection technique concerns the size of the edge masks. Selecting an appropriate size for the edge mask is important. The use of larger masks allows a larger range of edge directions. Also, larger masks are less sensitive to noise than smaller masks. The disadvantage of larger masks is that they are less sensitive to detail. The choice of mask size is application dependent.

Figure 3.10.3. Edge points with their associated directions.

3.11. Product of the Difference of Averages

If the neighborhoods over which differencing takes place are small, edge detectors based on discrete differencing will be sensitive to noise. However, the ability to precisely locate an edge suffers as the size of the neighborhood used for differencing increases in size. The *product of the differences of averages* technique [17] integrates both large and small differencing neighborhoods into its scheme to produce an edge detector that is insensitive to noise and that allows precise location of edges.

We first restrict our attention to the detection of horizontal edges. Let $\mathbf{a} \in \mathbb{R}^{\mathbf{X}}$, where $\mathbf{X} = \mathbb{Z}_m \times \mathbb{Z}_n$, be the source image. For $k \in K \subset \mathbb{Z}^{+}$ define the image $\mathbf{v}_k$ by

$$\mathbf{v}_k(x_1, x_2) = \frac{1}{k}|\mathbf{a}(x_1 + k - 1, x_2) + \mathbf{a}(x_1 + k - 2, x_2) + \cdots + \mathbf{a}(x_1, x_2)$$
$$-\mathbf{a}(x_1 - 1, x_2) - \mathbf{a}(x_1 - 2, x_2) - \cdots - \mathbf{a}(x_1 - k, x_2)|.$$

The product

$$\mathbf{v}(x_1, x_2) = \mathbf{v}_1(x_1, x_2) \cdot \mathbf{v}_2(x_1, x_2) \cdot \mathbf{v}_3(x_1, x_2) \cdots \mathbf{v}_p(x_1, x_2)$$

will be large only if each of its factors is large. The averaging that takes place over the large neighborhoods makes the factors with large k less likely to pick up false edges due to noise. The factors from large neighborhoods contribute to the product by cancelling out false edge readings that may be picked up by the factors with small k. As (x_1, x_2) moves farther from an edge point, the factors with small k get smaller. Thus, the factors in the product for small neighborhoods serve to pinpoint the edge point's location.

The above scheme can be extended to two dimensions by defining another image $\mathbf{h}$ to be

$$\mathbf{h}(x_1, x_2) = \mathbf{h}_1(x_1, x_2) \cdot \mathbf{h}_2(x_1, x_2) \cdot \mathbf{h}_3(x_1, x_2) \cdots \mathbf{h}_p(x_1, x_2),$$

where

$$\mathbf{h}_k(x_1, x_2) = \frac{1}{k}|\mathbf{a}(x_1, x_2 + k - 1) + \mathbf{a}(x_1, x_2 + k - 2) + \cdots + \mathbf{a}(x_1, x_2)$$
$$-\mathbf{a}(x_1, x_2 - 1) - \mathbf{a}(x_1, x_2 - 2) - \cdots - \mathbf{a}(x_1, x_2 - k)|.$$

The image $\mathbf{h}$ is sensitive to vertical edges. To produce an edge detector that is sensitive to both vertical and horizontal edges, one can simply take the maximum of $\mathbf{h}$ and $\mathbf{v}$ at each point in $\mathbf{X}$.

Figure 3.11.1 compares the results of applying discrete differencing and the product of the difference of averages techniques. The source image (labeled 90/10) in the top left corner of the figure is of a circular region whose pixel values have a 90% probability of being white against a background whose pixel values have a 10% probability of being white. The corresponding regions of the image in the upper right corner have probabilities of white pixels equal to 70% and 30%. The center images show the result of taking the maximum from discrete differencing along horizontal and vertical directions. The images at the bottom of the figure show the results of the product of the difference of average algorithm. Specifically, the figure's edge enhanced image e produced by using the product of the difference of averages is given by

$$\mathbf{e}(x_1, x_2) = max\{\mathbf{h}_1(x_1, x_2) \cdot \mathbf{h}_2(x_1, x_2) \cdot \mathbf{h}_4(x_1, x_2) \cdot \mathbf{h}_8(x_1, x_2) \cdot \mathbf{h}_{16}(x_1, x_2),$$
$$\mathbf{v}_1(x_1, x_2) \cdot \mathbf{v}_2(x_1, x_2) \cdot \mathbf{v}_4(x_1, x_2) \cdot \mathbf{v}_8(x_1, x_2) \cdot \mathbf{v}_{16}(x_1, x_2)\}.$$

Image Algebra Formulation

Let $\mathbf{a} \in \mathbb{R}^{\mathbf{X}}$ be the source image. Define the parameterized templates $\mathbf{h}(k)$ and $\mathbf{v}(k)$ to be

$$\mathbf{h}(k)_{\mathbf{y}}(\mathbf{x}) = \begin{cases} 1 & \text{if } \mathbf{x} = (y_1, y_2 + i) \text{ and } i = 0, 1, 2, \ldots, k - 1 \\ -1 & \text{if } \mathbf{x} = (y_1, y_2 - i) \text{ and } i = 1, 2, \ldots, k \\ 0 & \text{otherwise} \end{cases}$$

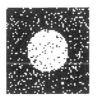

90/10 Source image

70/30 Source image

Discrete differencing
applied to 90/10 source
image

Discrete differencing
applied to 70/30 source
image

Product of the difference
of averages applied to
90/10 source image

Product of the difference
of averages applied to
70/30 source image

Figure 3.11.1. The product of the difference of averages
edge detection method vs. discrete differencing.

and

$$\mathbf{v}(k)_{\mathbf{y}}(\mathbf{x}) = \begin{cases} 1 & \text{if } \mathbf{x} = (y_1 + i, y_2) \text{ and } i = 0, 1, 2, \ldots, k - 1 \\ -1 & \text{if } \mathbf{x} = (y_1 - i, y_2) \text{ and } i = 1, 2, \ldots, k \\ 0 & \text{otherwise}, \end{cases}$$

where $(y_1, y_2) = \mathbf{y}$. The edge enhanced image e produced by the product of the difference of averages algorithm is given by the image algebra statement

$$\text{e} := \left(\prod_{k \in K} \frac{1}{k} \cdot |(\mathbf{a} \oplus \mathbf{h}(k))| \right) \bigvee \left(\prod_{k \in K} \frac{1}{k} \cdot |(\mathbf{a} \oplus \mathbf{v}(k))| \right).$$

The templates used in the formulation above are designed to be sensitive to vertical and horizontal edges.

3.12. Canny Edge Detection

The method of edge detection described by John Canny in 1986 [18] has been extremely influential. Numerous implementations of edge detectors based on Canny's ideas have been developed. Some written works give the misimpression that Canny developed a single *optimal* edge detector. In fact, Canny described a method of generating edge detectors using an optimization approach and showed how to use the technique to generate a robust detector for step edges. This section presents the method of Canny and discusses some issues associated with its implementation.

Canny's starting point is to note that the three most important performance criteria for an edge detector are that it should i) provide good *detection* with high probability of reporting an edge where there is an edge and low probability of report where there is no edge; ii) provide good *localization*, that is, the points marked as edge points should be close to the actual location of the edge; and iii) respond only once to a single edge (in a one-dimensional signal).

In looking at criteria i) and ii), Canny notes that an uncertainty principle relates them. Improving the detection of a filter—as measured by signal-to-noise ratio (SNR)—will reduce its localization. Conversely, improving the localization of a detector will hurt its detection probability. Canny shows how to develop a linear filter that will optimize the product of SNR and localization in response to a step edge in an image with Gaussian additive noise. The filter that optimizes this product is the edge itself (unsurprisingly). This filter does not, however, perform well with respect to the third criterion (which was to have only a single response to a single edge) in the presence of noise. Canny used constrained optimization to solve the problem of finding a detector for step edges that satisfies criteria i) and ii) under the constraint of yielding a single response. After solving the constrained optimization problem, Canny notes that the optimal filter can be approximated by the first derivative of the Gaussian with only a 20% reduction in performance on criteria i) and ii), and a 10% performance reduction in the single response measure. He further notes that the derivative of the Gaussian is identical to a one-dimensional Marr-Hildreth edge detector described in Section 3.14.

When finding edges in a two or higher dimensional image, one must employ some mechanism to find directional edges. Canny rejects the technique of sampling edges in orthogonal directions and using gradient to determine the correct orientation on the grounds that the uncertainty in the edge direction and the response of the filter will degrade the result. He notes that by using six different operator directions, the error in orientation of the best matching directional filter will be 15 degrees and that the operator output will fall to about 85% of its maximum value. Canny finds this magnitude of error acceptable.

After filtering edges, Canny applies a thresholding operation with hysteresis. This technique is employed to attempt to reduce the impact of noise in the edge image on the final result. Any point above a high threshold (a *primary point*) is marked in the result image, as is any group of points connected to a primary point, lying on the same contour, and having values above a lower threshold.

Image Algebra Formulation

Let $\mathbf{a} \in \mathbb{R}^{\mathbf{X}}$ be the source image whose edges are to be mapped.

Select separated Gaussian filtering templates $\mathbf{s}_1$ and $\mathbf{s}_2$ using an appropriate choice for σ. These will be used to generate a smoothed version of the source image.

Find a sequence of directional derivative templates $\mathbf{t}_0, \ldots \mathbf{t}_{n-1}$ using the method described in Section 11.5. (Canny chose $n = 6$.) These will be used to calculate edge magnitude.

Using a method similar to that for constructing directional derivative templates, construct $2n$ radial spoke templates $\mathbf{r}_0, \ldots, \mathbf{r}_{2n-1}$. Let

$$\mathbf{a}_l(\mathbf{x}_1, \mathbf{x}_2) = \begin{cases} 1 & if\ \mathbf{x}_1 = 0\ and\ \mathbf{x}_2 > 0 \\ 0 & otherwise \end{cases}$$

and let $\hat{f}_i$ be the rotation function for θ corresponding to direction i, as defined in Section 11.5. Templates $\mathbf{r}_i$ and $\mathbf{r}_{2i}$ are oriented normal to the edge responding with greatest intensity to template $\mathbf{t}_i$. Then let each template $\mathbf{r}_{i(0,0)} = \mathbf{a}'_l \circ \hat{f}$. These will be used to find the local maxima in the edge magnitude image.

Let $\mathbf{t}_c$ represent the cumulative histogram template defined in Section 10.11.

Using a method similar to that used to construct templates $\mathbf{t}_i$ and $\mathbf{r}_i$, construct n connector neighborhoods $N_0, \ldots, N_{n-1}$ each containing those points lying along the edge direction at the target point. These will be used to find low-threshold edge points connected to identified edge points.

The Canny edge detection algorithm is described in image algebra as follows:

$$\mathbf{b} := (\mathbf{a} \oplus \mathbf{s}_1) \oplus \mathbf{s}_2$$
$$\mathbf{e} := 0$$
$$\mathbf{d} := -1$$
for i in $0 .. n$ loop
$\quad \mathbf{e}_i := |\mathbf{b} \oplus \mathbf{t}_i|$
$\quad \mathbf{d} := \mathbf{d} \vee i \cdot \chi_{>\mathbf{e}}(\mathbf{e}_i)$
$\quad \mathbf{e} := \mathbf{e} \vee \mathbf{e}_i$
end loop
for i in $0 .. n - 1$ loop
$\quad \mathbf{c} := \mathbf{c} \vee \chi_{=i}\mathbf{d} \wedge \chi_{\geq(\mathbf{e} \oplus \mathbf{r}_i)}\mathbf{e} \wedge \chi_{\geq(\mathbf{e} \oplus \mathbf{r}_{i+n})}\mathbf{e}$
end loop
$\mathbf{h}_c = \Sigma \mathbf{t}_c(\mathbf{e})$
$\tau_{low} := \mathbf{h}'_c(0.4)$
$\tau_{high} := \mathbf{h}'_c(0.8)$
$\mathbf{c}_{low} := \mathbf{c} \wedge \chi_{\geq \mathbf{e}}\tau_{low}$
$\mathbf{c}_{new} := \mathbf{c} \wedge \chi_{\geq \mathbf{e}}\tau_{high}$
loop
$\quad \mathbf{c} := \mathbf{c}_{new}$
$\quad$ for i in $0 .. n - 1$ loop
$\quad\quad \mathbf{c}_{new} := \mathbf{c}_{new} \vee \mathbf{c}_{low} \wedge (\mathbf{c} \wedge \chi_{=i}\mathbf{d}) \bigvee N_i$
$\quad\quad$ exit when $\mathbf{c} = \mathbf{c}_{new}$
$\quad$ end loop
end loop.

Comments and Observations

Figures 3.12.1 through 3.12.4 show the results of applying the Canny edge detector with $\sigma = 1.5$ to the motorcycle image. Note the edge points added by hysteresis thresholding (Figure 3.12.3). Production of these extra edge points is a relatively time-consuming process but important in preserving contours.

Figure 3.12.1. Source Image (left) with Gaussian smoothing applied (right).

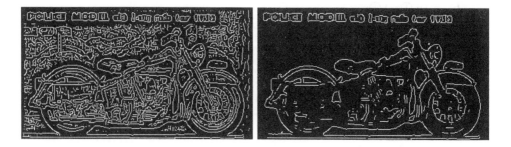

Figure 3.12.2. Edge maxima (left) and high threshold edge points (right).

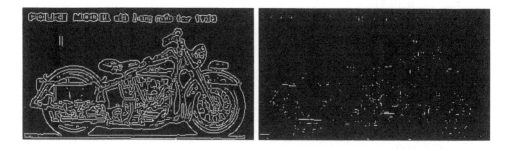

Figure 3.12.3. Low threshold points (left) and
points added by hysteresis thresholding (right).

Figure 3.12.4. Canny edge image.

Although Canny's detector calculates first derivative edge magnitudes, its attenuation of locally non-maximal edge points makes it essentially a second derivative technique.

Thus, it bears closer resemblance to the Marr-Hildreth detector (Section 3.14) than any of the first derivative techniques presented here. In comparing the final result to the Marr-Hildreth edge detector, we see that Canny's detector yields what seem to be the most intuitively important edges. But this is provided at the expense of using arbitrary threshold settings and a rather time-consuming hysteresis thresholding operation.

3.13.   Crack Edge Detection

Crack edge detection techniques comprise the family of algorithms that associate edge values with the points lying between neighboring image pixels [19]. More precisely, if one considers a 2-dimensional image with 4-neighbor connectivity, each spatial location lying exactly between two horizontally or vertically neighboring pixels corresponds to a crack. Each such point is assigned an edge value. The name *crack* edge stems from the view of pixels as square regions with cracks in between them.

Any of the edge detection methods presented in this chapter can be converted into a crack edge detector. Below, we consider the particular case of discrete differencing.

For a given source image $\mathbf{a} \in \mathbb{R}^{\mathbf{X}}$, *pixel edge* horizontal differencing will yield an image $\mathbf{b} \in \mathbb{R}^{\mathbf{X}}$ such that $\mathbf{b}(i,j) = \mathbf{a}(i,j) - \mathbf{a}(i,j+1)$ for all $(i,j) \in \mathbf{X}$. Thus, if there is a change in value from point (i,j) to point $(i,j+1)$, the difference in those values is associated with point (i,j).

The horizontal crack edge differencing technique, given $\mathbf{a} \in \mathbb{R}^{\mathbf{X}}$, will yield a result image $\mathbf{b} \in \mathbb{R}^{\mathbf{W}}$ where $\mathbf{W} = \left\{(i,r) : \left(i, r - \frac{1}{2}\right), \left(i, r + \frac{1}{2}\right) \in \mathbf{X}\right\}$, that is, $\mathbf{b}$'s point set corresponds to the cracks that lie between horizontally neighboring points in $\mathbf{X}$. The values of $\mathbf{b}$ are given by

$$\mathbf{b}(i,r) = \mathbf{a}(i, \lfloor r \rfloor) - \mathbf{a}(i, \lceil r \rceil).$$

One can compute the vertical differences as well, and store both horizontal and vertical differences in a single image with scalar edge values. (This cannot be done using pixel edges since each pixel is associated with both a horizontal and vertical edge.) That is, given $\mathbf{a} \in \mathbb{R}^{\mathbf{X}}$, we can compute $\mathbf{b} \in \mathbb{R}^{\mathbf{Y}}$ where

$$\mathbf{Y} = \left\{(y_1, y_2) : \left(y_1, y_2 - \frac{1}{2}\right), \left(y_1, y_2 + \frac{1}{2}\right), \left(y_1 - \frac{1}{2}, y_2\right), \text{ or } \left(y_1 + \frac{1}{2}, y_2\right) \in \mathbf{X}\right\}$$

and

$$\mathbf{b}(i,r) = \mathbf{a}(\lfloor i \rfloor, \lfloor r \rfloor) - \mathbf{a}(\lceil i \rceil, \lceil r \rceil).$$

Image Algebra Formulation

Given source image $\mathbf{a} \in \mathbb{R}^{\mathbf{X}}$, define spatial functions f_1, f_2, f_3, f_4 on $\mathbb{R}^2$ as follows:

$$f_1(x_1, x_2) = \left(x_1, x_2 + \frac{1}{2}\right)$$

$$f_2(x_1, x_2) = \left(x_1, x_2 - \frac{1}{2}\right)$$

$$f_1(x_1, x_2) = \left(x_1 + \frac{1}{2}, x_2\right)$$

$$f_1(x_1, x_2) = \left(x_1 - \frac{1}{2}, x_2\right).$$

Next construct the point set

$$\mathbf{Y} = (f_1(\mathbf{X}) \cap f_2(\mathbf{X})) \cup (f_3(\mathbf{X}) \cap f_4(\mathbf{X})),$$

and define $\mathbf{t} \in \left(\mathbb{R}^X\right)^Y$ by

$$\mathbf{t}_{(y_1, y_2)}(x_1, x_2) = \begin{cases} 1 & \text{if } (x_1, x_2) = (\lfloor y_1 \rfloor, \lfloor y_2 \rfloor) \\ -1 & \text{if } (x_1, x_2) = (\lceil y_1 \rceil, \lceil y_2 \rceil) \\ 0 & \text{otherwise.} \end{cases}$$

For a point with integral first coordinate, $\mathbf{t}$ can be pictured as follows:

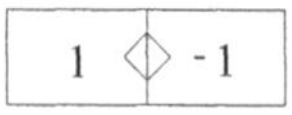

and for a point with integral second coordinate, $\mathbf{t}$ can be pictured as

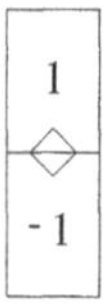

The crack edge image $\mathbf{b} \in \mathbb{R}^Y$ is given by

$$\mathbf{b} := \mathbf{a} \oplus \mathbf{t}.$$

Alternate Image Algebra Formulation

Some implementations of image algebra will not permit one to specify points with non-integral coordinates as are required above. In those cases, several different techniques can be used to represent the crack edges as images. One may want to map each point (y_1, y_2) in $\mathbf{Y}$ to the point $(2y_1, 2y_2)$. In such a case, the domain of the crack edge image does not cover a rectangular subset of $\mathbb{Z}^2$, all points involving two odd or two even coordinates are missing.

Another technique that can be used to transform the set of crack edges onto a less sparse subset of $\mathbb{Z}^2$ is to employ a spatial transformation that will rotate the crack edge points by angle $\pi/4$ and shift and scale appropriately to map them onto points with integral coordinates. The transformation $f : \mathbf{X} \to \mathbf{Z}$ defined by

$$f(y_1, y_2) = \left(y_1 - \frac{1}{2} + y_2, y_2 - y_1 + \frac{1}{2}\right) \quad \forall (y_1, y_2) \in \mathbf{Y}$$

$$f^{-1}(z_1, z_2) = \left(\frac{z_1 - z_2 + 1}{2}, \frac{z_1 + z_2}{2}\right) \quad \forall (z_1, z_2) \in \mathbf{Z}$$

is just such a function. The effect of applying f to a set of edge points is shown in Figure 3.13.1 below.

Horizontal Neighborhood

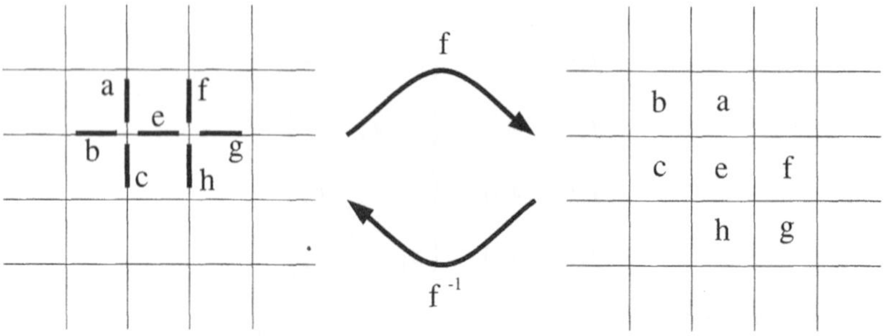

Figure 3.13.1. The effect of applying f to a set of edge points.

Comments and Observations

Crack edge finding can be used to find edges using a variety of underlying techniques. The primary benefit it provides is the ability to detect edges of features of single pixel width.

Representation of crack edges requires one to construct images over pointsets that are either sparse, rotated, or contain fractional coordinates. Such images may be difficult to represent efficiently.

3.14. Marr-Hildreth Edge Detection

The Marr-Hildreth edge detection approach [20] is motivated by biological studies of mammalian vision systems. Marr and Hildreth proposed a method that deals with an image separately at different resolutions, each resolution's edge map being generated by finding the zero crossings in a second derivative operator. The method for each resolution can be described succinctly as follows:

(a) Convolve the source image with a two-dimensional Gaussian.

(b) Calculate the Laplacian of the resulting image.

(c) Find the zero-crossings of that image.

Marr notes that the zero-crossings may be detected economically at each given scale by searching for the zero values in the resulting image, however, searching for zero values will not generally yield connected contours. We discuss several alternate and more robust methods by which zero-crossings can be found.

Image Algebra Formulation

Let $\mathbf{a} \in \mathbb{R}^X$ be the source image, and let $\mathbf{t}_\sigma \in \left(\mathbb{R}^X\right)^X$ be a Gaussian smoothing template with variance σ^2, constructed as described in Section 2.6. We can construct a template to calculate the Laplacian by adding together two orthogonal one-dimensional second derivative templates. A discrete, partial second derivative template can be constructed by calculating the first differences to the left and right of a point x, namely $\frac{f(x_1)-f(x_1-\Delta x)}{\Delta x}$ and $\frac{f(x_1)-f(x_1-\Delta x)}{\Delta x}$, then taking the first difference of these two, $\frac{f(x_1+\Delta x)-f(x_1)}{\Delta x} - \frac{f(x_1)-f(x_1-\Delta x)}{\Delta x} = \frac{f(x_1+\Delta x)-2f(x_1)+f(x_1-\Delta x)}{\Delta x}$. Letting $\Delta x = 1$ leads us to a template of the following form:

1	-2	1

The Laplacian is formed by adding the partial derivatives with respect to both x and y. Using the fact that $\mathbf{a} \oplus \mathbf{t}_1 + \mathbf{a} \oplus \mathbf{t}_2 = \mathbf{a} \oplus (\mathbf{t}_1 + \mathbf{t}_2)$, we note that the Laplacian of an image $\mathbf{a}$ is given by

$$\mathbf{b} = \mathbf{a} \oplus \mathbf{t}$$

where t has the form:

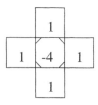

The finding of zero-crossings is a matter that is rarely discussed in any detail in connection with Laplacian of Gaussian techniques such as the Marr-Hildreth edge detector, however, it is not a trivial task. Huertas and Medioni [21] present a relatively complicated method of deriving zero-crossings accurate to the subpixel level. The marching cubes algorithm of Lorensen [22], which was developed to find surfaces of constant density in three-dimensional medical data, has been recast as the marching squares algorithm for application to two-dimensional images. Any technique that can be used to find isoelevation lines in digital elevation maps can be employed to find zero-crossings.

We present a morphological technique for identifying zero-crossings. While not as accurate as the other methods discussed, this technique is perfectly reasonable for generating an edge map for shape-based recognition tasks.

Note that to identify zero-crossings, we must distinguish the boundaries between regions of negative values and those of positive values. Let M be the Moore neighborhood. We can find the positive region image

$$\mathbf{c}_1 = \chi_{>0}\mathbf{b}$$

then calculate its outer boundary

$$\mathbf{d}_1 = \mathbf{c}_1 \oslash M \wedge \mathbf{c}_1^c.$$

We can also find the negative regions

$$\mathbf{c}_2 = \chi_{<0}\mathbf{b}$$

and calculate its inner boundary

$$\mathbf{d}_2 = \mathbf{c_2} \wedge \mathbf{c}_2^c \oslash M.$$

The conjunction of these two boundary images,

$$\mathbf{d} = \mathbf{d}_1 \vee \mathbf{d}_2$$

is an image that has a nonzero pixel neighboring every crossing from positive-to-negative, positive-to-zero, or negative-to-zero.

Comments and Observations

Figures 3.14.1 through 3.14.2 show the application of the Marr-Hildreth edge detector to the motorcycle image at several scales.

Figure 3.14.1. Laplacian of Gaussian with $\sigma = 2$ (left) and zero-crossings (right).

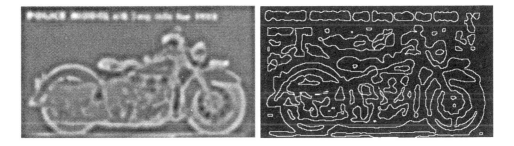

Figure 3.14.2. Laplacian of Gaussian with $\sigma = 4$ (left) and zero-crossings (right).

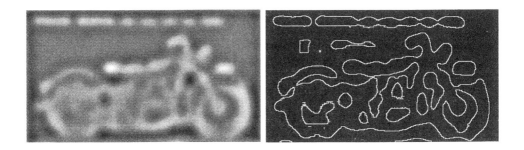

Figure 3.14.3. Laplacian of Gaussian with $\sigma = 8$ (left) and zero-crossings (right).

The complexity of the images constructed with smaller values of σ is striking. Like the local maximum image of the Canny detector, there are numerous edges that are not readily apparent to the human observer.

One can use a first-derivative edge detector to condition the results of the Marr-Hildreth algorithm, providing a more Canny-like edge image without the carrying out the hysteresis thresholding technique. If we perform Sobel edge detection (Section 3.6) on the Gaussian smoothed image and remove any zero-crossings with less than median edge strength then we get the results shown in Figure 3.14.4. This simple technique is significantly less complex than Canny's hysteresis thresholding yet yields similar, if not superior, qualitative results.

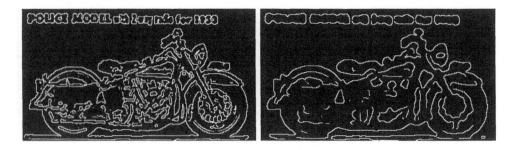

Figure 3.14.4. Zero-crossing images with low Sobel edge strength points removed for $\sigma = 2$ (left) and $\sigma = 4$ (right).

3.15. Local Edge Detection in Three-Dimensional Images

This technique detects surface elements in three-dimensional data. It consists of approximating the surface normal vector at a point and thresholding the magnitude of the normal vector. In three-dimensional data, boundaries of objects are surfaces and the image gradient is the local surface normal [11, 23, 24].

For $\mathbf{w} = (x, y, z)$, let $\|\mathbf{w}\| = \sqrt{(x^2 + y^2 + z^2)}$, and define $g_1(\mathbf{w}) = x/\|\mathbf{w}\|$, $g_2(\mathbf{w}) = y/\|\mathbf{w}\|$, and $g_3(\mathbf{w}) = z/\|\mathbf{w}\|$.

Suppose $\mathbf{a}$ denotes the source image and

$$M(\mathbf{w}_0) = \left\{ (x_0 \pm i, y_0 \pm j, z_0 \pm k) \ : \ 0 \le i, j, k \le \frac{m-1}{2} \right\}$$

denotes a three-dimensional m × m × m neighborhood about the point $\mathbf{w}_0 = (x_0, y_0, z_0)$. The components of the surface normal vector of the point $\mathbf{w}_0$ are given by

$$n_i = \sum_{\mathbf{w} \in M(\mathbf{w}_0)} \mathbf{a}(\mathbf{w}) \cdot g_i(\mathbf{w} - \mathbf{w}_0), \quad i = 1, 2, 3.$$

Thus, the surface normal vector of $\mathbf{w}_0$ is

$$\mathbf{n}(\mathbf{w}_0) = (n_1, n_2, n_3).$$

The edge image e for a given threshold T_0 is given by

$$e(\mathbf{w}) = \begin{cases} \|\mathbf{n}(\mathbf{w})\| & \text{if } \|\mathbf{n}(\mathbf{w})\| \ge T_0 \\ 0 & \text{otherwise.} \end{cases}$$

Image Algebra Formulation

Let $\mathbf{a} \in \mathbb{R}^{\mathbf{X}}$ be the source image. The edge image e is defined as

$$\mathbf{e} = \chi_{\ge T_0} \left(\sqrt{(\mathbf{a} \oplus \mathbf{g}_1)^2 + (\mathbf{a} \oplus \mathbf{g}_2)^2 + (\mathbf{a} \oplus \mathbf{g}_3)^2} \right),$$

where $\mathbf{g}_{i\,\mathbf{w}_0}(\mathbf{w}) = g_i(\mathbf{w} - \mathbf{w}_0)$ if $\mathbf{w} \in M(\mathbf{w}_0)$. For example, consider $M(\mathbf{w}_0)$ a $3 \times 3 \times 3$ domain. The figure below (Figure 3.15.1) shows the domain of the $\mathbf{g}_i$'s.

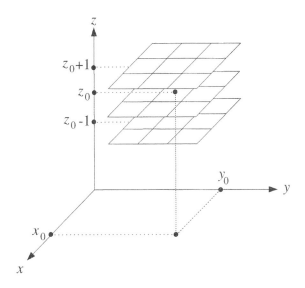

Figure 3.15.1. Illustration of a three-dimensional $3 \times 3 \times 3$ neighborhood.

Fixing z to have value z_0, we obtain

$$\mathbf{g}_{3\mathbf{w}_0}(x,y,z_0) = \begin{array}{|c|c|c|} \hline 0 & 0 & 0 \\ \hline 0 & 0 & 0 \\ \hline 0 & 0 & 0 \\ \hline \end{array}$$

Fixing z to have value $z_0 + 1$, we obtain

$$\mathbf{g}_{3\mathbf{w}_0}(x,y,z_0+1) = \begin{array}{|c|c|c|} \hline \frac{\sqrt{3}}{3} & \frac{\sqrt{2}}{2} & \frac{\sqrt{3}}{3} \\ \hline \frac{\sqrt{2}}{2} & 1 & \frac{\sqrt{2}}{2} \\ \hline \frac{\sqrt{3}}{3} & \frac{\sqrt{2}}{2} & \frac{\sqrt{3}}{3} \\ \hline \end{array}$$

Fixing z to have value $z_0 - 1$, we obtain

$$\mathbf{g}_{3\mathbf{w}_0}(x,y,z_0\text{-}1) = \begin{array}{|c|c|c|} \hline -\frac{\sqrt{3}}{3} & -\frac{\sqrt{2}}{2} & -\frac{\sqrt{3}}{3} \\ \hline -\frac{\sqrt{2}}{2} & -1 & -\frac{\sqrt{2}}{2} \\ \hline -\frac{\sqrt{3}}{3} & -\frac{\sqrt{2}}{2} & -\frac{\sqrt{3}}{3} \\ \hline \end{array}$$

Comments and Observations

The derivation uses a continuous model and the method minimizes the mean-square error with an ideal step edge.

3.16. Hierarchical Edge Detection

In this edge detection scheme, a pyramid structure is developed and used to detect edges. The idea is to consolidate neighborhoods of certain pixels into one pixel with a value obtained by measuring some local features, thus forming a new image. The new image will have lower resolution and fewer pixels. This procedure is iterated to obtain a *pyramid* of images. To detect an edge, one applies a local edge operator to an intermediate resolution image. If the magnitude of the response exceeds a given threshold at a pixel then one proceeds to the next higher resolution level and applies the edge operator to all the corresponding pixels in that level. The procedure is repeated until the source image resolution level is reached [25, 11].

For $k \in \{0, 1, \ldots, l-1\}$, let $\mathbf{a}_k \in \mathbb{R}^{\mathbf{X}_k}$ denote the images in a pyramid, where $\mathbf{X}_k = \{(i,j) : 0 \leq i, j \leq 2^k - 1\}$. The source image is $\mathbf{a}_{l-1} \in \mathbb{R}^{\mathbf{X}_{l-1}}$. Given $\mathbf{a}_{l-1}$, one can construct $\mathbf{a}_0, \ldots, \mathbf{a}_{l-2}$ in the following manner:

$$\mathbf{a}_k(i,j) = \frac{1}{4} \cdot [\mathbf{a}_{k+1}(2i, 2j) + \mathbf{a}_{k+1}(2i+1, 2j) + \mathbf{a}_{k+1}(2i, 2j+1) + \mathbf{a}_{k+1}(2i+1, 2j+1)].$$

The algorithm starts at some intermediate level, k, in the pyramid. This level is typically chosen to be $k = \lceil \frac{2l}{3} \rceil$.

The image $\mathbf{b}_k \in \mathbb{R}^{\mathbf{X}_k}$ containing boundary pixels at level k is formed by

$$\mathbf{b}_k(i,j) = |\mathbf{a}_k(i, j-1) - \mathbf{a}_k(i, j+1)| + |\mathbf{a}_k(i+1, j) - \mathbf{a}_k(i-1, j)|$$

for all $(i,j) \in \mathbf{X}_k$.

If, for some given threshold T, $\mathbf{b}_k(i,j) > T$, then apply the boundary pixel formation technique at points $(2i, 2j), (2i+1, 2j), (2i, 2j+1), (2i+1, 2j+1)$ in level $k+1$. All other points at level $k+1$ are assigned value 0. Boundary formation is repeated until level $l-1$ is reached.

Image Algebra Formulation

For any non-negative integer k, let $\mathbf{X}_k$ denote the set $\{(i,j) : 0 \leq i, j \leq 2^k - 1\}$. Given a source image $\mathbf{a}_{l-1} \in \mathbb{R}^{\mathbf{X}_{l-1}}$ construct a pyramid of images $\mathbf{a}_k \in \mathbb{R}^{\mathbf{X}_k}$ for $k \in \{0, 1, \ldots, l-2\}$ as follows:

$$\mathbf{a}_k := \frac{1}{4} \mathbf{a}_{k+1} \oplus \mathbf{s},$$

where

$$\mathbf{s}_{(i,j)}(\mathbf{x}) = \begin{cases} 1 & \text{if } \mathbf{x} \in \{(2i, 2j), (2i+1, 2j), (2i, 2j+1), (2i+1, 2j+1)\} \\ 0 & \text{otherwise.} \end{cases}$$

For $k = \lceil \frac{2l}{3} \rceil$, construct the boundary images $\mathbf{b}_k, \ldots, \mathbf{b}_{l-1}$ using the statement:

$$\mathbf{b}_k := abs(\mathbf{a}_k \oplus \mathbf{t}_1) + abs(\mathbf{a}_k \oplus \mathbf{t}_2),$$

where

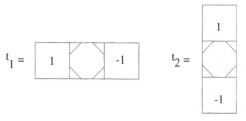

To construct the boundary images at level m where $k < m \leq l - 1$, first define

$$\mathbf{Y}_m := domain(\mathbf{b}_{m-1}\|_{>T}).$$

Next let $g : \mathbf{X}_{m-1} \rightarrow 2^{\mathbf{X}_m}$ be defined by

$$g(i, j) = \{(2i, 2j), (2i + 1, 2j), (2i, 2j + 1), (2i + 1, 2j + 1)\}$$

and $\breve{g} : 2^{\mathbf{X}_{m-1}} \rightarrow 2^{\mathbf{X}_m}$ be defined by

$$\breve{g}(\mathbf{W}) = \bigcup_{(i,j)\in\mathbf{W}} g(i, j), \text{ for } \mathbf{W} \subset \mathbf{X}_{j-1}.$$

The effect of this mapping is shown in Figure 3.16.1.

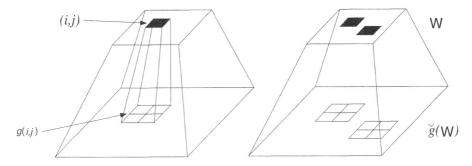

Figure 3.16.1. Illustration of the effect of the pyramidal map $\breve{g} : 2^{\mathbf{X}_{m-1}} \rightarrow 2^{\mathbf{X}_m}$.

The boundary image $\mathbf{b}_m$ is computed by

$$\mathbf{b}_m := \left[(abs(\mathbf{a}_m \oplus \mathbf{t}_1) + abs(\mathbf{a}_m \oplus \mathbf{t}_2))|_{\breve{g}(\mathbf{Y}_m)}\right]|^{\mathbf{0}},$$

where $\mathbf{0}$ denotes the zero image on the point set $\mathbf{X}_m$.

Alternate Image Algebra Formulation

The above image algebra formulation gives rise to a massively parallel process operating at each point in the pyramid. One can restrict the operator to only those points in levels $m > k$ where the boundary threshold is satisfied as described below.

Let $g_1(i, j) = \{(i, j), (i - 1, j), (i + 1, j), (i, j - 1), (i, j + 1)\}$. This maps a point (i, j) into its von Neumann neighborhood. Let $\breve{g}_1 : 2^{\mathbf{X}_m} \rightarrow 2^{\mathbf{X}_m}$ be defined by

$$\breve{g}_1(\mathbf{W}) = \bigcup_{(i,j)\in\mathbf{W}} g_1(i, j), \text{ for } \mathbf{W} \subset 2^{\mathbf{X}_m}.$$

Note that in contrast to the functions g and $\breve{g}$, g_1 and $\breve{g}_1$ are functions defined on the same pyramid level. We now compute $\mathbf{b}_m$ by

$$\mathbf{b}_m := \left[\left(abs\left(\mathbf{a}_m|_{\breve{g}_1(\breve{g}(\mathbf{Y}_m))} \oplus \mathbf{t}_1\right) + abs\left(\mathbf{a}_m|_{\breve{g}_1(\breve{g}(\mathbf{Y}_m))} \oplus \mathbf{t}_2\right)\right)|_{\breve{g}(\mathbf{Y}_m)}\right]|^{\mathbf{0}}.$$

Comments and Observations

This technique ignores edges that appear at high-resolution levels which do not appear at lower resolutions.

The method of obtaining intermediate-resolution images could easily blur, or wipe out edges at high resolutions.

3.17. Edge Detection Using K-Forms

The K-forms technique encodes the local intensity difference information of an image. The codes are numbers expressed either in ternary or decimal form. Local intensity differences are calculated and are labeled 0, 1, or 2 depending on the value of the difference. The K-form is a linear function of these labels, where k denotes the neighborhood size. The specific formulation, first described in Kaced [26], is as follows.

Let a_0 be a pixel value and let $a_1, \ldots, a_8$ be the pixel values of the 8-neighbors of a_0 represented pictorially as

a_4	a_3	a_2
a_5	a_0	a_1
a_6	a_7	a_8

Set

$$e_l = a_l - a_0$$

for all $l \in \{1, .., 8\}$ and for some positive threshold number T define $p : \mathbb{R} \to \{0, 1, 2\}$ as

$$p(r) = \begin{cases} 0 & \text{if } r < -T \\ 1 & \text{if } |r| \le T \\ 2 & \text{if } r > T. \end{cases}$$

Some common K-form neighborhoods are pictured in Figure 3.17.1. We use the notation

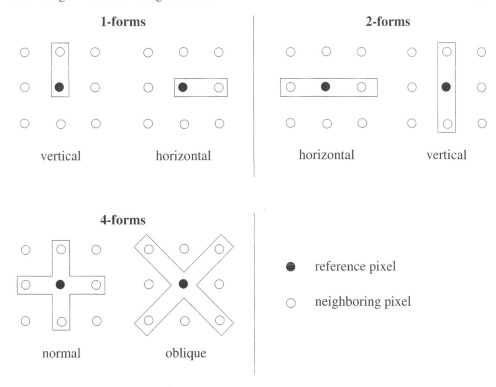

Figure 3.17.1. Common K-forms in a 3×3 window.

f_b^{nc} to denote the K-form with representation in base b, having a neighborhood containing n pixel neighbors of the reference pixel, and having orientation characterized by the roman letter c. Thus the horizontal decimal 2-form shown in Figure 3.17.1 is denoted f_{10}^{2h}, having base 10, 2 neighbors, and horizontal shape. Its value is given by

$$f_{10}^{2h}(a_0) = 3p(e_5) + p(e_1).$$

The value of f_{10}^{2h} is a single-digit decimal number encoding the topographical characteristics of the point in question. Figure 3.17.2 graphically represents each of the possible values for the horizontal 2-form showing a pixel and the orientation of the surface to left and right of that pixel. (The pixel is represented by a large dot, and the surface orientation on either side is shown by the orientation of the line segments to the left and right of the dot.)

horizontal 2-form	f_3	f_{10}	topographical analogy
	00	0	north-south ridge
	01	1	west plateau edge
	02	2	western slope
	10	3	east plateau edge
	11	4	plain or plateau
	12	5	foot of western slope
	20	6	eastern slope
	21	7	foot of eastern slope
	22	8	north-south valley

Figure 3.17.2. The nine horizontal 2-forms and their meaning.

The vertical decimal 2-form is defined to be

$$f_{10}^{2v}(a_0) = 3p(e_7) + p(e_3).$$

The ternary K-forms assign a ternary number to each topographical configuration. The ternary horizontal 2-form is given by

$$f_3^{2h}(a_0) = 10p(e_5) + p(e_1),$$

and the ternary normal 4-form is

$$f_3^4 = 1000p(e_7) + 100p(e_3) + 10p(e_5) + p(e_1) = 100f_3^{2v} + f_3^{2h}.$$

Note that the multiplicative constants are represented base 3, thus these expressions form the ternary number yielded by concatenating the results of application of p.

The values of other forms are calculated in a similar manner.

Three different ways of using the decimal 2-forms to construct a binary edge image are the following:

(a) An edge *in relief* is associated with any pixel whose K-form value is 0, 1, or 3.

(b) An edge *in depth* is associated with any pixel whose K-form value is 5, 7, or 8.

(c) An edge *by gradient threshold* is associated with a pixel whose K-form is not 0, 4, or 8.

Image Algebra Formulation

Let $\mathbf{a} \in \mathbb{R}^{X}$ be the source image. We can compute images $\mathbf{e}_l \in \mathbb{R}^{X}$ for $l \in \{1, ..., 8\}$, representing the fundamental edge images from which the K-forms are computed, as follows:

$$\mathbf{e}_l := \mathbf{a} \oplus \mathbf{t}(l),$$

where the $\mathbf{t}(l) \in \left(\mathbb{R}^{X}\right)^{X}$ are defined as shown in Figure 3.17.3.

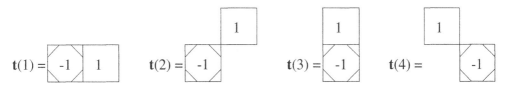

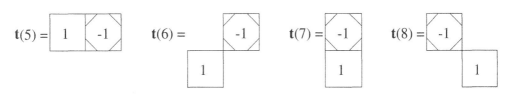

Figure 3.17.3. K-form fundamental edge templates.

Thus, for example, $\mathbf{t}(1)$ is defined by

$$\mathbf{t}(1)_{(x,y)}(i,j) = \begin{cases} -1 & \text{if } (x,y) = (i,j) \\ 1 & \text{if } (x,y) = (i,j+1) \\ 0 & \text{otherwise.} \end{cases}$$

Let the function p be defined as in the mathematical formulation above. The decimal horizontal 2-form is defined by

$$\mathbf{f}_{10}^{2h} = 3 \cdot p(\mathbf{e}_5) + p(\mathbf{e}_1).$$

Other K-forms are computed in a similar manner.

Edge *in relief* is computed as

$$\mathbf{r} := \chi_{\{0,1,3\}}\left(\mathbf{f}_{10}^{2h}\right).$$

Edge *in depth* is computed as

$$\mathbf{d} := \chi_{\{5,7,8\}}\left(\mathbf{f}_{10}^{2h}\right).$$

Edge *in threshold* is computed as

$$\mathbf{h} := \chi_{\{1,2,3,5,6,7\}}\left(\mathbf{f}_{10}^{2h}\right).$$

Comments and Observations

The K-forms technique captures the qualitative notion of topographical changes in an image surface.

The use of K-forms requires more computation than gradient formation to yield the same information as multiple forms must be computed to extract edge directions.

3.18. Hueckel Edge Operator

The Hueckel edge detection method is based on fitting image data to an ideal two-dimensional edge model [1, 27, 28]. In the one-dimensional case, the image **a** is fitted to a step function

$$\mathbf{s}(x) = \begin{cases} b & \text{if } x < x_0 \\ b+h & \text{if } x \geq x_0. \end{cases}$$

If the fit is sufficiently accurate at a given location, an edge is assumed to exist with the same parameters as the ideal edge model. (See Figure 3.18.1.) An edge is assumed if

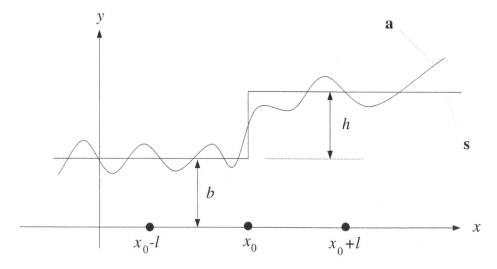

Figure 3.18.1. One-dimensional edge fitting.

the mean-square error

$$E = \int_{x_0-l}^{x_0+l} [\mathbf{a}(x) - \mathbf{s}(x)]^2 dx$$

is below some threshold value.

In the two-dimensional formulation the ideal step edge is defined as

$$\mathbf{s}(x,y) = \begin{cases} b & \text{if } x\cos\theta + y\sin\theta < \rho \\ b+h & \text{if } x\cos\theta + y\sin\theta \geq \rho, \end{cases}$$

where ρ represents the distance from the center of a test disk D of radius R to the ideal step edge $(\rho \leq R)$, and θ denotes the angle of the normal to the edge as shown in Figure 3.18.2.

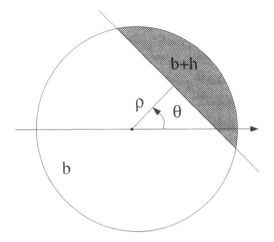

Figure 3.18.2. Two-dimensional edge fitting.

The edge fitting error is

$$E = \int \int_D [\mathbf{a}(x,y) - \mathbf{s}(x,y)]^2 dx dy. \tag{3.18.1}$$

In Hueckel's method, both image data and the ideal edge model are expressed as vectors in the vector space of continuous functions over the unit disk $D = \{(x,y) : x^2 + y^2 \leq 1\}$. A basis for this vector space — also known as a *Hilbert space* — is any complete sequence of orthonormalized continuous functions $\{h_i : i = 0, 1, \ldots\}$ with domain D. For such a basis, $\mathbf{a}$ and $\mathbf{s}$ have vector form $\mathbf{a} = (a_1, a_2, \ldots)$ and $\mathbf{s} = (s_1, s_2, \ldots)$, where

$$a_i = \int \int_D h_i(x,y)\mathbf{a}(x,y) dx dy, \quad i = 0, 1, \ldots$$

and

$$s_i = \int \int_D h_i(x,y)\mathbf{s}(x,y) dx dy, \quad i = 0, 1, \ldots \ .$$

For application purposes, only a finite number of basis functions can be used. Hueckel truncates the infinite Hilbert basis to only eight functions, $h_0, h_1, \ldots, h_7$. This provides for increased computational efficiency. Furthermore, his sub-basis was chosen so as to have a lowpass filtering effect for inherent noise smoothing. Although Hueckel provides an explicit formulation for his eight basis functions, their actual derivation has never been published!

Having expressed the signal $\mathbf{a}$ and edge model $\mathbf{s}$ in terms of Hilbert vectors, minimization of the mean-square error of Equation 3.18.1 can be shown to be equivalent to minimization of $\sum_{i=0}^{7} (a_i - s_i)^2$. Hueckel has performed this minimization by using some simplifying approximations. Also, although $\mathbf{a}$ is expressed in terms of vector components $a_0, \ldots, a_7$, $\mathbf{s}(x,y)$ is defined parametrically in terms of the parameters (b, h, ρ, θ). The exact discrete formulation is given below.

Definition of the eight basis functions. Let $D(x_0, y_0)$ be a disk with center (x_0, y_0) and radius R. For each $(x, y) \in D(x_0, y_0)$, define

$$r = \frac{1}{R}\sqrt{(x - x_0)^2 + (y - y_0)^2}$$
$$Q(r) = (1 - r^2)^{1/2},$$

and

$$H_0(x, y) = \frac{5}{6}\pi Q(r) \cdot (1 + 2r^2)$$

$$H_1(x, y) = \left(\frac{8}{27}\pi\right)^{1/2} \cdot Q(r)(5r^2 - 2)$$

$$H_2(x, y) = \left(\frac{3}{\pi}\right)^{1/2} \cdot Q(r)\frac{(x - x_0)}{R}(4r^2 - 1)$$

$$H_3(x, y) = \left(\frac{3}{\pi}\right)^{1/2} \cdot Q(r)\frac{(y - y_0)}{R}(4r^2 - 1)$$

$$H_4(x, y) = \left(\frac{3}{\pi}\right)^{1/2} \cdot Q(r)\frac{(x - x_0)}{R}(3 - 4r^2)$$

$$H_5(x, y) = \left(\frac{3}{\pi}\right)^{1/2} \cdot Q(r)\frac{(y - y_0)}{R}(3 - 4r^2)$$

$$H_6(x, y) = \left(\frac{32}{3\pi}\right)^{1/2} \cdot Q(r)\left[\left(\frac{(x - x_0)}{R}\right)^2 - \left(\frac{(y - y_0)}{R}\right)^2\right]$$

$$H_7(x, y) = \left(\frac{32}{3\pi}\right)^{1/2} \cdot Q(r) \cdot 2\frac{(x - x_0)}{R}\frac{(y - y_0)}{R}.$$

Then

$$h_0 = \left(\frac{24\pi}{25}\right)^{1/2} H_0$$

$$h_1 = \left(\frac{9}{2}\right)^{1/2} H_1$$

$$h_2 = H_2 + H_4$$

$$h_3 = H_3 + H_5$$

$$h_4 = \frac{3}{2} H_6$$

$$h_5 = \frac{3}{2} H_7$$

$$h_6 = \left(\frac{5}{4}\right)^{1/2} (H_2 - H_4)$$

$$h_7 = \left(\frac{5}{4}\right)^{1/2} (H_3 - H_5).$$

Note that when $\sqrt{(x - x_0)^2 + (y - y_0)^2} = R$, then $Q(r) = 0$. Thus, on the boundary of D each of the functions h_i is 0. In fact, the functions h_i intersect the disk D as shown in the following figure:

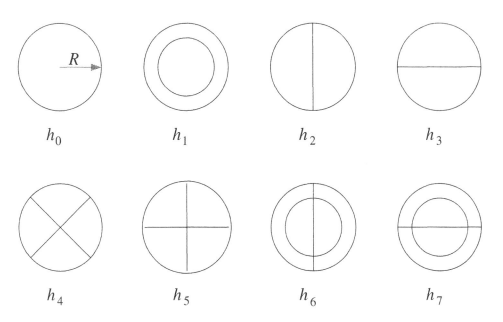

Figure 3.18.3. The intersection $D \cap \mathrm{graph}(h_i)$.

In his algorithm, however, Hueckel uses the functions H_i $(i = 0, 1, \cdots, 7)$ instead of h_i $(i = 0, 1, \cdots, 7)$ in order to increase computation efficiency. This is allowable since the H_i's are also linearly independent and span the same eight-dimensional subspace.

As a final remark, we note that the reason for normalizing r — i.e., $r = \frac{1}{R}\sqrt{(x - x_0)^2 + (y - y_0)^2}$ instead of $r = \sqrt{(x - x_0)^2 + (y - y_0)^2}$ forcing $0 \le r \le 1$ — is to scale the disk $D(x_0, y_0)$ back to the size of the unit disk, so that $\mathbf{a}$ restricted to $D(x_0, y_0)$ can be expressed as a vector in the Hilbert space of continuous functions on D.

Hueckel's algorithm. The algorithm proceeds as follows: Choose a digital disk of radius R (R is usually 4, 5, 6, 7, or 8, depending on the application) and a finite number of directions θ_j $(j = 1, \ldots, m)$ — usually $\theta = 0°, 45°, 90°, 135°$.

Step 1. For $i = 0, 1, \ldots, 7$, compute $a_i = \displaystyle\sum_{(x,y) \in D(x_0, y_0)} \mathbf{a}(x, y) H_i(x, y)$;

Step 2. For $j = 0, 1, \cdots, m$, compute $e_0(\theta_j) = a_2 cos\theta_j + a_3 sin\theta_j$;

Step 3. For $j = 0, 1, \cdots, m$, compute $e_1(\theta_j) = a_4 cos\theta_j + a_5 sin\theta_j$;

Step 4. For $j = 0, 1, \cdots, m$, compute $e_2(\theta_j) = a_1 + a_6\left(cos^2\theta_j - sin^2\theta_j\right) + a_7 2 sin\theta_j cos\theta_j$;

Step 5. For $j = 0, 1, \cdots, m$, compute $u(\theta_j) = \frac{e_0(\theta_j)}{|e_0(\theta_j)|}\left[e_1^2(\theta_j) + e_2^2(\theta_j)\right]^{1/2}$;

Step 6. For $j = 0, 1, \cdots, m$, compute $\Lambda(\theta_j) = e_0(\theta_j) + u(\theta_j)$;

Step 7. Find k such that $|\Lambda(\theta_k)| \ge |\Lambda(\theta_j)|$, $j = 0, 1, \cdots, m$;

Step 8. Calculate

$$\rho = e_2(\theta_k)/\left(\sqrt{2}[u(\theta_k) + e_1(\theta_k)]\right),$$

$$h = 4\Lambda(\theta_k)/\left[(3\pi)^{1/2}(1-\rho^2)^2(1+2\rho^2)\right],$$

$$b = a_0 - h[4 + \rho(3 + \rho(2 + \rho))](1-\rho)^2/8;$$

Step 9. Set

$$K = \left|\frac{\Lambda(\theta_k)}{[6a_1^2 + 2(a_2^2+a_3^2+a_4^2+a_5^2) + 3(a_6^2+a_7^2)]^{1/2}}\right|$$

and define

$$\mathbf{e}(x_0, y_0) = \begin{cases} (b, h, \rho, \theta_k) & \text{if } K > T \\ 0 & \text{otherwise.} \end{cases}$$

Remarks.   The edge image e has either pixel value zero or the parameter (b, h, ρ, θ_k). Thus, with each edge pixel we obtain parameters describing the nature of the edge. These parameters could be useful in further analysis of the edges. If only the existence of an edge pixel is desired, then the algorithm can be shortened by dropping STEP 8 and defining

$$\mathbf{e}(x_0, y_0) = \begin{cases} 1 & \text{if } K > T \\ 0 & \text{otherwise.} \end{cases}$$

Note that in this case we also need only compute $a_1, \dots, a_7$ since a_0 is only used in the calculation of **b**.

It can be shown that the constant K in STEP 9 is the cosine of the angle between the vectors $\mathbf{a} = (a_0, a_1, \dots, a_7)$ and $\mathbf{s} = (s_0, s_1, \dots, s_7)$. Thus, if the two vectors match exactly, then the angle is zero and $K = 1$; i.e., **a** is already an ideal edge. If $K < 0.9$, then the fit is not a good fit in the mean-square sense. For this reason the threshold T is usually chosen in the range $0.9 \leq T < 1$. In addition to thresholding with T, a further test is often made to determine if the edge constant factor h is greater than a threshold factor.

As a final observation we note that K would have remained the same whether the basis H_i or h_i had been used to compute the components a_i.

Image Algebra Formulation

In the algorithm, the user specifies the radius R of the disk over which the ideal step edge is to be fitted to the image data, a finite number of directions $\theta_0, \theta_1, \cdots, \theta_m$, and a threshold T.

Let $\mathbf{a} \in \mathbb{R}^X$ denote the source image. Define a parameterized template $\mathbf{t}(i) :$ $\mathbb{Z}^2 \to \mathbb{R}^{\mathbb{Z}^2}$, where $i = 0, 1, \dots, 7$, by

$$\mathbf{t}(i)_{\mathbf{y}}(\mathbf{x}) = \begin{cases} H_i(\mathbf{x}) & \text{if } \|\mathbf{x} - \mathbf{y}\| \leq R \\ 0 & \text{otherwise.} \end{cases}$$

Here $\mathbf{y} = (y_1, y_2)$ corresponds to the center of the disk $D(\mathbf{y})$ of radius R and $\mathbf{x} = (x_1, x_2)$. For j in $\{0, 1, \dots, m\}$, define $f_j : \mathbb{R} \to \mathbb{R}^3$ by $f_j(r) = (r, |r|, \theta_j)$.

The algorithm now proceeds as follows:

> for i in $0..7$ loop
>
> $\quad \mathbf{a}_i := \mathbf{a} \oplus \mathbf{t}(i)$
>
> end loop
>
> for j in $0..m$ loop
>
> $\quad \mathbf{e}_{0,j} := \mathbf{a}_2 \cdot \cos\theta_j + \mathbf{a}_3 \cdot \sin\theta_j$
>
> $\quad \mathbf{e}_{1,j} := \mathbf{a}_4 \cdot \cos\theta_j + \mathbf{a}_5 \cdot \sin\theta_j$
>
> $\quad \mathbf{e}_{2,j} := \mathbf{a}_1 + \mathbf{a}_6 \cdot \left(\cos^2\theta_j - \sin^2\theta_j\right) + \mathbf{a}_7 \cdot 2 \cdot \sin\theta_j \cdot \cos\theta_j$
>
> $\quad \mathbf{u}_j := \dfrac{\mathbf{e}_{0,j}}{|\mathbf{e}_{0,j}|} \cdot \left(\mathbf{e}_{1,j}^2 + \mathbf{e}_{2,j}^2\right)^{\frac{1}{2}}$
>
> $\quad \lambda_j := \mathbf{e}_{0,j} + \mathbf{u}_j$
>
> end loop
>
> $\lambda := \left(\vee|_2\right)_{j=0}^{m} f_j(\lambda_j)$
>
> $\mathbf{c} := \cos(p_3(\lambda))$
>
> $\mathbf{s} := \sin(p_3(\lambda))$
>
> $\mathbf{e}_0 := \mathbf{a}_2 \cdot \mathbf{c} + \mathbf{a}_3 \cdot \mathbf{s}$
>
> $\mathbf{e}_1 := \mathbf{a}_4 \cdot \mathbf{c} + \mathbf{a}_5 \cdot \mathbf{s}$
>
> $\mathbf{e}_2 := \mathbf{a}_1 + \mathbf{a}_6 \cdot \left(\mathbf{c}^2 - \mathbf{s}^2\right) + \mathbf{a}_7 \cdot 2 \cdot \mathbf{s} \cdot \mathbf{c}$
>
> $\mathbf{u} := \dfrac{\mathbf{e}_0}{|\mathbf{e}_0|} \cdot \left(\mathbf{e}_1^2 + \mathbf{e}_2^2\right)^{\frac{1}{2}}$
>
> $\mathbf{r} := \mathbf{e}_2 / \left(\sqrt{2}(\mathbf{u} + \mathbf{e}_1)\right)$
>
> $\mathbf{h} := 4 \cdot p_1(\lambda) / \left[\sqrt{3 \cdot \pi} \cdot \left(1 - \mathbf{r}^2\right)^2 \left(1 + 2 \cdot \mathbf{r}^2\right)\right]$
>
> $\mathbf{b} := \mathbf{a}_0 - \mathbf{h} \cdot [4 + \mathbf{r} \cdot (3 + \mathbf{r} \cdot (2 + \mathbf{r}))] \cdot (1 - \mathbf{r})^2 / 8$
>
> $\mathbf{k} := p_2(\lambda) / \left[6 \cdot \mathbf{a}_1^2 + 2 \cdot \left(\mathbf{a}_2^2 + \mathbf{a}_3^2 + \mathbf{a}_4^2 + \mathbf{a}_5^2\right) + 3 \cdot \left(\mathbf{a}_6^2 + \mathbf{a}_7^2\right)\right]^{\frac{1}{2}}$
>
> $\mathbf{e} := \chi_{>T}(\mathbf{k}) \cdot (\mathbf{b}, \mathbf{h}, \mathbf{r}, p_3(\lambda))$.

Remarks. Note that e is a vector valued image of four values. If we only seek a Boolean edge image, then the formulation can be greatly simplified. The first two loops remain the same and after the second loop the algorithm is modified as follows:

$$\lambda := \bigvee_{j=0}^{m} |\lambda_j|$$

$$\mathbf{k} := \lambda / \left[6 \cdot \mathbf{a}_1^2 + 2 \cdot \left(\mathbf{a}_2^2 + \mathbf{a}_3^2 + \mathbf{a}_4^2 + \mathbf{a}_5^2\right) + 3 \cdot \left(\mathbf{a}_6^2 + \mathbf{a}_7^2\right)\right]^{\frac{1}{2}}$$

$$\mathbf{e} := \chi_{>T}(\mathbf{k}).$$

Furthermore, one convolution can be saved as $\mathbf{a}_0$ need not be computed.

Comments and Observations

Experimental results indicate that the Hueckel operator performs well in noisy and highly textured environments. The operator not only provides edge strength, but also information as to the height of the step edge, orientation, and distance from the edge pixel.

The Hueckel operator is highly complex and computationally very intensive. The complexity renders it difficult to analyze the results theoretically.

3.19. Divide-and-Conquer Boundary Detection

Divide-and-conquer boundary detection is used to find a boundary between two known edge points. If two edge points are known to be on a boundary, then one searches along the perpendiculars of the line segment joining the two points, looking for an edge point. The method is recursively applied to the resulting 2 pairs of points. The result is an ordered list of points approximating the boundary between the two given points.

The following method was described in [11]. Suppose $a \in \{0,1\}^X$ is an edge image defined on a point set X, p and q denote two edge points, and $L(p, q)$ denotes the line segment joining them. Given $D > 0$, define a rectangular point set S of size $2D \times \|p - q\|$ as shown in Figure 3.19.1.

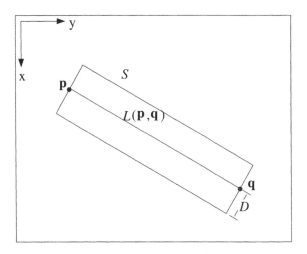

Figure 3.19.1. Divide-and-conquer region of search.

Let Y be the set of all the edge points in the rectangular point set S. Next, choose point r from Y such that the distance from r to $L(p, q)$ is maximal. Apply the above procedure to the pairs (p, r) and (r, q).

Image Algebra Formulation

Let $a \in \{0,1\}^X$ be an edge image, and $p = (p_1, p_2)$, $q = (q_1, q_2) \in X$ be the two edge points of a. Let $L(x, y)$ denote the line segment connecting points x and y, and let $d(x, L)$ denote the distance between point x and line L.

Find a rectangular point set S with width $2D$ about the line segment connecting p and q. Assume without loss of generality that $p_1 \leq q_1$. Let $\theta = sin^{-1}\left(\frac{q_2 - p_2}{\|q - p\|}\right)$. Let points s, t, u, v be defined as follows:

$$s = p + D \cdot \left(cos\left(\theta + \frac{\pi}{2}\right), sin\left(\theta + \frac{\pi}{2}\right)\right)$$

$$t = p + D \cdot \left(cos\left(\theta + \frac{3\pi}{2}\right), sin\left(\theta + \frac{3\pi}{2}\right)\right)$$

$$u = q + D \cdot \left(cos\left(\theta + \frac{\pi}{2}\right), sin\left(\theta + \frac{\pi}{2}\right)\right)$$

$$v = q + D \cdot \left(cos\left(\theta + \frac{3\pi}{2}\right), sin\left(\theta + \frac{3\pi}{2}\right)\right).$$

Figure 3.19.2 shows the relationships of $\mathbf{s}, \mathbf{t}, \mathbf{u}, \mathbf{v}$, and S to $\mathbf{p}$ and $\mathbf{q}$.

Let $f(j, \mathbf{x}, \mathbf{y})$, where $\mathbf{x} = (x_1, x_2)$ and $\mathbf{y} = (y_1, y_2)$, denote the first coordinate of the closest point to $\mathbf{x}$ on $L(\mathbf{x}, \mathbf{y})$ having second coordinate j, that is,

$$f(j, \mathbf{x}, \mathbf{y}) = \begin{cases} \frac{j(x_1 - y_1) + y_1(x_2 - y_2) - y_2(x_1 - y_1)}{x_2 - y_2} & \text{if } x_2 \neq y_2 \\ x_1 & \text{if } x_2 = y_2. \end{cases}$$

Note that f is partial in that it is defined only if there is a point on $L(\mathbf{x}, \mathbf{y})$ with second coordinate j. Furthermore, let $F(j, \mathbf{w}, \mathbf{x}, \mathbf{y}, \mathbf{z})$ denote the set $\{(i, j) : i \in \mathbb{Z}, f(j, \mathbf{w}, \mathbf{x}) \leq i \leq f(j, \mathbf{z}, \mathbf{y})\}$; that is, the set of points with second coordinate j bounded by $L(\mathbf{w}, \mathbf{x})$ and $L(\mathbf{y}, \mathbf{z})$.

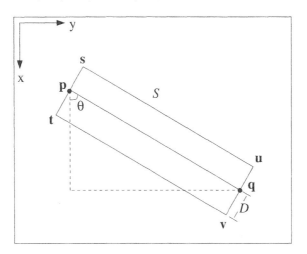

Figure 3.19.2. Variables characterizing the region of search.

We can then compute the set S containing all integral points in the rectangle described by corner points $\mathbf{s}, \mathbf{t}, \mathbf{u}$, and $\mathbf{v}$ as follows:

$$S := \varnothing;$$

if $s_2 \leq u_2$ then

    for j in $\lfloor t_2 \rfloor .. \lceil u_2 \rceil$ loop

        if $j > s_2$ then

            if $j > v_2$ then

                $S := S \cup F(j, \mathbf{u}, \mathbf{s}, \mathbf{u}, \mathbf{v});$

            else

                $S := S \cup F(j, \mathbf{u}, \mathbf{s}, \mathbf{v}, \mathbf{t});$

            end if;

        else $--j \leq s_2$

            if $j > v_2$ then

                $S := S \cup F(j, \mathbf{s}, \mathbf{t}, \mathbf{u}, \mathbf{v});$

            else

                $S := S \cup F(j, \mathbf{s}, \mathbf{t}, \mathbf{t}, \mathbf{v});$

            end if;

        end if;

    end loop;

else $--s_2 > u_2$

    for j in $\lfloor v_2 \rfloor .. \lceil s_2 \rceil$ loop

        if $j > t_2$ then

            if $j > u_2$ then
                $S \;:=\; S \cup F(j, \mathbf{s}, \mathbf{t}, \mathbf{s}, \mathbf{u});$
            else
                $S \;:=\; S \cup F(j, \mathbf{s}, \mathbf{t}, \mathbf{u}, \mathbf{v});$
            end if;

        else $--j \leq t_2$

            if $j > u_2$ then
                $S \;:=\; S \cup F(j, \mathbf{t}, \mathbf{v}, \mathbf{s}, \mathbf{u});$
            else
                $S \;:=\; S \cup F(j, \mathbf{t}, \mathbf{v}, \mathbf{v}, \mathbf{u});$
            end if;

        end if;

    end loop;

  end if.

Let $\mathbf{Y} = domain((\mathbf{a}|\mathbf{s})\|_{=1})$. If $card(\mathbf{Y}) = 0$, the line segment joining $\mathbf{p}$ and $\mathbf{q}$ defines the boundary between $\mathbf{p}$ and $\mathbf{q}$, and our work is finished.

If $card(\mathbf{Y}) > 0$, define image $\mathbf{b} \in \mathbb{R}^{\mathbf{Y}}$ by $\mathbf{b}(\mathbf{x}) = d(\mathbf{x}, L(\mathbf{p}, \mathbf{q}))$. Let $\mathbf{r} = choice(domain(\mathbf{b}\|_{=\vee \mathbf{b}}))$. Thus, as shown in Figure 3.19.3, $\mathbf{r}$ will be an edge point in set S farthest from line segment $L(\mathbf{p}, \mathbf{q})$. The algorithm can be repeated with points $\mathbf{p}$ and $\mathbf{r}$, and with points $\mathbf{r}$ and $\mathbf{q}$.

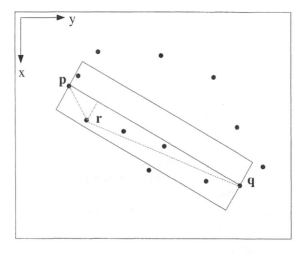

Figure 3.19.3. Choice of edge point $\mathbf{r}$.

Comments and Observations

This technique is useful in the case that a low curvature boundary is known to exist between edge elements and the noise levels in the image are low. It could be used for filling in gaps left by an edge detecting-thresholding-thinning operation.

Difficulties arise in the use of this technique when there is more than one candidate point for the new edge point, or when these points do not lie on the boundary of the object being segmented. In our presentation of the algorithm, we make a non-deterministic choice of the new edge point if there are several candidates. We assume that the edge point selection method — which is used as a preprocessing step to boundary detection — is robust and accurate, and that all edge points fall on the boundary of the object of interest. In practice, this does not usually occur, thus the sequence of boundary points detected by this technique will often be inaccurate.

3.20. Edge Following as Dynamic Programming

The purpose of this particular technique is to extract *one* high-intensity line or curve of *fixed* length and locally low-curvature from an image using dynamic programming. Dynamic programming in this particular case involves the definition and evaluation of a *figure-of-merit* (FOM) function that embodies a notion of "best curvature." The evaluation of the FOM function is accomplished by using the multistage optimization process described below (see also Ballard and Brown [11]).

Let $\mathbf{X} \subset \mathbb{Z}^2$ be a finite point set and suppose g is a real-valued function of n discrete variables with each variable a point of $\mathbf{X}$. That is,

$$g : \mathbf{X}^n \to \mathbb{R}$$
$$(\mathbf{x}_1, \ldots, \mathbf{x}_n) \mapsto g(\mathbf{x}_1, \ldots, \mathbf{x}_n).$$

The objective of the optimization process is to find an n-tuple $(\bar{\mathbf{x}}_1, \ldots, \bar{\mathbf{x}}_n) \in \mathbf{X}^n$ such that

$$g(\bar{\mathbf{x}}_1, \ldots, \bar{\mathbf{x}}_n) = max\{g(\mathbf{x}_1, \ldots, \mathbf{x}_n) : (\mathbf{x}_1, \ldots, \mathbf{x}_n) \in \mathbf{X}^n\}.$$

It can be shown that if g can be expressed as a sum of functions

$$g(\mathbf{x}_1, \ldots, \mathbf{x}_n) = g_1(\mathbf{x}_1, \mathbf{x}_2) + g_2(\mathbf{x}_2, \mathbf{x}_3) + \cdots + g_{n-1}(\mathbf{x}_{n-1}, \mathbf{x}_n),$$

then a multistage optimization procedure using the following recursion process can be applied:

1. Define a sequence of functions $f_k : \mathbf{X} \to \mathbb{R}$ recursively by setting

$$f_1(\mathbf{x}) = 0 \ \forall \mathbf{x} \in \mathbf{X}$$
$$f_{k+1}(\mathbf{y}) = max\{g_k(\mathbf{x}, \mathbf{y}) + f_k(\mathbf{y}) : \mathbf{x} \in \mathbf{X}\} \ \forall \mathbf{y} \in \mathbf{X},$$

for $k = 1, \ldots, n - 1$.

2. For each $\mathbf{y} \in \mathbf{X}$, determine the point $m_{k+1}(\mathbf{y}) = \mathbf{x}_k \in \mathbf{X}$ such that

$$f_{k+1}(\mathbf{y}) = g_k(\mathbf{x}_k, \mathbf{y}) + f_k(\mathbf{x}_k);$$

that is, find the point $\mathbf{x}_k$ for which the maximum of the function

$$\psi_{\mathbf{y},k}(\mathbf{x}) = g_k(\mathbf{x},\mathbf{y}) + f_k(\mathbf{x})$$

is achieved.

The n-tuple $(\bar{\mathbf{x}}_1, \ldots, \bar{\mathbf{x}}_n)$ can now be computed by using the recursion formula:

3. Find $\bar{\mathbf{x}}_n \in \mathbf{X}$ such that $f_n(\bar{\mathbf{x}}_n) = max\{f_n(\mathbf{y}) : \mathbf{y} \in \mathbf{X}\}$.

4. Set $\bar{\mathbf{x}}_k = m_{k+1}(\bar{\mathbf{x}}_{k+1})$ for $k = n-1, \ldots, 1$.

The input to this algorithm is an edge enhanced image with each pixel location $\mathbf{x}$ having edge magnitude $e(\mathbf{x}) \geq 0$ and edge direction $d(\mathbf{x}) \in \mathbb{Z}_8$. Thus, only the eight octagonal edge directions $0, 1, \ldots, 7$ are allowed (see Figure 3.20.2).

In addition, an *a priori* determined integer n, which represents the (desired) length of the curve, must be supplied by the user as input.

For two pixel locations $\mathbf{x}, \mathbf{y}$ of the image, the *directional difference* $q(\mathbf{x},\mathbf{y})$ is defined by

$$q(\mathbf{x},\mathbf{y}) = \begin{cases} |d(\mathbf{x}) - d(\mathbf{y})| & \text{if } |d(\mathbf{x}) - d(\mathbf{y})| \neq 7 \\ 1 & \text{otherwise.} \end{cases}$$

The directional difference q is a parameter of the FOM function g which is defined as

$$g(\mathbf{x}_1, \ldots, \mathbf{x}_n) = \sum_{k=1}^{n} e(\mathbf{x}_k) - \sum_{k=1}^{n-1} q(\mathbf{x}_k, \mathbf{x}_{k+1})$$

subject to the constraints

(i) $e(\mathbf{x}_k) > 0$ (or $e(\mathbf{x}_k) > k_0$ for some threshold k_0),

(ii) $\|\mathbf{x}_k - \mathbf{x}_{k+1}\| \leq \sqrt{2}$, and

(iii) $q(\mathbf{x}_k, \mathbf{x}_{k+1}) \leq 1$.

The solution $(\bar{\mathbf{x}}_1, \ldots, \bar{\mathbf{x}}_n)$ obtained through the multistage optimization process will be the optimal curve with respect to the FOM function g.

Note that constraint (i) restricts the search to actual edge pixels. Constraint (ii) further restricts the search by requiring the point $\mathbf{x}_k$ to be an 8-neighbor of the point $\mathbf{x}_{k+1}$. Constraint (iii), the low-curvature constraint, narrows the search for adjacent points even further by restricting the search to three points for each point on the curve. For example, given $\mathbf{x}_{k+1}$ with $d(\mathbf{x}_{k+1}) = 0$, then the point $\mathbf{x}_k$ preceding $\mathbf{x}_{k+1}$ can only have directional values $0, 1$, or 7. Thus, $\mathbf{x}_k$ can only be one of the following 8-neighbors of $\mathbf{x}_{k+1}$ shown in Figure 3.20.1.

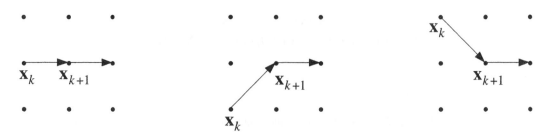

Figure 3.20.1. The three possible predecessors of $\mathbf{x}_{k+1}$.

In this example, the octagonal directions are assumed to be oriented as shown in Figure 3.20.2.

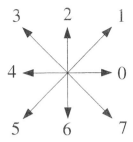

Figure 3.20.2. The eight possible directions $d(\mathbf{x})$.

In the multistage optimization process discussed earlier, the FOM function g had no constraints. In order to incorporate the constraints (i), (ii), and (iii) into the process, we define two step functions s_1 and s_2 by

$$s_1(r) = \begin{cases} \infty & \text{if } r \le 0 \\ 0 & \text{if } r > 0 \end{cases}$$

and

$$s_2(r) = \begin{cases} \infty & \text{if } r > 1 \\ 0 & \text{if } r \le 1. \end{cases}$$

Redefining g by

$$g(\mathbf{x}_1, \ldots, \mathbf{x}_n) = g_1(\mathbf{x}_1, \mathbf{x}_2) + g_2(\mathbf{x}_2, \mathbf{x}_3) + \cdots + g_{n-1}(\mathbf{x}_{n-1}, \mathbf{x}_n),$$

where

$$g_k(\mathbf{x}_k, \mathbf{x}_{k+1}) = [e(\mathbf{x}_k) - s_1(e(\mathbf{x}_k))] - [q(\mathbf{x}_k, \mathbf{x}_{k+1}) + s_2(q(\mathbf{x}_k, \mathbf{x}_{k+1}))]$$

for $k = 1, \ldots, n-1$, results in a function which has constraints (i), (ii), and (iii) incorporated into its definition and has the required format for the optimization process. The algorithm reduces now to applying steps 1 through 4 of the optimization process to the function g. The output $(\bar{\mathbf{x}}_1, \ldots, \bar{\mathbf{x}}_n)$ of this process represents the optimal curve of length n.

Image Algebra Formulation

The input consists of an edge magnitude image e, the corresponding direction image d, and a positive integer n representing the desired curve length. The edge magnitude/direction image (e, d) could be provided by the directional edge detection algorithm of Section 3.10.

Suppose $(\mathbf{e}, \mathbf{d}) \in (\mathbb{R}, \mathbb{Z}_8)^{\mathbf{X}}$. Define a parameterized template $\mathbf{t} : (\mathbb{R}, \mathbb{Z}_8)^{\mathbf{X}} \to (\mathbb{R}^{\mathbf{X}})^{\mathbf{X}}$ by

$$\mathbf{t}(\mathbf{e}, \mathbf{d})_{\mathbf{y}}(\mathbf{x}) = \begin{cases} e(\mathbf{x}) - q(\mathbf{x}, \mathbf{y}) & \text{if } \|\mathbf{x} - \mathbf{y}\| \le \sqrt{2}, e(\mathbf{x}) > 0, \text{ and } q(\mathbf{x}, \mathbf{y}) \le 1 \\ -\infty & \text{otherwise} \end{cases}$$

and a function $h : \mathbb{R}^{\mathbf{X}} \to \mathbf{X}$ by

$$h(\mathbf{a}) = choice(domain(\mathbf{a} \| \vee \mathbf{a})).$$

The multistage edge following process can then be expressed as

$$\mathbf{f}_1 := 0 \quad (\mathbf{f}_1 \in \mathbb{R}^{\mathbf{X}})$$

for k in $1..n-1$ loop

$\qquad \mathbf{f}_{k+1} := \mathbf{f}_k \boxtimes \mathbf{t}(\mathbf{e}, \mathbf{d})$

$\qquad \mathbf{m}_{k+1} := h \circ (\mathbf{t}(\mathbf{e}, \mathbf{d}) + \mathbf{f}_k) \qquad$ (note that $h \circ (\mathbf{t}(\mathbf{e}, \mathbf{d}) + \mathbf{f}_k) \in \mathbf{X}^{\mathbf{X}}$)

end loop

$\bar{\mathbf{x}}_n := h(\mathbf{f}_n)$

for k in $n-1..1$ loop

$\qquad \bar{\mathbf{x}}_k := \mathbf{m}_{k+1}(\bar{\mathbf{x}}_{k+1})$

end loop

Note that by restricting $\mathbf{f}_1$ to $\mathbf{Y} = domain(\mathbf{e}\|_{>0})$ one may further reduce the search area for the optimal curve $(\bar{\mathbf{x}}_1, \ldots, \bar{\mathbf{x}}_n)$.

Comments and Observations

The algorithm provides the optimal curve of length n that satisfies the FOM function. The method is very flexible and general. For example, the length n of the curve could be incorporated as an additional constraint in the FOM function or other desirable constraints could be added.

The method has several deficiencies. Computation time is much greater than many other local boundary following techniques. Storage requirements (tables for m_{k+1}) are high. The length n must somehow be predetermined. Including the length as an additional constraint into the FOM function drastically increases storage and computation requirements.

3.21. Exercises

1. Note the dates on the references to papers by Roberts [5], Prewitt [7], Sobel [9], Marr and Hildreth [20], and Canny [18]. Consider the use of multiple edge directions, smoothing, calculation of the first derivative, and calculation of the second derivative. Describe the trends that you see in edge detection over the period of 21 years covered by those papers.

2. Consider the Frei-Chen orthogonal basis templates shown in Section 3.8. Verify that the basis is orthogonal. How easy would it be to define a new basis with different weights for the edge subspace? Design such a basis with templates that more closely approximate those of the Sobel edge detector.

3.* Use the technique described in Section 11.5 to create a program written using the image algebra C++ library to implement Canny edge detection using just four directional derivatives (at 45° increments) rather than six as shown in Section 3.12. Compare the results with the six direction version of the algorithm. Discuss whether or not using more directions provided better results.

4. Consider the image algebra formulation for the Laplacian shown in Section 3.14. Reformulate this as the difference between an image and a smoothing of that image with a von Neumann neighborhood.

5.   Suppose you want to find edges at subpixel resolution. Which of the detectors presented in this book are amenable to such a modification? Using the image algebra C++ library, implement a modification of the Marr-Hildreth edge detector using bilinear interpolation to report edges at points in an image having four times the resolution of the original image. Test your algorithm using synthetic images containing step edges. Down-sample your images by a factor of 16 (four times both horizontally and vertically) before testing.

6.   Investigate the behavior of the Sobel, Canny, and Marr-Hildreth edge detectors to the following image:

$$\mathbf{a} \in \mathbb{R}^{\mathbb{Z}_{32} \times \mathbb{Z}_{32}}, \ \mathbf{a}(x_1, x_2) = \begin{cases} 0 & if \ x_1 < 8 \\ x_1 - 8 & if \ 8 \le x_1 \le 24 \\ 16 & if \ x_1 > 24 \end{cases}.$$

Discuss where you believe edges should appear in this image and where they do appear in the different algorithms. Discuss which of these edge algorithms represents a first-derivative and which is a second-derivative operator.

7.   Both the Canny and Marr-Hildreth detectors can be used at a sequence of scales to provide better locality for edges found at a lower scale. Design an algorithm to track edge points from a low resolution image (large σ) through a sequence of higher resolution images (with successively smaller σ values) to improve its location. What problems must you face in designing any such algorithm?

3.22. References

[1]   W. Pratt, *Digital Image Processing.* New York: John Wiley, 1978.

[2]   R. Gonzalez and P. Wintz, *Digital Image Processing.* Reading, MA: Addison-Wesley, second ed., 1987.

[3]   A. Rosenfeld and A. Kak, *Digital Image Processing.* New York: Academic Press, 1976.

[4]   R. Haralick and L. Shapiro, *Computer and Robot Vision*, vol. I. New York: Addison-Wesley, 1992.

[5]   L. G. Roberts, "Machine perception of three-dimensional solids," in *Optical and Electro-Optical Information Processing* (J. Tippett, ed.), pp. 159–197, Cambridge, MA: MIT Press, 1965.

[6]   M. Lineberry, "Image segmentation by edge tracing," *Applications of Digital Image Processing IV*, vol. 359, 1982.

[7]   J. Prewitt, "Object enhancement and extraction," in *Picture Processing and Psychopictorics* (B. Lipkin and A. Rosenfeld, eds.), New York: Academic Press, 1970.

[8]   A. Rosenfeld and A. Kak, *Digital Picture Processing.* New York, NY: Academic Press, 2nd ed., 1982.

[9]   I. Sobel, *Camera Models and Machine Perception.* Ph.D. thesis, Stanford University, Stanford, CA, 1970.

[10] R. Duda and P. Hart, *Pattern Classification and Scene Analysis.* New York, NY: John Wiley & Sons, 1973.

[11] D. Ballard and C. Brown, *Computer Vision.* Englewood Cliffs, NJ: Prentice-Hall, 1982.

[12] W. Frei and C. Chen, "Fast boundary detection: A generalization and a new algorithm," *IEEE Transactions on Electronic Computers*, vol. C-26, Oct. 1977.

[13] R. Gonzalez and R. Woods, *Digital Image Processing*. Reading, MA: Addison-Wesley, 1992.

[14] R. Kirsch, "Computer determination of the constituent structure of biological images," *Computers and Biomedical Research*, vol. 4, pp. 315–328, June 1971.

[15] R. Nevatia and K. Babu, "Linear feature extraction and description," *Computer Graphics and Image Processing*, vol. 13, 1980.

[16] G. Ritter, J. Wilson, and J. Davidson, "Image algebra: An overview," *Computer Vision, Graphics, and Image Processing*, vol. 49, pp. 297–331, Mar. 1990.

[17] A. Rosenfeld, "A nonlinear edge detection technique," *Proceedings of the IEEE, Letters*, vol. 58, May 1970.

[18] J. Canny, "A computational approach to edge-detection," *IEEE Transactions on Pattern Analysis and Machine Intelligence*, vol. 8, pp. 679–700, 1986.

[19] J. M. Prager, "Extracting and labeling boundary segments in natural scenes," *IEEE Transactions on Pattern Analysis and Machine Intelligence*, vol. 1, Jan. 1980.

[20] D. Marr and E. Hildreth, "Theory of edge detection," in *Proceedings of the Royal Society of London B*, vol. 207, pp. 187–217, 1980.

[21] A. Huertas and G. Medioni, "Detection of intensity changes with subpixel accuracy using Laplacian-Gaussian masks," *IEEE Transactions on Pattern Analysis and Machine Intelligence*, vol. 8, pp. 651–665, May 1986.

[22] W. Lorensen and H. Cline, "Marching cubes: A high resolution 3D surface construction algorithm," *Computer Graphics*, vol. 21, pp. 163–169, July 1987.

[23] S. Zucker and R. Hummel, "A three-dimensional edge operator," Tech. Rep. TR-79-10, McGill University, 1979.

[24] H. Liu, "Two- and three dimensional boundary detection," *Computer Graphics and Image Processing*, vol. 6, no. 2, 1977.

[25] S. Tanimoto and T. Pavlidis, "A hierarchical data structure for picture processing," *Computer Graphics and Image Processing*, vol. 4, no. 2, pp. 104–119, 1975.

[26] A. Kaced, "The K-forms: A new technique and its applications in digital image processing," in *IEEE Proceedings 5th International Conference on Pattern Recognition*, 1980.

[27] M. Hueckel, "An operator which locates edges in digitized pictures," *Journal of the ACM*, vol. 18, pp. 113–125, Jan. 1971.

[28] M. Hueckel, "A local visual operator which recognizes edges and lines," *Journal of the ACM*, vol. 20, pp. 634–647, Oct. 1973.

CHAPTER 4
THRESHOLDING TECHNIQUES

4.1. Introduction

This chapter describes several standard thresholding techniques. Thresholding is one of the simplest and most widely used image segmentation techniques. The goal of thresholding is to segment an image into regions of interest and to remove all other regions deemed inessential. The simplest thresholding methods use a single threshold in order to isolate objects of interest. In many cases, however, no single threshold provides a good segmentation result over an entire image. In such cases variable and multilevel threshold techniques based on various statistical measures are used. The material presented in this chapter provides some insight into different strategies of threshold selection.

4.2. Global Thresholding

The global thresholding technique is used to isolate objects of interest having values different from the background. Each pixel is classified as either belonging to an object of interest or to the background. This is accomplished by assigning to a pixel the value 1 if the source image value is within a given threshold range and 0 otherwise [1]. A sampling of algorithms used to determine threshold levels can be found in Sections 4.6, 4.7, and 4.8.

The global thresholding procedure is straightforward. Let $\mathbf{a} \in \mathbb{R}^{\mathbf{X}}$ be the source image and $[h, k]$ be a given threshold range. The thresholded image $\mathbf{b} \in \{0, 1\}^{\mathbf{X}}$ is given by

$$\mathbf{b}(\mathbf{x}) = \begin{cases} 1 & \text{if } h \leq \mathbf{a}(\mathbf{x}) \leq k \\ 0 & \text{otherwise} \end{cases}$$

for all $\mathbf{x} \in \mathbf{X}$.

Two special cases of this methodology are concerned with the isolation of uniformly high values or uniformly low values. In the first the thresholded image $\mathbf{b}$ will be given by

$$\mathbf{b}(\mathbf{x}) = \begin{cases} 1 & \text{if } \mathbf{a}(\mathbf{x}) \geq k \\ 0 & \text{otherwise} \end{cases}$$

while in the second case

$$\mathbf{b}(\mathbf{x}) = \begin{cases} 1 & \text{if } \mathbf{a}(\mathbf{x}) \leq k \\ 0 & \text{otherwise,} \end{cases}$$

where k denotes the suitable threshold value.

Image Algebra Formulation

Let $\mathbf{a} \in \mathbb{R}^{\mathbf{X}}$ be the source image and $[h, k]$ be a given threshold range. The thresholded result image $\mathbf{b} \in \{0, 1\}^{\mathbf{X}}$ can be computed using the characteristic function

$$\mathbf{b} := \chi_{[h,k]}(\mathbf{a}).$$

The characteristic functions

$$\mathbf{b} := \chi_{\geq k}(\mathbf{a})$$

and

$$\mathbf{b} := \chi_{\leq k}(\mathbf{a})$$

can be used to isolate object of high values and low values, respectively.

Comments and Observations

Global thresholding is effective in isolating objects of uniform value placed against a background of different values. Practical problems occur when the background is non-uniform or when the object and background assume a broad range of values. Note also that

$$\mathbf{b} := \chi_{[h,k]}(\mathbf{a}) = (\chi_{\geq h}(\mathbf{a})) \cdot (\chi_{\leq k}(\mathbf{a})).$$

4.3. Semithresholding

Semithresholding is a useful variation of global thresholding [1]. Pixels whose values lie within a given threshold range retain their original values. Pixels with values lying outside of the threshold range are set to 0. For a source image $\mathbf{a} \in \mathbb{R}^X$ and a threshold range $[h, k]$, the semithresholded image $\mathbf{b} \in \mathbb{R}^X$ is given by

$$\mathbf{b}(\mathbf{x}) = \begin{cases} \mathbf{a}(\mathbf{x}) & \text{if } h \leq \mathbf{a}(\mathbf{x}) \leq k \\ 0 & \text{otherwise} \end{cases}$$

for all $\mathbf{x} \in \mathbf{X}$.

Regions of high values can be isolated using

$$\mathbf{b}(\mathbf{x}) := \begin{cases} \mathbf{a}(\mathbf{x}) & \text{if } \mathbf{a}(\mathbf{x}) \geq k \\ 0 & \text{otherwise} \end{cases}$$

and regions of low values can be isolated using

$$\mathbf{b}(\mathbf{x}) := \begin{cases} \mathbf{a}(\mathbf{x}) & \text{if } \mathbf{a}(\mathbf{x}) \leq k \\ 0 & \text{otherwise.} \end{cases}$$

Image Algebra Formulation

The image algebra formulation for the semithresholded image $\mathbf{b} \in \mathbb{R}^X$ over the range of values $[h, k]$ is

$$\mathbf{b} := \mathbf{a} \cdot \chi_{[h,k]}(\mathbf{a}).$$

The images semithresholded over the unbounded sets $[k, \infty)$ and $(-\infty, k]$ are given by

$$\mathbf{b}(\mathbf{x}) := \mathbf{a} \cdot \chi_{\geq k}(\mathbf{a})$$

and

$$\mathbf{b}(\mathbf{x}) := \mathbf{a} \cdot \chi_{\leq k}(\mathbf{a}),$$

respectively.

Alternate Image Algebra Formulation

The semithresholded image can also be derived by restricting the source image to those points whose values lie in the threshold range, and then extending the restriction to $\mathbf{X}$ with value 0. The image algebra formulation for this method of semithresholding is

$$\mathbf{b} := \left(\mathbf{a}\|_{[h,k]}\right)|^{0}.$$

If appropriate, instead of constructing a result over $\mathbf{X}$, one can construct the subimage $\mathbf{c}$ of $\mathbf{a}$ containing only those pixels lying in the threshold range, that is,

$$\mathbf{c} := \mathbf{a}\|_{[h,k]}.$$

Comments and Observations

Figures 4.3.2 and 4.3.3 below show the thresholded and semithresholded images of the original image of the Thunderbirds in Figure 4.3.1.

Figure 4.3.1. Image of Thunderbirds $\mathbf{a}$.

Figure 4.3.2. Thresholded image of Thunderbirds $\mathbf{b} := \chi_{\leq 190}(\mathbf{a})$.

Figure 4.3.3. Semithresholded image of Thunderbirds $\mathbf{b} := \mathbf{a} \cdot \chi_{\leq 190}(\mathbf{a})$.

4.4. Multilevel Thresholding

Global thresholding and semithresholding techniques (Sections 4.2 and 4.3) seg-ment an image based on the assumption that the image contains only two types of regions. Certainly, an image may contain more than two types of regions. Multilevel thresholding is an extension of the two earlier thresholding techniques that allows for segmentation of pixels into multiple classes [1].

For example, if the image histogram contains three peaks, then it is possible to segment the image using two thresholds. These thresholds divide the value set into three nonoverlapping ranges, each of which can be associated with a unique value in the resulting image. Methods for determining such threshold values are discussed in Sections 4.6, 4.7, and 4.8.

Let $\mathbf{a} \in \mathbb{R}^{\mathbf{X}}$ be the source image, and let $k_1, \cdots, k_n$ be threshold values satisfying $k_1 > k_2 > \cdots > k_n$. These values partition $\mathbb{R}$ into $n+1$ intervals which are associated with values $v_1, \cdots, v_{n+1}$ in the thresholded result image. A typical sequence of result values might be $1, \frac{n-1}{n}, \cdots, \frac{1}{n}, 0$. The thresholded image $\mathbf{b} \in \mathbb{R}^{\mathbf{X}}$ is defined by

$$
\mathbf{b}(\mathbf{x}) = \begin{cases} v_1 & \text{if } k_1 < \mathbf{a}(\mathbf{x}) \\ v_i & \text{if } k_i < \mathbf{a}(\mathbf{x}) \leq k_{i-1} \\ v_{n+1} & \text{if } \mathbf{a}(\mathbf{x}) \leq k_n. \end{cases}
$$

Image Algebra Formulation

Define the function $f : \mathbb{R} \to \mathbb{R}$ by

$$
f(r) = \begin{cases} v_1 & \text{if } k_1 < r \\ v_i & \text{if } k_i < r \leq k_{i-1} \\ v_{n+1} & \text{if } r \leq k_n. \end{cases}
$$

The thresholded image $\mathbf{b} \in \mathbb{R}^{\mathbf{X}}$ can be computed by composing f with $\mathbf{a}$, i.e.,

$$
\mathbf{b} := f \circ \mathbf{a}.
$$

4.5. Variable Thresholding

No single threshold level may produce good segmentation results over an entire image. Variable thresholding allows different threshold levels to be applied to different regions of a image.

For example, objects may contrast with the background throughout an image, but due to uneven illumination, objects and background may have lower values on one side of the image than on the other. In such instances, the image may be subdivided into smaller regions. Thresholds are then established for each region and global (or other) thresholding is applied to each subimage corresponding to a region [1].

The exact methodology is as follows. Let $\mathbf{a} \in \mathbb{R}^{\mathbf{X}}$ be the source image, and let image $\mathbf{d} \in \mathbb{R}^{\mathbf{X}}$ denote the region threshold value associated with each point in $\mathbf{X}$, that is, $\mathbf{d}(\mathbf{x})$ is the threshold value associated with the region in which point $\mathbf{x}$ lies. The thresholded image $\mathbf{b} \in \{0,1\}^{\mathbf{X}}$ is defined by

$$\mathbf{b}(\mathbf{x}) = \begin{cases} 1 & \text{if } \mathbf{a}(\mathbf{x}) \geq \mathbf{d}(\mathbf{x}) \\ 0 & \text{if } \mathbf{a}(\mathbf{x}) < \mathbf{d}(\mathbf{x}). \end{cases}$$

Image Algebra Formulation

The thresholded image $\mathbf{b}$ can be computed as follows:

$$\mathbf{b} := \chi_{\geq \mathbf{d}}(\mathbf{a}).$$

Comments and Observations

Variable thresholding is effective for images with locally bimodal histograms. This method will produce the desired results if the objects are relatively small and are not clustered too close together. The subimages should be large enough to ensure that they contain both background and object pixels.

The same problems encountered in global thresholding can occur on a local level in variable thresholding. Thus, if an image has locally nonuniform background, or large ranges of values in some regions, or if the multimodal histogram does not distinguish object from background, the technique will fail. Also, it is difficult to define the image $\mathbf{d}$ without some *a priori* information.

4.6. Threshold Selection Using Mean and Standard Deviation

In this section we present the first of three automatic thresholding techniques. The particular threshold derived here is a linear combination, $k_1\mu + k_2\sigma$, of the mean and standard deviation of the source image and was proposed by Hamadani [2]. The mean and standard deviation are intrinsic properties of the source image. The weights k_1 and k_2 are pre-selected, based on image type information, in order to optimize performance.

For a given image $\mathbf{a} \in \mathbb{R}^{\mathbf{X}}$, where $\mathbf{X}$ is an $m \times n$ grid, the mean μ and standard deviation σ of $\mathbf{a}$ are given by

$$\mu = \frac{1}{mn} \sum_{i=1}^{m} \sum_{j=1}^{n} \mathbf{a}(i,j)$$

and

$$\sigma = \sqrt{\frac{1}{mn}\sum_{i=1}^{m}\sum_{j=1}^{n}\left(\mathbf{a}(i,j)-\mu\right)^{2}},$$

respectively. The threshold level τ is set at

$$\tau = k_1\mu + k_2\sigma,$$

where the constants k_1 and k_2 are image type dependent.

Image Algebra Formulation

Let $\mathbf{a} \in \mathbb{R}^{\mathbf{X}}$, where $\mathbf{X} = \mathbb{Z}_m \times \mathbb{Z}_n$. The mean and standard deviation of $\mathbf{a}$ are given by the image algebra statements

$$\mu := \frac{1}{mn}\sum\mathbf{a}$$

and

$$\sigma := \sqrt{\frac{1}{mn}\sum\left(\mathbf{a}-\mu\right)^{2}}.$$

The threshold level is given by

$$\tau := k_1\mu + k_2\sigma.$$

Comments and Observations

For typical low-resolution IR images $k_1 = k_2 = 1$ seems to work fairly well for extracting "warm" objects. For higher resolution $k_1 = 1$ or $k_1 = 1.5$ and $k_2 = 2$ may yield better results.

Figure 4.6.1 shows an original IR image of a jeep (top) and three binary images that resulted from thresholding the original based on its mean and standard deviation for various k_1 and k_2.

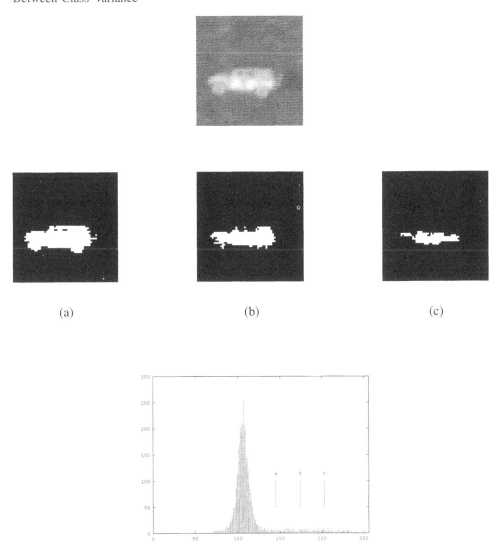

Figure 4.6.1. Thresholding an image based on threshold level
$\tau = k_1\mu + k_2\sigma$ for various k_1 and k_2: (a) $k_1 = 1, k_2 = 1, \tau = 145$,
(b) $k_1 = 1, k_2 = 2, \tau = 174$, (c) $k_1 = 1.5, k_2 = 1, \tau = 203$. At the
bottom of the figure is the histogram of the original image. The threshold
levels corresponding to images (a), (b), and (c) have been marked with arrows.

4.7. Threshold Selection by Maximizing Between-Class Variance

In this section we present the method of Otsu [3, 4] for finding threshold
parameters to be used in multithresholding schemes. For a given k the method finds
thresholds $0 \le \tau_1 < \tau_2 < \cdots < \tau_{k-1} < l - 1$ for partitioning the pixels of $\mathbf{a} \in (\mathbb{Z}_l)^{\mathbf{X}}$

into the classes

$$C_0 = \{(\mathbf{x}, \mathbf{a}(\mathbf{x})) : 0 \leq \mathbf{a}(\mathbf{x}) \leq \tau_1\}$$

$$\vdots$$

$$C_i = \{(\mathbf{x}, \mathbf{a}(\mathbf{x})) : \tau_i < \mathbf{a}(\mathbf{x}) \leq \tau_{i+1}\}$$

$$\vdots$$

$$C_{k-1} = \{(\mathbf{x}, \mathbf{a}(\mathbf{x})) : \tau_{k-1} < \mathbf{a}(\mathbf{x}) \leq l - 1\}$$

by maximizing the separability between classes. *Between-class variance* is used as the measure of separability between classes. Its definition, which follows below, uses histogram information derived from the source image. After threshold values have been determined, they can be used as parameters in the multithresholding algorithm of Section 4.4.

Let $\mathbf{a} \in (\mathbb{Z}_l)^{\mathbf{X}}$ and let $\overline{\mathbf{h}}$ be the normalized histogram of $\mathbf{a}$ (Section 10.10). The pixels of $\mathbf{a}$ are to be partitioned into k classes $C_0, C_1, \ldots, C_{k-1}$ by selecting the τ_i's as stipulated below.

The probabilities of class occurrence $Pr(C_i)$ are given by

$$Pr(C_0) = \omega_0 = \sum_{j=0}^{\tau_1} \overline{\mathbf{h}}(j) = \omega(\tau_1)$$

$$\vdots$$

$$Pr(C_i) = \omega_i = \sum_{j=\tau_i+1}^{\tau_{i+1}} \overline{\mathbf{h}}(j) = \omega(\tau_{i+1}) - \omega(\tau_i)$$

$$\vdots$$

$$Pr(C_{k-1}) = \omega_{k-1} = \sum_{j=\tau_{k-1}+1}^{l-1} \overline{\mathbf{h}}(j) = 1 - \omega(\tau_{k-1}),$$

where $\omega(\tau_i) = \sum_{j=0}^{\tau_i} \overline{\mathbf{h}}(j)$ is the 0th-order cumulative moment of the histogram evaluated up to the τ_ith level. The class mean levels are given by

$$\mu_0 = \sum_{j=0}^{\tau_1} \frac{j \cdot \overline{\mathbf{h}}(j)}{\omega_0} = \frac{\mu(\tau_1)}{\omega(\tau_1)}$$

$$\vdots$$

$$\mu_i = \sum_{j=\tau_i+1}^{\tau_{i+1}} \frac{j \cdot \overline{\mathbf{h}}(j)}{\omega_i} = \frac{\mu(\tau_{i+1}) - \mu(\tau_i)}{\omega(\tau_{i+1}) - \omega(\tau_i)}$$

$$\vdots$$

$$\mu_{k-1} = \sum_{j=\tau_{k-1}+1}^{l-1} \frac{j \cdot \overline{\mathbf{h}}(j)}{\omega_{k-1}} = \frac{\mu_\tau - \mu(\tau_{k-1})}{1 - \omega(\tau_{k-1})},$$

where $\mu(\tau_i) = \sum_{j=0}^{\tau_i} j \cdot \overline{\mathbf{h}}(j)$ is the 1st-order cumulative moment of the histogram up to the τ_ith level and $\mu = \sum_{j=0}^{l-1} j \cdot \overline{\mathbf{h}}(j)$ is the total mean level of $\mathbf{a}$.

In order to evaluate the goodness of the thresholds at levels $0 \leq \tau_1 < \tau_2 < \cdots < \tau_{k-1} < l - 1$, the between-class variance is used as discriminant criterion measure of class separability. This between-class variance σ_b^2 is defined as

$$\sigma_b^2 = \omega_0(\mu_0 - \mu)^2 + \omega_1(\mu_1 - \mu)^2 \cdots + \omega_{k-1}(\mu_{k-1} - \mu)^2$$

or, equivalently,

$$\sigma_b^2 = \omega_0\mu_0^2 + \omega_1\mu_1^2 + \cdots + \omega_{k-1}\mu_{k-1}^2 - \mu^2.$$

Note that σ_b^2 is a function of $\tau_1, \tau_2, \ldots, \tau_{k-1}$ (if we substitute the corresponding expressions involving the τ_i's). The problem of determining the goodness of the τ_i's reduces to optimizing σ_b^2, i.e., to finding the maximum of $\sigma_b^2(\tau_1, \tau_2, \ldots, \tau_{k-1})$ or

$$max\left\{\omega_0\mu_0^2 + \omega_1\mu_1^2 + \cdots + \omega_{k-1}\mu_{k-1}^2 - \mu^2 : 0 \leq \tau_1 < \tau_2 < \cdots < \tau_{k-1} < l - 1\right\}.$$

To the left in Figure 4.7.1 is the blurred image of three nested squares with pixel values 80, 160, and 240. Its histogram below in Figure 4.7.2 shows the two threshold levels, 118 and 193, that result from maximizing the variance between three classes. The result of thresholding the blurred image at the levels above is seen in the image to the right of Figure 4.7.1.

Figure 4.7.1. Blurred image (left) and the result of thresholding it (right).

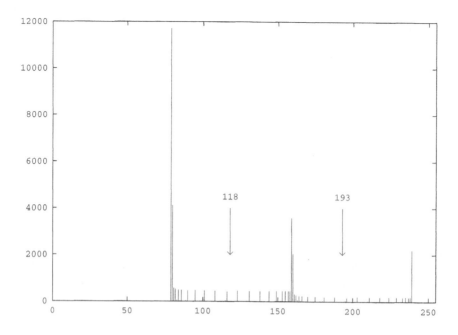

Figure 4.7.2. Histogram of blurred image with threshold levels marked.

Image Algebra Formulation

We will illustrate the image algebra formulation of the algorithm for $k = 3$ classes. It is easily generalized for other k with $1 \leq k < l - 1$.

Let $\mathbf{a} \in (\mathbb{Z}_l)^{\mathbf{X}}$ be the source image. Let $\overline{\mathbf{h}}$ and $\overline{\mathbf{c}}$ be the normalized histogram and normalized cumulative histogram of $\mathbf{a}$, respectively (Sections 10.10 and 10.11). The 1st order cumulative moment image $\mathbf{u}$ of $\overline{\mathbf{h}}$ is given by

$$\mathbf{u} := \sum \mathbf{t}(\overline{\mathbf{h}}),$$

where $\mathbf{t}$ is the parameterized template defined by

$$\mathbf{t}(\mathbf{a})_j(i) = \begin{cases} j \cdot \mathbf{a}(j) & \text{if } j \leq i \\ 0 & \text{otherwise.} \end{cases}$$

The image algebra pseudocode for the threshold finding algorithm for $k = 3$ is given by the following formulation:

$$m := 0$$
$$\tau_1 := 0$$
$$\tau_2 := 0$$
$$\mu := \mathbf{u}(l - 1)$$

for i in $0..l - 2$ loop
   for j in $i + 1..l - 1$ loop

$$\omega_0 := \overline{\mathbf{c}}(i)$$
$$\omega_1 := \overline{\mathbf{c}}(j) - \overline{\mathbf{c}}(i)$$
$$\omega_2 := 1 - \overline{\mathbf{c}}(j)$$
$$\mu_0 := \frac{\mathbf{u}(i)}{\omega_0}$$
$$\mu_1 := \frac{\mathbf{u}(j) - \mathbf{u}(i)}{\omega_1}$$
$$\mu_2 := \frac{\mu - \mathbf{u}(j)}{\omega_2}$$
$$\sigma_b^2 := \omega_0 \mu_0^2 + \omega_1 \mu_1^2 + \omega_2 \mu_2^2 - \mu^2$$

if $\sigma_b^2 > m$ then
$$\tau_1 := i$$
$$\tau_2 := j$$
$$m := \sigma_b^2$$
end if

   end loop
end loop.

When the algorithm terminates, the desired thresholds are τ_1 and τ_2.

Comments and Observations

This algorithm is an unsupervised algorithm. That is, it has the advantageous property of not requiring human intervention. For large k the method may fail unless the classes are extremely well separated.

In the figures below, the directional edge detection algorithm (Section 3.10) has been applied to the source image in Figure 4.7.3. The magnitude image produced by the directional edge detection algorithm is represented by Figure 4.7.4. The directional edge magnitude image has 256 gray levels. Its histogram can be seen in Figure 4.7.6.

Otsu's algorithm has been applied to find a bi-level threshold τ_1 that can be used to clean up Figure 4.7.4. The threshold level is determined to be 105, which lies in the valley between the broad peak to the left of the histogram and the spiked peak at 255.

The threshold is marked by an arrow in Figure 4.7.6. The thresholded image can be seen in Figure 4.7.5.

Figure 4.7.3. Source image of a causeway with bridge.

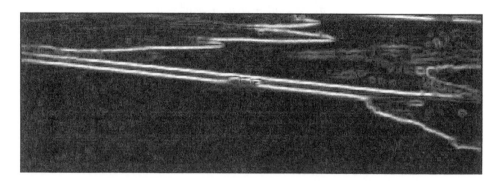

Figure 4.7.4. Magnitude image after applying directional edge detection to Figure 4.7.3.

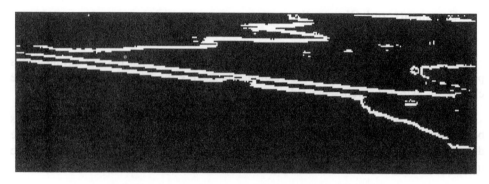

Figure 4.7.5. Result of thresholding Figure 4.7.4 at
level specified by maximizing between-class variance.

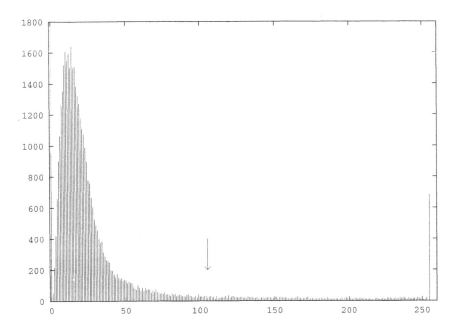

Figure 4.7.6. Histogram of edge magnitude image (Figure 4.7.4).

4.8. Threshold Selection Using a Simple Image Statistic

The *simple image statistic* (SIS) algorithm [5, 4, 6] is an automatic bi-level threshold selection technique. The method is based on simple image statistics that do not require computing a histogram for the image. Let $\mathbf{a} \in \mathbb{R}^{\mathbf{X}}$ be the source image over an $m \times n$ array $\mathbf{X}$. The SIS algorithm assumes that $\mathbf{a}$ is an imperfect representation of an object and its background. The ideal object is composed of pixels with gray level a and the ideal background has gray level b.

Let $e(i, j)$ be the maximum in the absolute sense of the gradient masks $\mathbf{s}$ and $\mathbf{t}$ (below) applied to $\mathbf{a}$ and centered at $(i, j) \in \mathbf{X}$.

$$\mathbf{s} = \begin{array}{|c|} \hline 1 \\ \hline \\ \hline -1 \\ \hline \end{array} \qquad \mathbf{t} = \begin{array}{|c|c|c|} \hline -1 & & 1 \\ \hline \end{array}$$

In [5], it is shown that

$$\frac{\sum\limits_{i=1}^{n} \sum\limits_{j=1}^{m} |\mathbf{a}(i,j) \cdot e(i,j)|}{\sum\limits_{i=1}^{n} \sum\limits_{j=1}^{m} |e(i,j)|} = \frac{a+b}{2}.$$

The fraction on the right-hand side of the equality is the midpoint between the gray level of the object and background. Intuitively, this midpoint is an appealing level for thresholding. Experimentation has shown that reasonably good performance has been achieved by thresholding at this midpoint level [6]. Thus, the SIS algorithm sets its threshold τ equal to

$$\tau = \frac{\displaystyle\sum_{i=1}^{n}\sum_{j=1}^{m}|\mathbf{a}(i,j)\cdot\mathbf{e}(i,j)|}{\displaystyle\sum_{i=1}^{n}\sum_{j=1}^{m}|\mathbf{e}(i,j)|}.$$

The image to the left in Figure 4.8.1 is that of a square region with pixel value 192 set against a background of pixel value 64 that has been blurred with a smoothing filter. Its histogram (seen in Figure 4.8.2) is bimodal. The threshold value 108 selected based on the SIS method divides the two modes of the histogram. The result of thresholding the blurred image at level 108 is seen in the image to the right of Figure 4.8.1.

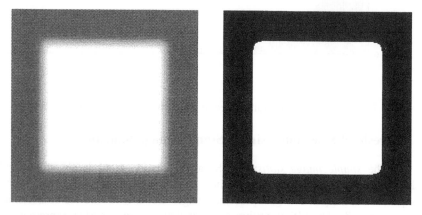

Figure 4.8.1. Blurred image (left) and the result of thresholding (right) using the threshold level determined by the SIS method.

Image Algebra Formulation

Let $\mathbf{a} \in \mathbb{R}^{\mathbf{X}}$ be the source image, where $\mathbf{X} = \mathbb{Z}_m \times \mathbb{Z}_n$. Let $\mathbf{s}$ and $\mathbf{t}$ be the templates pictured below.

$$\mathbf{s} = \quad \mathbf{t} =$$

Let $\mathbf{e} := |\mathbf{a} \oplus \mathbf{s}| \vee |\mathbf{a} \oplus \mathbf{t}|$. The SIS threshold is given by the image algebra statement

$$\tau := \frac{\sum(\mathbf{e}\cdot|\mathbf{a}|)}{\sum\mathbf{e}}.$$

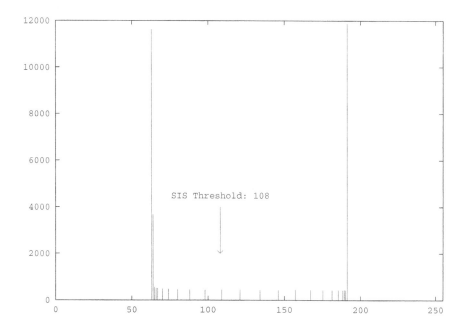

Figure 4.8.2. Histogram of blurred image.

Comments and Observations

Figures 4.8.4 and 4.8.5 are the results of applying Otsu's (Section 4.7) and the SIS algorithms, respectively, to the source image in Figure 4.8.3. The histogram of the source image with the threshold levels marked by arrows is shown in Figure 4.8.6. Note that the maximum between class variance for Otsu's threshold occurred at gray level values 170 through 186. Thus, the smallest gray level value at which the maximum between-class variance occurred was chosen as the Otsu threshold level for our illustration.

Figure 4.8.3. Original image.

Figure 4.8.4. Bi-level threshold image using Otsu's algorithm.

Figure 4.8.5. Bi-level threshold image using the SIS algorithm.

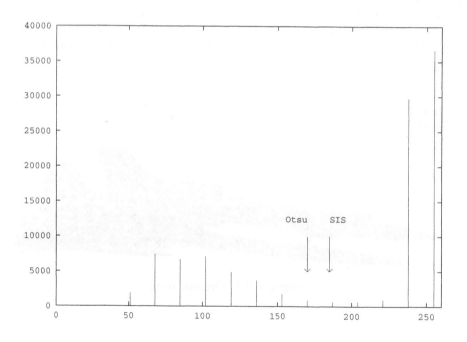

Figure 4.8.6. Histogram of original image with threshold levels marked by arrows.

4.9. Exercises

1.* Write a program that will compute the histogram of an image, automatically locate a major dip in the histogram for threshold selection, and threshold the image to produce a binary image. Express your program in image algebra pseudocode and test it on suitable images.

2. Develop a program in image algebra pseudocode that will generate a digital image containing several 80 to 240 gray-level-tall bell-shaped towers on a background of specified gray level. Implement the threshold selection techniques described in this chapter and test them on your artificial images.

3. Develop your own adaptive threshold program that can set the threshold for each tower of the image generated in Exercise 2. Your program should be capable of isolating any one (and only one) of the gray-level towers.

4. Threshold levels can be varied to suit local or neighborhood input image gray-level variations. The intuitive result is the preservation of local contrast, thus yielding an output image with less loss of information and a subjectively more pleasing appearance. Techniques implementing such local thresholds are commonly known as *locally adaptive thresholding*. Locally adaptive image thresholding techniques include line- or edge-sensitive thresholding and average gray-level based thresholding. An example of edge-sensitive thresholding was given in Section 4.5. In average gray-level based thresholding, the threshold is determined as a function of the average input image gray-value in a local region (usually a 3×3 or 5×5 neighborhood). Pixel intensities that differ significantly from this average are assumed to contain local contrast information and are thus distinguished in an output image. Develop an algorithm that implements locally adaptive image thresholding based on average gray-level thresholding. Write your algorithm in image algebra pseudocode. Implement your algorithm on suitable images.

4.10. References

[1] A. Rosenfeld and A. Kak, *Digital Picture Processing*. New York, NY: Academic Press, 2nd ed., 1982.

[2] N. Hamadani, "Automatic target cueing in IR imagery," Master's thesis, Air Force Institute of Technology, WPAFB, Ohio, Dec. 1981.

[3] N. Otsu, "A threshold selection method from gray-level histograms," *IEEE Transactions on Systems, Man, and Cybernetics*, vol. SMC-9, pp. 62–66, Jan. 1979.

[4] P. K. Sahoo, S. Soltani, A. C. Wong, and Y. C. Chen, "A survey of thresholding techniques," *Computer Vision, Graphics, and Image Processing*, vol. 41, pp. 233–260, 1988.

[5] J. Kittler, J. Illingworth, and J. Foglein, "Threshold selection based on a simple image statistic," *Computer Vision, Graphics, and Image Processing*, vol. 30, pp. 125–147, 1985.

[6] S. U. Lee, S. Y. Chung, and R. H. Park, "A comparative performance study of several global thresholding techniques for segmentation," *Computer Vision, Graphics, and Image Processing*, vol. 52, pp. 171–190, 1990.

CHAPTER 5
THINNING AND SKELETONIZING

5.1. Introduction

Thick objects in a discrete binary image are often reduced to thinner representations called skeletons, which are similar to stick figures. Most skeletonizing algorithms iteratively erode the contours in a binary image until thin skeletons or single pixels remain. These algorithms typically examine the neighborhood of each contour pixel and identify those pixels that can be deleted and those that can be classified as skeletal pixels.

Thinning and skeletonizing algorithms have been used extensively for processing thresholded images, data reduction, pattern recognition, and counting and labeling of connected regions. Another thinning application is edge thinning which is an essential step in the description of objects where boundary information is vital. Algorithms given in Chapter 3 describe how various transforms and operations applied to digital images yield primitive edge elements. These edge detection operations typically produce a number of undesired artifacts including parallel edge pixels which result in thick edges. The aim of edge thinning is to remove the inherent edge broadening in the gradient image without destroying the edge continuity of the image.

5.2. Pavlidis Thinning Algorithm

The Pavlidis thinning transform is a simple thinning transform [1]. It provides an excellent illustration for translating set theoretic notation into image algebra formulation.

Let $\mathbf{a} \in \{0,1\}^{\mathbf{X}}$ denote the source image and let $\mathbf{A}$ denote the support of $\mathbf{a}$; i.e., $\mathbf{A} = \{\mathbf{x} \in \mathbf{X} : \mathbf{a}(\mathbf{x}) = 1\}$. The inside boundary of $\mathbf{A}$ is denoted by $\mathrm{IB}(\mathbf{A})$. The set $\mathrm{C}(\mathbf{A})$ consists of those points of $\mathrm{IB}(\mathbf{A})$ whose only neighbors are in $\mathrm{IB}(\mathbf{A})$ or $\mathbf{A}'$.

The algorithm proceeds by starting with $n = 1$, setting $\mathbf{A}^0$ equal to $\mathbf{A}$, and iterating the statement

$$\mathbf{A}^n := \mathrm{C}(\mathbf{A}^{n-1}) \cup (\mathbf{A}^{n-1} \setminus \mathrm{IB}(\mathbf{A}^{n-1})) \tag{5.2.1}$$

until $\mathbf{A}^n = \mathbf{A}^{n-1}$.

It is important to note that the algorithm described may result in the thinned region having disconnected components. This situation can be remedied by replacing Eq. 5.2.1 with

$$\mathbf{A}^n = \mathrm{C}(\mathbf{A}^{n-1}) \cup (\mathbf{A}^{n-1} \setminus \mathrm{IB}(\mathbf{A}^{n-1})) \cup (\mathrm{OB}(\mathrm{C}(\mathbf{A}^{n-1})) \cap \mathbf{A}^{n-1}),$$

where $\mathrm{OB}(\mathbf{R})$ is the outside boundary of region $\mathbf{R}$. The trade-off for connectivity is higher computational cost and the possibility that the thinned image may not be reduced as much. Definitions for the various boundaries of a Boolean image and boundary detection algorithms can be found in Section 3.2.

Figure 5.2.1 below shows the disconnected and connected skeletons of the SR71 superimposed over the original image.

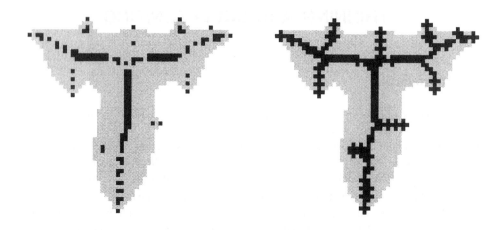

Figure 5.2.1. Disconnected and connected Pavlidis skeletons of the SR71.

Image Algebra Formulation

Let M be the Moore neighborhood and M_0 be the Moore neighborhood with its center pixel deleted as shown in Figure 5.2.2. The image algebra formulation of the

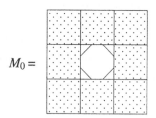

$$M_0 =$$

Figure 5.2.2. The deleted Moore neighborhood M_0.

Pavlidis thinning algorithm is now as follows:

$$
\begin{aligned}
&\mathbf{b} := 0 \\
&\texttt{while } \mathbf{b} \neq \mathbf{a} \texttt{ loop} \\
&\quad \mathbf{b} := \mathbf{a} \\
&\quad \mathbf{e} := \mathbf{a} \,\boxtimes\, M \\
&\quad \mathbf{i} := \mathbf{e}^c \wedge \mathbf{a} \\
&\quad \mathbf{c} := \chi_{=0}((\mathbf{a} \wedge \mathbf{i}^c) \oplus M_0) \wedge \mathbf{i} \\
&\quad \mathbf{a} := \mathbf{c} \vee \mathbf{e} \\
&\texttt{end loop.}
\end{aligned}
$$

When the loop terminates the thinned image will be contained in the image variable $\mathbf{a}$. The correspondences between the images in the code above and the mathematical formulation are

$$\mathbf{e} \equiv \text{eroded } \mathbf{a},$$
$$\mathbf{e}^c \equiv \mathbf{1} - \mathbf{e},$$
$$\mathbf{i} \equiv \text{inside boundary of } \mathbf{A} \equiv \text{IB}(\mathbf{A}),$$
$$\mathbf{i}^c \equiv \mathbf{1} - \mathbf{i}, \text{ and}$$
$$\mathbf{c} \equiv \text{C}(\mathbf{A}).$$

If connected components are desired then the last statement in the loop should be changed to

$$\mathbf{a} := \mathbf{c} \vee \mathbf{e} \vee (\mathbf{c}^c \wedge (\mathbf{c} \boxdot M) \wedge \mathbf{a}).$$

Comments and Observations

There are two approaches to this thinning algorithm. Note that the mathematical formulation of the algorithm is in terms of the underlying set (or domain) of the image $\mathbf{a}$, while the image algebra formulation is written in terms of image operations. Since image algebra also includes the set theoretic operations of union and intersection, the image algebra formulation could just as well have been expressed in terms of set theoretic notation; i.e., $\mathbf{A} := \text{domain}(\mathbf{a}\|_1)$, $\mathbf{A}' := \text{domain}(\mathbf{a}\|_0)$, and so on. However, deriving the sets $\text{IB}(\mathbf{A})$ and $\text{C}(\mathbf{A})$ would involve neighborhood or template operations, thus making the algorithm less translucent and more cumbersome. The image-based algorithm is far simpler and more translucent.

5.3. Medial Axis Transform (MAT)

Let $\mathbf{X} \subset \mathbb{R}^2$, $\mathbf{a} \in \{0,1\}^{\mathbf{X}}$, and let $\mathbf{A}$ denote the support of $\mathbf{a}$. The *medial axis* is the set, $\mathbf{M} \subseteq \mathbf{A}$, consisting of those points $\mathbf{x}$ for which there exists a ball of radius $r_{\mathbf{x}}$, centered at $\mathbf{x}$, that is contained in $\mathbf{A}$ and intersects the boundary of $\mathbf{A}$ in at least two distinct points. The dotted line (center dot for the circle) of Figure 5.3.1 represents the medial axis of some simple regions.

Figure 5.3.1. Medial axes.

The *medial axis transform* $\mathbf{m}$ is a gray level image defined over $\mathbf{X}$ by

$$\mathbf{m}(\mathbf{x}) = \begin{cases} r_{\mathbf{x}} & \text{if } \mathbf{x} \in \mathbf{M} \\ 0 & \text{otherwise.} \end{cases}$$

The medial axis transform is unique. The original image can be reconstructed from its medial axis transform.

The medial axis transform evolved through Blum's work on animal vision systems [1–10]. His interest involved how animal vision systems extract geometric shape information. There exists a wide range of applications that finds a minimal representation of an image useful [7]. The medial axis transform is especially useful for image compression since reconstruction of an image from its medial axis transform is possible.

Let $\mathbf{B}_r(\mathbf{x})$ denote the closed ball of radius r and centered at $\mathbf{x}$. The medial axis of region $\mathbf{A}$ is the set

$$\mathbf{M} = \{\mathbf{x} \in \mathbf{A} \ : \ \exists\, r_{\mathbf{x}} \in \mathbb{R}^+ \ni \mathbf{B}_{r_{\mathbf{x}}}(\mathbf{x}) \subset \mathbf{A}, \text{ and } card(\mathbf{B}_{r_{\mathbf{x}}}(\mathbf{x}) \cap \partial\mathbf{A}) \geq 2\}.$$

The medial axis transform is the function $\mathbf{m} : \mathbf{A} \rightarrow \mathbb{R}^+$ defined by

$$\mathbf{m}(\mathbf{x}) = \begin{cases} r_{\mathbf{x}} & \text{if } \mathbf{x} \in \mathbf{M} \\ 0 & \text{otherwise.} \end{cases}$$

The reconstruction of the domain of the original image $\mathbf{a}$ in terms of the medial axis transform $\mathbf{m}$ is given by

$$\mathbf{A} = \bigcup_{\mathbf{x} \in \mathbf{M}} \mathbf{B}_{\mathbf{m}(\mathbf{x})}(\mathbf{x}).$$

The original image $\mathbf{a}$ is then

$$\mathbf{a}(\mathbf{x}) = \begin{cases} 1 & \text{if } \mathbf{x} \in \mathbf{A} \\ 0 & \text{if } \mathbf{x} \in \mathbf{X} \setminus \mathbf{A}. \end{cases}$$

Image Algebra Formulation

Let $\mathbf{a} \in \{0,1\}^{\mathbf{X}}$ denote the source image. Usually, the neighborhood N is a digital disk of a specified radius. The shape of the digital disk will depend on the distance function used. In our example N is the von Neumann neighborhood.

The following image algebra pseudocode will generate the medial axis transform of $\mathbf{a}$. When the loop terminates the medial axis transform will be stored in the image variable $\mathbf{m}$.

$$i := 0$$
$$\mathbf{m} := 0$$
$$\texttt{while } \mathbf{a} \neq 0 \texttt{ loop}$$
$$\quad \mathbf{b} := \mathbf{a} \cdot \chi_{=0}((\mathbf{a} \boxminus N) \boxplus N)$$
$$\quad i := i + 1$$
$$\quad \mathbf{m} := \mathbf{m} + (i \cdot \mathbf{b})$$
$$\quad \mathbf{a} := \mathbf{a} \boxminus N$$
$$\texttt{end loop.}$$

The result of applying the medial axis transform to the SR71 can be seen in Figure 5.3.2. The medial axis transformation is superimposed over the original image of the SR71 which is represented by dots. Hexadecimal numbering is used for the display of the medial axis transform.

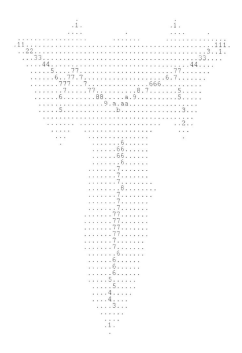

Figure 5.3.2. MAT superimposed over the SR71.

The original image **a** can be recovered from the medial axis transform **m** using the image algebra expression

$$\mathbf{a} := \bigvee_{k=1}^{i} \left(\chi_{=k}(\mathbf{m}) \ \boxtimes \ N^{k-1} \right).$$

Because each medial axis value is one greater than the associated neighborhood radius, the transform encodes isolated points (with value 1) and yields exact reconstruction.

Comments and Observations

The medial axis transform of a connected region may not be connected. Preserving connectivity is a desirable property for skeletonizing transforms.

The application of the medial axis on a discretized region may not represent a good approximation of the medial axis on the continuous region.

Different distance functions may be used for the medial axis transform. The choice of distance function is reflected by the template or neighborhood employed in this algorithm.

5.4.   Distance Transforms

A *distance transform* assigns to each feature pixel of a binary image a value equal to its distance to the nearest non-feature pixels. The algorithm may be performed in parallel or sequentially [11]. In either the parallel or sequential case, global distances are propagated using templates that reflect local distances to the target point of the template [9, 11, 12, 13].

A thinned subset of the original image can be derived from the distance transform by extracting the image that consists of the local maxima of the distance transform. This derived subset is called the *distance skeleton*. The original distance transform, and thus the original binary image, can be reconstructed from the distance skeleton. The distance skeleton can be used as a method of image compression.

Let **a** be a binary image defined on $\mathbf{X} \subset \mathbb{Z}^2$ with ∞ for feature pixels and 0 for non-feature pixels. The SR71 of Figure 5.4.1 will serve as our source image for illustrations purposes. An asterisk represents a pixel value of ∞. Non-feature pixel values are not displayed.

Figure 5.4.1. Source binary image.

The distance transform of **a** is a gray level image $\mathbf{b} \in \mathbb{R}^{\mathbf{X}}$ such that pixel value $\mathbf{b}(i, j)$ is the distance between the pixel location (i, j) and the nearest non-feature pixels. The distance can be measured in terms of Euclidean distance, city block, or chess board, etc., subject to the application's requirements. The result of applying the city block distance transform to the image of Figure 5.4.1 can be seen in Figure 5.4.2. Note that hexadecimal numbering is used in the figure.

The distance skeleton **c** of the distance transform **b** is the image whose nonzero pixel values consist of local maxima of **b**. Figure 5.4.3 represents the distance skeleton extracted from Figure 5.4.2 and superimposed over the original SR71 image.

The *restoration transform* is used to reconstruct the distance transform **b** from the distance skeleton **c**.

```
                                          1                              1
                                        121                            121
                                       1221                            1221        1
                           11111111111123321          1112111        123321111121111
                           1222222222223443211111112223222111111234432222322221
                           12333333333345543222222233343332222223455433333343221
                           12344444444566543333333444544433333345665444443211
                           12345555556776544444455565555444444567765554321
                           12345666678876555555566676665555555567887654321
                           12345677788787666666777877766666667787654321
                           12345678887678777778889888777776667654321
                           1234567876567888888999a999887665556654321
                           1234567765456789999aaabaa9876554445654321
                           1234566543456789aabbbba9876544333454321
                           12345654323456789abcba98765433222344321
                           1234543212345678 9aba98765432211123321
                           1234321 123456789a9876543211      12321
                             12321  12345678987654321          121
                              121    123456787654321            1
                               1      12345677654321
                                      12345677654321
                                      12345677654321
                                      12345677654321
                                      123456787654321
                                      123456787654321
                                      123456787654321
                                      123456788654321
                                      12345678987654321
                                      1234567887654321
                                      123456787654321
                                      123456787654321
                                      1234567887654321
                                      1234567887654321
                                      1234567887654321
                                      1234567887654321
                                      12345678787654321
                                      123456787654321
                                      12345677654321
                                      1234567654321
                                      1234567654321
                                      1234567654321
                                      123456654321
                                       12345654321
                                       1234554321
                                       123454321
                                        12344321
                                         123321
                                          1221
                                          121
                                           1
```

Figure 5.4.2. Distance transform image.

Image Algebra Formulation

Let $\mathbf{X} \subset \mathbb{Z}^2$, and $\mathbf{a} \in \{0,1\}^{\mathbf{X}}$ be the source image. Note that the templates $\mathbf{s}$, $\mathbf{t} = (\mathbf{t}_{\nearrow}, \mathbf{t}_{\prec})$, and $\mathbf{u}$ may be defined according to the specific distance measure being used. The templates used for our city block distance example are defined pictorially below in Figure 5.4.4.

Distance transform

The distance transform $\mathbf{b}$ of $\mathbf{a}$, computed recursively, is given by

$$\mathbf{b} := (\mathbf{a}\,\boxslash\,_{\prec}\mathbf{t})\,\boxslash\,_{\succ}\mathbf{t}',$$

where $\prec$ is the forward scanning order defined on $\mathbf{X}$ and $\succ$ is the backward scanning order defined on $\mathbf{X}$. The parallel image algebra formulation for the distance transform is given by

$$\mathbf{b} := 0$$

```
while a ≠ b loop
   b := a
   a := (a ⊠ u)
end loop.
```

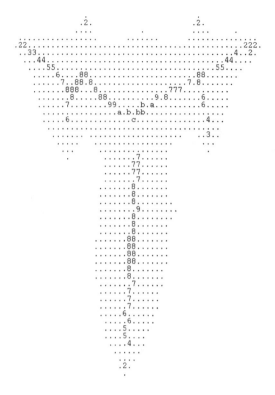

Figure 5.4.3. Distance skeleton.

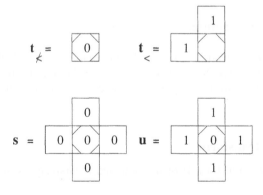

Figure 5.4.4. Templates used for city block distance image algebra formulation.

Distance skeleton transform

The distance skeleton transform is formulated nonrecursively as

$$\mathbf{c} := \mathbf{b} \cdot \chi_{=0}((\mathbf{b} \boxtimes \mathbf{s}) - \mathbf{b}).$$

Restoration transform

The restoration of $\mathbf{b}$ from $\mathbf{c}$ can be done recursively by

$$\mathbf{b} := (\mathbf{c} \boxtimes {}_{\prec}(-\mathbf{t})) \boxtimes {}_{\succ} \mathbf{t}^{*}.$$

5.5.   Zhang-Suen Skeletonizing

The *Zhang-Suen transform* is one of many derivatives of Rutovitz' thinning algorithm [6, 7, 14, 15, 16]. This class of thinning algorithms repeatedly removes boundary points from a region in a binary image until an irreducible skeleton remains. Deleting or removing a point in this context means to change its pixel value from 1 to 0. Not every boundary point qualifies for deletion. The 8-neighborhood of the boundary point is examined first. Only if the configuration of the 8-neighborhood satisfies certain criteria will the boundary point be removed.

It is the requirements placed on its 8-neighborhood that qualify a point for deletion together with the order of deletion that distinguish the various modifications of Rutovitz' original algorithm. Some of the derivations of the original algorithm are applied sequentially, and some are parallel algorithms. One iteration of a parallel algorithm may consist of several subiterations, targeting different boundary points on each subiteration. It is the order of removal and the configuration of the 8-neighborhood that qualify a boundary point for deletion that ultimately determine the topological properties of the skeleton that is produced.

The Zhang-Suen skeletonizing transform is a parallel algorithm that reduces regions of a Boolean image to an 8-connected skeletons of unit thickness. Each iteration of the algorithm consists of two subiterations. The first subiteration examines a 3×3 neighborhood of every southeast boundary point and northwest corner point. The configuration of its 3×3 neighborhood will determine whether the point can be deleted without corrupting the ideal skeleton. On the second subiteration a similar process is carried out to select northwest boundary points and southeast corner points that can be removed without corrupting the ideal skeleton. What is meant by corrupting the ideal skeleton and a discussion of the neighborhood configurations that make a contour point eligible for deletion follows next.

Let $\mathbf{a} \in \{0,1\}^{\mathbf{X}}$ be the source image. Select contour points are removed iteratively from $\mathbf{a}$ until a skeleton of unit thickness remains. A contour (boundary) point is a point that has pixel value 1 and has at least one 8-neighbor whose pixel value is 0. In order to preserve 8-connectivity, only contour points that will not disconnect the skeleton are removed.

Each iteration consists of two subiterations. The first subiteration removes southeast boundary points and northwest corner points. The second iteration removes northwest boundary points and southeast corner points. The conditions that qualify a contour point for deletion are discussed next.

Let $\mathbf{p}_1, \mathbf{p}_2, ..., \mathbf{p}_8$ be the 8-neighbors of $\mathbf{p}$. (See Figure 5.5.1 below.) Recalling our convention that boldface characters are used to represent points and italics are used for pixel values, let $p_i = \mathbf{a}(\mathbf{p}_i)$, $i \in \{1,\dots,8\}$.

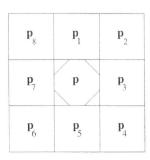

Figure 5.5.1. The 8-neighbors of $\mathbf{p}$.

Let $B(\mathbf{p})$ denote the number of nonzero 8-neighbors of $\mathbf{p}$ and let $A(\mathbf{p})$ denote the number of zero-to-one transitions in the ordered sequence $\mathbf{p}_1, \mathbf{p}_2, ..., \mathbf{p}_8, \mathbf{p}_1$. If $\mathbf{p}$ is a contour point and its 8-neighbors satisfy the following four conditions listed below, then $\mathbf{p}$ will be removed on the first subiteration. That is, the new value associated with $\mathbf{p}$ will be 0. The conditions for boundary pixel removal are

(a) $2 \leq B(\mathbf{p}) \leq 6$

(b) $A(\mathbf{p}) = 1$

(c) $p_1 \cdot p_3 \cdot p_5 = 0$

(d) $p_3 \cdot p_5 \cdot p_7 = 0$

In the example shown in Figure 5.5.2, $\mathbf{p}$ is not eligible for deletion during the first subiteration. With this configuration $B(\mathbf{p}) = 5$ and $A(\mathbf{p}) = 3$.

0	1	1
1	**p**	0
1	0	1

Figure 5.5.2. Example pixel values of the 8-neighborhood about $\mathbf{p}$.

Condition (a) insures that endpoints are preserved. Condition (b) prevents the deletion of points of the skeleton that lie between endpoints. Conditions (c) and (d) select southeast boundary points and northwest corner points for the first subiteration.

A contour point will be subject to deletion during the second subiteration provided its 8-neighbors satisfy conditions (a) and (b) above, and conditions (c') and (d') are given by

(c') $p_1 \cdot p_3 \cdot p_7 = 0$

(d') $p_1 \cdot p_5 \cdot p_7 = 0$

Iteration continues until either subiteration produces no change in the image. Figure 5.5.3 shows the Zhang-Suen skeleton superimposed over the original image of the SR71.

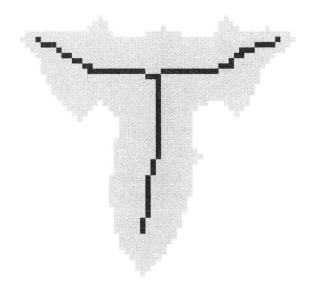

Figure 5.5.3. Zhang-Suen skeleton of the SR71.

Image Algebra Formulation

Let $\mathbf{a} \in \{0,1\}^{\mathbf{X}}$ be the source image. The template $\mathbf{t}$, defined pictorially in

t =

128	1	2
64		4
32	16	8

Figure 5.5.4. Census template used for Zhang-Suen thinning.

Figure 5.5.4, is a census template used in conjunction with characteristic functions defined on the sets

$$\begin{aligned} S_1 = \{&3, 6, 7, 12, 14, 15, 24, 28, 30, 48, 56, 60, 62, 96,\\ &112, 120, 129, 131, 135, 143, 192, 193, 195, 199,\\ &207, 224, 225, 227, 231, 240, 241, 243, 248, 249\} \end{aligned}$$

and

$$\begin{aligned} S_2 = \{&3, 6, 7, 12, 14, 15, 24, 28, 30, 31, 48, 56, 60, 62\\ &63, 96, 112, 120, 124, 126, 129, 131, 135, 143,\\ &159, 192, 193, 195, 224, 225, 227, 240, 248, 252\} \end{aligned}$$

to identify points that satisfy the conditions for deletion. S_1 targets points that satisfy (a) to (d) for the first subiteration and S_2 targets points that satisfy (a), (b), (c'), and (d') for the second subiteration.

The image algebra pseudocode for the Zhang-Suen skeletonizing transform is

$$\mathbf{b} := \mathbf{0}$$

$$\texttt{while } (\mathbf{a} \neq \mathbf{b}) \texttt{ loop}$$

$$\mathbf{b} := \mathbf{a}$$

$$\mathbf{a} := \mathbf{a} \cdot (1 - \chi_{S_1}(\mathbf{a} \oplus \mathbf{t}))$$

$$\mathbf{a} := \mathbf{a} \cdot (1 - \chi_{S_2}(\mathbf{a} \oplus \mathbf{t}))$$

$$\texttt{end loop}.$$

It may be more efficient to use lookup tables rather than the characteristic functions χ_{S_1}, and χ_{S_2}. The lookup tables that correspond to χ_{S_1}, and χ_{S_2} are defined by

$$f(r) = \begin{cases} 1 & \text{if } r \in S_1 \\ 0 & \text{otherwise,} \end{cases} \quad \text{and}$$

$$g(r) = \begin{cases} 1 & \text{if } r \in S_2 \\ 0 & \text{otherwise,} \end{cases}$$

respectively.

Comments and Observations

The Zhang-Suen algorithm reduces the image to a skeleton of unit pixel width. Endpoints are preserved. The Zhang-Suen algorithm is also immune to contour noise. The transform does not allow reconstruction of the original image from the skeleton.

Two-pixel-wide diagonal lines may become excessively eroded. 2×2 blocks will be completely removed.

Finite sets A and B are homotopic if there exists a continuous Euler number preserving one-to-one correspondence between the connected components of A and B. The fact that 2×2 blocks are removed by the Zhang-Suen transform means that the transform does not preserve homotopy. Preservation of homotopy is considered to be an important topological property. We will provide an example of a thinning algorithm that preserves homotopy in Section 5.6.

5.6.   Zhang-Suen Transform — Modified to Preserve Homotopy

The Zhang-Suen transform of Section 5.5 is an effective thinning algorithm. However, the fact that 2×2 blocks are completely eroded means that the transform does not preserve homotopy. Preservation of homotopy is a desirable property of a thinning algorithm. We take this opportunity to discuss what it means for a thinning transform to preserve homotopy.

Two finite sets A and B are homotopic if there exists a continuous Euler number preserving one-to-one correspondence between the connected components of A and B.

Figure 5.6.1 shows the results of applying the two versions of the Zhang-Suen thinning algorithm to an image. The original image is represented in gray. The largest portion of the original image is the SR71. A 3×3-pixel square hole has been punched out near the top center. An X has been punched out near the center of the SR71. A 2×2-pixel square has been added to the lower right-hand corner. The original Zhang-Suen transform has been applied to the image on the left. The Zhang-Suen transform modified to preserve homotopy has been applied to the image on the right. The resulting skeletons are represented in black.

The connectivity criterion for feature pixels is 8-connectivity. For the complement 4-connectivity is the criterion. In analyzing the image, note that the white X in the center of the SR71 is not one 8-connected component of the complement of the image, but rather nine 4-connected components.

The feature pixels of the original image are contained in two 8-connected components. Its complement consists of eleven 4-connected components. The skeleton image produced by the original Zhang-Suen algorithm (left Figure 5.6.1) has one 8-connected components. Its complement has eleven 4-connected components. Therefore, there is not a one-to-one correspondence between the connected components of the original image and its skeleton.

Notice that the modified Zhang-Suen transform (right Figure 5.6.1) does preserve homotopy. This is because the 2×2 in the lower right-hand corner of the original image was shrunk to a point rather than being erased.

To preserve homotopy, the conditions that make a point eligible for deletion must be made more stringent. The conditions that qualify a point for deletion in the original Zhang-Suen algorithm remain in place. However, the 4-neighbors (see Figure 5.5.1) of the point $\mathbf{p}$ are examined more closely. If the target point $\mathbf{p}$ has none or one 4-neighbor that has pixel value 1, then no change is made to the existing set of criteria for deletion. If $\mathbf{p}$ has two or three 4-neighbors with pixel value 1, it can be deleted on the first pass provided $p_3 \cdot p_5 = 0$. It can be deleted on the second pass if $p_1 \cdot p_7 = 0$. These changes insure that 2×2 blocks do not get completely removed.

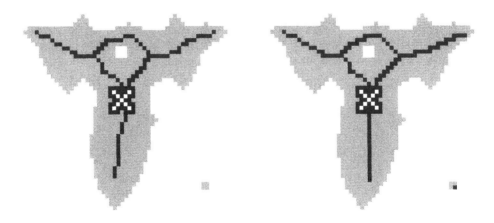

Figure 5.6.1. Original Zhang-Suen transform
(left) and modified Zhang-Suen transform (right).

Image Algebra Formulation

As probably anticipated, the only effect on the image algebra formulation caused by modifying the Zhang-Suen transform to preserve homotopy shows up in the sets S_1 and

S_2 of Section 5.5. The sets S'_1 and S'_2 that replace them are

$$S'_1 = S_1 \setminus \{28, 30, 60, 62\}$$
and
$$S'_2 = S_2 \setminus \{193, 195, 225, 227\}.$$

5.7.   Thinning Edge Magnitude Images

The edge thinning algorithm discussed here uses edge magnitude and edge direction information to reduce the edge elements of an image to a set that is only one pixel wide. The algorithm was originally proposed by Nevatia and Babu [17]. The algorithm presented here is a variant of Nevatia and Babu's algorithm. It was proposed in Ritter, Gader, and Davidson [18] and was implemented by Norris [19].

The input to the algorithm is an image of edge magnitudes and directions. The input image is also assumed to have been thresholded. The edge thinning algorithm retains a point as an edge point if its magnitude and direction satisfy certain heuristic criteria in relation to the magnitudes and directions of two of its 8-neighbors. The criteria that a point must satisfy to be retained as an edge point will be presented in the next subsection.

The edge image that was derived in Section 3.10 will serve as our example source image. Recall that even after thresholding the thickness of the resulting edges was undesirable. Let $\mathbf{a} = (\mathbf{m}, \mathbf{d}) \in \left(\mathbb{R}^2\right)^{Z^2}$ be an image with edge magnitudes $\mathbf{m} \in \mathbb{R}^{Z^2}$ and directions $\mathbf{d} \in \mathbb{R}^{Z^2}$. The range of $\mathbf{d}$ in this variation of the edge thinning algorithm is the set (in degrees), $\{0, 30, 60, \ldots, 330\} \subset \mathbb{R}$.

To understand the requirements that a point must satisfy to be an edge point, it is necessary to know what is meant by the expression "the two 8-neighbors normal to the direction at a point." The two 8-neighbors that comprise the normal to a given horizontal or vertical edge direction are apparent. The question then arises about points in the 8-neighborhood of (i, j) that lie on the normal to, for instance, a 30° edge direction. For non-vertical and non-horizontal edge directions, the diagonal elements of the 8-neighborhood are assumed to make up the normals. For example, the points $(i - 1, j - 1)$ and $(i + 1, j + 1)$ are on the normal to a 30° edge direction. These two points are also on the normal for edge directions 60°, 210°, and 240° (see Figure 5.7.1). We are now ready to present the conditions a point must satisfy to qualify as an edge point.

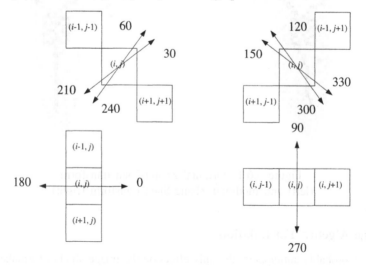

Figure 5.7.1. Directions and their 8-neighborhood normals about (i, j).

An edge element is deemed to exist at the point (i, j) if any one of the following sets of conditions hold:

(1) The edge magnitudes of points on the normal to $\mathbf{d}(i, j)$ are both zero.

(2) The edge directions of the points on the normal to $\mathbf{d}(i, j)$ are both within $30°$ of $\mathbf{d}(i, j)$, and $\mathbf{m}(i, j)$ is greater than the magnitudes of its neighbors on the normal.

(3) One neighbor on the normal has direction within $30°$ of $\mathbf{d}(i, j)$, the other neighbor on the normal has direction within $30°$ of $\mathbf{d}(i, j) + 180°$, and $\mathbf{m}(i, j)$ is greater than the magnitude of the former of its two neighbors on the normal.

The result of applying the edge thinning algorithm to Figure 3.10.3 can be seen in Figure 5.7.2.

Image Algebra Formulation

Let $\mathbf{a} = (\mathbf{m}, \mathbf{d}) \in (\mathbb{R}^2)^{\mathbf{X}}$ be an image on the point set $\mathbf{X}$, where $\mathbf{m} \in \mathbb{R}^{\mathbf{X}}$ is the edge magnitude image, and $\mathbf{d} \in \mathbb{R}^{\mathbf{X}}$ is the edge direction image.

The parameterized template $\mathbf{t}(i)$ has support in the three-by-three 8-neighborhood about its target point. The ith neighbor (ordered clockwise starting from bottom center) has value 1. All other neighbors have value 0. Figure 5.7.3 provides an illustration for the case $i = 4$. The small number in the upper right-hand corner of the template cell indicates the point's position in the neighborhood ordering.

The function $g : \{0, 30, \ldots, 330\} \rightarrow \{0, 1, \ldots, 7\}$ defined by

$$g(r) = \left\lfloor \frac{2r + 30}{90} \right\rfloor$$

is used to defined the image $\mathbf{i} = g(\mathbf{d}) \in \{0, \ldots, 7\}^{\mathbf{X}}$. The pixel value $\mathbf{i}(\mathbf{x})$ is equal to the positional value of one of the 8-neighbors of $\mathbf{x}$ that is on the normal to $\mathbf{d}(\mathbf{x})$.

The function f is used to discriminate between points that are to remain as edge points and those that are to be deleted (have their magnitude set to 0). It is defined by

$$f(m, m_1, m_2, d, d_1, d_2) = \begin{cases} (m, d) & \text{if } m_1 = m_2 = 0 \\[1.5em] \cdot & \text{if } m > m_1, m > m_2, \\ \cdot & \quad |d - d_1| \le 30, \\ \cdot & \quad \text{and } |d - d_2| \le 30 \\[1.5em] & \text{if } m > m_1, |d - d_1| \le 30, \\ & \quad \text{and } |180 + d - d_2| \le 30 \\[1.5em] & \text{if } m > m_2, |d - d_2| \le 30, \\ & \quad \text{and } |180 + d - d_1| \le 30 \\[1.5em] (0, d) & \text{otherwise.} \end{cases}$$

The edge thinned image $\mathbf{b} \in (\mathbb{R}^2)^{\mathbf{X}}$ can now be expressed as

$$\mathbf{b} := f(\mathbf{m}, \mathbf{m} \oplus \mathbf{t}(\mathbf{i}), \mathbf{m} \oplus \mathbf{t}((\mathbf{i} + 4) \bmod 8), \mathbf{d}, \mathbf{d} \oplus \mathbf{t}(\mathbf{i}), \mathbf{d} \oplus \mathbf{t}((\mathbf{i} + 4) \bmod 8)).$$

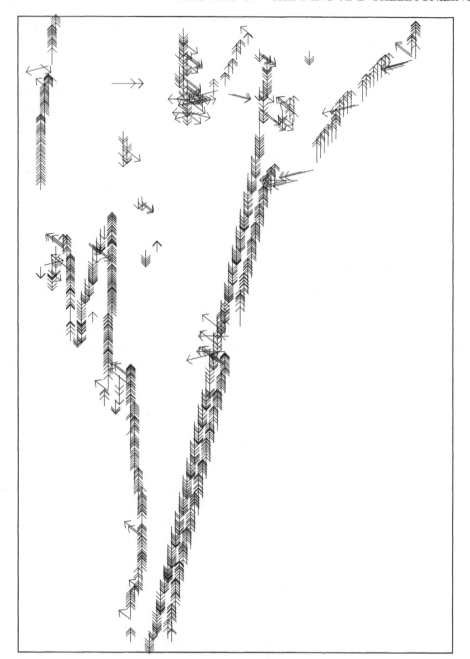

Figure 5.7.2. Thinned edges with direction vectors.

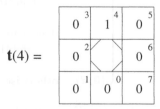

$\mathbf{t}(4) =$

0 [3]	1 [4]	0 [5]
0 [2]		0 [6]
0 [1]	0 [0]	0 [7]

Figure 5.7.3. Parameterized template used for edge thinning.

5.8. Exercises

1. Give a set theoretic image algebra formulation of the Pavlidis thinning algorithm.

2. a. Provide an image algebra formulation of the medial axis transform using the chessboard distance. Implement the algorithm on a Boolean image containing irregularly shaped objects.
   b. Repeat 2.a using the city block distance. Compare and analyze the results with those obtained in 2.a.

3. Explore and discuss how a skeletonized object can be used as a primitive for pattern recognition.

4. The medial axis transform has been successfully applied in path planning for robotic applications. Given a robot of circumference d, construct a building floor with hallways of different widths, some wider than d and others narrower than d. Find the medial axis of the hallways and eliminate those that cannot be navigated by the robot. Given your map of the building floor as the input image, write an algorithm (in image algebra) that will compute the paths the robot can travel.

5. a. Provide an image algebra formulation of the distance transform using the Euclidean distance. Implement the algorithm on a Boolean image containing irregularly shaped objects.
   b. Repeat 5.a using the chessboard distance. Compare and analyze the results with those obtained in 5.a.

6. Give a non-recursive image algebra formulation of a restoration transform that restores the distance skeletons obtained from exercises 5.a and 5.b.

7. Let $N_r : \mathbf{X} \to 2^{\mathbf{X}}$ be defined by $N_r(\mathbf{x}) = \{\mathbf{y} \in \mathbf{X} : d(\mathbf{x}, \mathbf{y}) < r\}$, $\mathbf{a} \in \{0,1\}^{\mathbf{X}}$, and let A denote the support of $\mathbf{a}$.
   a. Show that $d\left(\mathbf{x}, \tilde{A}\right) > r$ if and only if $\mathbf{x} \in B$ where B denotes the support of $\mathbf{a} \boxtimes N_r$.
   b. Show that $d(\mathbf{x}, A) < r$ if and only if $\mathbf{x} \in C$ where C denotes the support of $\mathbf{a} \boxtimes N_r$.

8. Derive a less cumbersome algorithm that yields a one-pixel-wide directional edge image than the one given in Section 5.7. Express your algorithm in image algebra.

5.9. References

[1] T. Pavlidis, *Structural Pattern Recognition*. New York: Springer-Verlag, 1977.

[2] H. Blum, "An associative machine for dealing with the visual field and some of its biological implications," in *Biological Prototypes and Synthetic Systems* (Bernard and Kare, eds.), vol. 1, New York: Plenum Press, 1962.

[3] H. Blum, "A transformation for extracting new descriptors of shape," in *Symposium on Models for Perception of Speech and Visual Form* (W. Whaten-Dunn, ed.), Cambridge, MA: MIT Press, 1967.

[4] H. Blum, "Biological shape and visual science (part I)," *Journal of Theoretical Biology*, vol. 38, 1973.

[5] J. Davidson, "Thinning and skeletonizing: A tutorial and overview," in *Digital Image Processing: Fundamentals and Applications* (E. Dougherty, ed.), New York: Marcel Dekker, Inc., 1991.

[6] B. Jang and R. Chin, "Analysis of thinning algorithms using mathematical morphology," *IEEE Transactions on Pattern Analysis and Machine Intelligence*, vol. 12, pp. 541–551, June 1990.

[7] L. Lam, S. Lee, and C. Suen, "Thinning methodologies — a comprehensive survey," *IEEE Transactions on Pattern Analysis and Machine Intelligence*, vol. 14, pp. 868–885, June 1992.

[8] D. Marr, "Representing visual information," tech. rep., MIT AI Lab, 1977. AI Memo 415.

[9] A. Rosenfeld and J. Pfaltz, "Sequential operators in digital picture processing," *Journal of the ACM*, vol. 13, 1966.

[10] A. Rosenfeld and A. Kak, *Digital Picture Processing*. New York, NY: Academic Press, 2nd ed., 1982.

[11] G. Borgefors, "Distance transformations in digital images," *Computer Vision, Graphics, and Image Processing*, vol. 34, pp. 344–371, 1986.

[12] D. Li and G. Ritter, "Recursive operations in image algebra," in *Image Algebra and Morphological Image Processing*, vol. 1350 of *Proceedings of SPIE*, (San Diego, CA), July 1990.

[13] G. Ritter, "Recent developments in image algebra," in *Advances in Electronics and Electron Physics* (P. Hawkes, ed.), vol. 80, pp. 243–308, New York, NY: Academic Press, 1991.

[14] U. Eckhardt, "A note on Rutovitz' method for parallel thinning," *Pattern Recognition Letters*, vol. 8, no. 1, pp. 35–38, 1988.

[15] D. Rutovitz, "Pattern recognition," *Journal of the Royal Statistical Society*, vol. 129, Series A, pp. 504–530, 1966.

[16] T. Zhang and C. Suen, "A fast parallel algorithm for thinning digital patterns," *Communications of the ACM*, vol. 27, pp. 236–239, Mar. 1984.

[17] R. Nevatia and K. Babu, "Linear feature extraction and description," *Computer Graphics and Image Processing*, vol. 13, 1980.

[18] G. Ritter, P. Gader, and J. Davidson, "Automated bridge detection in FLIR images," in *Proceedings of the Eighth International Conference on Pattern Recognition*, (Paris, France), 1986.

[19] K. Norris, "An image algebra implementation of an autonomous bridge detection algorithm for forward looking infrared (flir) imagery," Master's thesis, University of Florida, Gainesville, FL, 1992.

CHAPTER 6
CONNECTED COMPONENT ALGORITHMS

6.1. Introduction

A wide variety of techniques employed in computer vision reduce gray level images to binary images. These binary images usually contain only objects deemed interesting and worthy of further analysis. Objects of interest are analyzed by computing various geometric properties such as size, shape, or position. Before such an analysis is done, it is often necessary to remove various undesirable artifacts such as feelers from components through *pruning* processes. Additionally, it may be desirable to label the objects so that each object can be analyzed separately and its properties listed under its' label. As such, component labeling can be considered a fundamental segmentation technique.

The labeling of connected components also provides for the number of components in an image. However, there are often faster methods for determining the number of components. For example, if components contain small holes, the holes can be rapidly filled. An application of the Euler characteristic to an image containing components with no holes provides one fast method for counting components. This chapter provides some standard labeling algorithms as well as examples of pruning, hole filling, and component counting algorithms.

6.2. Component Labeling for Binary Images

Component labeling algorithms segment the domain of a binary image into partitions that correspond to connected components. Component labeling is one of the fundamental segmentation techniques used in image processing. As well as being an important segmentation technique in itself, it is often an element of more complex image processing algorithms [1].

The set of interest for component labeling of an image, $a \in \{0,1\}^X$, is the set of points $Y \subseteq X$ that have pixel value 1. This set is partitioned into disjoint connected subsets. The partitioning is represented by an image b in which all points of Y that lie in the same connected component have the same pixel value. Distinct pixel values are assigned to distinct connected components.

Either 4- or 8-connectivity can be used for component labeling. For example, let the image to the left in Figure 6.2.1 be the binary source image (pixel values equal to 1 are shown in black). The image in the center represents the labeling of 4-components. The image to the right represents the 8-components. Different colors (or gray levels) are often used to distinguish connected components. This is why component labeling is also referred to as *blob coloring*. From the different gray levels in Figure 6.2.1, we see that the source image has five 4-connected components and one 8-connected component.

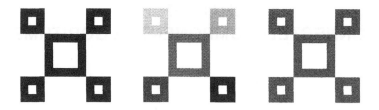

Figure 6.2.1. Original image, labeled 4-component image, and labeled 8-component image.

Image Algebra Formulation

Let $a \in \{0,1\}^{\mathbf{X}}$ be the source image, where $\mathbf{X}$ is an $m \times n$ grid. Let the image $\mathbf{d}$ be defined by

$$\mathbf{d}(i,j) = (i-1)n + j, \text{where } 1 \leq i \leq m \text{ and } 1 \leq j \leq n.$$

The algorithm starts by assigning each black pixel a unique label and uses $\mathbf{c}$ to keep track of component labels as it proceeds. The neighborhood N is either the von Neumann neighborhood for 4-connectivity or the Moore neighborhood for 8-connectivity.

When the loop of the pseudocode below terminates, $\mathbf{b}$ will be the image that contains connected component information. That is, points of $\mathbf{X}$ that have the same pixel value in the image $\mathbf{b}$ lie in the same connected component of $\mathbf{X}$.

$$\mathbf{b} := 0$$
$$\mathbf{c} := \mathbf{d} \cdot \mathbf{a}$$
$$\texttt{while } \mathbf{b} \neq \mathbf{c} \texttt{ loop}$$
$$\mathbf{b} := \mathbf{c}$$
$$\mathbf{c} := (\mathbf{b} \boxtimes N) \cdot \mathbf{a}$$
$$\texttt{end loop}.$$

Initially, each feature pixel of $\mathbf{a}$ has a corresponding unique label in the image $\mathbf{c}$. The algorithm works by propagating the maximum label within a connected component to every point in the connected component.

Alternate Image Algebra Formulations

Let $\mathbf{d}$ be as defined above. Define the neighborhood function $N(\mathbf{a})$ as follows:

$$[N(\mathbf{a})](\mathbf{y}) = \{\mathbf{y}\} \cup \{\mathbf{x} = \mathbf{y} + (0,i) \text{ or } \mathbf{y} + (i,0) : \mathbf{a}(\mathbf{x}) \neq 0, i = 1 \text{ or } -1\}.$$

The components of $\mathbf{a}$ are labeled using the image algebra pseudocode that follows. When the loop terminates the image with the labeled components will be contained in image variable $\mathbf{b}$.

$$\mathbf{b} := 0$$
$$\mathbf{c} := \mathbf{d} \cdot \mathbf{a}$$
$$\texttt{while } \mathbf{c} \neq \mathbf{b} \texttt{ loop}$$
$$\mathbf{b} := \mathbf{c}$$
$$\mathbf{c} := \mathbf{b} \boxtimes N(\mathbf{a})$$
$$\texttt{end loop}.$$

Note that this formulation propagates the minimum value within a component to every point in the component. The variant neighborhood N as defined above is used to label 4-connected components. It is easy to see how the transition from invariant neighborhood to variant has been made. The same transition can be applied to the invariant neighborhood used for labeling 8-components.

Variant neighborhoods are not as efficient to implement as invariant neighborhoods. However, in this implementation of component labeling there is one less image multiplication for each iteration of the while loop.

Although the above algorithms are simple, their computational cost is high. The number of iterations is directly proportional to mn. A faster alternate labeling algorithm [2] proceeds in two phases. The number of iterations is only directly proportional to $m+n$. However, the price for decreasing computational complexity is an increase in space complexity.

In the first phase, the alternate algorithm applies the shrinking operation developed in Section 6.4 to the source image $m + n$ times. The first phase results in $m + n$ binary images $a_0, a_1, \ldots, a_{m+n-1}$. Each a_k represents an intermediate image in which connected components are shrinking toward a unique isolated pixel. This isolated pixel is the top left corner of the component's bounding rectangle, thus it may not be in the connected component. The image a_{m+n-1} consists of isolated black pixels.

In the second phase, the faster algorithm assigns labels to each of the isolated black pixels of the image a_{m+n-1}. These labels are then propagated to the pixels connected to them in a_{m+n-2} by applying a label propagating operation, then to a_{m+n-3}, and so on until a_0 is labeled. In the process of propagating labels, new isolated black pixels may be encountered in the intermediate images. When this occurs, the isolated pixels are assigned unique labels and label propagation continues.

Note that the label propagating operations are applied to the images in the reverse order they are generated by the shrinking operations. More than one connected component may be shrunk to the same isolated pixel, but at different iterations. A unique label assigned to an isolated pixel should include the iteration number at which it is generated.

The template s used for image shrinking and the neighborhood function P used for label propagation are defined in following figure. The faster algorithm for labeling

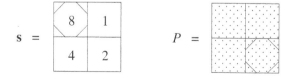

8-components is as follows:

$$\mathbf{a}_0 := \mathbf{a}$$

$$k := 0$$

$$\text{while } \mathbf{a}_k \neq 0 \text{ loop}$$

$$\mathbf{a}_{k+1} := \chi_{\{5,7,9,10,11,12,13,14,15\}}(\mathbf{a}_k \oplus \mathbf{s})$$

$$k := k + 1$$

$$\text{end loop}$$

$$l := k - 1$$

$$\mathbf{c} := \mathbf{a}_l$$

$$\text{while } l > 0 \text{ loop}$$

$$\mathbf{c} := (\mathbf{d} \cdot (m + n) + l) \cdot \chi_{=1}(\mathbf{c}) \vee \mathbf{c}$$

$$\mathbf{c} := \mathbf{c} \vee \mathbf{a}_{l-1}$$

$$\mathbf{c} := (\mathbf{c} \boxdot P) \cdot \mathbf{a}_{l-1}$$

$$l := l - 1$$

$$\text{end loop.}$$

When the last loop terminates the labeled image will be contain in the variable c. If 4-component labeling is preferred, the characteristic function in line 4 should be replaced with

$$\chi_{\{7,9,11,12,13,14,15\}} \cdot$$

Comments and Observations

The alternate algorithm must store $m + n$ intermediate images. There are other fast algorithms proposed to reduce this storage requirement [2, 3, 4, 5]. It may also be more efficient to replace the characteristic function $\chi_{\{5,7,9,...,15\}}$ by a lookup table.

6.3.   Labeling Components with Sequential Labels

The component labeling algorithms of Section 6.2 began by assigning each point in the domain of the source image the number that represents its position in a forward scan of the domain grid. The label assigned to a component is either the maximum or minimum from the set of point numbers encompassed by the component (or its bounding box). Consequently, label numbers are not used sequentially, that is, in $1, 2, 3, \ldots, n$ order.

In this section two algorithms for labeling components with sequential labels are presented. The first locates connected components and assigns labels sequentially, and thus takes the source image as input. The second takes as input labeled images, such as those produced by the algorithms of Section 6.2, and reassigns new labels sequentially.

It is easy to determine the number of connected components in an image from its corresponding sequential label image; simply find the maximum pixel value of the label image. The set of label numbers is also known, namely $\{1, 2, 3, \ldots, v\}$, where v is the maximum pixel value of the component image. If component labeling is one step in larger image processing regimen, this information may facilitate later processing steps.

Labeling with sequential labels also offers a savings in storage space. Suppose one is working with 17×17 gray level images whose pixel values have eight bit representations. Up to 255 labels can be assigned if labeling with sequential labels is used. It may not be possible to represent the label image at all if labeling with sequential labels is not used and a label value greater than 255 is assigned.

Image Algebra Formulation

Let $\mathbf{a} \in \{0,1\}^{\mathbf{X}}$ be the source image, where $\mathbf{X}$ is an $m \times n$ grid. The neighborhood N used is either the von Neumann neighborhood for labeling 4-connected components, or the Moore neighborhood for labeling 8-connected components.

When the algorithm below ends, the sequential label image will be contained in image variable $\mathbf{b}$.

$$l := 0$$
$$\mathbf{b} := \mathbf{a}$$
$$\mathbf{c} := \mathbf{a}$$
```
for i in 1..m loop
  for j in 1..n loop
    if b(i, j) = 1 then
      l := l + 1
      b(i, j) := b(i, j) + l
      while b ≠ c loop
        c := b
        b := (c ☑ N) · a
      end loop
    end if
  end loop
end loop
b := b − a.
```

Alternate Image Algebra Formulation

The alternate image algebra algorithm takes as input a label image $\mathbf{a}$ and reassigns labels sequentially. When the `while` loop in the pseudocode below terminates, the image variable $\mathbf{b}$ will contain the sequential label image. The component connectivity criterion (four or eight) for $\mathbf{b}$ will be the same that was used in generating the image $\mathbf{a}$.

$$i := 0$$
$$\mathbf{b} := 0$$
$$m := \vee \mathbf{a}$$
```
while m ≠ 0 loop
  i := i + 1
  b := i · χ_{=m}(a) + b
  a := (χ_{<m}(a)) · a
  m := ∨ a
end loop.
```

Comments and Observations

Figure 6.3.1 shows a binary image (left) whose feature pixels are represented by an asterisk. The pixel values assigned to its 4- and 8-components after sequential labeling are seen in the center and right images, respectively, of Figure 6.3.1.

```
. . . . . . . . . . . . . . . . . .     . . . . . . . . . . . . . . . . . .     . . . . . . . . . . . . . . . . . .
. .***********.     ..222222222222.     ..111111111111.
.*............*.     .1...........2.     .1...........1.
.*............*.     .1...........2.     .1...........1.
.*............*.     .1...........2.     .1...........1.
.*..***..***..*.     .1..666..333..2.     .1..444..222..1.
.*..***..***..*.     .1..666..333..2.     .1..444..222..1.
.*..***..***..*.     .1..666..333..2.     .1..444..222..1.
.*..***....**.*.     .1..666....33.2.     .1..444....22.1.
.*.....***.**.*.     .1.....555.33.2.     .1.....444.22.1.
.*.**..***.**.*.     .1.44..555.33.2.     .1.33..444.22.1.
.*.**..***.**.*.     .1.44..555.33.2.     .1.33..444.22.1.
.*.**..***.**.*.     .1.44..555.33.2.     .1.33..444.22.1.
.*.**......**.*.     .1.44......33.2.     .1.33......22.1.
.*............*.     .1...........2.     .1...........1.
.*************..     .111111111111..     .111111111111..
. . . . . . . . . . . . . . . . . .     . . . . . . . . . . . . . . . . . .     . . . . . . . . . . . . . . . . . .
```

Figure 6.3.1. A binary image, its 4-labeled image (center) and 8-labeled image (right).

6.4. Counting Connected Components by Shrinking

The purpose of the shrinking algorithm presented in this section is to count the connected components in a Boolean image. The idea, based on Levialdi's approach [6], is to shrink each component to a point and then count the number of points obtained. To express this idea explicitly, let $\mathbf{a} \in \{0, 1\}^{\mathbf{X}}$ denote the source image and let h denote the Heaviside function defined by

$$h(t) = \begin{cases} 0 & \text{if } t \leq 0 \\ 1 & \text{if } t > 0. \end{cases}$$

One of four windows may be chosen as the shrinking pattern. Each is distinguished by the direction it compresses the image.

Shrinking toward top right —

$$\mathbf{a}(i, j) := h[h(\mathbf{a}(i, j - 1) + \mathbf{a}(i, j) + \mathbf{a}(i + 1, j) - 1) + h(\mathbf{a}(i, j) + \mathbf{a}(i + 1, j - 1) - 1)].$$

Shrinking toward top left —

$$\mathbf{a}(i, j) := h[h(\mathbf{a}(i, j) + \mathbf{a}(i, j + 1) + \mathbf{a}(i + 1, j) - 1) + h(\mathbf{a}(i, j) + \mathbf{a}(i + 1, j + 1) - 1)].$$

Shrinking toward bottom left —

$$\mathbf{a}(i, j) := h[h(\mathbf{a}(i, j) + \mathbf{a}(i - 1, j) + \mathbf{a}(i, j + 1) - 1) + h(\mathbf{a}(i, j) + \mathbf{a}(i - 1, j + 1) - 1)].$$

Shrinking toward bottom right —

$$\mathbf{a}(i, j) := h[h(\mathbf{a}(i, j) + \mathbf{a}(i, j - 1) + \mathbf{a}(i - 1, j) - 1) + h(\mathbf{a}(i, j) + \mathbf{a}(i - 1, j - 1) - 1)].$$

Each iteration of the algorithm consists of applying, in parallel, the selected shrinking window on every element of the image. Iteration continues until the original image has been reduced to the zero image. After each iteration the number of isolated points is counted and added to a running total of isolated points. Each isolated point corresponds to an 8-connected component of the original image.

This shrinking process is illustrated in the series (in left to right and top to bottom order) of images in Figure 6.4.1. The original image (in the top left corner) consists of five 8-connected patterns. By the fifth iteration the component in the upper left corner has been reduced to an isolated pixel. Then the isolated pixel is counted and removed in the following iteration. This process continues until each component has been reduced to an isolated pixel, counted, and then removed.

Figure 6.4.1. The parallel shrinking process.

Image Algebra Formulation

Let $\mathbf{a} \in \{0,1\}^{\mathbf{X}}$ be the source image. We will illustrate the image algebra formulation using the shrinking window that compresses toward the top right. This choice of shrinking window dictates the form of neighborhoods M and N below. The template $\mathbf{u}$ is used to find isolated points.

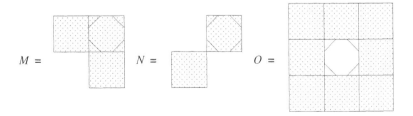

The image variable $\mathbf{a}$ is initialized with the original image. The integer variable p is initially set to zero. The algorithm consists of executing the following loop until the image variable **a** is reduced to the zero image.

$$\text{while } a \neq 0 \text{ loop}$$
$$p := p + \sum \left(\mathbf{a} \wedge \left(1 - \left(\mathbf{a} \, \oslash O \right) \right) \right)$$
$$\mathbf{a} := \chi_{>0}(\chi_{>1}(\mathbf{a} \oplus M) + \chi_{>1}(\mathbf{a} \oplus N))$$
$$\text{end loop.}$$

When the algorithm terminates, the value of p will represent the number of 8-connected components of the original image.

Alternate Image Algebra Formulation

The above image algebra formulation closely parallels the formulation presented in Levialdi [6]. However, it involves three convolutions in each iteration. We can reduce number of convolutions in each iteration to only one by using the following census template. A binary census template (whose weights are powers of 2), when applied to a binary image, encodes the neighborhood configurations of target pixels. From the configuration of a pixel, we can decide whether to change it from 1 to 0 or from 0 to 1, and we can also decide whether it is an isolated pixel or not.

$$
\mathbf{s} =
\begin{array}{|c|c|c|}
\hline
16 & 32 & 64 \\
\hline
8 & 1 & 128 \\
\hline
4 & 2 & 256 \\
\hline
\end{array}
$$

The algorithm using the binary census template is expressed as follows:

$$\text{while } a \neq 0 \text{ loop}$$
$$\mathbf{a} := \mathbf{a} \oplus \mathbf{s}$$
$$p := p + \Sigma(\chi_{=1}(\mathbf{a}))$$
$$\mathbf{a} := f(\mathbf{a})$$
$$\text{end loop.}$$

In order to count 8-components, the function f should be defined as

$$f(r) = \begin{cases} 1 & \text{if } r \bmod 16 \in \{3, 5, 7, 9, 10, 11, 13, 14, 15\} \\ 0 & \text{otherwise.} \end{cases}$$

The 4-components of the image can be counted by defining f to be

$$f(r) = \begin{cases} 1 & \text{if } r \bmod 16 \in \{3, 7, 9, 11, 13, 14, 15\} \\ 0 & \text{otherwise.} \end{cases}$$

In either case, the movement of the shrinking component is toward the pixel at the top right of the component.

Comments and Observations

The maximum number of iterations required to shrink a component to its corresponding isolated point is equal to the d_1 distance of the element of the region farthest from top rightmost corner of the rectangle circumscribing the region.

6.5. Pruning of Connected Components

Pruning of connected components is a common step in removing various undesirable artifacts created during preceding image processing steps. Pruning usually removes thin objects such as feelers from thick, blob-shaped components. There exists a wide variety of pruning algorithms. Most of these algorithms are usually structured to solve a particular task or problem. For example, after edge detection and thresholding, various objects of interest may be reduced to closed contours. However, in real situations there may also be many unwanted edge pixels left. In this case a pruning algorithm can be tailored so that only closed contours remain.

In this section we present a pruning algorithm that removes feelers and other objects that are of single-pixel width while preserving closed contours of up to single-pixel width as well as 4-connectivity of components consisting of thick bloblike objects that may be connected by thin lines. In this algorithm, a pixel is removed from an object if it contains fewer than two 4-neighbors. The procedure is iterative and continues until every object pixel has two or more 4-neighbors.

Image Algebra Formulation

Let $a \in \{0,1\}^X$ be the source image and N denote the von Neumann neighborhood. The following simple algorithm will produce the pruned version of a described above:

$$\mathbf{b} := 0$$
$$\text{while } \mathbf{b} \neq \mathbf{a} \text{ loop}$$
$$\quad \mathbf{b} := \mathbf{a}$$
$$\quad \mathbf{a} := \chi_{\geq 3}[\mathbf{a} \cdot (\mathbf{a} \oplus N)]$$
$$\text{end loop}.$$

The results of applying this pruning algorithm are shown in Figure 6.5.1.

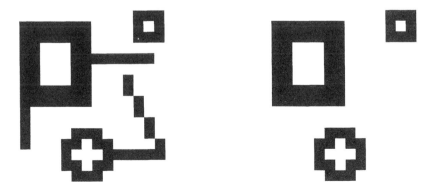

Figure 6.5.1. The source image $\mathbf{a}$ is shown at
the left and the pruned version of $\mathbf{a}$ on the right.

Comments and Observations

The algorithm presented above was first used in an autonomous target detection scheme for detecting tanks and military vehicles in infrared images [7]. As such, it represents a specific pruning algorithm that removes only thin feelers and *non-closed* thin lines. Different size neighborhoods must be used for the removal of larger artifacts.

Since the algorithm presented here is an iterative algorithm that uses a small neighborhood, removal of long feelers may increase computational time requirements beyond acceptable levels, especially if processing is done on sequential architectures. In such cases, morphological filters may be more appropriate.

6.6. Hole Filling

As its name implies, the hole filling algorithm fills holes in binary images. A hole in this context is a region of 0-valued pixels bounded by an 8-connected set of 1-valued pixels. A hole is filled by changing the pixel value of points in the hole from 0 to 1. See Figure 6.6.1 for an illustration of hole filling.

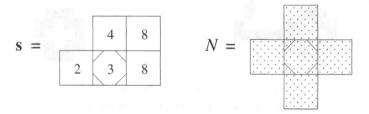

Figure 6.6.1. Original image (left) and filled image (right).

Image Algebra Formulation

Let $a \in \{0, 1\}^{\mathbf{X}}$ be the source image, and let the template s and the neighborhood N be pictured below. The hole filling image algebra code below is from the code derived by Ritter [8].

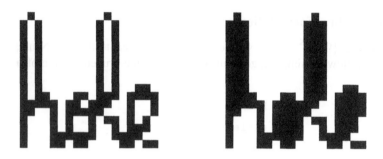

$$c := 0$$
$$b := a$$
$$\texttt{while } b \neq c \texttt{ loop}$$
$$c := b;$$
$$b = b + \chi_{\{14,22\}}(b \oplus s)$$
$$\texttt{end loop}$$
$$c := 0$$
$$\texttt{while } b \neq c \texttt{ loop}$$
$$c := b$$
$$b := (b \boxtimes N) \vee a$$
$$\texttt{end loop}.$$

When the first **while** loop terminates, the image variable **b** will contain the objects in the original image **a** filled with 1's and some extra 1's attached to their exteriors. The second **while** loop "peels off" the extraneous 1's to produce the final output image **b**. The algorithm presented above is efficient for filling small holes. The alternate version presented next is more efficient for filling large holes.

Alternate Image Algebra Formulation

Let $a \in \{0,1\}^X$ be the source image and N the neighborhood as previously defined. The holes in **a** are filled using the image algebra pseudocode below.

$$b := 0$$
$$c := 1$$
$$\texttt{while } c \neq b \texttt{ loop}$$
$$b := c$$
$$c := (b \boxtimes N) \vee a$$
$$\texttt{end loop}.$$

When the loop terminates the "filled" image will be contained in the image variable **c**. The second method fills the whole domain of the source image with 1's. Then, starting at the edges of the domain of the image, extraneous 1's are peeled away until the exterior boundary of the objects in the source image are reached. Depending on how $\boxtimes$ is implemented, it may be necessary to add the line

$$c(0,0) = 0$$

immediately after the statement that initializes **c** to **1**. This is because template **t** needs to encounter a 0 before the peeling process can begin.

6.7. Exercises

1.   Rewrite the connected component labeling algorithms using the operations of $\bigotimes$ and $\bigwedge$ instead of $\boxtimes$ and $\boxtimes$.

2.   As mentioned in Section 6.2, there exist other fast connected component labeling algorithms. Using the references provided, give an image algebra specification of one such algorithm.

3.   Given a Boolean image $\mathbf{a} \in \{0,1\}^{\mathbf{X}}$, let $\mathbf{Y} \subseteq \mathbf{X}$ be the set of points having pixel value 1 in $\mathbf{a}$. Let $\mathbf{b} \in \{0,1,2\}^{\mathbf{X}}$ be defined by

$$\mathbf{b}(\mathbf{x}) = \begin{cases} 2 & if\ M(\mathbf{x}) \subset \mathbf{Y} \\ \mathbf{a}(\mathbf{x}) & otherwise. \end{cases}$$

Specify an algorithm in image algebra that computes $\mathbf{b}$.

4.   A binary image over a rectangular $m \times n$ array is equivalent to a binary $m \times n$ matrix. Both the image and the matrix are bulky and can be easily compressed. For example, given an 8×8 boolean image of the form

$$\begin{bmatrix} 0 & 1 & 0 & 0 & 0 & 1 & 0 & 0 \\ 0 & 0 & 1 & 0 & 0 & 1 & 0 & 0 \\ 0 & 0 & 0 & 0 & 1 & 0 & 0 & 0 \\ 0 & 0 & 1 & 1 & 1 & 1 & 0 & 0 \\ 0 & 0 & 1 & 0 & 0 & 1 & 0 & 0 \\ 0 & 0 & 1 & 0 & 0 & 1 & 0 & 0 \\ 0 & 0 & 1 & 1 & 1 & 1 & 0 & 0 \\ 0 & 0 & 1 & 1 & 1 & 1 & 0 & 0 \end{bmatrix}$$

then the first row is equivalent to the binary number 01000100 = 68. Hence the matrix can be written as a single column vector

$$\begin{bmatrix} 68 \\ 36 \\ 8 \\ 60 \\ 36 \\ 36 \\ 60 \\ 60 \end{bmatrix}$$

It follows that the binary image can be written as a simple string "68 36 8 60 36 36 60 60" and the image value $\mathbf{a}(i,j)$ can be easily recovered from the condensed string by using a decimal-to-binary conversion function. Specify an image algebra algorithm that compresses a binary image into a column vector or string. Provide a decompression algorithm specified in image algebra.

5.   Let $\mathbf{a} \in \{0,1\}^{\mathbf{X}}$ and $\mathbf{Y} \subseteq \mathbf{X}$ denote the set of points corresponding to one of the connected components $\mathbf{b}$ of $\mathbf{a}$. Note that $\mathbf{b} \equiv \mathbf{a}|_{\mathbf{Y}}$. Let $N(\mathbf{Y}) \subseteq \mathbf{X}$ be the smallest rectangular array containing $\mathbf{Y}$. A pixel $(\mathbf{y}, \mathbf{b}(\mathbf{y}))$ is called an *extremal point* of the component $\mathbf{b}$ of $\mathbf{a}$ if $\mathbf{y}$ is on the boundary of $N(\mathbf{Y})$. Provide an algorithm that labels the extremal points of connected components of $\mathbf{a}$. Specify your algorithm in image algebra. Can your algorithm be implemented in parallel?

6.   As mentioned in this chapter, component labeling algorithms form a computational bottleneck in many algorithms. Can you develop a more efficient algorithm to label connected components than the ones presented in this chapter? Can you specify a parallel algorithm?

7. Specify an alternate pruning algorithm. Can you specify a pruning algorithm that removes long feelers without significantly increasing computation time? Can you specify a parallel pruning algorithm?

6.8. References

[1] D. Ballard and C. Brown, *Computer Vision*. Englewood Cliffs, NJ: Prentice-Hall, 1982.

[2] R. Cypher, J. Sanz, and L. Snyder, "Algorithms for image component labeling on SIMD mesh connected computers," *IEEE Transactions on Computers*, vol. 39, no. 2, pp. 276–281, 1990.

[3] H. Alnuweiri and V. Prasanna, "Fast image labeling using local operators on mesh-connected computers," *IEEE Transactions on Pattern Analysis and Machine Intelligence*, vol. PAMI-13, no. 2, pp. 202–207, 1991.

[4] H. Alnuweiri and V. Prasanna, "Parallel architectures and algorithms for image component labeling," *IEEE Transactions on Pattern Analysis and Machine Intelligence*, vol. PAMI-14, no. 10, pp. 1014–1034, 1992.

[5] H. Shi and G. Ritter, "Image component labeling using local operators," in *Image Algebra and Morphological Image Processing IV*, vol. 2030 of *Proceedings of SPIE*, (San Diego, CA), pp. 303–314, July 1993.

[6] S. Levialdi, "On shrinking binary picture patterns," *Communications of the ACM*, vol. 15, no. 1, pp. 7–10, 1972.

[7] N. Hamadani, "Automatic target cueing in IR imagery," Master's thesis, Air Force Institute of Technology, WPAFB, Ohio, Dec. 1981.

[8] G. Ritter, "Boundary completion and hole filling," Tech. Rep. TR 87–05, CCVR, University of Florida, Gainesville, 1987.

CHAPTER 7
MORPHOLOGICAL TRANSFORMS AND TECHNIQUES

7.1. Introduction

Mathematical morphology is that part of digital image processing that is concerned with image filtering and geometric analysis by structuring elements. It grew out of the early work of Minkowski and Hadwiger on geometric measure theory and integral geometry [1, 2, 3], and entered the modern era through the work of Matheron and Serra of the Ecole des Mines in Fontainebleau, France [4, 5]. Matheron and Serra not only formulated the modern concepts of morphological image transformations, but also designed and built the *Texture Analyzer System*, a parallel image processing architecture based on morphological operations [6]. In the U.S., research into mathematical morphology began with Sternberg at ERIM. Serra and Sternberg were the first to unify morphological concepts and methods into a coherent algebraic theory specifically designed for image processing and image analysis. Sternberg was also the first to use the term *image algebra* [7, 8].

Initially the main use of mathematical morphology was to describe Boolean image processing in the plane, but Sternberg and Serra extended the concepts to include gray valued images using the cumbersome notion of an *umbra*. During this time a divergence of definition of the basic operations of *dilation* and *erosion* occurred, with Sternberg adhering to Minkowski's original definition. Sternberg's definitions have been used more regularly in the literature, and, in fact, are used by Serra in his book on mathematical morphology[5].

Since those early days, morphological operations and techniques have been applied from low-level, to intermediate, to high-level vision problems. Among some recent research papers on morphological image processing are Crimmins and Brown [9], Haralick, et al. [10, 11], Maragos and Schafer [12, 13, 14], Davidson [15, 16, 17], Dougherty [18, 19], Koskinen and Astola [20], and Sivakumar and Goutsias [21]. The rigorous mathematical foundation of morphology in terms of lattice algebra was independently established by Davidson and Heijmans [22, 23, 24]. Davidson's work differs from that of Heijmans' in that the foundation provided by Davidson is more general and extends classical morphology by allowing for *shift variant* structuring elements. Furthermore, Davidson's work establishes a connection between morphology, minimax algebra, and a subalgebra of image algebra.

7.2. Basic Morphological Operations: Boolean Dilations and Erosions

Dilation and erosion are the two fundamental operations that define the algebra of mathematical morphology. These two operations can be applied in different combinations in order to obtain more sophisticated operations. Noise removal in binary images provides one simple application example of the operations of dilation and erosion (Section 7.4).

The language of Boolean morphology is that of set theory. Those points in a set being morphologically transformed are considered the selected set of points, and those in the complement set are considered to be not selected. In Boolean (binary) images the set of pixels selected is the foreground and the set of pixels not selected is the background. The selected set of pixels is viewed as a set in Euclidean 2-space. For example, the set of all *black* pixels in a Boolean image constitutes a complete description of the image and is viewed as a set in Euclidean 2-space.

A dilation is a morphological transformation that combines two sets by using vector addition of set elements. In particular, a dilation of the set of black pixels in a binary image by another set (usually containing the origin), say B, is the set of all points obtained by adding the points of B to the points in the underlying point set of the black pixels. An erosion can be obtained by dilating the complement of the black pixels and then taking the complement of the resulting point set.

Dilations and erosions are based on the two classical operations of *Minkowski addition* and *Minkowski subtraction* of integral geometry. For any two sets $\mathbf{A} \subset \mathbb{R}^n$ and $\mathbf{B} \subset \mathbb{R}^n$, Minkowski addition is defined as

$$\mathbf{A} \times \mathbf{B} = \{\mathbf{a} + \mathbf{b} : \mathbf{a} \in \mathbf{A}, \mathbf{b} \in \mathbf{B}\}$$

and Minkowski subtraction as

$$\mathbf{A}/\mathbf{B} = (\mathbf{A}' \times \mathbf{B}^*)',$$

where $\mathbf{B}^* = \{-\mathbf{b} : \mathbf{b} \in \mathbf{B}\}$ and $\mathbf{A}' = \{\mathbf{x} \in \mathbb{R}^n : \mathbf{x} \notin \mathbf{A}\}$; i.e., $\mathbf{B}^*$ denotes the reflection of $\mathbf{B}$ across the origin $\mathbf{0} = (0, 0, \ldots, 0) \in \mathbb{R}^n$, while $\mathbf{A}'$ denotes the complement of $\mathbf{A}$. Here we have used the original notation employed in Hadwiger's book [3].

Defining $\mathbf{A}_\mathbf{b} = \{\mathbf{a} + \mathbf{b} : \mathbf{a} \in \mathbf{A}\}$, one also obtains the relations

$$\mathbf{A} \times \mathbf{B} = \bigcup_{\mathbf{b} \in \mathbf{B}} \mathbf{A}_\mathbf{b} = \{\mathbf{p} : \mathbf{B}_\mathbf{p}^* \cap \mathbf{A} \neq \varnothing\} \qquad (7.2.1)$$

and

$$\mathbf{A}/\mathbf{B} = \bigcap_{\mathbf{b} \in \mathbf{B}^*} \mathbf{A}_\mathbf{b} = \{\mathbf{p} : \mathbf{B}_\mathbf{p} \subset \mathbf{A}\}, \qquad (7.2.2)$$

where $\mathbf{B}_\mathbf{p} = \{\mathbf{b} + \mathbf{p} : \mathbf{b} \in \mathbf{B}\}$. It is these last two equations that makes morphology so appealing to many researchers. Equation 7.2.1 is the basis of morphological *dilations*, while Equation 7.2.2 provides the basis for morphological *erosions*. Suppose $\mathbf{B}$ contains the origin $\mathbf{0}$. Then Equation 7.2.1 says that $\mathbf{A} \times \mathbf{B}$ is the set of all points $\mathbf{p}$ such that the translate of $\mathbf{B}^*$ by the vector $\mathbf{p}$ intersects $\mathbf{A}$. Figures 7.2.1 and 7.2.2 illustrate this situation.

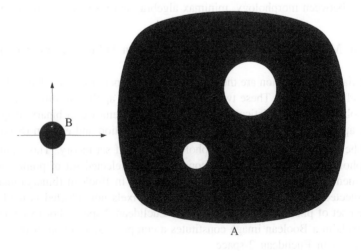

Figure 7.2.1. The set $\mathbf{A} \subset \mathbb{R}^2$ with structuring element B.

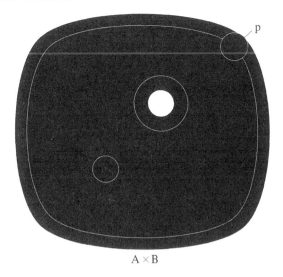

$$A \times B$$

Figure 7.2.2. The dilated set $\mathbf{A} \times \mathbf{B}$.
Note: original set boundaries shown with thin white lines.

The set $\mathbf{A/B}$, on the other hand, consists of all points p such that the translate of $\mathbf{B}$ by the vector $\mathbf{p}$ is completely contained inside $\mathbf{A}$. This situation is illustrated in Figure 7.2.3.

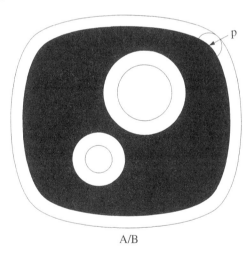

$$A/B$$

Figure 7.2.3. The eroded set $\mathbf{A}/\mathbf{B}$.
Note: original set boundaries shown with thin black lines.

In the terminology of mathematical morphology when doing a dilation or erosion *of* $\mathbf{A}$ *by* $\mathbf{B}$, it is assumed that $\mathbf{A}$ is the set to be analyzed and that $\mathbf{B}$ is the measuring stick, called a *structuring element*. To avoid anomalies without practical interest, the structuring element $\mathbf{B}$ is assumed to include the origin, and both $\mathbf{A}$ and $\mathbf{B}$ are assumed to be compact. The set $\mathbf{A}$ corresponds to either the support of a binary image $\mathbf{a} \in \{0,1\}^{\mathbf{X}}$ or to the complement of the support.

The dilation of the *image* $\mathbf{a}$ using the structuring element $\mathbf{B}$ results in another binary image $\mathbf{b} \in \{0,1\}^{\mathbf{X}}$ which is defined as

$$\mathbf{b}(\mathbf{x}) = \begin{cases} 1 & \text{if } \mathbf{x} \in \mathbf{X} \cap (\mathbf{A} \times \mathbf{B}) \\ 0 & \text{otherwise.} \end{cases} \tag{7.2.3}$$

Similarly, the erosion of the *image* $\mathbf{a} \in \{0,1\}^{\mathbf{X}}$ by $\mathbf{B}$ is the binary image $\mathbf{b} \in \{0,1\}^{\mathbf{X}}$ defined by

$$\mathbf{b}(\mathbf{x}) = \begin{cases} 1 & \text{if } \mathbf{x} \in \mathbf{A}/\mathbf{B} \\ 0 & \text{otherwise.} \end{cases} \tag{7.2.4}$$

The dilation and erosion of images as defined by Equations 7.2.3 and 7.2.4 can be realized by using max and min operations or, equivalently, by using OR and AND operations. In particular, it follows from Equations 7.2.3 and 7.2.1 that the dilated image $\mathbf{b}$ obtained from $\mathbf{a}$ is given by

$$\mathbf{b}(\mathbf{x}) = max\{\mathbf{a}(\mathbf{x} + \mathbf{y}) : \mathbf{y} \in \mathbf{B}^*\}.$$

Similarly, the eroded image $\mathbf{b}$ obtained from $\mathbf{a}$ is given by

$$\mathbf{b}(\mathbf{x}) = min\{\mathbf{a}(\mathbf{x} + \mathbf{y}) : \mathbf{y} \in \mathbf{B}\}.$$

Image Algebra Formulation

Let $\mathbf{a} \in \{0,1\}^{\mathbf{X}}$ denote a source image (usually, $\mathbf{X} \subset \mathbb{Z}^2$ is a rectangular array) and suppose we wish to dilate $\mathbf{A} \subset \mathbf{X}$, where $\mathbf{A}$ denotes the support of $\mathbf{a}$, using a structuring element $\mathbf{B}$ containing the origin.

Define a neighborhood $N : \mathbf{X} \to 2^{\mathbf{X}}$ by

$$N(\mathbf{y}) = \{\mathbf{x} \in \mathbf{X} : \mathbf{x} - \mathbf{y} \in \mathbf{B}^*\}.$$

The image algebra formulation of the dilation of the image $\mathbf{a}$ by the structuring element $\mathbf{B}$ is given by

$$\mathbf{b} := \mathbf{a} \boxtimes N.$$

The image algebra equivalent of the erosion of $\mathbf{a}$ by the structuring element $\mathbf{B}$ is given by

$$\mathbf{b} := \mathbf{a} \boxtimes N^*.$$

Replacing $\mathbf{a}$ by its Boolean complement $\mathbf{a}^c$ in the above formulation for dilation (erosion) will dilate (erode) the complement of the support of $\mathbf{a}$.

Alternate Image Algebra Formulations

We present an alternate formulation in terms of image-template operations. The main rationale for the alternate formulation is its easy extension to the gray level case (see Section 7.6).

Let $\mathbb{F}$ denote the 3-element bounded subgroup $\{-\infty, 0, \infty\}$ of $\mathbb{R}_{\pm\infty}$. Define a template $\mathbf{t} \in \left(\mathbb{F}^{\mathbb{Z}^2}\right)^{\mathbb{Z}^2}$ by

$$\mathbf{t}_{\mathbf{y}}(\mathbf{x}) = \begin{cases} 1 & \text{if } \mathbf{x} = \mathbf{y} + \mathbf{p}, \ \mathbf{p} \in \mathbf{B}^* \\ -\infty & \text{otherwise.} \end{cases}$$

The image algebra formulation of the dilation of the image **a** by the structuring element **B** is given by

$$\mathbf{b} := \mathbf{a} \boxtimes \mathbf{t}.$$

The image algebra equivalent of the erosion of **a** by the structuring element **B** is given by

$$\mathbf{b} := \mathbf{a} \boxtimes \mathbf{t}^{*}.$$

With a minor change in template definition, binary dilations and erosions can just as well be accomplished using the lattice convolution operations $\boxed{\vee}$ and $\boxed{\wedge}$, respectively. In particular, let $\mathbb{F} = \{0, 1, \infty\}$ and define $\mathbf{t} \in \left(\mathbb{F}^{\mathbb{Z}^2} \right)^{\mathbb{Z}^2}$ by

$$\mathbf{t}_{\mathbf{y}}(\mathbf{x}) = \begin{cases} 1 & \text{if } \mathbf{x} = \mathbf{y} + \mathbf{p}, \ \mathbf{p} \in \mathbf{B}^{*} \\ 0 & \text{otherwise.} \end{cases}$$

The dilation of the image **a** by the structuring element **B** is now given by

$$\mathbf{b} := \mathbf{a} \boxed{\vee} \mathbf{t}$$

and the erosion of **a** by **B** is given by

$$\mathbf{b} := \mathbf{a} \boxed{\wedge} \overline{\mathbf{t}}.$$

Comments and Observations

It can be shown that

$$\mathbf{A} \subset (\mathbf{A} \times \mathbf{B})/\mathbf{B},$$
$$(\mathbf{A}/\mathbf{B}) \times \mathbf{B} \subset \mathbf{A},$$
$$(\mathbf{A}/\mathbf{B})/\mathbf{C} = \mathbf{A}/(\mathbf{B} \times \mathbf{C}),$$
$$\mathbf{A} \times (\mathbf{B}/\mathbf{C}) \subset (\mathbf{A} \times \mathbf{B})/\mathbf{C},$$
$$\mathbf{A} \times (\mathbf{B} \cup \mathbf{C}) = (\mathbf{A} \times \mathbf{B}) \cup (\mathbf{A} \times \mathbf{C}), \text{ and}$$
$$\mathbf{A}/(\mathbf{B} \cup \mathbf{C}) = (\mathbf{A}/\mathbf{B}) \cap (\mathbf{A}/\mathbf{C}).$$

These equations constitute the basic laws that govern the algebra of mathematical morphology and provide the fundamental tools for geometric shape analysis of images.

The current notation used in morphology for a dilation of **A** by **B** is $\mathbf{A} \oplus \mathbf{B}$, while an erosion of A by B is denoted by $\mathbf{A} \ominus \mathbf{B}$. In order to avoid confusion with the linear image-template product $\oplus$, we use Hadwiger's notation [3] in order to describe morphological image transforms. In addition, we use the now more commonly employed definitions of Sternberg for dilation and erosion. A comparison of the different notation and definitions used to describe erosions and dilations is provided by Table 7.2.1.

Table 7.2.1 Notation and Definitions Used to Describe Dilations and Erosions

Minkowski	Addition $$\mathbf{A} \times \mathbf{B} = \bigcup_{b \in B} \mathbf{A_b}$$	Subtraction $$\mathbf{A}/\mathbf{B} = \bigcap_{b \in B^*} \mathbf{A_b} = (\mathbf{A}' \times \mathbf{B}^*)'$$
Serra Maragos	Dilation of $\mathbf{A}$ by $\mathbf{B}$ $$\mathbf{A} \oplus \mathbf{B}^* = \bigcup_{b \in B^*} \mathbf{A_b}$$	Erosion of $\mathbf{A}$ by $\mathbf{B}$ $$\mathbf{A} \ominus \mathbf{B}^* = \bigcap_{b \in B^*} \mathbf{A_b} = (\mathbf{A}' \oplus \mathbf{B})'$$
Sternberg	Dilation of $\mathbf{A}$ by $\mathbf{B}$ $$\mathbf{A} \oplus \mathbf{B} = \bigcup_{b \in B} \mathbf{A_b}$$	Erosion of $\mathbf{A}$ by $\mathbf{B}$ $$\mathbf{A} \ominus \mathbf{B} = \bigcap_{b \in B^*} \mathbf{A_b} = (\mathbf{A}' \oplus \mathbf{B}^*)'$$

7.3.   Opening and Closing

Dilations and erosions are usually employed in pairs; a dilation of an image is usually followed by an erosion of the dilated result or vice versa. In either case, the result of successively applied dilations and erosions results in the elimination of specific image detail smaller than the structuring element without the global geometric distortion of unsuppressed features.

An *opening* of an image is obtained by first eroding the image with a structuring element and then dilating the result using the *same* structuring element. The *closing* of an image is obtained by first dilating the image with a structuring element and then eroding the result using the *same* structuring element. The next section shows that opening and closing provide a particularly simple mechanism for shape filtering.

The operations of opening and closing are idempotent; their reapplication effects no further changes to the previously transformed results. In this sense openings and closings are to morphology what orthogonal projections are to linear algebra. An orthogonal projection operator is idempotent and selects the part of a vector that lies in a given subspace. Similarly, opening and closing provide the means by which given subshapes or supershapes of a complex geometric shape can be selected.

The opening of $\mathbf{A}$ by $\mathbf{B}$ is denoted by $\mathbf{A} \circ \mathbf{B}$ and defined as

$$\mathbf{A} \circ \mathbf{B} = (\mathbf{A}/\mathbf{B}) \times \mathbf{B}.$$

The closing of $\mathbf{A}$ by $\mathbf{B}$ is denoted by $\mathbf{A} \bullet \mathbf{B}$ and defined as

$$\mathbf{A} \bullet \mathbf{B} = (\mathbf{A} \times \mathbf{B})/\mathbf{B}.$$

Image Algebra Formulation

Let $\mathbf{a} \in \{0,1\}^{\mathbf{X}}$ denote a source image and $\mathbf{B}$ the desired structuring element containing the origin. Define $N : \mathbf{X} \to 2^{\mathbf{X}}$ is defined by

$$N(\mathbf{y}) = \{\mathbf{x} \in \mathbf{X} : \mathbf{x} - \mathbf{y} \in \mathbf{B}^*\}.$$

The image algebra formulation of the opening of the image $\mathbf{a}$ by the structuring element $\mathbf{B}$ is given by

$$\mathbf{b} := (\mathbf{a} \boxminus N^*) \boxplus N.$$

The image algebra equivalent of the closing of $\mathbf{a}$ by the structuring element $\mathbf{B}$ is given by

$$\mathbf{b} := (\mathbf{a} \boxplus N) \boxminus N^*.$$

Comments and Observations

It follows from the basic theorems that govern the algebra of erosions and dilations that $\mathbf{A} \circ \mathbf{B} \subset \mathbf{A}$, $(\mathbf{A} \circ \mathbf{B}) \circ \mathbf{B} = \mathbf{A} \circ \mathbf{B}$, $\mathbf{A} \subset \mathbf{A} \bullet \mathbf{B}$, and $(\mathbf{A} \bullet \mathbf{B}) \bullet \mathbf{B} = \mathbf{A} \bullet \mathbf{B}$. This shows the analogy between the morphological operations of opening and closing and the specification of a filter by its bandwidth. Morphologically filtering an image by an opening or closing operation corresponds to the ideal nonrealizable bandpass filters of conventional linear filters. Once an image is ideal bandpass filtered, further ideal bandpass filtering does not alter the result.

7.4. Salt and Pepper Noise Removal

Opening an image with a disk-shaped structuring element smooths the contours, breaks narrow isthmuses, and eliminates small islands. Closing an image with a disk structuring element smooths the contours, fuses narrow breaks and long thin gulfs, eliminates small holes, and fills gaps in contours. Thus, a combination of openings and closings can be used to remove small holes and small speckles or islands in a binary image. These small holes and islands are usually caused by factors such as system noise, threshold selection, and preprocessing methodologies, and are referred to as salt and pepper noise.

Consider the image shown in Figure 7.4.1. Let $\mathbf{A}$ denote the set of all black pixels. Choosing the structuring element $\mathbf{B}$ shown in Figure 7.4.2, the opening of $\mathbf{A}$ by $\mathbf{B}$

$$\mathbf{A} \circ \mathbf{B} = (\mathbf{A}/\mathbf{B}) \times \mathbf{B},$$

removes all the pepper noise (small black areas) from the input images. Doing a closing on $\mathbf{A} \circ \mathbf{B}$,

$$(\mathbf{A} \circ \mathbf{B}) \bullet \mathbf{B} = [(\mathbf{A} \circ \mathbf{B}) \times \mathbf{B}]/\mathbf{B},$$

closes the small white holes (salt noise) and results in the image shown in Figure 7.4.3.

Figure 7.4.1. SR71 with salt and pepper noise.

Figure 7.4.2. The structuring element B.

Figure 7.4.3. SR71 with salt and pepper noise removed.

Image Algebra Formulation

Let $\mathbf{a} \in \{0,1\}^{\mathbf{X}}$ denote a source image and N the von Neumann neighborhood. The image $\mathbf{b}$ derived from $\mathbf{a}$ using the morphological salt and pepper noise removal technique is given by

$$\mathbf{b} := (((\mathbf{a} \boxtimes N) \boxtimes N) \boxtimes N) \boxtimes N \, .$$

Comments and Observations

Salt and pepper noise removal can also be accomplished with the appropriate median filter. In fact, there is a close relationship between the morphological operations of opening and closing (gray level as well as binary) and the median filter. Images that remain unchanged after being median filtered are called *median root images*. To obtain the median root image of a given input image one simply repeatedly median filters the given image until there is no change. An image that is both opened and closed with respect to the same structuring element is a median root image.

7.5. The Hit-and-Miss Transform

The hit-and-miss transform (HMT) is a natural operation to select *out* pixels that have certain geometric properties such as corner points, isolated points, boundary points, etc. In addition, the HMT performs template matching, thinning, thickening, and centering.

Since an erosion or a dilation can be interpreted as special cases of the hit-and-miss transform, the HMT is considered to be the most general image transform in mathematical morphology. This transform is often viewed as the universal morphological transformation upon which mathematical morphology is based.

Let $\mathbf{B} = (\mathbf{D}, \mathbf{E})$ be a pair of structuring elements. Then the hit-and-miss transform of the set $\mathbf{A}$ is given by the expression

$$\mathbf{A} \circledast \mathbf{B} = \{\mathbf{p} : \mathbf{D_p} \subset \mathbf{A} \text{ and } \mathbf{E_p} \subset \mathbf{A'}\}. \tag{7.5.1}$$

For practical applications it is assumed that $\mathbf{D} \cap \mathbf{E} = \varnothing$. The erosion of $\mathbf{A}$ by $\mathbf{D}$ is obtained by simply letting $\mathbf{E} = \varnothing$, in which case Equation 7.5.1 becomes

$$\mathbf{A} \circledast \mathbf{B} = \mathbf{A}/\mathbf{D}.$$

Since a dilation can be obtained from an erosion via the duality $\mathbf{A} \times \mathbf{B} = (\mathbf{A'}/\mathbf{B^*})'$, it follows that a dilation is also a special case of the HMT.

Image Algebra Formulation

Let $\mathbf{a} \in (\mathbb{Z}_2)^{\mathbf{X}}$ and $\mathbf{B} = (\mathbf{D}, \mathbf{E})$. Define $\mathbf{t} \in \left(\mathbb{F}^{\mathbb{Z}^2}\right)^{\mathbb{Z}^2}$, where $\mathbb{F} = \mathbb{R}_{\pm\infty}$, by

$$\mathbf{t_y}(\mathbf{x}) = \begin{cases} 0 & \text{if } \mathbf{x} = \mathbf{y} + \mathbf{p}, \ \mathbf{p} \in \mathbf{D^*} \\ 1 & \text{if } \mathbf{x} = \mathbf{y} + \mathbf{p}, \ \mathbf{p} \in \mathbf{E^*}. \end{cases}$$

The image algebra equivalent of the hit-and-miss transform applied to the image $\mathbf{a}$ using the structuring element $\mathbf{B}$ is given by

$$\mathbf{c} := \mathbf{a} \, \widetilde{\boxtimes} \, \mathbf{t}.$$

Alternate Image Algebra Formulation

Let the neighborhoods $N, M : \mathbf{X} \to 2^{\mathbf{X}}$ be defined by

$$N(\mathbf{y}) = \{\mathbf{x} \in \mathbf{X} : \mathbf{x} - \mathbf{y} \in \mathbf{D^*}\}$$

and

$$M(\mathbf{y}) = \{\mathbf{x} \in \mathbf{X} : \mathbf{x} - \mathbf{y} \in \mathbf{E^*}\}.$$

An alternate formulation image algebra for the hit-and-miss transform is given by

$$\mathbf{c} := (\mathbf{a} \boxtimes N^*) \cdot (\chi_0(\mathbf{a}) \boxtimes M^*).$$

Comments and Observations

Davidson proved that the HMT can also be accomplished using a linear convolution followed by a simple threshold [22]. Let $\{\mathbf{x}_1(\mathbf{y}), \mathbf{x}_2(\mathbf{y}), \ldots, \mathbf{x}_n(\mathbf{y})\} = S_\infty(\mathbf{t}_\mathbf{y}^*) \cup S_\infty(\mathbf{s}_\mathbf{y}^*)$, where the enumeration is such that $\{\mathbf{x}_1(\mathbf{y}), \mathbf{x}_2(\mathbf{y}), \ldots, \mathbf{x}_k(\mathbf{y})\} = S_\infty(\mathbf{t}_\mathbf{y}^*)$ and $\{\mathbf{x}_{k+1}(\mathbf{y}), \mathbf{x}_{k+2}(\mathbf{y}), \ldots, \mathbf{x}_n(\mathbf{y})\} = S_\infty(\mathbf{s}_\mathbf{y}^*)$. Define an integer-valued template $\mathbf{r}$ from $\mathbb{Z}^2$ to $\mathbb{Z}^2$ by

$$\mathbf{r}_\mathbf{y}(\mathbf{x}) = \begin{cases} 2^{i-1} & \text{if } \mathbf{x} = \mathbf{x}_i(\mathbf{y}) \\ 0 & \text{if } \mathbf{x} \neq \mathbf{x}_i(\mathbf{y}) \ \ i = 1, \ldots, n \,. \end{cases}$$

Then

$$\mathbf{c} := \chi_m(\mathbf{a} \oplus \mathbf{r}), \quad \text{where } m = \sum_{i=1}^{k} 2^{i-1}$$

is another image algebra equivalent formulation of the HMT.

Figure 7.5.1 shows how the hit-and-miss transform can be used to locate square regions of a certain size. The source image is to the left of the figure. The structuring element $\mathbf{B} = (\mathbf{D}, \mathbf{E})$ is made up of the 3×3 solid square $\mathbf{D}$ and the 9×9 square border $\mathbf{E}$. In this example, the hit-and-miss transform is designed to "hit" regions that cover $\mathbf{D}$ and "miss" $\mathbf{E}$. The two smaller square regions satisfy the criteria of the example design. This is seen in the image to the right of Figure 7.5.1.

The template used for the image algebra formulation of the example HMT is shown in Figure 7.5.2. In a similar fashion, templates can be designed to locate any of the region configuration seen in the source image.

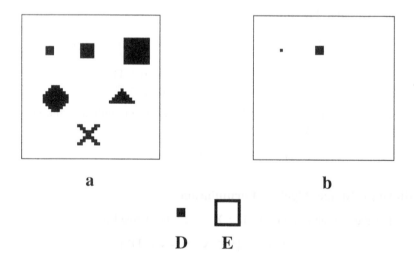

a   b

D   E

Figure 7.5.1. Hit-and-miss transform used to locate square regions. Region **A** (corresponding to source image **a**) is transformed to region B (image **b**) with the morphological hit-and-miss transform using structuring element $\mathbf{F} = (\mathbf{D}, \mathbf{E})$ by $\mathbf{B} = \mathbf{A} \circledast \mathbf{F}$. In image algebra notation, $\mathbf{b} = \mathbf{a} \,\tilde{\boxtimes}\, \mathbf{t}$ with $\mathbf{t}$ as shown in Figure 7.5.2.

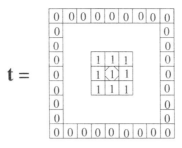

$$\mathbf{t} =$$

Figure 7.5.2. Template used for the image algebra formulation of the hit-and-miss transform designed to locate square regions.

7.6. Gray Value Dilations, Erosions, Openings, and Closings

Although morphological operations on binary images provide useful analytical tools for image analysis and classification, they play only a very limited role in the processing and analysis of gray level images. In order to overcome this severe limitation, Sternberg and Serra extended binary morphology in the early 1980s to gray scale images via the notion of an *umbra*. As in the binary case, dilations and erosions are the basic operations that define the algebra of gray scale morphology.

While there have been several extensions of the Boolean dilation to the gray level case, Sternberg's formulae for computing the gray value dilation and erosion are the most straightforward even though the underlying theory introduces the somewhat extraneous concept of an *umbra*. Let $\mathbf{X} \subset \mathbb{R}^n$ and $f : \mathbf{X} \to \mathbb{R}$ be a function. Then the *umbra* of f, denoted by $\mathcal{U}(f)$, is the set $\mathcal{U}(f) \subset \mathbb{R}^{n+1}$, defined by

$$\mathcal{U}(f) = \left\{ (x_1, x_2, \ldots, x_n, x_{n+1}) \in \mathbb{R}^{n+1} : \begin{array}{l} (x_1, x_2, \ldots, x_n) \in \mathbf{X} \text{ and} \\ x_{n+1} \leq f(x_1, x_2, \ldots, x_n) \end{array} \right\}.$$

Note that the notion of an unbounded set is exhibited in this definition; the value of x_{n+1} can approach $-\infty$.

Since $\mathcal{U}(f) \subset \mathbb{R}^{n+1}$, we can dilate $\mathcal{U}(f)$ by any other subset of $\mathbb{R}^{n+1}$. This observation provides the clue for dilation of gray-valued images. In general, the dilation of a function $f : \mathbb{R}^n \to \mathbb{R}$ by a function $g : \mathbf{X} \to \mathbb{R}$, where $\mathbf{X} \subset \mathbb{R}^n$, is defined through the dilation of their umbras $\mathcal{U}(f) \times \mathcal{U}(g)$ as follows. Let $\mathbf{x} = (x_1, \ldots, x_n) \in \mathbb{R}^n$ and define a function $d : \mathbb{R}^n \to \mathbb{R}$ by $d(\mathbf{x}) = max\{z \in \mathbb{R} : (\mathbf{x}, z) \in \mathcal{U}(f) \times \mathcal{U}(g)\}$. We now define $f \times g \equiv d$. The erosion of f by g is defined as the function $f/g \equiv e$, where $e = -[(-f) \times \hat{g}]$ and $\hat{g}(\mathbf{x}) = g(-\mathbf{x})$.

Openings and closings of f by g are defined in the same manner as in the binary case. Specifically, the *opening* of f by g is defined as $f \circ g = (f/g) \times g$ while the *closing* of f by g is defined as $f \bullet g = (f \times g)/g$.

When calculating the new functions $d = f \times g$ and $e = f/g$, the following formulas for two-dimensional dilations and erosions are actually being used:

$$d(\mathbf{y}) = \max_{\mathbf{x} \in \mathbf{X}} \{f(\mathbf{y} + \mathbf{x}) + g(\mathbf{x})\}$$

for the dilation, and

$$e(\mathbf{y}) = \min_{\mathbf{x} \in \mathbf{X}'} \{f(\mathbf{x} + \mathbf{y}) - g^*(\mathbf{x})\} \tag{7.6.1}$$

for the erosion, where $g^* : \mathbf{X}^* \to \mathbb{R}$ is defined by $g^*(\mathbf{x}) = g(-\mathbf{x}) \ \forall \mathbf{x} \in \mathbf{X}^*$. In practice, the function f represents the image, while g represents the structuring element, where the support of g corresponds to a binary structuring element. That is, one starts by defining a binary structuring element $\mathbf{B}$ in $\mathbb{Z}^2$ about the origin, and then adds gray values to the cells in $\mathbf{B}$. This defines a real-valued function g whose support is $\mathbf{B}$. Also, the support of g is, in general, much smaller than the array on which f is defined. Thus, in practice, the notion of an umbra need not be introduced at all.

Since there are slight variations in the definitions of dilation and erosion in the literature, we again remind the reader that we are using formulations that coincide with Minkowski's addition and subtraction.

Image Algebra Formulation

Let $\mathbf{a} \in \mathbb{R}^{\mathbf{X}}$ denote the gray scale source image and g the structuring element whose support is $\mathbf{B}$. Define an extended real-valued template $\mathbf{t}$ from $\mathbb{Z}^2$ to $\mathbb{Z}^2$ by

$$\mathbf{t_y}(\mathbf{x}) = \begin{cases} g(\mathbf{y} - \mathbf{x}) & \text{if } \mathbf{y} - \mathbf{x} \in \mathbf{B}^* \\ -\infty & \text{otherwise.} \end{cases}$$

Note that the condition $\mathbf{y} - \mathbf{x} \in \mathbf{B}^*$ is equivalent to $-\mathbf{x} \in \mathbf{B}^*_{\mathbf{y}}$, where $\mathbf{B}^*_{\mathbf{y}}$ denotes the translation of $\mathbf{B}^*$ by the vector $\mathbf{y}$.

The image algebra equivalents of a gray scale dilation and a gray scale erosion of $\mathbf{a}$ by the structuring element g are now given by

$$\mathbf{b} := \mathbf{a} \boxedvee \mathbf{t}$$

and

$$\mathbf{b} := \mathbf{a} \boxedwedge \mathbf{t}^*,$$

respectively.   Here

$$\mathbf{t}^*_{\mathbf{y}}(\mathbf{x}) = \begin{cases} -g(\mathbf{y} - \mathbf{x}) & \text{if } \mathbf{y} - \mathbf{x} \in \mathbf{B} \\ +\infty & \text{otherwise.} \end{cases}$$

The opening of $\mathbf{a}$ by g is given by

$$\mathbf{b} := (\mathbf{a} \boxedwedge \mathbf{t}^*) \boxedvee \mathbf{t}$$

and the closing of $\mathbf{a}$ by g is given by

$$\mathbf{b} := (\mathbf{a} \boxedvee \mathbf{t}) \boxedwedge \mathbf{t}^*.$$

Comments and Observations

Davidson has shown that a subalgebra of the full image algebra, namely $\mathcal{A} = \left(\mathbb{R}^{\mathbf{X}}_{\pm\infty}, \left(\mathbb{R}^{\mathbf{X}}_{\pm\infty}\right)^{\mathbf{Y}}; \ \boxedvee, \ \boxedwedge, \vee, \wedge, +, - \right)$, contains the algebra of gray scale mathematical morphology as a special case [22]. It follows that all morphological transformations can be easily expressed in the language of image algebra.

That the algebra $\mathcal{A}$ is more general than mathematical morphology should come as no surprise as templates are more general objects than structuring elements. Since structuring elements correspond to translation invariant templates, morphology lacks the ability to implement translation variant lattice transforms effectively. In order to implement such transforms effectively, morphology needs to be extended to include the notion of translation variant structuring elements. Of course, this extension is already a part of image algebra.

7.7. The Rolling Ball Algorithm

As was made evident in the previous sections, morphological operations and transforms can be expressed in terms of image algebra by using the operators ☑ and ☒. Conversely, any image transform that is based on these operators and uses only invariant templates can be considered a morphological transform. Thus, many of the transforms listed in this synopsis such as skeletonizing and thinning, are morphological image transforms. We conclude this chapter by providing an additional morphological transform known as the *rolling ball algorithm*.

The rolling ball algorithm, also known as the *top hat transform*, is a geometric shape filter that corresponds to the residue of an opening of an image [5, 7, 8]. The *ball* used in this image transformation corresponds to a structuring element (shape) that does not fit into the geometric shape (mold) of interest, but fits well into the background clutter. Thus, by removing the object of interest, complementation will provide its location.

In order to illustrate the basic concept behind the rolling ball algorithm, let **a** be a surface in 3-space. For example, **a** could be a function whose values **a(x)** represent some physical measurement such as reflectivity at points $\mathbf{x} \in \mathbb{R}^2$. Figure 7.7.1 represents a one-dimensional analogue in terms of a two-dimensional slice perpendicular to the (x, y)-plane. Now suppose **s** is a ball of some radius r. Rolling this ball beneath the surface **a** in such a way that the boundary of the ball, and only the boundary of the ball, always touches **a**, results in another surface **b** which is determined by the set of points consisting of all possible locations of the center of the ball as it rolls below **a**. The tracing of the ball's center locations is illustrated in Figure 7.7.2.

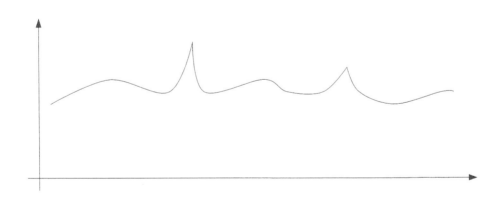

Figure 7.7.1. The signal **a**.

It may be obvious that the generation of the surface **b** is equivalent to an erosion. To realize this equivalence, let $s : \mathbf{X} \to \mathbb{R}$ be defined by $s(\mathbf{x}) = \sqrt{r^2 - \|\mathbf{x}\|^2}$, where $\mathbf{X} = \left\{ \mathbf{x} \in \mathbb{R}^2 : \|\mathbf{x}\| \leq r \right\}$. Obviously, the graph of s corresponds to the upper surface (upper hemisphere) of a ball **s** of radius r with center at the origin of $\mathbb{R}^3$. Using Equation 7.6.1, it is easy to show that $\mathbf{b} = \mathbf{a}/\mathbf{s}$.

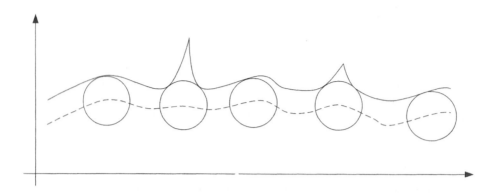

Figure 7.7.2. The surface generated by the center of a rolling ball.

Next, let **s** roll *in* the surface **b** such that the center of **s** is always a point of **b** and such that every point of **b** gets hit by the center of **s**. Then the *top* point of **s** traces another surface **c** *above* **b**. If **a** is flat or smooth, with local curvature never less than r, then **c** = **a**. However, if **a** contains crevasses into which **s** does *not* fit — that is, locations at which **s** is not tangent to **a** or points of **a** with curvature less than r, etc. — then **c** $\neq$ **a**. Figure 7.7.3 illustrates this situation. It should also be clear by now that **c** corresponds to a dilation of **b** by s. Therefore, **c** is the opening of **a** by s, namely

$$\mathbf{c} = (\mathbf{a}/s) \times s = \mathbf{a} \circ s .$$

Hence, **c** $\leq$ **a**.

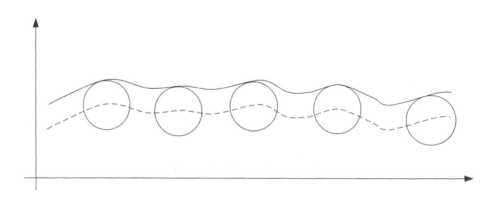

Figure 7.7.3. The surface generated by the top of a rolling ball.

In order to remove the background of **a** — that is, those locations where **s** fits well beneath **a** — one simply subtracts **c** from **a** in order to obtain the image **d** = **a** − **c** containing only areas of interest (Figure 7.7.4).

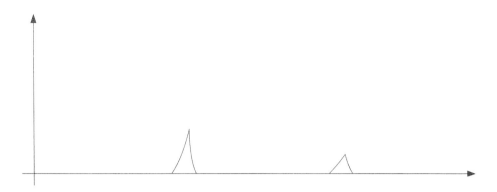

Figure 7.7.4. The result of the rolling ball algorithm.

Let $\mathbf{a} \in \mathbb{R}^{\mathbf{X}}$ denote the digital source image and s a structuring element represented by digital hemisphere of some desired radius r about the origin. Note that the support of s is a digital disk $\mathbf{S}$ of radius r. The rolling ball algorithm is given by the transformation $\mathbf{a} \rightarrow \mathbf{d}$ which is defined by

$$\mathbf{d} = \mathbf{a} - [(\mathbf{a}/s) \times s].$$

Image Algebra Formulation

Let $\mathbf{a} \in \mathbb{R}^{\mathbf{X}}$ denote the source image. Define an extended real-valued template s from $\mathbb{Z}^2$ to $\mathbb{Z}^2$ by

$$\mathbf{s_y}(\mathbf{x}) = \begin{cases} s(\mathbf{y} - \mathbf{x}) & \text{if } \mathbf{y} - \mathbf{x} \in \mathbf{S}^* \\ -\infty & \text{otherwise.} \end{cases}$$

The image algebra formulation of the rolling ball algorithm is now given by

$$\mathbf{d} := \mathbf{a} - (\mathbf{a} \boxtimes \mathbf{s}^*) \boxtimes \mathbf{s}.$$

Comments and Observations

It should be clear from Figure 7.7.4 that such an algorithm could be used for the easy and fast detection of *hot spots* in IR images. There are many variants of this basic algorithm. Also, in many applications it is advantageous to use different size balls for the erosion and dilation steps, or to use a sequence of balls (i.e., a sequence of transforms $\mathbf{a} \rightarrow \mathbf{d_i}$, $i = 1, 2, \ldots, k$) in order to obtain different regions (different in shape) of interest. Furthermore, there is nothing to prevent an algorithm developer from using shapes different from disks or balls in the above algorithm. The shape of the structuring element is determined by *a priori* information and/or by the objects one seeks.

7.8.   Exercises

1.   Prove the six basic laws of the algebra of mathematical morphology (Section 7.2).

2.   a.   Express the algorithm $(\mathbf{a} \boxtimes \mathbf{t}) \boxtimes \mathbf{s}$ in terms of a single template $\mathbf{r}$. How is $\mathbf{r}$ defined?
     b.   Can the algorithm $(\mathbf{a} \boxtimes \mathbf{t}) \boxtimes \mathbf{s}$ be expressed in terms of a single template $\mathbf{r}$? Explain your answer.

3.   Let $\mathbf{a}, \mathbf{b} \in \mathbb{R}^X$ and $\mathbf{c} = \mathbf{a} - \mathbf{b}$. Define a template $\mathbf{s} \in \left(\mathbb{R}^X_{-\infty}\right)$ so that $\mathbf{a} \boxdot \mathbf{s} = \mathbf{c}$.

4.   In Chapter 3 (Section 3.2) morphological operations were used to find edge in binary images. Specify a morphology-based edge detector for gray-level images. Compare the performance of your edge detector with the performance of the Prewitt and Sobel edge detectors.

5.   Repeat Exercise 4 using the operations of $\vee$ and $\wedge$ instead of $\boxdot$ and $\boxminus$.

6.   Specify a morphology-based hole filling algorithm that is different from those presented in Chapter 6.

7.   Construct several binary images containing various digitized objects of interest such as coins, rectangles, and pencils as well as other objects of no interest such as amorphous blobs, noise, trapezoids, etc. Specify a morphology-based recognition algorithm that will identify all objects of interest and ignore or eliminate all objects of no interest.

8.   Let $\mathbf{a}$ be a 32×32 binary image containing a single black pixel.

   a.   Define a template $\mathbf{t}$ such that the iteration $(\ldots((\mathbf{a} \boxdot \mathbf{t}) \boxdot \mathbf{t}) \ldots \boxdot \mathbf{t})$ results in a chessboard image (i.e., each white pixel's horizontal and vertical neighbors are black and vice versa).
   b.   Specify an iterative morphology-based algorithm such that application to image $\mathbf{a}$ results in a herringbone pattern (i.e., alternating zigzag lines of black and white pixels).

9.   The medial axis transform (Section 5.3) is a morphological transform based on erosions and dilations using a ball shaped neighborhood. Implement the medial axis transform using different template shapes such as rectangles, triangles, and line segments. Compare the results for various objects.

7.9.   References

[1]   H. Minkowski, "Volumen und oberflache," *Mathematische Annalen*, vol. 57, pp. 447–495, 1903.

[2]   H. Minkowski, *Gesammelte Abhandlungen*. Leipzig-Berlin: Teubner Verlag, 1911.

[3]   H. Hadwiger, *Vorlesungen Über Inhalt, Oberflæche und Isoperimetrie*. Berlin: Springer-Verlag, 1957.

[4]   G. Matheron, *Random Sets and Integral Geometry*. New York: Wiley, 1975.

[5]   J. Serra, *Image Analysis and Mathematical Morphology*. London: Academic Press, 1982.

[6]   J. Klein and J. Serra, "The texture analyzer," *Journal of Microscopy*, vol. 95, 1972.

[7]   S. Sternberg, "Biomedical image processing," *Computer*, vol. 16, Jan. 1983.

[8]   S. Sternberg, "Overview of image algebra and related issues," in *Integrated Technology for Parallel Image Processing* (S. Levialdi, ed.), London: Academic Press, 1985.

[9]   T. Crimmins and W. Brown, "Image algebra and automatic shape recognition," *IEEE Transactions on Aerospace and Electronic Systems*, vol. AES-21, pp. 60–69, Jan. 1985.

[10] R. Haralick, L. Shapiro, and J. Lee, "Morphological edge detection," *IEEE Journal of Robotics and Automation*, vol. RA-3, pp. 142–157, Apr. 1987.

[11] R. Haralick, S. Sternberg, and X. Zhuang, "Image analysis using mathematical morphology: Part I," *IEEE Transactions on Pattern Analysis and Machine Intelligence*, vol. 9, pp. 532–550, July 1987.

[12] P. Maragos and R. Schafer, "Morphological filters Part I: Their set-theoretic analysis and relations to linear shift-invariant filters," *IEEE Transactions on Acoustics, Speech, and Signal Processing*, vol. ASSP-35, pp. 1153–1169, Aug. 1987.

[13] P. Maragos and R. Schafer, "Morphological filters Part II : Their relations to median, order-statistic, and stack filters," *IEEE Transactions on Acoustics, Speech, and Signal Processing*, vol. ASSP-35, pp. 1170–1184, Aug. 1987.

[14] P. Maragos, *A Unified Theory of Translation-Invariant Systems with Applications to Morphological Analysis and Coding of Images*. Ph.D. dissertation, Georgia Institute of Technology, Atlanta, 1985.

[15] J. Davidson, "Simulated annealing and morphological neural networks," in *Image Algebra and Morphological Image Processing III*, vol. 1769 of *Proceedings of SPIE*, (San Diego, CA), pp. 119–127, July 1992.

[16] J. Davidson and A. Talukder, "Template identification using simulated annealing in morphology neural networks," in *Second Annual Midwest Electro-Technology Conference*, (Ames, IA), pp. 64–67, IEEE Central Iowa Section, Apr. 1993.

[17] J. Davidson and F. Hummer, "Morphology neural networks: An introduction with applications," *IEEE Systems Signal Processing*, vol. 12, no. 2, pp. 177–210, 1993.

[18] E. Dougherty, "Unification of nonlinear filtering in the context of binary logical calculus, part ii: Gray-scale filters," *Journal of Mathematical Imaging and Vision*, vol. 2, pp. 185–192, Nov. 1992.

[19] E. Dougherty, "Optimal binary morphological bandpass filters induced by granulometric spectral representation," *Journal of Mathematical Imaging and Vision*, vol. 7, pp. 175–191, Mar. 1997.

[20] L. Koskinen and J. Astola, "Asymptotic behaviour of morphological filters," *Journal of Mathematical Imaging and Vision*, vol. 2, pp. 117–136, Nov. 1992.

[21] K. Sivakumar and J. Goutsias, "Morphologically constrained grfs: Applications to texture synthesis and analysis," *IEEE Transactions on Pattern Analysis and Machine Intelligence*, vol. 21, pp. 99–131, Feb. 1999.

[22] J. Davidson, *Lattice Structures in the Image Algebra and Applications to Image Processing*. Ph.D. thesis, University of Florida, Gainesville, FL, 1989.

[23] J. Davidson, "Foundation and applications of lattice transforms in image processing," in *Advances in Electronics and Electron Physics* (P. Hawkes, ed.), vol. 84, pp. 61–130, New York, NY: Academic Press, 1992.

[24] H. Heijmans, "Theoretical aspects of gray-level morphology," *IEEE Transactions on Pattern Analysis and Machine Intelligence*, vol. 13(6), pp. 568–582, 1991.

CHAPTER 8
LINEAR IMAGE TRANSFORMS

8.1. Introduction

A large class of image processing transformations is linear in nature; an output image is formed from linear combinations of pixels of an input image. Such transforms include convolutions, correlations, and unitary transforms. Applications of linear transforms in image processing are numerous. Linear transforms have been utilized to enhance images and to extract various features from images. For example, the Fourier transform is used in highpass and lowpass filtering (Chapter 2) as well as in texture analysis. Another application is image coding in which bandwidth reduction is achieved by deleting low-magnitude transform coefficients. In this chapter we provide some typical examples of linear transforms and their reformulations in the language of image algebra.

8.2. Fourier Transform

The *Fourier transform* is one of the most useful tools in image processing. It provides a realization of an image that is a composition of sinusoidal functions over an infinite band of frequencies. This realization facilitates many image processing techniques. Filtering, enhancement, encoding, restoration, texture analysis, feature classification, and pattern recognition are but a few of the many areas of image processing that utilize the Fourier transform.

The Fourier transform is defined over a function space. A function in the domain of the Fourier transform is said to be defined over a *spatial domain*. The corresponding element in the range of the Fourier transform is said to be defined over a *frequency domain*. We will discuss the significance of the frequency domain after the one-dimensional Fourier transform and its inverse have been defined.

The Fourier transform $\mathfrak{F}(f) = \hat{f}$ of the function $f \in C(\mathbb{R}^1)$ is defined by

$$\hat{f}(u) = [\mathfrak{F}(f)](u) = \int_{-\infty}^{\infty} f(x)e^{-2\pi iux}dx,$$

where $i = \sqrt{-1}$. Given $\hat{f}$, then f can be recovered by using the *inverse Fourier transform* $\mathfrak{F}^{-1}$ which is given by the equation

$$f(x) = [\mathfrak{F}^{-1}(\mathfrak{F}(f))](x) = \int_{-\infty}^{\infty} \hat{f}(u)e^{2\pi iux}du.$$

The functions f and $\hat{f}$ are called a *Fourier transform pair*.

Substituting the integral definitions of the Fourier transform and its inverse into the above equation, the following equality is obtained

$$f(x) = \int_{-\infty}^{\infty} \left[\int_{-\infty}^{\infty} f(x)e^{-2\pi iux}dx \right] e^{2\pi iux}du.$$

The inner integral (enclosed by square brackets) is the Fourier transform of f. It is a function of u alone. Replacing the integral definition of the Fourier transform of f by $\hat{f}(u)$ we get

$$f(x) = \int\limits_{-\infty}^{\infty} \hat{f}(u)e^{2\pi i u x}\,du.$$

Euler's formula allows $e^{2\pi i u x}$ to be expressed as $cos(2\pi u x) + i\,sin(2\pi u x)$. Thus, $e^{2\pi i u x}$ is a sum of a real and complex sinusoidal function of frequency u. The integral is the continuous analog of summing over all frequencies u. The Fourier transform evaluated at u, $\hat{f}(u)$, can be viewed as a weight applied to the real and complex sinusoidal functions at frequency u in a continuous summation.

Combining all these observations, the original function $f(x)$ is seen as a continuous weighted sum of sinusoidal functions. The weight applied to the real and complex sinusoidal functions of frequency u is given by the Fourier transform evaluated at u. This explains the use of the term "frequency" as an adjective for the domain of the Fourier transform of a function.

For image processing, an image can be mapped into its frequency domain representation via the Fourier transform. In the frequency domain representation the weights assigned to the sinusoidal components of the image become accessible for manipulation. After the image has been processed in the frequency domain representation, the representation of the enhanced image in the spatial domain can be recovered using the inverse Fourier transform.

The discrete equivalent of the one-dimensional continuous Fourier transform pair is given by

$$\hat{\mathbf{a}}(j) = [\mathfrak{F}(\mathbf{a})](j) = \sum_{k=0}^{n-1} \mathbf{a}(k)e^{-2\pi j k i/n}$$

and

$$\mathbf{a}(k) = \left[\mathfrak{F}^{-1}(\hat{\mathbf{a}})\right](k) = \frac{1}{n}\sum_{j=0}^{n-1} \hat{\mathbf{a}}(j)e^{2\pi k j i/n},$$

where $\mathbf{a} \in \mathbb{R}^{\mathbb{Z}_n}$ and $\hat{\mathbf{a}} \in \mathbb{C}^{\mathbb{Z}_n}$. In digital signal processing, $\mathbf{a}$ is usually viewed as having been obtained from a continuous function $f \in C(\mathbb{R}^1)$ by sampling f at some finite number of uniformly spaced points $\{x_0, x_1, \ldots, x_{n-1}\} \subset \mathbb{R}$ and setting $\mathbf{a}(k) = f(x_k)$.

For $f \in C(\mathbb{R}^2)$, the *two-dimensional* continuous Fourier transform pair is given by

$$\hat{f}(u, v) = \int\limits_{-\infty}^{\infty} \int\limits_{-\infty}^{\infty} f(x, y)e^{-2\pi i(ux+vy)}\,dx\,dy$$

and

$$f(x, y) = \int\limits_{-\infty}^{\infty} \int\limits_{-\infty}^{\infty} \hat{f}(u, v)e^{2\pi i(xu+yv)}\,du\,dv.$$

For discrete functions $\mathbf{a} : \mathbb{Z}_m \times \mathbb{Z}_n \to \mathbb{R}$ we have the two-dimensional discrete Fourier transform

$$\hat{\mathbf{a}}(u, v) = [\mathfrak{F}(\mathbf{a})](u, v) = \sum_{k=0}^{n-1}\sum_{j=0}^{m-1} \mathbf{a}(j, k)e^{-2\pi i\left(j\frac{u}{m}+k\frac{v}{n}\right)},$$

with the inverse transform specified by

$$\mathbf{a}(i,j) = \left[\mathfrak{F}^{-1}(\hat{\mathbf{a}})\right](i,j) = \frac{1}{nm}\sum_{v=0}^{n-1}\sum_{u=0}^{m-1}\hat{\mathbf{a}}(u,v)e^{2\pi i\left(u\frac{j}{m}+v\frac{k}{n}\right)}.$$

Figure 8.2.1 shows the image of a jet and its Fourier transform image. The Fourier transform image is complex-valued and, therefore, difficult to display. The value at each point in Figure 8.2.1 is actually the magnitude of its corresponding complex pixel value in the Fourier transform image. Figure 8.2.1 does show a full period of the transform, however the origin of the transform does not appear at the center of the display. A representation of one full period of the transform image with its origin shifted to the center of the display can be achieved by multiplying each point $\mathbf{a}(x,y)$ by $(-1)^{x+y}$ before applying the transform. The jet's Fourier transform image shown with the origin at the center of the display is seen in Figure 8.2.2.

Figure 8.2.1. The image of a jet (left) and its Fourier transform image (right).

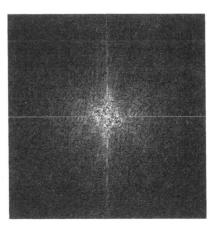

Figure 8.2.2. Fourier transform of jet with origin at center of display.

Image Algebra Formulation

The discrete Fourier transform of the one-dimensional image $\mathbf{a} \in \mathbb{R}^{\mathbb{Z}_n}$ is given by the image algebra expression

$$\hat{\mathbf{a}} = \mathbf{a} \oplus \mathbf{f},$$

where $\mathbf{f} \in \left(\mathbb{C}^{\mathbb{Z}_n}\right)^{\mathbb{Z}_n}$ is the template defined by

$$\mathbf{f}_j(k) = e^{-2\pi k j i / n}.$$

The template $\mathbf{f}$ is called the *one-dimensional Fourier template*. It follows directly from the definition of $\mathbf{f}$ that $\mathbf{f}' = \mathbf{f}$ and $(\mathbf{f}^*)' = \mathbf{f}^*$, where $\mathbf{f}^*$ denotes the complex conjugate of $\mathbf{f}$ defined by $\mathbf{f}_j^*(k) = (\mathbf{f}_j(k))^* = e^{2\pi k j i / n}$. Hence, the equivalent of the discrete inverse Fourier transform of $\hat{\mathbf{a}}$ is given by

$$\mathbf{a} = \frac{1}{n}(\hat{\mathbf{a}} \oplus \mathbf{f}^*).$$

The image algebra equivalent formulation of the two-dimensional discrete Fourier transform pair is given by

$$\hat{\mathbf{a}} = \mathbf{a} \oplus \mathbf{f}$$

and

$$\mathbf{a} = \frac{1}{nm}(\hat{\mathbf{a}} \oplus \mathbf{f}^*),$$

where the *two-dimensional Fourier template* $\mathbf{f}$ is defined by

$$\mathbf{f}_{(u,v)}(j,k) = e^{-2\pi i \left(j \frac{u}{m} + k \frac{v}{n}\right)},$$

and $\mathbf{X} = \mathbb{Z}_m \times \mathbb{Z}_n$.

8.3.   Centering the Fourier Transform

Centering the Fourier transform is a common operation in image processing. Centered Fourier transforms are useful for displaying the Fourier spectra as intensity functions, for interpreting Fourier spectra, and for the design of filters.

The discrete Fourier transform and its inverse exhibit the periodicity

$$\hat{\mathbf{a}}(u,v) = \hat{\mathbf{a}}(u+m,v) = \hat{\mathbf{a}}(u,v+n) = \hat{\mathbf{a}}(u+m,v+n)$$

for an $m \times n$ image $\mathbf{a}$. Additionally, since $\hat{\mathbf{a}}(u,v) = \hat{\mathbf{a}}^*(-u,-v)$, the magnitude of the Fourier transform exhibits the symmetry

$$|\hat{\mathbf{a}}(u,v)| = |\hat{\mathbf{a}}(-u,-v)|.$$

Periodicity and symmetry are the key ingredients for understanding the need for centering the Fourier transform for interpretation purposes. The periodicity property indicates that $\hat{\mathbf{a}}(u,v)$ has period of length m in the u direction and of length n in the v direction, while the symmetry shows that the magnitude is centered about the origin. This is shown in Figure 8.2.1 where the origin $(0,0)$ is located at the upper left hand corner of the image. Since the discrete Fourier transform has been formulated for values of u in the interval $[0, m-1]$ and values v in the interval $[0, n-1]$, the result of this formulation yields two half periods in these intervals that are back to back. This is illustrated in Figure 8.3.1 (a), which shows a one-dimensional slice along the u-axis of the magnitude function $|\hat{a}(u, 0)|$. Therefore, in order to display one full period in both the u and v directions, it is necessary to shift the origin to the midpoint $\left(\frac{m}{2}, \frac{n}{2}\right)$ and add points that shift to the outside of the image domain back into the image domain using modular arithmetic. Figures 8.2.2 and 8.3.1 illustrate the effect of this centering method. The reason for using modular arithmetic is that, in contrast to Figure 8.3.1 (a), when Fourier transforming $a \in \mathbb{R}^{\mathbf{X}}$, where $\mathbf{X} = \mathbb{Z}_m \times \mathbb{Z}_n$, there is no information available outside the intervals $[0, m-1]$ and $[0, n-1]$. Thus, in order to display the full periodicity centered at the midpoint in the interval $[0, m-1]$, the intervals $\left[0, \frac{m}{2}\right]$ and $\left[\frac{m}{2}+1, m-1\right]$ need to be flipped and their ends glued together. The same gluing procedure needs to be performed in the v direction. This amounts to doing modular arithmetic. Figure 8.3.2 illustrates this process.

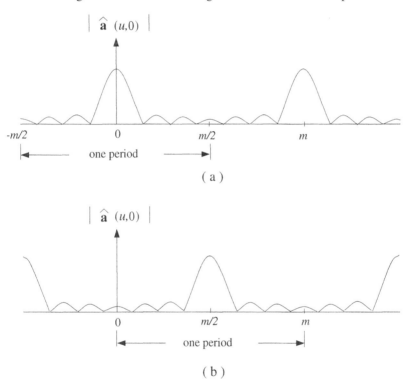

Figure 8.3.1. Periodicity of the Fourier transform. Figure (a) shows the two half periods, one on the interval $\left[0, \frac{m}{2}\right]$ and the other on the interval $\left[\frac{m}{2}, m\right]$. Figure (b) shows the full period of the shifted Fourier transform on the interval $[0, m]$.

Image Algebra Formulation

Let $\hat{a} \in \mathbb{C}^{\mathbf{X}}$, where $\mathbf{X} = \mathbb{Z}_m \times \mathbb{Z}_n$ and m and n are even integers. Define the centering function $center : \mathbf{X} \to \mathbf{X}$ by

$$center(\mathbf{p}) = \left[\mathbf{p} + \left(\frac{m}{2}, \frac{n}{2}\right)\right] mod(m, n).$$

Note that the centering function is its own inverse since $center(center(\mathbf{p})) = \mathbf{p}$. Therefore, if $\hat{c}$ denotes the centered version of $\hat{a}$, then $\hat{c}$ is given by

$$\hat{c} := \hat{a} \circ center.$$

Figure 8.3.2 illustrates the mapping of points when applying *center* to an array **X**.

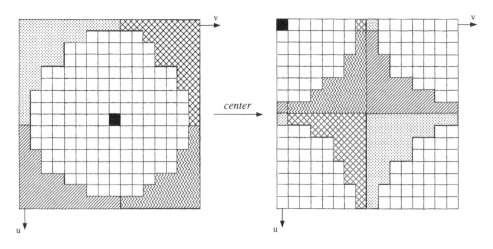

Figure 8.3.2. The centering function applied to a 16×16 array. The different shadings indicate the mapping of points under the center function.

Comments and Observations

If **X** is some arbitrary rectangular subset of $\mathbb{Z}_2$, then the centering function needs to take into account the location of the midpoint of **X** with respect to the minimum of **X**, whenever $min(\mathbf{X}) \neq (0, 0)$. In particular, by defining the function

$$mid(\mathbf{X}) = \frac{max(\mathbf{X}) - min(\mathbf{X}) + 1}{2},$$

the centering function now becomes

$$center(\mathbf{p}) = [\mathbf{p} - min(\mathbf{X}) + mid(\mathbf{X})]mod(2mid(\mathbf{X})) + min(\mathbf{X}).$$

Note that $mid(\mathbf{X})$ does not correspond to the midpoint of **X**, but gives the midpoint of the set $\mathbf{X} - min(\mathbf{X})$. The actual midpoint of **X** is given by $center(min(\mathbf{X}))$. This is illustrated by the example shown in Figure 8.3.3. In this example, $min(\mathbf{X}) = (4, 5)$, $max(\mathbf{X}) = (13, 18)$, and $mid(\mathbf{X}) = (5, 7)$. Also, if $\mathbf{X} = \mathbb{Z}_m \times \mathbb{Z}_n$, then $min(\mathbf{X}) = (0, 0)$, $max(\mathbf{X}) = (m - 1, n - 1)$, and $mid(\mathbf{X}) = \left(\frac{m}{2}, \frac{n}{2}\right)$. This shows that the general center function reduces to the previous center function.

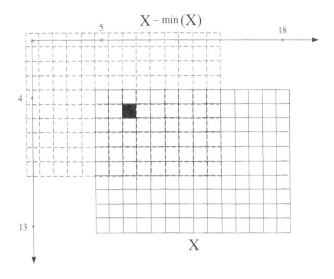

Figure 8.3.3. The shift $\mathbf{X} - min(\mathbf{X})$ of a 10×14
array $\mathbf{X}$. The point $mid(\mathbf{X}) = (5,7)$ is shown in black.

If $\mathbf{X}$ is a square array of form $\mathbf{X} = \mathbb{Z}_n \times \mathbb{Z}_n$, then centering can be accomplished
by multiplying the image $\mathbf{a}$ by the image $\mathbf{b}$ defined by $\mathbf{b}(x,y) = (-1)^{x+y}$ prior to taking
the Fourier transform. This follows from the simple fact expressed by the equation

$$e^{2\pi i (u_0 x + v_0 y)/n} = e^{\pi i (x+y)} = (-1)^{x+y},$$

which holds whenever $(u_0, v_0) = \left(\frac{n}{2}, \frac{n}{2}\right)$. Thus, in this case we can compute $\hat{\mathbf{c}}$ by using
the formulation

$$\hat{\mathbf{c}} := \mathfrak{F}(\mathbf{a} \cdot \mathbf{b}).$$

8.4. Fast Fourier Transform

In this section we present the image algebra formulation of the *Cooley-Tukey*
radix-2 fast Fourier transform (FFT) [1, 2]. The image algebra expression for the Fourier
transform of $\mathbf{a} \in \mathbb{C}^{\mathbb{Z}_n}$ is given in Section 8.2 by

$$\hat{\mathbf{a}} = \mathbf{a} \oplus \mathbf{f},$$

where $\mathbf{f}$ is the Fourier template defined by

$$\mathbf{f}_j(k) = e^{-2\pi k j i / n}.$$

Expanding $\mathbf{a} \oplus \mathbf{f}$, we get the following equation for $\hat{\mathbf{a}}$:

$$\hat{\mathbf{a}}(j) = \sum_{k=0}^{n-1} \mathbf{a}(k) e^{-2\pi i j k / n}.$$

In this form it is easy to see that $O(n^2)$ complex adds and multiplications are required to
compute the Fourier transform using the formulation of Section 8.2. For each $0 \leq j \leq n-1$,
n complex multiplications of $\mathbf{a}(k)$ by $e^{-2\pi i j k / n}$ are required for a total of n^2 complex

multiplications. For each $0 \leq j \leq n-1$ the sum $\sum\limits_{k=0}^{n-1}$ requires $n-1$ complex adds for a total of $n^2 - n$ complex adds. The complex arithmetic involved in the computation of the $e^{-2\pi i j k/n}$ terms does not enter into the measure of computational complexity that is optimized by the fast Fourier algorithm.

The number of complex adds and multiplications can be reduced to $O(n\, log_2\, n)$ by incorporating the Cooley-Tukey radix-2 fast Fourier algorithm into the image algebra formulation of the Fourier transform. It is assumed for the FFT optimization that n is a power of 2.

The separability of the Fourier transform can be exploited to obtain a similar computational savings on higher dimensional images. Separability allows the Fourier transform to be computed over an image as a succession of one-dimensional Fourier transforms along each dimension of the image. Separability will be discussed in more detail later.

For the mathematical foundation of the Cooley-Tukey fast Fourier algorithm the reader is referred to Cooley, et al. [1, 2, 3]. A detailed exposition on the integration of the Cooley-Tukey algorithm into the image algebra formulation of the FFT can be found in Ritter [4]. The separability of the Fourier transform and the definition of the permutation function $\rho_n : \mathbb{Z}_n \rightarrow \mathbb{Z}_n$ used in the image algebra formulation of the FFT will be discussed here.

The separability of the Fourier transform is key to the decomposition of the two-dimensional FFT into two successive one-dimensional FFTs. The structure of the image algebra formulation of the two-dimensional FFT will reflect the utilization of separability.

The discrete Fourier transform, as defined in Section 8.2, is given by

$$\hat{\mathbf{a}}(u,v) = \sum_{k=0}^{n-1} \sum_{j=0}^{m-1} \mathbf{a}(j,k) e^{-2\pi i \left(j \frac{u}{m} + k \frac{v}{n} \right)}.$$

The double summation can be rewritten as

$$\hat{\mathbf{a}}(u,v) = \sum_{k=0}^{n-1} \left[\left(e^{-2\pi i k \frac{v}{n}} \right) \sum_{j=0}^{m-1} \mathbf{a}(j,k) e^{-2\pi i j \frac{u}{m}} \right]$$

or

$$\hat{\mathbf{a}}(u,v) = \sum_{k=0}^{n-1} \hat{\mathbf{a}}(u,k) e^{-2\pi i k \frac{v}{n}},$$

where

$$\hat{\mathbf{a}}(u,k) = \sum_{j=0}^{m-1} \mathbf{a}(j,k) e^{-2\pi i j \frac{u}{m}}.$$

For each $0 \leq k \leq n-1$, $\hat{\mathbf{a}}(u,k)$ is the one-dimensional Fourier transform of the kth column of $\mathbf{a}$. For each $0 \leq u \leq m-1$,

$$\hat{\mathbf{a}}(u,v) = \sum_{k=0}^{n-1} \hat{\mathbf{a}}(u,k) e^{-2\pi i k \frac{v}{n}}$$

is the one-dimensional Fourier transform of the uth row of $\hat{\mathbf{a}}(u,k)$.

From the above it is seen that a two-dimensional Fourier transform can be computed by a two-stage application of one-dimensional Fourier transforms. First, a one-dimensional Fourier transform is applied to the columns of the image. In the second stage, another one-dimensional Fourier transform is applied to the rows of the result of the first stage. The same result is obtained if the one-dimensional Fourier transform is first applied to the rows of the image, then the columns. Computing the two-dimensional Fourier transform in this way has "separated" it into two one-dimensional Fourier transforms. By using separability, any saving in the computational complexity of the one-dimensional FFT can be passed on to higher dimensional Fourier transforms.

For $n = 2^k$, the permutation $\rho_n : \mathbb{Z}_n \to \mathbb{Z}_n$ is the function that reverses the bit order of the k-bit binary representation of its input, and outputs the decimal representation of the value of the reversed bit order. For example, the table below represents $\rho_8 : \mathbb{Z}_8 \to \mathbb{Z}_8$. The permutation function will be used in the image algebra formulation of the FFT. The algorithm used to compute the permutation function ρ_n is presented next.

Table 8.4.1 The Permutation ρ_8 and Its Corresponding Binary Evaluations

i	Binary i	Reversed binary i	$\rho_8(i)$
0	000	000	0
1	001	100	4
2	010	010	2
3	011	110	6
4	100	001	1
5	101	101	5
6	110	011	3
7	111	111	7

For $n = 2^k$ and $i = 0, 1, \ldots, n-1$ compute $\rho_n(i)$ as follows:

```
begin
    j := 0
    m := i
    for l := 0 to log₂(n) − 1 loop
```
$$h := \left\lfloor \frac{m}{2} \right\rfloor$$
$$j := 2j + (m - 2h)$$
$$m := h$$
```
    end loop
    return j
end.
```

Image Algebra Formulation

One-Dimensional FFT

Let $\mathbf{a} \in \mathbb{C}^{\mathbb{Z}_n}$, where $n = 2^k$ for some positive integer k. Let $P = \left\{ 2^i : i = 0, 1, \ldots, log_2 n - 1 \right\}$ and for $p \in P$, define the parameterized template $\mathbf{t}(p)$ by

$$\mathbf{t}(p)_j(l) = \begin{cases} 1 & \text{if } \lfloor j/p \rfloor \text{ is even and } l = j \\ w(j,p) & \text{if } \lfloor j/p \rfloor \text{ is even and } l = j + p \\ -w(j,p) & \text{if } \lfloor j/p \rfloor \text{ is odd and } l = j \\ 1 & \text{if } \lfloor j/p \rfloor \text{ is odd and } l = j - p \\ 0 & \text{otherwise,} \end{cases}$$

where $w(j,p) = e^{-\frac{\pi i (j \bmod p)}{p}}$. The following image algebra algorithm computes the Cooley-Tukey radix-2 FFT.

$$\mathbf{a} := \mathbf{a} \circ \rho_n$$
$$\text{for } i := 1 \text{ to } log_2 n \text{ loop}$$
$$\mathbf{a} := \mathbf{a} \bigoplus \mathbf{t}\left(2^{i-1}\right)$$
$$\text{end loop.}$$

The function ρ_n is the permutation function defined earlier.

By the definition of the generalized convolution operator, $\mathbf{a} \bigoplus \mathbf{t}\left(2^{i-1}\right)$ is equal to

$$\mathbf{a}(j) = \sum_{l \in S\left(\mathbf{t}(p)_j\right)} \mathbf{a}(l) \cdot \mathbf{t}(p)_j(l) .$$

Notice that from the definition of the template $\mathbf{t}$ there are at most 2 values of l in the support of $\mathbf{t}$ for every $0 \leq j \leq n - 1$. Thus, only $2n$ complex multiplications and n complex adds are required to evaluate $\mathbf{a} \bigoplus \mathbf{t}\left(2^{i-1}\right)$. Since the convolution $\mathbf{a} \bigoplus \mathbf{t}\left(2^{i-1}\right)$ is contained within a loop that consists of $log_2 n$ iterations, there are $O(n \, log \, n)$ complex adds and multiplications in the image algebra formulation of the FFT.

One-Dimensional Inverse FFT

The inverse Fourier transform can be computed in terms of the Fourier transform by simple conjugation. That is, $\mathbf{a} = \left(\frac{1}{n} \mathfrak{F}(\hat{\mathbf{a}}^*) \right)^*$. The following algorithm computes the inverse FFT of $\hat{\mathbf{a}} \in \mathbb{C}^{\mathbb{Z}_n}$ using the forward FFT and conjugation.

$$\mathbf{a} := \hat{\mathbf{a}}^* \circ \rho_n$$
$$\text{for } i := 1 \text{ to } log_2 n \text{ loop}$$
$$\mathbf{a} := \mathbf{a} \bigoplus \mathbf{t}\left(2^{i-1}\right)$$
$$\text{end loop}$$
$$\mathbf{a} := \left(\frac{1}{n} \mathbf{a} \right)^* .$$

Two-Dimensional FFT

In our earlier discussion of the separability of the Fourier transform, we noted that the two-dimensional DFT can be computed in two steps by successive applications of the one-dimensional DFT; first along each row followed by a one-dimensional DFT along each column. Thus, to obtain a fast Fourier transform for two-dimensional images, we need to apply the image algebra formulation of the one-dimensional FFT in simple succession. However, in order to perform the operations $\mathbf{a} \circ \rho_n$ and $\mathbf{a} \oplus \mathbf{t}(p)$ specified by the algorithm, it becomes necessary to extend the function ρ_n and the template $\mathbf{t}(p)$ to two-dimensional arrays. For this purpose, suppose that $\mathbf{X} = \mathbb{Z}_m \times \mathbb{Z}_n$, where $n = 2^h$ and $n = 2^k$, and assume without loss of generality that $n \leq m$.

Let $P = \{2^i : i = 0, 1, \ldots, log_2 m - 1\}$ and for $p \in P$ define the parameterized row template $\mathbf{t}(p) : \mathbf{X} \to \mathbb{C}^{\mathbf{X}}$ by

$$
\mathbf{t}(p)_{(u,v)}(x,y) = \begin{cases} 1 & \text{if } \lfloor u/p \rfloor \text{ is even and } (x,y) = (u,v) \\ w(u,p) & \text{if } \lfloor u/p \rfloor \text{ is even and } (x,y) = (u+p,v) \\ -w(u,p) & \text{if } \lfloor u/p \rfloor \text{ is odd and } (x,y) = (u,v) \\ 1 & \text{if } \lfloor u/p \rfloor \text{ is odd and } (x,y) = (u-p,v) \\ 0 & \text{otherwise,} \end{cases}
$$

where $w(u,p) = e^{-\frac{\pi i (u \bmod p)}{p}}$. Note that for each $p \in P$, $\mathbf{t}(p)$ is a row template which is essentially identical to the template used in the one-dimensional case.

The permutation ρ is extended to a function $r : \mathbf{X} \to \mathbf{X}$ in a similar fashion by restricting its actions to the rows of $\mathbf{X}$. In particular, define

$$
r_m : \mathbf{X} \to \mathbf{X}
$$
$$
\text{by } r_m(i,j) = (\rho_m(i), j).
$$

With the definitions of r and $\mathbf{t}$ completed, we are now in a position to specify the two-dimensional radix-2 FFT in terms of image algebra notation.

If $\mathbf{X}$, r_m and $\mathbf{t}$ are specified as above and $\mathbf{a} \in \mathbb{C}^{\mathbf{X}}$, then the following algorithm computes the two-dimensional fast Fourier transform of $\mathbf{a}$.

$$
\mathbf{a} := \mathbf{a} \circ r_m
$$

$$\text{for } i := 1 \text{ to } log_2 m \text{ loop}$$

$$\mathbf{a} := \mathbf{a} \oplus \mathbf{t}(2^{i-1})$$

$$\text{end loop}$$

$$\mathbf{a} := \mathbf{a}' \circ r_n$$

$$\text{for } i := 1 \text{ to } log_2 n \text{ loop}$$

$$\mathbf{a} := \mathbf{a} \oplus \mathbf{t}(2^{i-1})$$

$$\text{end loop}$$

$$\hat{\mathbf{a}} := \mathbf{a}'.$$

Two-Dimensional Inverse FFT

As in the one-dimensional case, the two-dimensional inverse FFT is also computed using the forward FFT and conjugation. Let $\hat{\mathbf{a}} \in \mathbb{C}^{\mathbf{X}}$. The inverse FFT is given by the following algorithm:

$$\mathbf{a} := \hat{\mathbf{a}}^*$$

Apply forward FFT

$$\mathbf{a} := \left(\frac{1}{mn}\mathbf{a}\right)^*.$$

Alternate Image Algebra Formulation

The formulation of the fast Fourier transform above assumes that the image-template operation $\oplus$ is only applied over the support of the template. If this is not the case, the "fast" formulation using templates will result in much poorer performance than the formulation of Section 8.2. The alternate image algebra formulation for the fast transform uses spatial transforms rather than templates. The alternate formulation is more appropriate if the implementation of $\oplus$ does not restrict its operations to the template's support.

For the alternate formulation, let the variables of the two-dimensional FFT above remain as defined. Define the functions $f_i, g_i : \mathbf{X} \to \mathbf{X}$ as follows:

$$f_i(x, y) = \begin{cases} \left(x + 2^{i-1}, y\right) & \text{if } \left\lfloor \frac{x}{2^{i-1}} \right\rfloor \text{ is even} \\ (x, y) & \text{otherwise} \end{cases}$$

$$g_i(x, y) = \begin{cases} \left(x - 2^{i-1}, y\right) & \text{if } \left\lfloor \frac{x}{2^{i-1}} \right\rfloor \text{ is odd} \\ (x, y) & \text{otherwise.} \end{cases}$$

The images $\mathbf{w}_i \in \mathbb{C}^{\mathbf{X}}$ used in the alternate formulation are defined as

$$\mathbf{w}_i(x, y) = \begin{cases} w\left(x, 2^{i-1}\right) & \text{if } \left\lfloor \frac{x}{2^{i-1}} \right\rfloor \text{ is even} \\ -w\left(x, 2^{i-1}\right) & \text{if } \left\lfloor \frac{x}{2^{i-1}} \right\rfloor \text{ is odd.} \end{cases}$$

The alternate formulation for the fast Fourier transform using spatial transforms is given by

$$\mathbf{a} := \mathbf{a} \circ r_m$$

for $i := 1$ to $log_2 m$ loop

$\quad \mathbf{a} := \mathbf{a} \circ g_i + \mathbf{w}_i \cdot (\mathbf{a} \circ f_i)$

end loop

$$\mathbf{a} := \mathbf{a}' \circ r_n$$

for $i := 1$ to $log_2 n$ loop

$\quad \mathbf{a} := \mathbf{a} \circ g_i + \mathbf{w}_i \cdot (\mathbf{a} \circ f_i)$

end loop

$$\hat{\mathbf{a}} := \mathbf{a}'.$$

Note that the use of the transpose $\mathbf{a}'$ before and after the second loop in the code above is used for notational simplicity, not computational efficiency. The use of transposes can be eliminated by using analogues of the functions f_i, $\bar{g}_i$, and $\mathbf{w}_i$ in the second loop that are functions of the column coordinate of $\mathbf{X}$.

Comments and Observations

The discussion presented here concerns a general algebraic optimization of the Fourier transform. Many important issues arise when optimizing the formulation for specific implementations and architectures [4].

The permutation function is not needed in the formulation of the Fourier transform of Section 8.2. The computation of the permutation function ρ_n does add to the computational cost of the FFT. For each $i \in \{0, 1, \ldots, n-1\}$, $\rho_n(i)$ requires $O(log_2 n)$ integer operations. Thus, the evaluation of ρ_n will require a total of $O(n log_2 n)$ integer operations. The amount of integer arithmetic involved in computing ρ_n is of the same order of magnitude as the amount of floating point arithmetic for the FFT. Hence the overhead associated with bit reversal is nontrivial in the computation of the FFT, often accounting for 10% to 30% of the total computation time.

8.5. Discrete Cosine Transform

Let $\mathbf{a} \in \mathbb{R}^{Z_n}$. The *one-dimensional discrete cosine transform* $\mathfrak{C}(\mathbf{a})$ is defined by

$$[\mathfrak{C}(\mathbf{a})](u) = \frac{2 \cdot c(u)}{n} \sum_{x=0}^{n-1} \mathbf{a}(x) \cdot cos\left(\frac{(2x+1)u\pi}{2n}\right) \qquad u = 0, 1, \ldots, n-1,$$

where

$$c(u) = \begin{cases} \frac{1}{\sqrt{2}} & \text{if } u = 0 \\ 1 & \text{if } u = 1, 2, \ldots, n-1. \end{cases}$$

The *inverse one-dimensional discrete cosine transform* $\mathfrak{C}^{-1}(\mathbf{a})$ is given by

$$[\mathfrak{C}^{-1}(\mathbf{a})](x) = \sum_{u=0}^{n-1} c(u) \cdot \mathbf{a}(u) \cdot cos\left(\frac{(2x+1)u\pi}{2n}\right).$$

The cosine transform provide the means of expressing an image as a weighted sum of cosine functions. The weights in the sum are given by the cosine transform. Figure 8.5.1 illustrates this by showing how the square wave

$$s(x) = \begin{cases} 1 & \text{if } 0 \leq x < 1 \\ -1 & \text{if } 1 \leq x \leq 2 \end{cases}$$

is approximated using the first five terms of the discrete cosine function.

For $\mathbf{a} \in \mathbb{R}^{Z_n \times Z_n}$ the two-dimensional cosine transform and its inverse are given by

$$[\mathfrak{C}(\mathbf{a})](u, v) = \frac{2 \cdot c(u) \cdot c(v)}{n} \sum_{x=0}^{n-1} \sum_{y=0}^{n-1} \mathbf{a}(x, y) \cdot cos\left(\frac{(2x+1)u\pi}{2n}\right) \cdot cos\left(\frac{(2y+1)v\pi}{2n}\right)$$

and

$$[\mathfrak{C}^{-1}(\mathbf{a})](x, y) = \frac{2}{n} \sum_{u=0}^{n-1} \sum_{v=0}^{n-1} c(u) \cdot c(v) \cdot \mathbf{a}(u, v) \cdot cos\left(\frac{(2x+1)u\pi}{2n}\right) \cdot cos\left(\frac{(2y+1)v\pi}{2n}\right),$$

respectively.

Figure 8.5.2 shows the image of a jet and the image which represents the pixel magnitude of its cosine transform image.

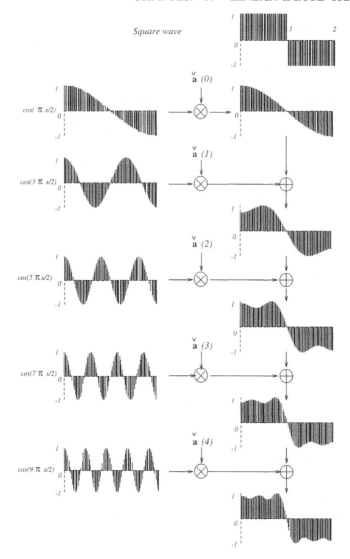

Figure 8.5.1. Approximation of a square wave using
the first five terms of the discrete cosine transform.

Image Algebra Formulation

The template $\mathbf{t}$ used for the two-dimensional cosine transform is specified by

$$\mathbf{t}_{(u,v)}(x,y) = c(u) \cdot c(v) \cdot cos\left(\frac{(2x+1)u\pi}{2n}\right) \cdot cos\left(\frac{(2y+1)v\pi}{2n}\right).$$

The image algebra formulations of the two-dimensional transform and its inverse are

$$\mathfrak{C}(\mathbf{a}) := \frac{2}{n}(\mathbf{a} \oplus \mathbf{t})$$

and

$$\mathfrak{C}^{-1}(\mathbf{a}) := \frac{2}{n}(\mathbf{a} \oplus \mathbf{t}'),$$

respectively.

Figure 8.5.2. Jet image (left) and its cosine transform image.

The one-dimensional transforms are formulated similarly. The template used for the one-dimensional case is

$$\mathbf{t}_u(x) = c(u) \cdot cos\left(\frac{(2x+1)u\pi}{2n}\right).$$

The image algebra formulations of the one-dimensional discrete cosine transform and its inverse are

$$\mathfrak{C}(\mathbf{a}) := \frac{2}{n}(\mathbf{a} \oplus \mathbf{t})$$

and

$$\mathfrak{C}^{-1}(\mathbf{a}) := (\mathbf{a} \oplus \mathbf{t}'),$$

respectively.

In [5], a method for computing the one-dimensional cosine transform using the fast Fourier transform is introduced. The method requires only a simple rearrangement of the input data into the FFT. The even and odd terms of the original image $\mathbf{a} \in \mathbb{R}^{\mathbb{Z}_n}$, $n = 2^k$ are rearranged to form a new image $\mathbf{b}$ according to the formula

$$\begin{array}{ll} \mathbf{b}(x) = \mathbf{a}(2x) & \\ \qquad \text{and} & x = 0, 1, \ldots, \dfrac{n}{2} - 1. \\ \mathbf{b}(n-1-x) = \mathbf{a}(2x+1) & \end{array}$$

Using the new arrangement of terms the cosine transform can be written as

$$[\mathfrak{C}(\mathbf{a})](u) = \frac{2c(u)}{n}[\sum_{x=0}^{\frac{n}{2}-1} \mathbf{b}(x) \cdot cos\left(\frac{(4x+1)u\pi}{2n}\right)$$

$$+ \sum_{x'=0}^{\frac{n}{2}-1} \mathbf{b}(n-1-x') \cdot cos\left(\frac{(4x'+3)u\pi}{2n}\right)].$$

Substituting $x = n - 1 - x'$ into the second sum and using the sum of two angles formula for cosine functions yields

$$[\mathfrak{C}(\mathbf{a})](u) = \frac{2c(u)}{n} \sum_{x=0}^{n-1} \mathbf{b}(x) \cdot cos\left(\frac{(4x+1)u\pi}{2n}\right).$$

The above can be rewritten in terms of the complex exponential function as

$$[\mathfrak{C}(\mathbf{a})](u) = \frac{2c(u)}{n} \sum_{x=0}^{n-1} \mathbf{b}(x) \cdot Re\left(e^{u\pi i/2n} \cdot e^{xu2\pi i/n}\right)$$

$$= 2c(u) \cdot Re\left(e^{u\pi i/2n} \cdot \frac{1}{n} \sum_{x=0}^{n-1} \left(\mathbf{b}(x) \cdot e^{ux2\pi i/n}\right)\right)$$

$$= 2c(u) \cdot Re\left(e^{u\pi i/2n} \cdot \left[\mathfrak{F}^{-1}(\mathbf{b})\right](u)\right),$$

where $\mathfrak{F}^{-1}$ denotes the inverse Fourier transform.

Let $\mathbf{d} = e^{u\pi i/2n} \cdot \mathfrak{F}^{-1}(\mathbf{b})$. Since $\mathbf{b}$ is real-valued we have that $\mathbf{d}(n-u) = i\mathbf{d}^*(u)$. Consequently,

$$\begin{array}{cc} [\mathfrak{C}(\mathbf{a})](u) = 2 \cdot c(u) \cdot Re(\mathbf{d}(u)) & \\ \text{and} & u = 0, 1, \ldots, \dfrac{n}{2}. \\ [\mathfrak{C}(\mathbf{a})](n-u) = 2 \cdot Im(\mathbf{d}(u)) & \end{array}$$

Therefore, it is only necessary to compute $\frac{n}{2}$ terms of the inverse Fourier transform in order to calculate $[\mathfrak{C}(\mathbf{a})](u)$ for $u = 0, 1, \ldots, n-1$.

The inverse cosine transform can be obtained from the real part of the inverse Fourier transform of

$$c(u) \cdot \mathbf{a}(u) \cdot e^{u\pi i/2n}.$$

Let the images $\mathbf{c}$ and $\mathbf{e}$ be defined by

$$\mathbf{c}(u) = c(u)$$
$$\mathbf{e}(u) = e^{u\pi i/2n}, \qquad u = 0, 1, \ldots, n-1.$$

The even indexed terms are calculated using

$$\left[\mathfrak{C}^{-1}(\mathbf{a})\right](2x) = Re\left[\sum_{u=0}^{n-1} \left(c(u) \cdot \mathbf{a}(u) \cdot e^{u\pi i/2n}\right) \cdot e^{ux2\pi i/n}\right]$$

$$= Re\left(\left[\mathfrak{F}^{-1}(\mathbf{c} \cdot \mathbf{a} \cdot \mathbf{e})\right](x)\right) \qquad x = 0, 1, \ldots, \frac{n}{2} - 1.$$

The odd indexed terms are given by

$$\left[\mathfrak{C}^{-1}(\mathbf{a})\right](2x+1) = Re\left[\sum_{u=0}^{n-1} \left(c(u) \cdot \mathbf{a}(u) \cdot e^{u\pi i/2n}\right) \cdot e^{ux2\pi i/n}\right]$$

$$= Re\left(\left[\mathfrak{F}^{-1}(\mathbf{c} \cdot \mathbf{a} \cdot \mathbf{e})\right](n-1-x)\right) \qquad x = 0, 1, \ldots, \frac{n}{2} - 1.$$

With the information above it is easy to adapt the formulations of the one-dimension fast Fourier transforms (Section 8.4) for the implementation of fast one-dimensional cosine transforms. The cosine transform and its inverse are separable. Therefore, fast two-dimensional transforms implementations are possible by taking one-dimensional transforms along the rows of the image, followed by one-dimensional transforms along the columns [5, 6, 7].

8.6.   Walsh Transform

The Walsh transform was first defined in 1923 by Walsh [8], although in 1893 Hadamard [9] had achieved a similar result by the application of certain orthogonal matrices, generally called Hadamard matrices, which contain only the entries +1 and –1.

In 1931 Paley provided an entirely different definition of the Walsh transform, which is the one used most frequently by mathematicians [10] and is the one used in this discussion.

The one-dimensional *Walsh transform* of $\mathbf{a} \in \mathbb{R}^{\mathbb{Z}_n}$, where $n = 2^k$, is given by

$$[\mathfrak{W}(\mathbf{a})](u) = \frac{1}{n} \sum_{x=0}^{n-1} \mathbf{a}(x) \prod_{i=0}^{k-1} (-1)^{b_i(x) b_{k-1-i}(u)}, \qquad 0 \le u < n.$$

The term $b_j(z)$ denotes the jth bit in the binary expansion of z. For example,

$$b_0(17) = 1, \ b_1(17) = 0, \ b_2(17) = 0, \ b_3(17) = 0, \ \text{and } b_4(17) = 1.$$

The inverse of the Walsh transform is

$$\left[\mathfrak{W}^{-1}(\mathbf{a})\right](x) = \sum_{u=0}^{n-1} \mathbf{a}(u) \prod_{i=0}^{k-1} (-1)^{b_i(x) b_{k-1-i}(u)}.$$

The set of functions created from the product

$$g_u(x) = \prod_{i=0}^{k-1} (-1)^{b_i(x) b_{k-1-i}(u)}$$

form the basis of the Walsh transform and for each pair (x, u), $g_u(x)$ represents the (x, u) entry of the Hadamard matrix mentioned above. Rewriting the expression for the inverse Walsh transform as

$$\mathbf{a}(x) = \sum_{u=0}^{n-1} [\mathfrak{W}(\mathbf{a})](u) \cdot g_u(x),$$

it is seen that $\mathbf{a}$ is a weighted sum of the basis functions $g_u(x)$. The weights in the sum are given by the Walsh transform. Figure 8.6.1 shows the Walsh basis functions for $n = 4$ and how an image is written as a weighted sum of the basis functions.

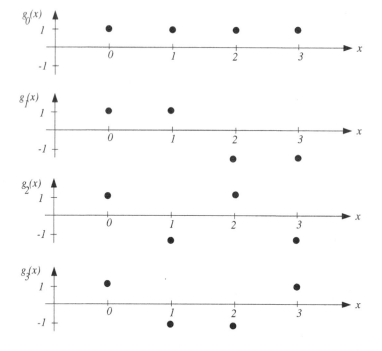

Figure 8.6.1. The Walsh basis for $n = 4$.

Thus, the Walsh transform and the Fourier transform are similar in that they both provide the coefficients for the representation of an image as a weighted sum of basis functions. The basis functions for the Fourier transform are sinusoidal functions of varying frequencies. The basis functions for the Walsh transform are the elements of $\{-1, 1\}^{\mathbb{Z}_n}$ defined above. The rate of transition from negative to positive value in the Walsh basis function is analogous to the frequency of the Fourier basis function. The frequencies of the basis functions for the Fourier transform increase as u increases. However, the rate at which Walsh basis functions change signs is not an increasing function of u.

For $\mathbf{a} \in \mathbb{R}^{\mathbb{Z}_n \times \mathbb{Z}_n}$, the forward and reverse Walsh transforms are given by

$$[\mathfrak{W}(\mathbf{a})](u, v) = \frac{1}{n} \sum_{x=0}^{n-1} \sum_{y=0}^{n-1} \mathbf{a}(x, y) \prod_{i=0}^{k-1} (-1)^{[b_i(x)b_{k-1-i}(u)+b_i(y)b_{k-1-i}(v)]}$$

and

$$\left[\mathfrak{W}^{-1}(\mathbf{a})\right](x, y) = \frac{1}{n} \sum_{u=0}^{n-1} \sum_{v=0}^{n-1} \mathbf{a}(u, v) \prod_{i=0}^{k-1} (-1)^{[b_i(x)b_{k-1-i}(u)+b_i(y)b_{k-1-i}(v)]},$$

respectively. Here again, $\mathbf{a}$ can be represented as a weighted sum of the basis functions

$$g_{(u,v)}(x, y) = \prod_{i=0}^{k-1} (-1)^{[b_i(x)b_{k-1-i}(u)+b_i(y)b_{k-1-i}(v)]}$$

with coefficients given by the Walsh transform. Figure 8.6.2 shows the two-dimensional Walsh basis for $n = 4$. The function $g_{(u,v)}(x, y)$ is represented by the image $\mathbf{g}_{uv}$ in which pixel values of 1 are white and pixel values of -1 are black.

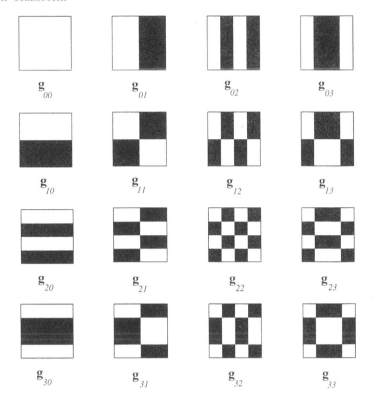

Figure 8.6.2. Two-dimensional Walsh basis for $n = 4$.

Figure 8.6.3 shows the magnitude image (right) of the two-dimensional Walsh transform of the image of a jet (left).

Figure 8.6.3. Jet image and the magnitude image of its Walsh transform image.

Image Algebra Formulation

The image algebra formulation of the fast Walsh transform is identical to that of the fast Fourier formulation (Section 8.4), with the exception that the template t used for the Walsh transform is

$$
\mathbf{t}(p)_{(u,v)}(x,y) = \begin{cases} 1 & \text{if } \lfloor u/p \rfloor \text{ is even and } (x,y) = (u,v) \\ 1 & \text{if } \lfloor u/p \rfloor \text{ is even and } (x,y) = (u+p,v) \\ -1 & \text{if } \lfloor u/p \rfloor \text{ is odd and } (x,y) = (u,v) \\ 1 & \text{if } \lfloor u/p \rfloor \text{ is odd and } (x,y) = (u-p,v) \\ 0 & \text{otherwise.} \end{cases}
$$

The Walsh transform shares the important property of separability with the Fourier transform. Thus, the two-dimensional Walsh transform can also be computed by taking the one-dimensional Walsh transforms along each row of the image, followed by another one-dimensional Walsh transform along the columns.

Alternate Image Algebra Formulations

If the convolution $\oplus$ does not restrict its operations to the template's support, the spatial transform approach will be much more efficient. The only change that needs to be made to the alternate fast Fourier transform of Section 8.2 is in the definition of the $\mathbf{w}_i$ images. For the spatial transform implementation of the fast Walsh transform, the images $\mathbf{w}_i \in \{-1,1\}^{\mathbf{X}}$ are defined as

$$
\mathbf{w}_i(x,y) = \begin{cases} 1 & \text{if } \left\lfloor \frac{x}{2^{i-1}} \right\rfloor \text{is even} \\ -1 & \text{if } \left\lfloor \frac{x}{2^{i-1}} \right\rfloor \text{is odd.} \end{cases}
$$

In [11, 12], Zhu provides a fast version of the Walsh transform in terms of the p-product. Zhu's method also eliminates the reordering process required in most fast versions of the Walsh transform. Specifically, given a one-dimensional signal $\mathbf{a} \in \mathbb{R}^{\mathbb{Z}_m}$, where $m = 2^k$, the Walsh transform of $\mathbf{a}$ is given by

$$
\mathfrak{W}(\mathbf{a}) := \frac{1}{m}[\mathbf{w}_1 \oplus_2 (\mathbf{w}_2 \oplus_2 (\cdots (\mathbf{w}_k \oplus_2 \mathbf{a}')))]',
$$

where $\mathbf{w}_i = (1,1,1,-1)$ for $i = 1, \ldots, k$.

Note that the 2-product formulation of the Walsh transform involves only the values +1 and –1. Therefore, there is no multiplication involved except for final multiplication by the quantity $\frac{1}{m}$.

For $\mathbf{a} \in \mathbb{R}^{\mathbb{Z}_m \times \mathbb{Z}_n}$, where $m = 2^k$ and $n = 2^l$, the two-dimensional Walsh transform of $\mathbf{a}$ is given by

$$
\mathbf{b} := \psi_m[\mathbf{w}_1 \oplus_2 (\mathbf{w}_2 \oplus_2 (\cdots (\mathbf{w}_k \oplus_2 \mathbf{a}')))]
$$
$$
\mathfrak{W}(\mathbf{a}) := \frac{1}{mn}\psi_n[\mathbf{w}_1 \oplus_2 (\mathbf{w}_2 \oplus_2 (\cdots (\mathbf{w}_l \oplus_2 \mathbf{b}')))]
$$

where $\varphi_h : \mathsf{F}^r \to \mathsf{F}_{q \times h}$, with $r = q \cdot h$, is defined by

$$
\varphi_h(f_1, f_2, \ldots, f_r) = \begin{pmatrix} f_1 & f_2 & \cdots & f_h \\ f_{h+1} & f_{h+2} & \cdots & f_{2h} \\ \vdots & \vdots & \ddots & \vdots \\ f_{(q-1)h+1} & f_{(q-1)h+2} & \cdots & f_{qh} \end{pmatrix}.
$$

Thus, the function φ_m converts the vector $[\mathbf{w}_1 \oplus_2 (\mathbf{w}_2 \oplus_2 (\cdots (\mathbf{w}_k \oplus_2 \mathbf{a}')))]$ back into matrix form.

Additional p-product formulations for signals whose lengths are not powers of two can be found in [11, 12].

8.7. The Haar Wavelet Transform

The simplest example of a family of functions yielding a multiresolution analysis of the space $L^2[\mathbb{R}]$ of all square-integrable functions on the real line is given by the Haar family of functions. These functions are well localized in space and are therefore appropriate for spatial domain analysis of signals and images. Furthermore, the Haar functions constitute an orthogonal basis. Hence, the discrete Haar transform benefits from the properties of orthogonal transformations. For example, these transformations preserve inner products when interpreted as linear operators from one space to another. The inverse of an orthogonal transformation is also particularly easy to implement, since it is simply the transpose of the direct transformation. Andrews [13] points out that orthogonal transformations are entropy preserving in an information theoretic sense. A particularly attractive feature of the Haar wavelet transform is its computational complexity, which is linear with respect to the length of the input signal. Moreover, this transform, as well as other wavelet representations of images (see Mallat [14]), discriminates oriented edges at different scales.

The appropriate theoretical background for signal and image analysis in the context of a multiresolution analysis is given by Mallat [14], where multiresolution wavelet representations are obtained by way of pyramidal algorithms. Daubechies provides a more general treatment of orthonormal wavelet bases and multiresolution analysis [15], where the Haar multiresolution analysis is presented. The Haar function, defined below, is the simplest example of an orthogonal wavelet. It may be viewed as the first of the family of compactly supported wavelets discovered by Daubechies.

The discrete *Haar wavelet transform* is a separable linear transform that is based on the scaling function

$$g(x) = \begin{cases} 1 & 0 \le x < 1 \\ 0 & \text{otherwise,} \end{cases}$$

and the dyadic dilations and integer translations of the Haar function

$$h(x) = \begin{cases} 1 & 0 \le x < \frac{1}{2} \\ -1 & \frac{1}{2} \le x < 1 \\ 0 & \text{otherwise.} \end{cases}$$

A matrix formulation of the Haar transform using orthogonal matrices follows. A convenient factorization of these matrices leads to a pyramidal algorithm, which makes use of the generalized matrix product of order two. The algorithm has complexity $O(n)$.

Let $n = 2^k, k \in \mathbb{Z}^+$, and $\mathbf{a} \in \mathbb{R}^{\mathbb{Z}_n}$ be a real-valued, one-dimensional signal. Associate with $\mathbf{a}$ the column vector $(a_0, a_1, \ldots, a_{n-1})'$, where $a_i = \mathbf{a}(i)$ for $i \in \mathbb{Z}_n$. To determine an orthonormal basis $\{\mathbf{u}_i \mid i \in \mathbb{Z}_n\}$ for the vector space $\mathbb{R}^{\mathbb{Z}_n}$, define

$$h_{pq}(x) \equiv h(2^p x - q)$$

for the range of integer indices $p \in \mathbb{N}$ and $0 \le q < 2^p$. For fixed p and q, the function $h_{pq}(x)$ is a translation by q of the function $h(2^p x)$, which is a dilation of the Haar wavelet function $h(x)$. Furthermore, $h_{pq}(x)$ is supported on an interval of length 2^{-p}. The infinite family of functions $\{h_{pq}(x) \mid p \in \mathbb{N}, 0 \le q < 2^p\}$ together with the scaling function $g(x)$ constitute an orthogonal basis, known as the *Haar basis*, for the space $L^2[0, 1]$ of square-integrable functions on the unit interval. This basis can be extended to a basis for $L^2[\mathbb{R}]$. The Haar basis was first described in 1910 [16].

To obtain a discrete version of the Haar basis, note that any positive integer i can be written uniquely as

$$i = 2^p + q,$$

where $p \in \mathbb{N}$ and $0 \le q < 2^p$. Using this fact, define

$$\mathbf{u}_0(j) \equiv \frac{1}{\sqrt{n}} \quad \text{for all } j \in \mathbb{Z}_n$$

and

$$\mathbf{u}_i(j) \equiv \sqrt{\frac{2^p}{n}} \cdot h_{pq}\left(\frac{j}{n}\right) = 2^{(p-k)/2} h_{pq}\left(2^{-k}j\right) \quad \text{for all } j \in \mathbb{Z}_n, \qquad (8.7.1)$$

where $i = 2^p + q$ for $i = 1, 2, \ldots, n - 1$. The factor $2^{(p-k)/2}$ in Equation 8.7.1 normalizes the Euclidean vector norm of the $\mathbf{u}_i$. Hence, $\|\mathbf{u}_i\|_2 = 1$ for all $i \in \mathbb{Z}_n$. The vector $\mathbf{u}_0$ is a normalized, discrete version of the scaling function $g(x)$. Similarly, the vectors $\mathbf{u}_i$ for $i = 1, 2, \ldots, n - 1$ are normalized, discrete versions of the wavelet functions $h_{pq}(x)$. Furthermore, these vectors are mutually orthogonal, i.e., the dot product

$$\mathbf{u}_i \bullet \mathbf{u}_j = 0 \quad \text{whenever } i \ne j.$$

Therefore, $\mathcal{B}_n = \{\mathbf{u}_i \mid i \in \mathbb{Z}_n\}$ is an orthonormal basis for the vector space $\mathbb{R}^{\mathbb{Z}_n}$ of one-dimensional signals of length n.

Now define the Haar matrices H_n for $n = 2^k$, k a positive integer, by letting $\mathbf{u}_i \in \mathcal{B}_n$ be the ith row vector of H_n. The orthonormality of the row vectors of H_n implies that H_n is an orthogonal matrix, i.e., $H'_n H_n = H_n H'_n = I_n$, where I_n is the $n \times n$ identity matrix. Hence, H_n is invertible with inverse $H_n^{-1} = H'_n$. Setting $\eta = 2^{-1/2}$, the normalization factor may be written as

$$2^{(p-k)/2} = 2^{-(k-p)/2} = \eta^{k-p}.$$

The Haar matrices for $n = 2, 4$, and 8 in terms of powers of η are

$$H_2 = \begin{pmatrix} \eta & \eta \\ \eta & -\eta \end{pmatrix},$$

$$H_4 = \begin{pmatrix} \eta^2 & \eta^2 & \eta^2 & \eta^2 \\ \eta^2 & \eta^2 & -\eta^2 & -\eta^2 \\ \eta & -\eta & 0 & 0 \\ 0 & 0 & \eta & -\eta \end{pmatrix},$$

and

$$H_8 = \begin{pmatrix} \eta^3 & \eta^3 & \eta^3 & \eta^3 & \eta^3 & \eta^3 & \eta^3 & \eta^3 \\ \eta^3 & \eta^3 & \eta^3 & \eta^3 & -\eta^3 & -\eta^3 & -\eta^3 & -\eta^3 \\ \eta^2 & \eta^2 & -\eta^2 & -\eta^2 & 0 & 0 & 0 & 0 \\ 0 & 0 & 0 & 0 & \eta^2 & \eta^2 & -\eta^2 & -\eta^2 \\ \eta & -\eta & 0 & 0 & 0 & 0 & 0 & 0 \\ 0 & 0 & \eta & -\eta & 0 & 0 & 0 & 0 \\ 0 & 0 & 0 & 0 & \eta & -\eta & 0 & 0 \\ 0 & 0 & 0 & 0 & 0 & 0 & \eta & -\eta \end{pmatrix}.$$

The signal $\mathbf{a}$ may be written as a linear combination of the basis vectors $\mathbf{u}_i$:

$$\mathbf{a} = \sum_{i=0}^{n-1} c_i \mathbf{u}_i, \qquad (8.7.2)$$

where the c_i (for all $i \in \mathbb{Z}_n$) are unknown coefficients to be determined. Equation 8.7.2 may be written as

$$\begin{aligned} \mathbf{a} &= [\mathbf{u}_0, \mathbf{u}_1, \ldots, \mathbf{u}_{n-1}]\mathbf{c} \\ &= H'_n \mathbf{c}, \end{aligned} \qquad (8.7.3)$$

where $\mathbf{c} \in \mathbb{R}^{\mathbb{Z}_n}$ is the vector of unknown coefficients, i.e., $\mathbf{c}(i) = c_i$ for each $i \in \mathbb{Z}_n$. Using $H_n^{-1} = H_n'$ and solving for $\mathbf{c}$, one obtains

$$\mathbf{c} = H_n \mathbf{a}. \qquad (8.7.4)$$

Equation 8.7.4 defines the Haar transform `HaarTransform1D` for one-dimensional signals of length $n = 2^k$, i.e.,

$$\mathtt{HaarTransform1D}(\mathbf{a}, n) \equiv H_n \mathbf{a}.$$

Furthermore, an image may be reconstructed from the vector $\mathbf{c}$ of Haar coefficients using Equation 8.7.3. Hence, define the inverse Haar transform `InverseHaarTransform1D` for one-dimensional signals by

$$\mathtt{InverseHaarTransform1D}(\mathbf{c}, n) \equiv H_n' \mathbf{c}.$$

To define the Haar transform for two-dimensional images, let $m = 2^k$ and $n = 2^l$, with $k, l \in \mathbb{Z}^+$, and let $\mathbf{a} \in \mathbb{R}^{\mathbb{Z}_m \times \mathbb{Z}_n}$ be a two-dimensional image. The Haar transform of $\mathbf{a}$ yields an $m \times n$ matrix $\mathbf{c} = [c_{ij}]$ of coefficients given by

$$\mathbf{c} = H_m [a_{ij}] H_n', \qquad (8.7.5)$$

where $a_{ij} = \mathbf{a}(i, j)$ for all $(i, j) \in \mathbb{Z}_m \times \mathbb{Z}_n$. Equation 8.7.5 defines the Haar transform `HaarTransform2D` of a two-dimensional $m \times n$ image:

$$\mathtt{HaarTransform2D}(\mathbf{a}, m, n) \equiv H_m [a_{ij}] H_n'.$$

The image $\mathbf{a}$ can be recovered from the $m \times n$ matrix $\mathbf{c}$ of Haar coefficients by the inverse Haar transform, `InverseHaarTransform2D`, for two-dimensional images:

$$\mathtt{InverseHaarTransform2D}(\mathbf{c}, m, n) \equiv H_m' [c_{ij}] H_n,$$

i.e.,

$$\mathbf{a} = H_m' [c_{ij}] H_n. \qquad (8.7.6)$$

Equation 8.7.6 is equivalent to the linear expansion

$$\mathbf{a} = \sum_{i=0}^{m-1} \sum_{j=0}^{n-1} c_{ij} \cdot \mathbf{u}_i \otimes \mathbf{v}_j,$$

where $\mathbf{u}_i \in \mathcal{B}_m$ for all $i \in \mathbb{Z}_m$ and $\mathbf{v}_j \in \mathcal{B}_n$ for all $j \in \mathbb{Z}_n$. The outer product $\mathbf{u}_i \otimes \mathbf{v}_j$ may be interpreted as the image of a two-dimensional, discrete Haar wavelet. The sum over all combinations of the outer products, appropriately weighted by the coefficients c_{ij}, reconstructs the original image $\mathbf{a}$.

Image Algebra Formulation

The image algebra formulation of the Haar transform is based on a convenient factorization of the Haar matrices [13]. To factor H_n for $n = 2, 4$, and 8, let

$$K_2 = \begin{pmatrix} \eta & \eta \\ \eta & -\eta \end{pmatrix},$$

$$K_4 = \begin{pmatrix} \eta & \eta & 0 & 0 \\ 0 & 0 & \eta & \eta \\ \eta & -\eta & 0 & 0 \\ 0 & 0 & \eta & -\eta \end{pmatrix},$$

and

$$K_8 = \begin{pmatrix} \eta & \eta & 0 & 0 & 0 & 0 & 0 & 0 \\ 0 & 0 & \eta & \eta & 0 & 0 & 0 & 0 \\ 0 & 0 & 0 & 0 & \eta & \eta & 0 & 0 \\ 0 & 0 & 0 & 0 & 0 & 0 & \eta & \eta \\ \eta & -\eta & 0 & 0 & 0 & 0 & 0 & 0 \\ 0 & 0 & \eta & -\eta & 0 & 0 & 0 & 0 \\ 0 & 0 & 0 & 0 & \eta & -\eta & 0 & 0 \\ 0 & 0 & 0 & 0 & 0 & 0 & \eta & -\eta \end{pmatrix}.$$

The Haar matrices H_2, H_4, and H_8 may be factored as follows:

$$H_2 = K_2,$$

$$H_4 = \begin{pmatrix} K_2 & O_2 \\ O_2 & I_2 \end{pmatrix} K_4,$$

and

$$H_8 = \begin{pmatrix} H_4 & O_4 \\ O_4 & I_4 \end{pmatrix} K_8$$

$$= \begin{pmatrix} \begin{matrix} K_2 & & O_2 \\ \\ O_2 & & I_2 \end{matrix} & \vdots & O_4 \\ - & - & - & + & - & - & - \\ & O_4 & & \vdots & I_4 \end{pmatrix} \begin{pmatrix} K_4 & O_4 \\ O_4 & I_4 \end{pmatrix} K_8.$$

From this factorization, it is clear that only pairwise sums and pairwise differences with multiplication by η are required to obtain the Haar wavelet transforms (direct and inverse) of an image. Furthermore, the one-dimensional Haar transform of a signal of length 2^k may be computed in k stages. The number of stages corresponds to the number of factors for the Haar matrix H_{2^k}. Computation of the Haar transform of a signal of length n requires $4(n - 1)$ multiplications and $2(n - 1)$ sums or differences. Hence, the computational complexity of the Haar transform is $O(n)$. Further remarks concerning the complexity of the Haar transform can be found in Andrews [13] and Strang [17].

The image algebra formulation of the Haar wavelet transform may be expressed in terms of the *normalization constant* $\eta = 2^{-1/2}$, the *Haar scaling vector* $\mathbf{g} = (\eta, \eta)'$, and the *Haar wavelet vector* $\mathbf{h} = (\eta, -\eta)'$, as described below.

One-Dimensional Haar Transform

Let $\mathbf{a} \in \mathbb{R}^{\mathbb{Z}_n}$, with $n = 2^k$, $k \in \mathbb{Z}^+$. The following pyramidal algorithm implements

$$\mathbf{c} := \texttt{HaarTransform1D}(\mathbf{a}, n),$$

where $\mathbf{a}$ is treated as a row vector.

$$\sigma^0 := \mathbf{a}$$
$$\text{for } i := 1 \text{ to } k \text{ loop}$$
$$\delta^i := \sigma^{i-1} \oplus_2 \mathbf{h}$$
$$\sigma^i := \sigma^{i-1} \oplus_2 \mathbf{g}$$
$$\text{end loop}$$
$$\mathbf{c} := \left(\sigma^k \mid \delta^k \mid \delta^{k-1} \mid \cdots \mid \delta^1\right).$$

Example: Let $\mathbf{a} = (a_0, a_1, \ldots, a_7)$. The Haar transform

$$\mathbf{c} = \left(\sigma^3 \mid \delta^3 \mid \delta^2 \mid \delta^1\right)$$
$$= \left(\sigma_0^3, \delta_0^3, \delta_0^2, \delta_1^2, \delta_0^1, \delta_1^1, \delta_2^1, \delta_3^1\right)$$

of $\mathbf{a}$ is computed in three stages, as shown in Figure 8.7.1.

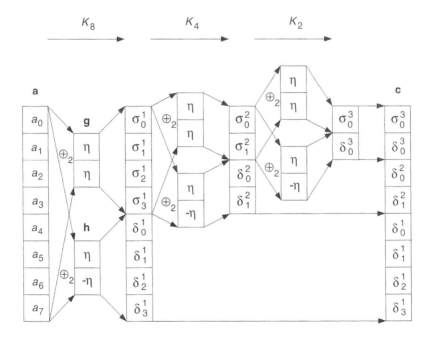

Figure 8.7.1. Pyramidal algorithm data flow.

Note that the first entry of the resulting vector is a scaled global sum, since

$$\sigma_0^3 = \eta^3 \sum \mathbf{a}.$$

Two-Dimensional Haar Transform

Let $m = 2^k$ and $n = 2^l$, with $k, l \in \mathbb{Z}^+$, and let $\mathbf{a} \in \mathbb{R}^{\mathbb{Z}_m \times \mathbb{Z}_n}$ be a two-dimensional image. The following algorithm implements

$$\mathbf{c} := \texttt{HaarTransform2D}(\mathbf{a}, m, n).$$

The one-dimensional Haar transform of each row of $\mathbf{a}$ is computed first as an intermediate image $\mathbf{b}$. The second loop of the algorithm computes the Haar transform of each column of $\mathbf{b}$. This procedure is equivalent to the computation $\mathbf{c} = H_m([a_{ij}]H_n')$.

$$\sigma^0 := \mathbf{a}$$

for $i := 1$ to l loop

$$\delta^i := \sigma^{i-1} \oplus_2 \mathbf{h}$$

$$\sigma^i := \sigma^{i-1} \oplus_2 \mathbf{g}$$

end loop

$$\mathbf{b} := \left(\sigma^l \mid \delta^l \mid \delta^{l-1} \mid \cdots \mid \delta^1 \right)$$

$$\sigma^0 := \mathbf{b}$$

for $j := 1$ to k loop

$$\delta^j := \mathbf{h}' \oplus_2' \sigma^{j-1}$$

$$\sigma^j := \mathbf{g}' \oplus_2' \sigma^{j-1}$$

end loop

$$\mathbf{c} := \begin{pmatrix} \sigma^k \\ - \\ \delta^k \\ - \\ \delta^{k-1} \\ - \\ \vdots \\ - \\ \delta^1 \end{pmatrix}.$$

Example: Consider the gray scale rendition of an input image of a 32×32 letter "A" as shown in Figure 8.7.2. In the gray scale images shown here, black corresponds to the lowest value in the image and white corresponds to the highest value. For the original image, black = 0 and white = 255.

Figure 8.7.2. Input image.

The row-by-row Haar transform of the original image is shown in Figure 8.7.3, and the column-by-column Haar transform of the original image is shown in Figure 8.7.4.

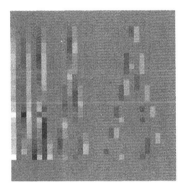

Figure 8.7.3. Row-by-row Haar transform of the input image.

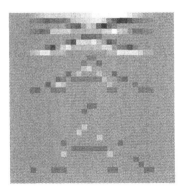

Figure 8.7.4. Column-by-column Haar transform of the input image.

Computing the column-by-column Haar transform of the image in Figure 8.7.3, or computing the row-by-row Haar transform of the image in Figure 8.7.4, yields the Haar transform of the original image. The result is shown in Figure 8.7.5. Interpretation of Figure 8.7.5 is facilitated by recognizing that the $2^5 \times 2^5$ matrix of coefficients is partitioned into a multiresolution grid consisting of $(5 + 1) \times (5 + 1)$ subregions, as shown in Figure 8.7.6. Each subregion corresponds to a particular resolution or scale along the x- and y-axes. For example, the labeled subregion $\delta^{2,3}$ contains a $2^2 \times 2^3 = 4 \times 8$ version of the original image at a scale of $1/8$ along the x-axis and a scale of $1/4$ along the y-axis.

Figure 8.7.5. Gray scale Haar transform of the input image.

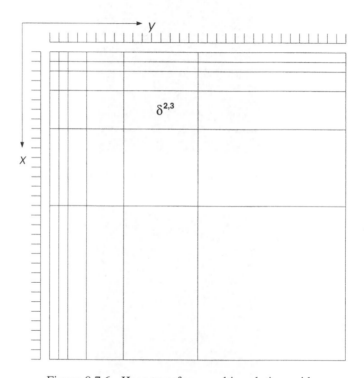

Figure 8.7.6. Haar transform multiresolution grid.

The large value in the upper left-hand pixel in Figure 8.7.5 corresponds to a scaled global sum. By eliminating the top row and the left column of pixels, the remaining pixels are rendered more clearly. The result is shown in Figure 8.7.7. The bright white pixel in the top row results from a dominant feature of the original image, namely, the right-hand diagonal of the letter "A." The Haar wavelet transform is therefore sensitive to the presence and orientation of edges at different scales in the image.

Figure 8.7.7. Gray scale Haar transform without top row and left column.

8.8. Daubechies Wavelet Transforms

The Daubechies family of orthonormal wavelet bases for the space $L^2[\mathbb{R}]$ of square-integrable functions generalize the Haar wavelet basis. Each member of this family consists of the dyadic dilations and integer translations

$$\psi_{m,n}(x) = \frac{1}{\sqrt{2^m}} \cdot \psi\left(\frac{1}{2^m}x - n\right), \quad m, n \in \mathbb{Z},$$

of a compactly supported *wavelet function* $\psi(x)$. The wavelet function in turn is derived from a *scaling function* $\varphi(x)$, which also has compact support. The scaling and wavelet functions have good localization in both the spatial and frequency domains [15, 18]. Daubechies was the first to describe compactly supported orthonormal wavelet bases [18].

The orthonormal wavelet bases were developed by Daubechies within the framework of a multiresolution analysis. Mallat [14] has exploited the features of multiresolution analysis to develop pyramidal decomposition and reconstruction algorithms for images. The image algebra algorithms presented here are based on these pyramidal schemes and build upon the work by Zhu and Ritter [12], in which the Daubechies wavelet transform and its inverse are expressed in terms of the *p*-product. The one-dimensional wavelet transforms (direct and inverse) described by Zhu and Ritter correspond to a single stage of the one-dimensional pyramidal image algebra algorithms. Computer routines and further aspects of the computation of the Daubechies wavelet transforms may be found in Press et al. [19].

For every $g \in \mathbb{Z}^+$, a discrete wavelet system, denoted by D_{2g}, is defined by a finite number of coefficients, $a_0, a_1, \ldots, a_{2g-1}$. These coefficients, known as *scaling* or *wavelet filter* coefficients, satisfy specific orthogonality conditions. The scaling function $\varphi(x)$ is defined to be a solution of

$$\varphi(x) = \sum_{k=0}^{2g-1} a_k \varphi(2x - k).$$

Moreover, the set of coefficients

$$b_k = (-1)^k a_{2g-1-k}, \quad k = 0, 1, \ldots, 2g - 1,$$

defines the wavelet function $\psi(x)$ associated with the scaling function of the wavelet system:

$$\psi(x) = \sum_{k=0}^{2g-1} b_k \varphi(2x - k).$$

To illustrate how the scaling coefficients are derived, consider the case $g = 2$. Let

$$\mathbf{s} = (a_0, a_1, a_2, a_3)$$

and

$$\mathbf{w} = (b_0, b_1, b_2, b_3)$$
$$= (a_3, -a_2, a_1, -a_0).$$

Note that $\mathbf{s}$ and $\mathbf{w}$ satisfy the orthogonality relation $\mathbf{s} \bullet \mathbf{w} = 0$. The coefficients $a_0, a_1, a_2,$ and a_3 are uniquely determined by the following equations:

$$a_0^2 + a_1^2 + a_2^2 + a_3^2 = 1 \qquad\qquad (8.8.1a)$$
$$a_0 a_2 + a_1 a_3 = 0 \qquad\qquad (8.8.1b)$$
$$a_3 - a_2 + a_1 - a_0 = 0 \qquad\qquad (8.8.1c)$$
$$0a_3 - 1a_2 + 2a_1 - 3a_0 = 0. \qquad\qquad (8.8.1d)$$

Equations 8.8.1a-d are equivalent to

$$\|\mathbf{s}\|_2 = 1 \qquad\qquad (8.8.2a)$$
$$(a_0, a_1, a_2, a_3, 0, 0) \bullet (0, 0, a_0, a_1, a_2, a_3) = 0 \qquad\qquad (8.8.2b)$$
$$\sum \mathbf{w} = 0 \qquad\qquad (8.8.2c)$$
$$(0, 1, 2, 3) \bullet \mathbf{w} = 0. \qquad\qquad (8.8.2d)$$

Solving for the four unknowns in the four Equations 8.8.1a-d yields

$$a_0 = \left(1 + \sqrt{3}\right)/4\sqrt{2}$$
$$a_1 = \left(3 + \sqrt{3}\right)/4\sqrt{2}$$
$$a_2 = \left(3 - \sqrt{3}\right)/4\sqrt{2} \qquad\qquad (8.8.3)$$
$$a_3 = \left(1 - \sqrt{3}\right)/4\sqrt{2}.$$

In general, the coefficients for the discrete wavelet system D_{2g} are determined by $2g$ equations in $2g$ unknowns. Daubechies has tabulated the coefficients for $g = 2, 3, \ldots, 10$ [15, 18]. Higher-order wavelets are smoother but have broader supports.

Image Algebra Formulation

Let $g \in \mathbb{Z}^+$. For the direct and inverse D_{2g} wavelet transforms, let

$$\mathbf{s} = \begin{pmatrix} a_0 & a_1 & \cdots & a_{2g-1} \end{pmatrix}'$$

and

$$\mathbf{w} = \begin{pmatrix} b_0 & b_1 & \cdots & b_{2g-1} \end{pmatrix}'$$

denote the *scaling vector* and the *wavelet vector*, respectively, of the wavelet transform. Furthermore, let

$$\mathbf{s}_0 = \begin{pmatrix} a_0 \\ a_1 \end{pmatrix}, \mathbf{s}_1 = \begin{pmatrix} a_2 \\ a_3 \end{pmatrix}, \ldots, \mathbf{s}_{g-1} = \begin{pmatrix} a_{2g-2} \\ a_{2g-1} \end{pmatrix},$$

and

$$\mathbf{w}_0 = \begin{pmatrix} b_0 \\ b_1 \end{pmatrix}, \mathbf{w}_1 = \begin{pmatrix} b_2 \\ b_3 \end{pmatrix}, \ldots, \mathbf{w}_{g-1} = \begin{pmatrix} b_{2g-2} \\ b_{2g-1} \end{pmatrix}.$$

In the computation of the wavelet transform of a signal or image, the scaling vector acts as lowpass filter, while the wavelet vector acts as a bandpass filter.

The algorithms in this section make use of specific column and row representations of matrices. For $\mathbf{X} = \mathbb{Z}_m \times \mathbb{Z}_n$ and $\mathbf{f} \in \mathbb{R}^\mathbf{X}$, the *column vector* of $\mathbf{f}$ is defined by

$$\mathrm{col}(\mathbf{f}) = (f_{0,0}, f_{0,1}, \ldots, f_{0,n-1}, f_{1,0}, f_{1,1}, \ldots, f_{1,n-1}, \ldots, f_{m-1,0}, f_{m-1,1}, \ldots, f_{m-1,n-1})',$$

and the *row vector* of $\mathbf{f}$ is defined by $\mathrm{row}(\mathbf{f}) = (\mathrm{col}(\mathbf{f}))'$.

When computing wavelet coefficients of a data vector near the end of the vector, the support of the scaling vector or wavelet vector may exceed the support of the data. In such cases, it is convenient to assume *periodic boundary conditions* on the data vector, i.e., the data vector is treated as a circular vector. Appropriate spatial maps are defined below to represent required shifts for data addressing.

One-Dimensional Daubechies Wavelet Transform

Let $\mathbf{X} = \mathbb{Z}_N$, with $N = 2^k$, $k \in \mathbb{Z}^+$, and let $\mathbf{f} \in \mathbb{R}^\mathbf{X}$ be a one-dimensional signal. Circular shifts of a row vector may be represented by composition with the spatial map $\kappa_q : \mathbf{X} \to \mathbf{X}$ defined by

$$\kappa_q : j \mapsto (j + q) mod(N),$$

where $q \in \mathbb{Z}$.

The following pyramidal algorithm computes the *one-dimensional wavelet transform* $\mathbf{c} \in \mathbb{R}^\mathbf{X}$ of $\mathbf{f}$, where $\mathbf{f}$ is treated as a row vector.

$$\sigma^0 := \mathbf{f}$$
$$R := k - (\lceil log_2(2g) \rceil - 1)$$
$$\text{for } i := 1 \text{ to } R \text{ loop}$$
$$\delta^i := \sigma^{i-1} \oplus_2 \mathbf{w}_0 + \sum_{p=1}^{g-1} (\sigma^{i-1} \circ \kappa_{2p}) \oplus_2 \mathbf{w}_p$$
$$\sigma^i := \sigma^{i-1} \oplus_2 \mathbf{s}_0 + \sum_{p=1}^{g-1} (\sigma^{i-1} \circ \kappa_{2p}) \oplus_2 \mathbf{s}_p$$
$$\text{end loop}$$
$$\mathbf{c} := (\sigma^R \,|\, \delta^R \,|\, \delta^{R-1} \,|\, \cdots \,|\, \delta^1).$$

Remarks

- The wavelet transform is computed in R stages. Each stage corresponds to a scale or level of resolution. The number of stages required to compute the wavelet transform is a function of the length $N = 2^k$ of the input data vector and the length $2g$ of the wavelet or scaling vector.

- σ^R and δ^R are row vectors of length $2^{\lceil log_2(2g) \rceil - 1}$ each.

- The case $g = 1$ yields the Haar wavelet transform with $\mathbf{s} = \mathbf{s}_0 = \mathbf{g}$ and $\mathbf{w} = \mathbf{w}_0 = \mathbf{h}$, where $\mathbf{g}$ and $\mathbf{h}$ are the Haar scaling and wavelet vectors, respectively.

The original signal $\mathbf{f}$ may be reconstructed from the vector $\mathbf{c}$ by the *inverse wavelet transform*. The following algorithm uses the tensor product to obtain the inverse wavelet transform of $\mathbf{c} \in \mathbb{R}^X$, where $\mathbf{c}$ is treated as a row vector.

$$L := 2^{(\lceil log_2(2g) \rceil - 1)}$$
$$\mathbf{d} := [c(0), c(1), \dots, c(L-1)]$$
$$R := k - (\lceil log_2(2g) \rceil - 1)$$
$$\textbf{for } i := 1 \textbf{ to } R \textbf{ loop}$$
$$\quad \mathbf{c}^s := \mathbf{d}$$
$$\quad \mathbf{c}^w := [c(L), c(L+1), \dots, c(2L-1)]$$
$$\quad \mathbf{d}^s := \mathbf{s}_0 \otimes \mathbf{c}^s + \sum_{p=1}^{g-1} \mathbf{s}_p \otimes (\mathbf{c}^s \circ \kappa_p)$$
$$\quad \mathbf{d}^w := \mathbf{w}_0 \otimes \mathbf{c}^w + \sum_{p=1}^{g-1} \mathbf{w}_p \otimes (\mathbf{c}^w \circ \kappa_p)$$
$$\quad \mathbf{d} := \mathbf{d}^s + \mathbf{d}^w$$
$$\quad \mathbf{d} := \text{row}(\mathbf{d}')$$
$$\quad L := 2L$$
$$\textbf{end loop.}$$

Remarks

- The reconstructed data vector $\mathbf{d}$ has dimensions $1 \times N$.

- $\mathbf{d}^s$ and $\mathbf{d}^w$ are matrices having dimensions $2 \times L$.

Example: Let $g = 2$. The scaling coefficients for D_4 are given by Equations 8.8.3. A typical wavelet in the D_4 wavelet basis may be obtained by taking the inverse wavelet transform of a canonical unit vector of length $N = 2^k$. For $k = 8$, the graph of the inverse D_4 wavelet transform of $\mathbf{e}_{11}$ is shown in Figure 8.8.1.

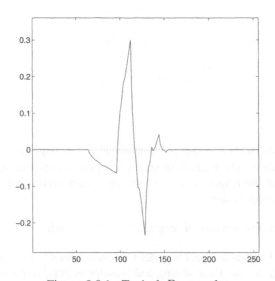

Figure 8.8.1. Typical D_4 wavelet.

Two-Dimensional Daubechies Wavelet Transform

Let $\mathbf{X} = \mathbb{Z}_M \times \mathbb{Z}_N$, with $M = 2^k$, $N = 2^l$, and $k, l \in \mathbb{Z}^+$. Circular column shifts of a two-dimensional image are easily represented by composition with the spatial map $\kappa_q : \mathbf{X} \to \mathbf{X}$ defined by

$$\kappa_q : (i, j) \mapsto (i, (j + q)mod(N)),$$

with $q \in \mathbb{Z}$. Analogously, circular row shifts of two-dimensional images may be represented by composition with the spatial map $\rho_q : \mathbf{X} \to \mathbf{X}$ defined by

$$\rho_q : (i, j) \mapsto ((i + q)mod(M), j),$$

with $q \in \mathbb{Z}$.

Let $\mathbf{f} \in \mathbb{R}^{\mathbf{X}}$ be a two-dimensional image. The following algorithm computes the *two-dimensional wavelet transform* $\mathbf{d} \in \mathbb{R}^{\mathbf{X}}$ of $\mathbf{f}$, where $\mathbf{f}$ is treated as an $M \times N$ matrix. The one-dimensional transform of each row of $\mathbf{f}$ is computed first as an intermediate image c. The second loop of the algorithm computes the one-dimensional transform of each column of c.

$$\sigma^0 := \mathbf{f}$$
$$R := l - (\lceil log_2(2g) \rceil - 1)$$
$$\text{for } i := 1 \text{ to } R \text{ loop}$$
$$\delta^i := \sigma^{i-1} \oplus_2 \mathbf{w}_0 + \sum_{p=1}^{g-1} (\sigma^{i-1} \circ \kappa_{2p}) \oplus_2 \mathbf{w}_p$$
$$\sigma^i := \sigma^{i-1} \oplus_2 \mathbf{s}_0 + \sum_{p=1}^{g-1} (\sigma^{i-1} \circ \kappa_{2p}) \oplus_2 \mathbf{s}_p$$
$$\text{end loop}$$
$$\mathbf{c} := \left(\sigma^R \,|\, \delta^R \,|\, \delta^{R-1} \,|\, \cdots \,|\, \delta^1 \right)$$
$$\sigma^0 := \mathbf{c}$$
$$S := k - (\lceil log_2(2g) \rceil - 1)$$
$$\text{for } j := 1 \text{ to } S \text{ loop}$$
$$\delta^j := \mathbf{w}_0' \oplus_2' \sigma^{j-1} + \sum_{p=1}^{g-1} \mathbf{w}_p' \oplus_2' (\sigma^{j-1} \circ \rho_{2p})$$
$$\sigma^j := \mathbf{s}_0' \oplus_2' \sigma^{j-1} + \sum_{p=1}^{g-1} \mathbf{s}_p' \oplus_2' (\sigma^{j-1} \circ \rho_{2p})$$
$$\text{end loop}$$

$$\mathbf{d} := \begin{pmatrix} \sigma^S \\ - \\ \delta^S \\ - \\ \delta^{S-1} \\ - \\ \vdots \\ - \\ \delta^1 \end{pmatrix}.$$

Example: Consider the gray scale rendition of an input image of a 32×32 letter "A" as shown in Figure 8.8.2. In the gray scale images shown here, black corresponds to the lowest value in the image and white corresponds to the highest value. For the original image, black = 0 and white = 255.

Figure 8.8.2. Input image.

Suppose $g = 2$. The row-by-row D_4 wavelet transform of the original image is shown in Figure 8.8.3, and the column-by-column D_4 wavelet transform of the original image is shown in Figure 8.8.4.

Figure 8.8.3. Row-by-row wavelet transform of the input image.

Figure 8.8.4. Column-by-column wavelet transform of the input image.

Computing the column-by-column wavelet transform of the image in Figure 8.8.3, or computing the row-by-row wavelet transform of the image in Figure 8.8.4, yields the D_4 wavelet transform of the original image. The result is shown in Figure 8.8.5. As in the Haar wavelet representation, the Daubechies wavelet representation discriminates the location, scale, and orientation of edges in the image.

Figure 8.8.5. Gray scale wavelet transform of the input image.

Alternate Image Algebra Formulation

The alternate image algebra formulation of the one-dimensional wavelet transform presented here uses the dual $\oplus_2'$ of the 2-product instead of the tensor product. Let

$$\mathbf{u}_0 = \begin{pmatrix} a_0 \\ b_0 \\ a_1 \\ b_1 \end{pmatrix}, \mathbf{u}_1 = \begin{pmatrix} a_2 \\ b_2 \\ a_3 \\ b_3 \end{pmatrix}, \dots, \mathbf{u}_{g-1} = \begin{pmatrix} a_{2g-2} \\ b_{2g-2} \\ a_{2g-1} \\ b_{2g-1} \end{pmatrix}.$$

The following algorithm computes the *inverse one-dimensional wavelet transform* $\mathbf{d} \in \mathbb{R}^{\mathbf{X}}$ of $\mathbf{c} \in \mathbb{R}^{\mathbf{X}}$, where $\mathbf{X} = \mathbb{Z}_N$, $N = 2^k$, and $k \in \mathbb{Z}^+$.

$$L := 2^{(\lceil log_2(2g)\rceil - 1)}$$
$$\mathbf{d} := [c(0), c(1), \dots, c(L-1)]$$
$$R := k - (\lceil log_2(2g)\rceil - 1)$$
$$\textbf{for } i := 1 \textbf{ to } R \textbf{ loop}$$
$$\quad \mathbf{c}^s := \mathbf{d}$$
$$\quad \mathbf{c}^w := [c(L), c(L+1), \dots, c(2L-1)]$$
$$\quad \mathbf{d} := (\mathbf{c}^s \mid \mathbf{c}^w) \oplus_2' \mathbf{u}_0 + \sum_{p=1}^{g-1} (\mathbf{c}^s \circ \kappa_p \mid \mathbf{c}^w \circ \kappa_p) \oplus_2' \mathbf{u}_p$$
$$\quad \mathbf{d} := \text{row}(\mathbf{d}')$$
$$\quad L := 2L$$
$$\textbf{end loop.}$$

8.9. Exercises

1. The Fourier spectrum of an image is complex-valued and usually displayed as a pair of images: either magnitude-phase or real-imaginary part.

    a.   Find a relationship between the two representations.
    b.   Give an interpretation of each.

2. Prove that a representation of one full period of the Fourier magnitude spectrum of an image, with the origin in the center of the display, can be achieved by multiplying $\mathbf{a}(x, y)$ by $(-1)^{x+y}$ before applying the transform.

3.   Give the image algebra formulations of the following properties of the discrete Fourier transform:

    a.   The addition theorem:

$$\mathfrak{F}(\mathbf{a} + \mathbf{b}) = \mathfrak{F}(\mathbf{a}) + \mathfrak{F}(\mathbf{b})$$

    where $\mathbf{a}, \mathbf{b}$ can be one-dimensional, $\mathbf{a}, \mathbf{b} \in \mathbb{R}^{\mathbb{Z}_n}$, or two-dimensional, $\mathbf{a}, \mathbf{b} \in \mathbb{R}^{\mathbb{Z}_m \times \mathbb{Z}_n}$.

    b.   The time (space) convolution theorem:

$$\mathfrak{F}(\mathbf{a} * \mathbf{b}) = \mathfrak{F}(\mathbf{a}) \cdot \mathfrak{F}(\mathbf{b}).$$

    c.   The frequency convolution theorem:

$$\mathfrak{F}(\mathbf{a} \cdot \mathbf{b}) = \mathfrak{F}(\mathbf{a}) * \mathfrak{F}(\mathbf{b}).$$

4.   Prove that the discrete Haar transform, as an orthogonal transformation, preserves inner products.

5.   Refer to the gray scale Haar transform depicted in Figures 8.7.5 and 8.7.6, and identify other features of the original image besides the one discussed at the end of Section 8.7.

6.   Explain why in the computation of the Daubechies wavelet transform of a signal or image, the scaling vector acts as a lowpass filter, while the wavelet vector acts as a bandpass filter.

7.   Discuss how the wavelet representations discriminate the location, scale, and orientation of edges in the original image.

8.   Discuss how the Haar transform can be used for finding and/or removing edges from an image.

9.   Give the image algebra formulation of the two-dimensional discrete sine transform and its inverse:

$$[\mathcal{S}(\mathbf{a})](u, v) = \frac{2}{n+1} \sum_{x=0}^{n-1} \sum_{y=0}^{n-1} \mathbf{a}(x, y) \cdot sin\left(\frac{(x+1)(u+1)\pi}{n+1}\right) \cdot sin\left(\frac{(y+1)(v+1)\pi}{n+1}\right)$$

$$[\mathcal{S}(\mathbf{a})](x, y) = \frac{2}{n+1} \sum_{u=0}^{n-1} \sum_{v=0}^{n-1} \mathbf{a}(u, v) \cdot sin\left(\frac{(x+1)(u+1)\pi}{n+1}\right) \cdot sin\left(\frac{(y+1)(v+1)\pi}{n+1}\right)$$

where $\mathbf{a} \in \mathbb{R}^{\mathbb{Z}_n \times \mathbb{Z}_n}$.

8.10.   References

[1] J. Cooley, P. Lewis, and P. Welch, "The fast Fourier transform and its applications," *IEEE Transactions on Education*, vol. E-12, no. 1, pp. 27–34, 1969.

[2] J. Cooley and J. Tukey, "An algorithm for the machine calculation of complex Fourier series," *Mathematics of Computation*, vol. 19, pp. 297–301, 1965.

[3]   R. Gonzalez and P. Wintz, *Digital Image Processing*. Reading, MA: Addison-Wesley, second ed., 1987.

[4]   G. Ritter, "Image algebra." Unpublished manuscript, available via anonymous ftp from `ftp://ftp.cise.ufl.edu/pub/src/ia/documents`, 1994.

[5]   M. Narasimha and A. Peterson, "On the computation of the discrete cosine transform," *IEEE Transactions on Communications*, vol. COM-26, no. 6, pp. 934–936, 1978.

[6]   W. Chen, C. Smith, and S. Fralick, "A fast computational algorithm for the discrete cosine transform," *IEEE Transactions on Communications*, vol. COM-25, pp. 1004–1009, 1977.

[7]   N. Ahmed, T. Natarajan, and K. Rao, "Discrete cosine transform," *IEEE Transactions on Computers*, vol. C-23, pp. 90–93, 1974.

[8]   J. Walsh, "A closed set of normal orthogonal functions," *American Journal of Mathematics*, vol. 45, no. 1, pp. 5–24, 1923.

[9]   M. J. Hadamard, "Resolution d'une question relative aux determinants," *Bulletin des Sciences Mathematiques*, vol. A17, pp. 240–246, 1893.

[10] R. E. A. C. Paley, "A remarkable series of orthogonal functions," *Proceedings of the London Mathematical Society*, vol. 34, pp. 241–279, 1932.

[11] H. Zhu, *The Generalized Matrix Product and its Applications in Signal Processing*. Ph.D. dissertation, University of Florida, Gainesville, 1993.

[12] H. Zhu and G. X. Ritter, "The p-product and its applications in signal processing," *SIAM Journal of Matrix Analysis and Applications*, vol. 16, no. 2, pp. 579–601, 1995.

[13] H. Andrews, *Computer Techniques in Image Processing*. New York: Academic Press, 1970.

[14] S. Mallat, "A theory for multiresolution signal decomposition: the wavelet representation," *IEEE Pattern Analysis and Machine Intelligence*, vol. 11, no. 7, pp. 674–693, 1989.

[15] I. Daubechies, *Ten Lectures on Wavelets*. CBMS-NSF Regional Conference Series in Applied Mathematics, Philadelphia, PA: SIAM, 1992.

[16] A. Haar, "Zur theorie der orthogonalen funktionensysteme," *Mathematische Annalen*, vol. 69, pp. 331–371, 1910.

[17] G. Strang, "Wavelet transforms versus fourier transforms," *Bulletin of the American Mathematical Society (new series)*, vol. 28, pp. 288–305, Apr. 1993.

[18] I. Daubechies, "Orthonormal bases of wavelets," *Communications on Pure and Applied Mathematics*, vol. 41, pp. 909–996, 1988.

[19] W. Press, S. Teukolsky, W. Vetterling, and B. Flannery, *Numerical Recipes in C*. Cambridge: Cambridge University Press, 1992.

CHAPTER 9
PATTERN MATCHING AND SHAPE DETECTION

9.1. Introduction

This chapter covers two related image analysis tasks: object detection by pattern matching and shape detection using Hough transform techniques. One of the most fundamental methods of detecting an object of interest is by pattern matching using templates. In template matching, a replica of an object of interest is compared to all objects in the image. If the pattern match between the template and an object in the image is sufficiently close (e.g., exceeding a given threshold), then the object is labeled as the template object.

The Hough transform provides for versatile methods for detecting shapes that can be described in terms of closed parametric equations or in tabular form. Examples of parameterizable shapes are lines, circles, and ellipses. Shapes that fail to have closed parametric equations can be detected by a generalized version of the Hough transform that employs lookup table techniques. The algorithms presented in this chapter address both parametric and non-parametric shape detection.

9.2. Pattern Matching Using Correlation

Pattern matching is used to locate an object of interest within a larger image. The pattern, which represents the object of interest, is itself an image. The image is scanned with the given pattern to locate sites on the image that *match* or bear a strong visual resemblance to the pattern. The determination of a good match between an image $\mathbf{a}$ and a pattern template $\mathbf{p}$ is usually given in terms of the metric

$$d = \Sigma(\mathbf{a} - \mathbf{p})^2 = \Sigma\mathbf{a}^2 - 2\Sigma\mathbf{ap} + \Sigma\mathbf{p}^2,$$

where the sum is over the support of $\mathbf{p}$. The value of d will be small when $\mathbf{a}$ and $\mathbf{p}$ are almost identical and large when they differ significantly. It follows that the term $\Sigma\mathbf{ap}$ will have to be large whenever d is small. Therefore, a large value of $\Sigma\mathbf{ap}$ provides a good measure of a match. Shifting the pattern template $\mathbf{p}$ over all possible locations of $\mathbf{a}$ and computing the match $\Sigma\mathbf{ap}$ at each location can therefore provide for a set candidate pixels of a good match. Usually, thresholding determines the final locations of a possible good match.

The method just described is known as *unnormalized correlation, matched filtering*, or *template matching*. There are several major problems associated with unnormalized correlation. If the values of $\mathbf{a}$ are large over the template support at a particular location, then it is very likely that $\Sigma\mathbf{ap}$ is also large at that location, even if no good match exists at that location. Another problem is in regions where $\mathbf{a}$ and $\mathbf{p}$ have a large number of zeros in common (i.e., a good match of zeros). Since zeros do not contribute to an increase of the value $\Sigma\mathbf{ap}$, a mismatch may be declared, even though a good match exists. To remedy this situation, several methods of *normalized correlation* have been proposed. One such method uses the formulation

$$\mathbf{c} = \frac{1}{\alpha}\Sigma\mathbf{ap},$$

where $\alpha = \Sigma a$ and the sum is over the region of the support of **p**. Another method uses the factor

$$\alpha = (\Sigma \mathbf{a})^{\frac{1}{2}} (\Sigma \mathbf{p})^{\frac{1}{2}}$$

which keeps the values of the normalized correlation between −1 and 1. Values closer to 1 represent better matches [1, 2].

The following figures illustrate pattern matching using normalized correlation. Figure 9.2.1 is an image of an industrial site with fuel storage tanks. The fuel storage tanks are the objects of interest for this example, thus the image of Figure 9.2.2 is used as the pattern. Figure 9.2.3 is the image representation for the values of positive normalized correlation between the storage tank pattern and the industrial site image for each point in the domain of the industrial site image. Note that the locations of the six fuel storage tanks show up as bright spots. There are also locations of strong correlation that are not locations of storage tanks. Referring back to the source image and pattern, it is understandable why there is relatively strong correlation at these false locations. Thresholding Figure 9.2.3 helps to pinpoint the locations of the storage tanks. Figure 9.2.4 represents the thresholded correlation image.

Figure 9.2.1. Industrial site with fuel storage tanks.

Figure 9.2.2. Pattern used to locate fuel storage tanks.

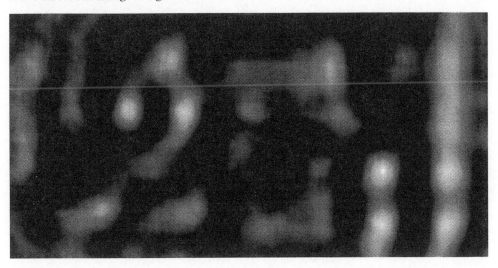

Figure 9.2.3. Image representation of positive normalized correlation resulting from applying the pattern of Figure 9.2.2 to the image of Figure 9.2.1.

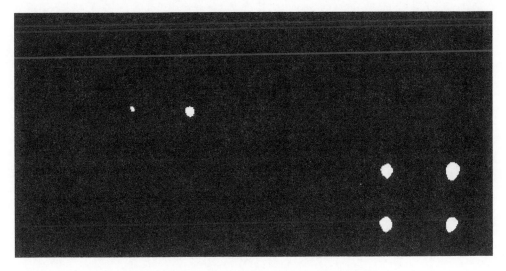

Figure 9.2.4. Thresholded correlation image.

Image Algebra Formulation

The exact formulation of a discrete correlation of an $M \times N$ image $\mathbf{a} \in \mathbb{R}^{\mathbf{X}}$ with a pattern $\mathbf{p}$ of size $(2m - 1) \times (2n - 1)$ centered at the origin is given by

$$\mathbf{c}(x, y) = \sum_{l=-(n-1)}^{n-1} \sum_{k=-(m-1)}^{m-1} \mathbf{a}(x + k, y + l) \cdot \mathbf{p}(k, l). \qquad (9.2.1)$$

For $(x + k, y + l) \notin \mathbf{X}$, one assumes that $\mathbf{a}(x + k, y + l) = 0$. It is also assumed that the pattern size is generally smaller than the sensed image size. Figure 9.2.5 illustrates the correlation as expressed by Equation 9.2.1.

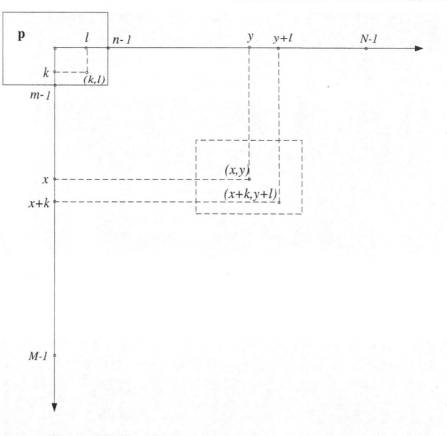

Figure 9.2.5. Computation of the correlation value $c(x, y)$ at a point $(x, y) \in \mathbf{X}$.

To specify template matching in image algebra, define an invariant pattern template $\mathbf{t}$, corresponding to the pattern $\mathbf{p}$ centered at the origin, by setting

$$\mathbf{t}_{(x,y)}(u, v) = \begin{cases} \mathbf{p}(u - x, v - y) & \text{if} \quad -(m-1) \le u - x \le m - 1 \\ & \text{and} \; -(n-1) \le v - y \le n - 1 \\ 0 & \text{otherwise.} \end{cases}$$

The unnormalized correlation algorithm is then given by

$$\mathbf{c} := \mathbf{a} \oplus \mathbf{t}.$$

The following simple computation shows that this agrees with the formulation given by Equation 9.2.1.

By definition of the operation $\oplus$, we have that

$$c(x, y) = \sum_{(u,v) \in \mathbf{X}} \mathbf{a}(u, v) \cdot \mathbf{t}_{(x,y)}(u, v). \tag{9.2.2}$$

Since $\mathbf{t}$ is translation invariant, $\mathbf{t}_{(x,y)}(u, v) = \mathbf{t}_{(0,0)}(u - x, v - y)$. Thus, Equation 9.2.2 can be written as

$$c(x, y) = \sum_{(u,v) \in \mathbf{X}} \mathbf{a}(u, v) \cdot \mathbf{t}_{(0,0)}(u - x, v - y). \tag{9.2.3}$$

Now $t_{(0,0)}(u - x, v - y) = 0$ unless $(u - x, v - y) \in S(t_{(0,0)})$ or, equivalently, unless $-(m - 1) \leq u - x \leq m - 1$ and $-(n - 1) \leq v - y \leq n - 1$. Changing variables by letting $k = u - x$ and $l = v - y$ changes Equation 9.2.3 to

$$
\begin{aligned}
c(x, y) &= \sum_{l=-(n-1)}^{n-1} \sum_{k=-(m-1)}^{m-1} a(x + k, y + l) \cdot t_{(0,0)}(k, l) \\
&= \sum_{l=-(n-1)}^{n-1} \sum_{k=-(m-1)}^{m-1} a(x + k, y + l) \cdot p(k, l) .
\end{aligned}
\tag{9.2.4}
$$

To compute the normalized correlation image c, let N denote the neighborhood function defined by $N(y) = S(t_y)$. The normalized correlation image is then computed as

$$
c := (a \oplus t)/(a \oplus N).
$$

An alternate normalized correlation image is given by the statement

$$
c := (a \oplus t)/(a \oplus N)^{\frac{1}{2}} (\Sigma t_{(0,0)})^{\frac{1}{2}}.
$$

Note that $\Sigma t_{(0,0)}$ is simply the sum of all pixel values of the pattern template at the origin.

Comments and Observations

To be effective, pattern matching requires an accurate pattern. Even if an accurate pattern exists, slight variations in the size, shape, orientation, and gray level values of the object of interest will adversely affect performance. For this reason, pattern matching is usually limited to smaller local features which are more invariant to size and shape variations of an object.

9.3. Pattern Matching in the Frequency Domain

The purpose of this section is to present several approaches to template matching in the spectral or Fourier domain. Since convolutions and correlations in the spatial domain correspond to multiplications in the spectral domain, it is often advantageous to perform template matching in the spectral domain. This holds especially true for templates with large support as well as for various parallel and optical implementations of matched filters.

It follows from the convolution theorem [3] that the spatial correlation $a \oplus t$ corresponds to multiplication in the frequency domain. In particular,

$$
a \oplus t = \mathfrak{F}^{-1}(\hat{a} \cdot \hat{t}^*),
\tag{9.3.1}
$$

where $\hat{a}$ denotes the Fourier transform of a, $\hat{t}^*$ denotes the complex conjugate of $\hat{t}$, and $\mathfrak{F}^{-1}$ the inverse Fourier transform. Thus, simple pointwise multiplication of the image $\hat{a}$ with the image $\hat{t}^*$ and Fourier transforming the result implements the spatial correlation $a \oplus t$.

One limitation of the matched filter given by Equation 9.3.1 is that the output of the filter depends primarily on the gray values of the image a rather than on its spatial structures. This can be observed when considering the output image and its corresponding gray value surface shown in Figure 9.3.2. For example, the letter E in the input image (Figure 9.3.1) produced a high-energy output when correlated with the pattern letter B shown in Figure 9.3.1. Additionally, the filter output is proportional to its autocorrelation, and the shape of the filter output around its maximum match is fairly broad. Accurately locating this maximum can therefore be difficult in the presence of noise. Normalizing the correlated image $\mathbf{a} \oplus \mathbf{t} = \mathfrak{F}^{-1}(\hat{\mathbf{a}} \cdot \hat{\mathbf{t}}^*)$, as done in the previous section, alleviates the problem for some mismatched patterns. Figure 9.3.3 provides an example of a normalized output. An approach to solving this problem in the Fourier domain is to use *phase-only* matched filters.

The transfer function of the phase-only matched filter is obtained by eliminating the amplitude of $\hat{\mathbf{t}}^*$ through factorization. As shown in Figure 9.3.4, the output of the phase-only matched filter $\mathfrak{F}^{-1}\left(\hat{\mathbf{a}} \cdot \dfrac{\hat{\mathbf{t}}^*}{|\hat{\mathbf{t}}^*|}\right)$ provides a much sharper peak than the simple matched filter since the spectral phase preserves the location of objects but is insensitive to the image energy [4].

Further improvements of the phase-only matched filter can be achieved by correlating the phases of both $\mathbf{a}$ and $\mathbf{t}$. Figure 9.3.5 shows that the image function $\mathfrak{F}^{-1}\left(\dfrac{\hat{\mathbf{a}}}{|\hat{\mathbf{a}}|} \cdot \dfrac{\hat{\mathbf{t}}^*}{|\hat{\mathbf{t}}^*|}\right)$ approximates the Dirac δ–function at the center of the letter B, thus providing even sharper peaks than the phase-only matched filter. Note also the suppression of the enlarged B in the left-hand corner of the image. This filtering technique is known as the *symmetric phase-only matched filter* or SPOMF [5].

Figure 9.3.1. The input image $\mathbf{a}$ is shown on the left and the pattern template $\mathbf{t}$ on the right.

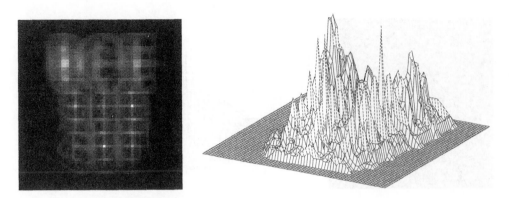

Figure 9.3.2. The correlated output image $\mathbf{c} = \mathfrak{F}^{-1}\left(\hat{\mathbf{a}} \cdot \hat{\mathbf{t}}^*\right)$ and its gray value surface.

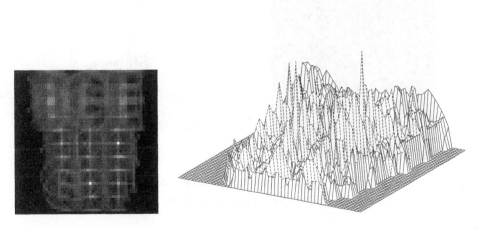

Figure 9.3.3. The normalized correlation image $\mathbf{c} = \mathfrak{F}^{-1}\left(\hat{\mathbf{a}} \cdot \hat{\mathbf{t}}^*\right) / (\mathbf{a} \oplus N)^{1/2} \left(\Sigma \mathbf{t}_{(0,0)}\right)^{1/2}$, where $N(\mathbf{y}) = S(\mathbf{t_y})$ and its gray value surface.

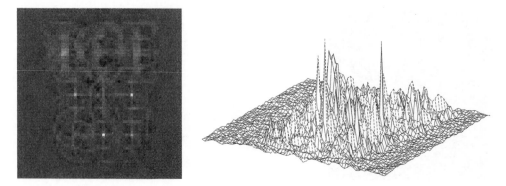

Figure 9.3.4. The phase-only correlation image $\mathfrak{F}^{-1}\left(\hat{\mathbf{a}} \cdot \frac{\hat{\mathbf{t}}^*}{|\hat{\mathbf{t}}^*|}\right)$ and its gray value surface.

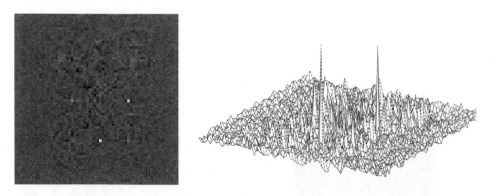

Figure 9.3.5. The symmetric phase-only correlation
image $\mathfrak{F}^{-1}\left(\frac{\hat{\mathbf{a}}}{|\hat{\mathbf{a}}|} \cdot \frac{\hat{\mathbf{t}}^*}{|\hat{\mathbf{t}}^*|}\right)$ and its gray value surface.

Image Algebra Formulation

Although the methods for matched filtering in the frequency domain are mathematically easily formulated, the exact digital specification is a little more complicated. The multiplication $\hat{\mathbf{a}} \cdot \hat{\mathbf{t}}^*$ implies that we must have two images $\hat{\mathbf{a}}$ and $\hat{\mathbf{t}}^*$ of the same size. Thus a first step is to create the *image* $\hat{\mathbf{t}}^*$ from the pattern *template* $\mathbf{t}$.

To create the image $\hat{\mathbf{t}}^*$, reflect the pattern template $\mathbf{t}$ across the origin by setting $\mathbf{p}(\mathbf{x}) = \mathbf{t}_{(0,0)}(-\mathbf{x})$. This will correspond to conjugation in the spectral domain. Since $\mathbf{t}$ is an invariant template defined on $\mathbb{Z}^2$, $\mathbf{p}$ is an image defined over $\mathbb{Z}^2$. However, what is needed is an image over $domain(\mathbf{a}) = \mathbf{X}$. As shown in Figure 9.2.5, the support of the pattern template intersects $\mathbf{X}$ in only the positive quadrant. Hence, simple restriction of $\mathbf{p}$ to $\mathbf{X}$ will not work. Additionally, the pattern image $\mathbf{p}$ needs to be *centered* with respect to the transformed image $\hat{\mathbf{a}}$ for the appropriate multiplication in the Fourier domain. This is achieved by translating $\mathbf{p}$ to the corners of $\mathbf{a}$ and then restricting to the domain of $\mathbf{a}$. This process is illustrated in Figure 9.3.6. Specifically, define

$$\mathbf{p} := (\mathbf{p} + [\mathbf{p} + (0, N)] + [\mathbf{p} + (M, 0)] + [\mathbf{p} + (M, N)])|\mathbf{x} \ .$$

Note that there are two types of additions in the above formula. One is image addition and the other is image-point addition which results in a shift of the image. Also, the translation of the image $\mathbf{p}$, achieved by vector addition, translates $\mathbf{p}$ one unit beyond the boundary of $\mathbf{X}$ in order to avoid duplication of the intersection of the boundary of $\mathbf{X}$ with $\mathbf{p}$ at the corner of the origin.

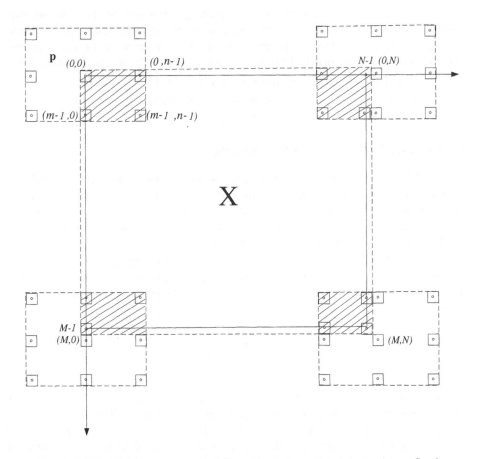

Figure 9.3.6. The image $\mathbf{p}$ created from the pattern template $\mathbf{t}$ using reflection, translations to the corners of the array $\mathbf{X}$, and restriction to $\mathbf{X}$. The values of $\mathbf{p}$ are zero except in the shaded area, where the values are equal the corresponding gray level values in the support of $\mathbf{t}$ at (0,0).

The image $\hat{\mathbf{t}}^*$ is now given by $\hat{\mathbf{p}}$, the Fourier transform of $\mathbf{p}$. The correlation image $\mathbf{c}$ can therefore be obtained using the following algorithm:

$$\mathbf{p}(\mathbf{x}) := \mathbf{t}_{(0,0)}(-\mathbf{x})$$
$$\mathbf{p} := (\mathbf{p} + [\mathbf{p} + (0, N)] + [\mathbf{p} + (M, 0)] + [\mathbf{p} + (M, N)])|\mathbf{x}$$
$$\mathbf{c} := \mathfrak{F}^{-1}(\hat{\mathbf{a}} \cdot \hat{\mathbf{p}}).$$

Using the image $\mathbf{p}$ constructed in the above algorithm, the phase-only filter and the symmetric phase-only filter have now the following simple formulation:

$$\mathbf{c} := \mathfrak{F}^{-1}\left(\hat{\mathbf{a}} \cdot \frac{\hat{\mathbf{p}}}{|\hat{\mathbf{p}}|}\right)$$

and

$$\mathbf{c} := \mathfrak{F}^{-1}\left(\frac{\hat{\mathbf{a}}}{|\hat{\mathbf{a}}|} \cdot \frac{\hat{\mathbf{p}}}{|\hat{\mathbf{p}}|}\right),$$

respectively.

Comments and Observations

In order to achieve the phase-only matching component to the matched filter approach we needed to divide the complex image $\hat{\mathbf{p}}$ by the amplitude image $|\hat{\mathbf{p}}|$. Problems can occur if some pixel values of $|\hat{\mathbf{p}}|$ are equal to zero. However, in the image algebra pseudocode of the various matched filters we assume that $\frac{\hat{\mathbf{p}}}{|\hat{\mathbf{p}}|} = \hat{\mathbf{p}} \cdot |\hat{\mathbf{p}}|^{-1}$, where $|\hat{\mathbf{p}}|^{-1}$ denotes the pseudoinverse of $|\hat{\mathbf{p}}|$. A similar comment holds for the quotient $\frac{\hat{\mathbf{a}}}{|\hat{\mathbf{a}}|}$.

Some further improvements of the symmetric phase-only matched filter can be achieved by processing the spectral phases [6, 7, 8, 9].

9.4.   Rotation Invariant Pattern Matching

In Section 9.2 we noted that pattern matching using simple pattern correlation will be adversely affected if the pattern in the image is different in size or orientation then the template pattern. Rotation invariant pattern matching solves this problem for patterns varying in orientation. The technique presented here is a digital adaptation of optical methods of rotation invariant pattern matching [10, 11, 12, 13, 14].

Computing the Fourier transform of images and ignoring the phase provides for a pattern matching approach that is insensitive to position (Section 9.3) since a shift in $\mathbf{a}(x, y)$ does not affect $|\hat{\mathbf{a}}(u, v)|$. This follows from the Fourier transform pair relation

$$\mathbf{a}(x - x_0, y - y_0) \Leftrightarrow \hat{\mathbf{a}}(u, v) = e^{-2\pi i(ux_0 + vy_0)/N}$$

which implies that

$$\left| \hat{\mathbf{a}}(u, v)e^{-2\pi i(ux_0 + vy_0)/N} \right| = |\hat{\mathbf{a}}(u, v)|,$$

where $x_0 = y_0 = N/2$ denote the midpoint coordinates of the $N \times N$ domain of $\hat{\mathbf{a}}$. However, rotation of $\mathbf{a}(x, y)$ rotates $|\hat{\mathbf{a}}(u, v)|$ by the same amount. This rotational effect can be taken care of by transforming $|\hat{\mathbf{a}}(u, v)|$ to polar form $(u, v) \mapsto (r, \theta)$. A rotation of $\mathbf{a}(x, y)$ will then manifest itself as a shift in the angle θ. After determining this shift, the pattern template can be rotated through the angle θ and then used in one of the standard correlation schemes in order to find the location of the pattern in the image.

The exact specification of this technique — which, in the digital domain, is by no means trivial — is provided by the image algebra formulation below.

Image Algebra Formulation

Let $\mathbf{t}$ denote the pattern template and let $\mathbf{a} \in \mathbb{R}^{\mathbb{Z}_N \times \mathbb{Z}_N}$, where $N = 2^n$, denote the image containing the rotated pattern corresponding to $\mathbf{t}$. The rotation invariant pattern scheme alluded to above can be broken down into seven basic steps.

Step 1. Extend the pattern template $\mathbf{t}$ to a pattern image over the domain $\mathbb{Z}_N \times \mathbb{Z}_N$ of $\mathbf{a}$.

This step can be achieved in a variety of ways. One way is to use the method defined in Section 9.3. Another method is to simply set

$$\mathbf{p} := \mathbf{t}_{(N/2, N/2)}|_{\mathbb{Z}_N \times \mathbb{Z}_N}.$$

This is equivalent to extending $\mathbf{t}_{(N/2, N/2)}$ outside of its support to the zero image on $\mathbb{Z}_N \times \mathbb{Z}_N$.

Step 2. Fourier transform $\mathbf{a}$ and $\mathbf{p}$.

$$\hat{\mathbf{a}} := \mathfrak{F}(\mathbf{a})$$
$$\hat{\mathbf{p}} := \mathfrak{F}(\mathbf{p}).$$

Step 3. Center the Fourier transformed images (see Section 8.3).

$$\hat{\mathbf{a}} := \hat{\mathbf{a}} \circ center$$
$$\hat{\mathbf{p}} := \hat{\mathbf{p}} \circ center.$$

Centering is a necessary step in order to avoid boundary effects when using bilinear interpolation at a subsequent stage of this algorithm.

Step 4. Scale the Fourier spectrum.

$$\hat{\mathbf{a}} := log(|\hat{\mathbf{a}}| + 1)$$
$$\hat{\mathbf{p}} := log(|\hat{\mathbf{p}}| + 1).$$

This step is vital for the success of the proposed method. Image spectra decrease rather rapidly as a function of increasing frequency, resulting in suppression of high-frequency terms. Taking the logarithm of the Fourier spectrum increases the amplitude of the side lobes and thus provides for more accurate results when employing the symmetric phase-only filter at a later stage of this algorithm.

Step 5. Convert $\hat{\mathbf{a}}$ and $\hat{\mathbf{p}}$ to *continuous* image.
The conversion of $\hat{\mathbf{a}}$ and $\hat{\mathbf{p}}$ to continuous images is accomplished by using bilinear interpolation. An image algebra formulation of bilinear interpolation can be found in Section 11.4. Note that because of Step 4, $\hat{\mathbf{a}}$ and $\hat{\mathbf{p}}$ are real-valued images. Thus, if $\hat{\mathbf{a}}_b$ and $\hat{\mathbf{p}}_b$ denote the interpolated images, then $\hat{\mathbf{p}}_b, \hat{\mathbf{a}}_b \in \mathbb{R}^{\mathbf{X}}$, where

$$\mathbf{X} = \{(x, y) : x, y \in \mathbb{R} \text{ and } 0 \leq x, y \leq N - 1\}.$$

That is, $\hat{\mathbf{a}}_b$ and $\hat{\mathbf{p}}_b$ are real-valued images over a point set $\mathbf{X}$ with real-valued coordinates.

Although nearest neighbor interpolation can be used, bilinear interpolation results in a more robust matching algorithm.

Step 6. Convert to polar coordinates.
Define the point set

$$\mathbf{Y} = \left\{ (r_i, \theta_j) : r_i = \frac{i}{N} \cdot \frac{N-1}{2}, \; \theta_j = \frac{2\pi j}{N} - \pi, \; i, j \in \mathbb{Z}_N \right\}$$

and a spatial function $f : \mathbf{Y} \rightarrow \mathbf{X}$ by

$$f(r_i, \theta) = (r_i \cos\theta_j + N/2, \ r_i \sin\theta_j + N/2).$$

Next compute the polar images.

$$\hat{\mathbf{a}} := \hat{\mathbf{a}}_b \circ f$$

$$\hat{\mathbf{p}} := \hat{\mathbf{p}}_b \circ f.$$

Step 7. Apply the SPOMF algorithm (Section 9.3).

Since the spectral magnitude is a periodic function of π and θ ranges over the interval $[-\pi = \theta_0, \ \theta_N = \pi]$, the output of the SPOMF algorithm will produce two peaks along the θ axis, θ_j and θ_k for some $j \in \mathbb{Z}_N$ and $k \in \mathbb{Z}_N$. Due to the periodicity, $|\theta_j| + |\theta_k| = \pi$ and, hence, $k = -(j + N/2)$. One of these two angles corresponds to the angle of rotation of the pattern in the image with respect to the template pattern. The complementary angle corresponds to the same image pattern rotated $180°$.

To find the location of the rotated pattern in the spatial domain image, one must rotate the pattern template (or input image) through the angle θ_j as well as the angle θ_k. The two templates thus obtained can then be used in one of the previous correlation methods. Pixels with the highest correlation values will correspond to the pattern location.

Comments and Observations

The following example will help to further clarify the algorithm described above. The pattern image $\mathbf{p}$ and input image $\mathbf{a}$ are shown in Figure 9.4.1. The exemplar pattern is a rectangle rotated through an angle of $15°$ while the input image contains the pattern rotated through an angle of $70°$. Figure 9.4.2 shows the output of Step 4 and Figure 9.4.3 illustrates the conversion to polar coordinates of the images shown in Figure 9.4.2. The output of the SPOMF process (before thresholding) is shown in Figure 9.4.4. The two high peaks appear on the θ axis ($r = 0$).

Figure 9.4.1. The input image $\mathbf{a}$ is shown on the left and the pattern template $\mathbf{p}$ on the right.

The reason for choosing grid spacing $r_i = \frac{i}{N} \frac{N-1}{2}$ in Step 6 is that the maximum value of r is $r_{N-1} = \frac{(N-1)^2}{2N}$ which prevents mapping the polar coordinates outside the set $\mathbf{X}$. Finer sampling grids will further improve the accuracy of pattern detection; however, computational costs will increase proportionally. A major drawback of this method is that it works best only when a single object is present in the image, and when the image and template backgrounds are identical.

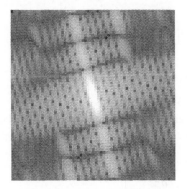 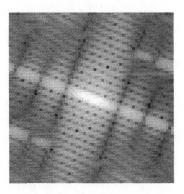

Figure 9.4.2. The log of the spectra of $\hat{\mathbf{a}}$ (left) and $\hat{\mathbf{p}}$ (right).

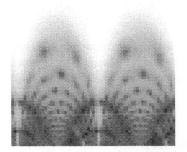 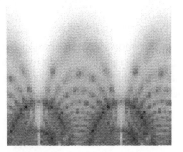

Figure 9.4.3. Rectangular to polar conversion of $\hat{\mathbf{a}}$ (left) and $\hat{\mathbf{p}}$ (right).

Figure 9.4.4. SPOMF of image and pattern shown in Figure 9.4.3.

9.5. Rotation and Scale Invariant Pattern Matching

In this section we discuss a method of pattern matching which is invariant with respect to both rotation and scale. The two main components of this method are the Fourier transform and the Mellin transform. Rotation invariance is achieved by using the approach described in Section 9.4. For scale invariance we employ the Mellin transform. Since the Mellin transform $\mathfrak{M}(\mathbf{a})$ of an image $\mathbf{a} \in \mathbb{R}^{\mathbb{Z}_N \times \mathbb{Z}_N}$ is given by

$$[\mathfrak{M}(\mathbf{a})](u,v) = \sum_{k=0}^{N-1} \sum_{j=0}^{N-1} \mathbf{a}(j,k) j^{-(iu+1)} k^{-(iv+1)},$$

it follows that if $\mathbf{b}(x, y) = \mathbf{a}(\alpha x, \alpha y)$, then

$$[\mathfrak{M}(\mathbf{b})](u, v) = \alpha^{-i(u+v)}[\mathfrak{M}(\mathbf{a})](u, v).$$

Therefore,

$$|[\mathfrak{M}(\mathbf{b})](u, v)| = |[\mathfrak{M}(\mathbf{a})](u, v)|,$$

which shows that the Mellin transform is scale invariant.

Implementation of the Mellin transform can be accomplished by use of the Fourier transform by rescaling the input function. Specifically, letting $\gamma = log x$ and $\beta = log y$ we have

$$x = e^\gamma \text{ and } dx = e^\gamma d\gamma$$
$$y = e^\beta \text{ and } dy = e^\beta d\beta.$$

Therefore,

$$[\mathfrak{M}(\mathbf{a})](u, v) = \int \int \mathbf{a}(x, y) x^{-(iu+1)} y^{-(iv+1)} dx dy$$
$$= \int \int \mathbf{a}(e^\gamma, e^\beta) e^{-i(u\gamma + v\beta)} d\gamma d\beta$$

which is the desired result.

It follows that combining the Fourier and Mellin transform with a rectangular to polar conversion yields a rotation and scale invariant matching scheme. The approach takes advantage of the individual invariance properties of these two transforms as summarized by the following four basic steps:

(1) Fourier transform

$$\mathbf{a}(x, y), \mathbf{p}(x, y) \rightarrow \hat{\mathbf{a}}(u, v), \hat{\mathbf{p}}(u, v)$$

(2) Rectangular to polar conversion

$$\hat{\mathbf{a}}(u, v), \hat{\mathbf{p}}(u, v) \rightarrow \hat{\mathbf{a}}(r, \theta), \hat{\mathbf{p}}(r, \theta)$$

(3) Logarithmic scaling of r

$$\hat{\mathbf{a}}(r, \theta), \hat{\mathbf{p}}(r, \theta) \rightarrow \hat{\mathbf{a}}(e^{log r}, \theta), \hat{\mathbf{p}}(e^{log r}, \theta)$$

(4) SPOMF

Image Algebra Formulation

The image algebra formulation of the rotation and scale invariant pattern matching algorithm follows the same steps as those listed in the rotation invariant pattern matching scheme in Section 9.4 with the exception of Step 6, in which the point set $\mathbf{Y}$ should be defined as follows:

$$\mathbf{Y} = \left\{ (r_i, \theta_j) : \ r_i = \frac{1}{2} N^{i/N}, \ \theta_j = \frac{2\pi j}{N} - \pi, \ i, j \in \mathbb{Z}_N \right\}.$$

The rescaling of r_i corresponds to the logarithmic scaling of r in the Fourier-Mellin transform.

As in the rotation invariant filter, the output of the SPOMF will result in two peaks which will be a distance π apart in the θ direction and will be offset by a factor proportional to the scaling factor from the θ axis. For $0 \le i < N/2$, this will correspond to an enlargement of the pattern, while for $N/2 < i \le N - 1$, the proportional scaling factor will correspond to a shrinking of the pattern [5]. The following example (Figures 9.5.1 to 9.5.3) illustrates the important steps of this algorithm.

Figure 9.5.1. The input image **a** is shown on the left and the pattern template **p** on the right.

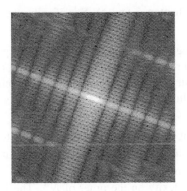

Figure 9.5.2. The log of the spectra of $\hat{\text{a}}$ (left) and $\hat{\text{p}}$ (right).

Figure 9.5.3. Rectangular to polar-log conversion of $\hat{\text{a}}$ (left) and $\hat{\text{p}}$ (right).

9.6. Line Detection Using the Hough Transform

The *Hough transform* is a mapping from $\mathbb{R}^2$ into the function space of sinusoidal functions. It was first formulated in 1962 by Hough [15]. Since its early formulation, this transform has undergone intense investigations which have resulted in several generalizations and a variety of applications in computer vision and image processing [1, 2, 16, 17, 18]. In this section we present a method for finding straight lines using the Hough transform. The input for the Hough transform is an image that has been preprocessed by

Figure 9.5.4. SPOMF of image and pattern shown in Figure 9.5.3.

some type of edge detector and thresholded (see Chapters 3 and 4). Specifically, the input should be a binary edge image.

A straight "line" in the sense of the Hough algorithm is a colinear set of points. Thus, the number of points in a straight line could range from one to the number of pixels along the diagonal of the image. The quality of a straight "line" is judged by the number of points in it. It is assumed that the natural straight lines in an image correspond to digitized straight "lines" in the image with relatively large cardinality.

A brute force approach to finding straight lines in a binary image with N feature pixels would be to examine all $\frac{N(N-1)}{2}$ possible straight lines between the feature pixels. For each of the $\frac{N(N-1)}{2}$ possible lines, $N-2$ tests for colinearity must be performed. Thus, the brute force approach has a computational complexity on the order of N^3. The Hough algorithm provides a method of reducing this computational cost.

To begin the description of the Hough algorithm, we first define the Hough transform and examine some of its properties. The Hough transform is a mapping h from $\mathbb{R}^2$ into the function space of sinusoidal functions defined by

$$h : (x, y) \rightarrow \rho = x \, cos(\theta) + y \, sin(\theta).$$

To see how the Hough transform can be used to find straight lines in an image, a few observations need to be made.

Any straight line l_0 in the xy-plane corresponds to a point (ρ_0, θ_0) in the $\rho\theta$-plane, where $\theta_0 \in [0, \pi)$ and $\rho_0 \in \mathbb{R}$. Let n_0 be the line normal to l_0 that passes through the origin of the xy-plane. The angle n_0 makes with the positive x-axis is θ_0. The distance from $(0, 0)$ to l_0 along n_0 is $|\rho_0|$. Figure 9.6.1 below illustrates the relation between l_0, n_0, θ_0, and ρ_0. Note that the x-axis in the figure corresponds to the point $(0, 0)$, while the y-axis corresponds to the point $(0, \pi/2)$.

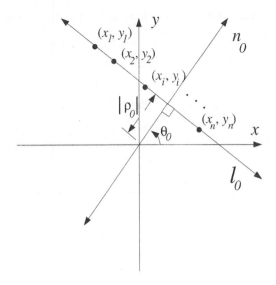

Figure 9.6.1. Relation of rectangular to polar representation of a line.

Suppose (x_i, y_i), $1 \leq i \leq n$, are points in the xy-plane that lie along the straight line l_0 (see Figure 9.6.1). The line l_0 has a representation (ρ_0, θ_0) in the $\rho\theta$-plane. The Hough transform takes each of the points (x_i, y_i) to a sinusoidal curve $\rho = x_i cos(\theta) + y_i sin(\theta)$ in the $\theta\rho$-plane. The property that the Hough algorithm relies on is that each of the curves $\rho = x_i cos(\theta) + y_i sin(\theta)$ have a common point of intersection, namely (ρ_0, θ_0). Conversely, the sinusoidal curve $\rho = x \, cos(\theta) + y \, sin(\theta)$ passes through the point (ρ_0, θ_0) in the $\rho\theta$-plane only if (x, y) lies on the line (ρ_0, θ_0) in the xy-plane.

As an example, consider the points $(1, 7)$, $(3, 5)$, $(5, 3)$, and $(6, 2)$ in the xy-plane that lie along the line l_0 with θ and ρ representation $\theta_0 = \frac{\pi}{4} \approx 0.7854$ and $\rho_0 = \sqrt{32} \approx 5.657$, respectively. Figure 9.6.2 shows these points and the line l_0.

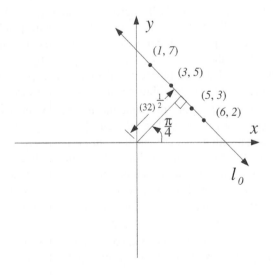

Figure 9.6.2. Polar parameters associated with points lying on a line.

The Hough transform maps the indicated points to the sinusoidal functions as follows:

$$h : (1, 7) \rightarrow \rho = \cos(\theta) + 7\sin(\theta)$$
$$h : (3, 5) \rightarrow \rho = 3\cos(\theta) + 5\sin(\theta)$$
$$h : (5, 3) \rightarrow \rho = 5\cos(\theta) + 3\sin(\theta)$$
$$h : (6, 2) \rightarrow \rho = 6\cos(\theta) + 2\sin(\theta) \ .$$

The graphs of these sinusoidal functions can be seen in Figure 9.6.3. Notice how the four sinusoidal curves intersect at $\theta = \frac{\pi}{4} \approx 0.7854$ and $\rho = \sqrt{32} \approx 5.657$.

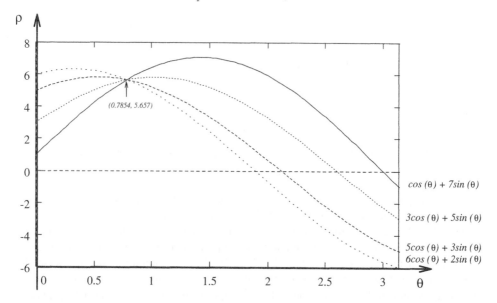

Figure 9.6.3. Sinusoidals in Hough space associated with points on a line.

Each point (x, y) at a feature pixel in the domain of the image maps to a sinusoidal function by the Hough transform. If the feature point (x_i, y_i) of an image lies on a line in the xy-plane parameterized by (ρ_0, θ_0) (θ_0, ρ_0), its corresponding representation as a sinusoidal curve in the $\rho\theta$-plane will intersect the point (ρ_0, θ_0). Also, the sinusoid $\rho = x\cos(\theta) + y\sin(\theta)$ will intersect (ρ_0, θ_0) only if the feature pixel location (x, y) lies on the line (ρ_0, θ_0) in the xy-plane. Therefore, it is possible to count the number of feature pixel points that lie along the line (ρ_0, θ_0) in the xy-plane by counting the number of sinusoidal curves in the $\rho\theta$ plane that intersect at the point (ρ_0, θ_0). This observation is the basis of the Hough line detection algorithm which is described next.

Obviously, it is impossible to count the number of intersection of sinusoidal curves at every point in the $\rho\theta$-plane. The $\rho\theta$-plane for $-R \leq \rho \leq R$, $0 \leq \theta < \pi$ must be quantized. This quantization is represented as an *accumulator array* $\mathbf{a}(i, j)$. Suppose that for a particular application it is decided that the $\rho\theta$-plane should be quantized into an $r \times c$ accumulator array. Each column of the accumulator represents a $\frac{\pi}{c}$ increment in the angle θ. Each row in the accumulator represents a $\frac{2R}{r}$ increment in ρ. The cell location $\mathbf{a}(i, j)$ of the accumulator is used as a counting bin for the point $\left((i - \frac{r}{2})\frac{2R}{r}, j\frac{\pi}{c}\right)$ in the $\rho\theta$-plane (and the corresponding line in the xy-plane).

Initially, every cell of the accumulator is set to 0. The value $\mathbf{a}(i, j)$ of the accumulator is incremented by 1 for every feature pixel (x, y) location at which the inequality

$$|\rho_i - (x\cos(\theta_j) + y\sin(\theta_j))| < \epsilon$$

is satisfied, where $(\rho_i, \theta_j) = \left(\left(i - \frac{r}{2}\right)\frac{2R}{r}, j\frac{\pi}{c}\right)$ and ϵ is an error factor used to compensate for quantization and digitization. That is, if the point (ρ_i, θ_j) lies on the curve $\rho = x\cos(\theta) + y\sin(\theta)$ (within a margin of error), the accumulator at cell location $\mathbf{a}(i, j)$ is incremented. Error analysis for the Hough transform is addressed in Shapiro's works [19, 20, 21].

When the process of incrementing cell values in the accumulator terminates, each cell value $\mathbf{a}(i, j)$ will be equal to the number of curves $\rho = x\cos(\theta) + y\sin(\theta)$ that intersect the point (ρ_i, θ_j) in the $\rho\theta$-plane. As we have seen earlier, this is the number of feature pixels in the image that lie on the line (ρ_i, θ_j).

The criterion for a good line in the Hough algorithm sense is a large number of colinear points. Therefore, the larger entries in the accumulator are assumed to correspond to lines in the image.

Image Algebra Formulation

Let $\mathbf{b} \in \{0, 1\}^X$ be the source image and let the accumulator image $\mathbf{a}$ be defined over $\mathbf{Y}$, where

$$\mathbf{Y} = \left\{(\rho_i, \theta_j) \;:\; \rho_i = \left(i - \frac{r}{2}\right)\frac{2R}{r}, \; \theta_j = \frac{j\pi}{c}, \; 0 \le i \le r, \; 0 \le j < c\right\}.$$

Define the parameterized template $\mathbf{t} \in \left(\mathbb{Z}^Y\right)^X$ by

$$\mathbf{t}(\mathbf{b})_{(x,y)}(\rho, \theta) = \begin{cases} 1 & \text{if } \mathbf{b}(x, y) = 1 \text{ and} \\ & \quad |\rho - (x\cos(\theta) + y\sin(\theta))| < \epsilon \\ 0 & \text{otherwise.} \end{cases}$$

The accumulator image is given by the image algebra expression

$$\mathbf{a} := \sum \mathbf{t}(\mathbf{b}).$$

Computation of this variant template sum is computationally intensive and inefficient. A more efficient implementation is given below.

Comments and Observations

For the c quantized values of θ in the accumulator, the computation of $x\cos(\theta) + y\sin(\theta)$ is carried out for each of the N feature pixel locations (x, y) in the image. Next, each of the rc cells of the accumulator are examined for high counts. The computational cost of the Hough algorithm is $Nc + rc$, or $O(N)$. This is a substantial improvement over the $O(N^3)$ complexity of the brute force approach mentioned at the beginning of this section. This complexity comparison may be a bit misleading. A true comparison would have to take into account the dimensions of the accumulator array. A smaller accumulator reduces computational complexity. However, better line detection performance can be achieved with a finer quantization of the $\rho\theta$-plane.

As presented above, the algorithm can increment more than one cell in the accumulator array for any choice of values for the point (x, y) and angle θ. We can insure that at most one accumulator cell will be incremented by calculating the unique value ρ as a function of (x, y) and quantized θ_j as follows:

$$\rho = [x\cos(\theta_j) + y\sin(\theta_j)],$$

where $[z]$ denotes the rounding of z to the nearest integer.

The equation for ρ can be used to define the neighborhood function

$$N : \mathbf{Y} \rightarrow 2^{\mathbf{X}}$$

by

$$N(\rho, \theta) = \{(x, y) \; : \; \rho = [x cos(\theta) + y sin(\theta)]\}.$$

The statement

$$\mathbf{b} \; := \; \mathbf{b}\|_1$$

$$\mathbf{a} \; := \; \mathbf{b} \oplus N$$

computes the accumulator image since

$$\mathbf{a}(i, j) = \sum_{(x,y) \in N(\rho_i, \theta_j)} \mathbf{b}(x, y) \, .$$

A straight line l_0 in the xy-plane can also be represented by a point (m_0, c_0), where m_0 is the slope of l_0 and c_0 is its y intercept. In the original formulation of the Hough algorithm [15], the Hough transform took points in the xy-plane to lines in the slope-intercept plane; i.e., $h : (x_i, y_i) \rightarrow y_i = mx_i + c$. The slope intercept representation of lines presents difficulties in implementation of the algorithm because both the slope and the y intercept of a line go to infinity as the line approaches the vertical. This difficulty is not encountered using the $\rho\theta$-representation of a line.

As an example, we have applied the Hough algorithm to the thresholded edge image of a causeway with a bridge (Figure 9.6.4). The $\rho\theta$-plane has been quantized using the 41×20 accumulator seen in Table 9.6.1. Accumulator values greater than 80 were deemed to correspond to lines. Three values in the accumulator satisfied this threshold; they are indicated within the accumulator by double underlining. The three detected lines are shown in Figure 9.6.5.

The lines produced by our example probably are not the lines that a human viewer would select. A finer quantization of θ and ρ would probably yield better results. All the parameters for our example were chosen arbitrarily. No conclusions on the performance of the algorithm should be drawn on the basis of our example. It serves simply to illustrate an accumulator array. However, it is instructive to apply a straight edge to the source image to see how the quantization of the $\rho\theta$-plane affected the accumulator values.

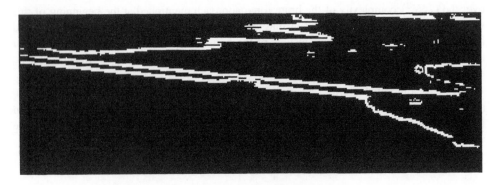

Figure 9.6.4. Source binary image.

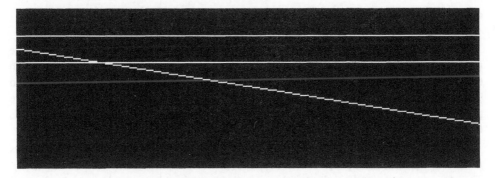

Figure 9.6.5. Detected lines.

Table 9.6.1 Hough Space Accumulator Values

ρ\θ	0	π/20	π/10	3π/20	π/5	π/4	3π/10	7π/20	2π/5	9π/20	π/2	11π/20	3π/5	13π/20	7π/10	3π/4	4π/5	17π/20	9π/10	19π/20
-200	0	0	0	0	0	0	0	0	0	0	0	0	0	0	0	0	0	0	0	0
-190	0	0	0	0	0	0	0	0	0	0	0	0	0	0	0	0	0	0	0	0
-180	0	0	0	0	0	0	0	0	0	0	0	0	0	0	0	0	0	0	2	2
-170	4	4	0	0	0	0	0	0	0	0	0	0	0	0	0	0	0	0	2	2	3
-160	6	4	3	0	0	0	0	0	0	0	0	0	0	0	0	0	3	2	5	11
-150	6	6	5	0	0	0	0	0	0	0	0	0	0	0	0	2	3	4	11	11
-140	7	7	6	3	0	0	0	0	0	0	0	0	0	0	0	2	4	9	8	8
-130	9	7	8	5	2	0	0	0	0	0	0	0	0	0	3	2	6	9	10	16
-120	7	9	8	8	4	0	0	0	0	0	0	0	0	0	0	4	11	17	14	8
-110	9	9	9	8	9	1	0	0	0	0	0	0	0	3	3	13	11	12	9	10
-100	8	8	8	9	11	8	0	0	0	0	0	0	0	0	2	4	16	16	9	9
-90	8	8	8	7	12	7	0	0	0	0	0	0	6	2	9	9	7	8	9	13
-80	7	8	10	11	11	8	7	0	0	0	0	0	2	5	18	10	9	12	7	5
-70	8	8	9	9	11	8	10	0	0	0	0	0	5	15	16	6	12	8	6	7
-60	7	8	8	11	11	8	13	1	0	0	0	0	7	12	13	8	9	7	5	4
-50	7	8	7	9	11	9	16	14	0	0	0	6	15	18	14	9	7	6	7	9
-40	9	7	8	11	10	16	14	22	0	0	17	6	15	13	8	14	8	7	8	13
-30	13	9	7	8	10	13	13	21	8	0	6	6	26	18	12	13	6	13	15	19
-20	8	12	10	9	9	8	15	16	21	0	14	26	14	18	15	6	11	18	18	15
-10	9	8	10	11	9	7	14	18	34	0	6	22	14	17	9	11	13	18	16	15
0	8	9	7	7	15	10	11	16	30	77	74	22	25	14	15	9	23	17	12	15
10	13	11	9	11	13	14	23	25	40	**144**	75	15	20	17	14	15	15	10	11	7
20	14	14	15	12	10	9	17	21	48	39	**84**	32	28	19	29	15	16	13	10	8
30	13	14	15	14	13	16	15	22	19	14	7	34	30	36	18	24	14	10	9	12
40	17	14	15	14	15	27	18	54	21	34	**97**	70	42	31	19	14	12	16	15	10
50	13	19	20	15	19	14	36	32	27	45	29	53	24	23	21	12	13	11	10	8
60	8	12	13	17	21	24	31	24	22	10	0	6	27	14	13	8	9	8	8	8
70	6	7	11	17	29	16	29	13	5	0	0	0	23	19	11	6	6	8	8	9
80	8	6	7	12	10	9	6	13	1	15	0	0	6	17	14	6	9	8	8	8
90	13	7	7	8	9	7	11	9	7	0	0	0	0	15	11	8	10	9	10	8
100	8	7	7	6	13	10	9	7	0	0	0	0	0	10	15	16	9	9	9	9
110	11	10	9	12	9	16	7	6	0	0	0	0	0	0	7	15	10	7	11	9
120	9	10	15	11	13	12	7	0	0	0	0	0	0	0	10	15	9	7	8	10
130	16	15	15	15	14	7	5	2	0	0	0	0	0	0	0	0	6	9	6	10	9
140	14	14	11	11	7	5	7	0	0	0	0	0	0	0	0	4	6	7	8	9
150	8	13	12	8	2	1	0	0	0	0	0	0	0	0	0	1	5	6	7	7
160	13	9	3	2	2	2	0	0	0	0	0	0	0	0	0	0	2	5	7	6
170	7	6	5	3	3	0	0	0	0	0	0	0	0	0	0	0	0	4	3	4
180	0	2	3	3	0	0	0	0	0	0	0	0	0	0	0	0	0	0	2	4
190	0	0	0	0	0	0	0	0	0	0	0	0	0	0	0	0	0	0	0	0
200	0	0	0	0	0	0	0	0	0	0	0	0	0	0	0	0	0	0	0	0

9.7.   Detecting Ellipses Using the Hough Transform

The Hough algorithm can be easily extended to finding any curve in an image that can be expressed analytically in the form $f(\mathbf{x}, \mathbf{p}) = 0$ [22]. Here, $\mathbf{x}$ is a point in the domain of the image and $\mathbf{p}$ is a parameter vector. For example, the lines of Section 9.6 can be expressed in analytic form by letting $g(\mathbf{x}, \mathbf{p}) = x\cos(\theta) + y\sin(\theta) - \rho$, where $\mathbf{p} = (\theta, \rho)$ and $\mathbf{x} = (x, y) \in \mathbb{R}^2$. We will first discuss how the Hough algorithm extends for any analytic curve using circle location to illustrate the method.

The circle $(x - \chi)^2 + (y - \psi)^2 = \rho^2$ in the xy-plane with center (χ, ψ) and radius ρ can be expressed as $f(\mathbf{x}, \mathbf{p}) = (x - \chi)^2 + (y - \psi)^2 - \rho^2 = 0$, where $\mathbf{p} = (\chi, \psi, \rho)$. Therefore, just as a line l_0 in the xy-plane can parameterized by an angle θ_0 and a directed distance ρ_0, a circle c_0 in the xy-plane can be parameterized by the location of its center (x_0, y_0) and its radius ρ_0.

The Hough transform used for circle detection is a map defined over feature points in the domain of the image into the function space of conic surfaces. The Hough transform h used for circle detection is the map

$$h : (x_0, y_0) \rightarrow (\chi - x_0)^2 + (\psi - y_0)^2 - \rho^2 = 0.$$

Note that the Hough transform is only applied to the feature points in the domain of the image. The surface $(\chi - x_0)^2 + (\psi - y_0)^2 - \rho^2 = 0$ is a cone in $\chi\psi\rho$ space with $\chi\psi$ intercept $(x_0, y_0, 0)$ (see Figure 9.7.1).

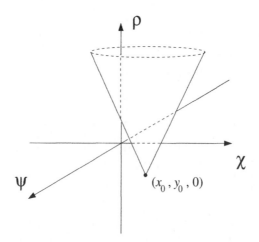

Figure 9.7.1. The surface $(\chi - x_0)^2 + (\psi - y_0)^2 - \rho^2 = 0$.

The point (x_i, y_i) lies on the circle $(x - \chi_0)^2 + (y - \psi_0)^2 - \rho_0^2 = 0$ in the xy-plane if and only if the conic surface $(\chi - x_i)^2 + (\psi - y_i)^2 - \rho^2 = 0$ intersects the point (χ_0, ψ_0, ρ_0) in $\chi\psi\rho$-space. For example, the points $(-1, 0)$ and $(1, 0)$ lie on the circle $f((x, y), (0, 0, 1)) = x^2 + y^2 - 1 = 0$ in the xy-plane. The Hough transform used for circle detection maps these points to conic surfaces in $\chi\psi\rho$-space as follows

$$h : (-1, 0) \rightarrow (\chi + 1)^2 + \psi^2 - \rho^2 = 0$$
$$h : (1, 0) \rightarrow (\chi - 1)^2 + \psi^2 - \rho^2 = 0.$$

These two conic surfaces intersect the point $(0, 0, 1)$ in $\chi\psi\rho$ space (see Figure 9.7.2).

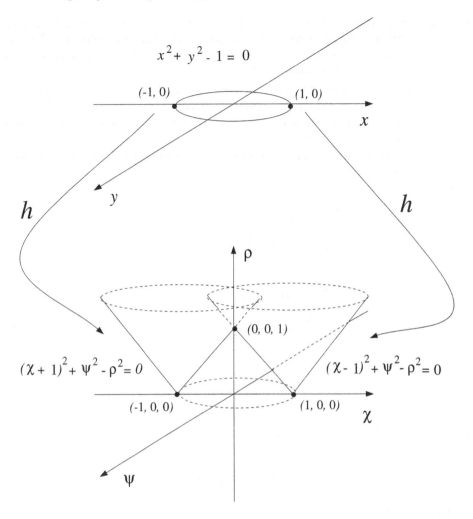

Figure 9.7.2. The two conics corresponding to the two points $(-1, 0)$ and $(1, 0)$ on the circle $x^2 + y^2 = 1$ under the Hough transform h.

The feature point (x_i, y_i) will lie on the circle (χ_0, ψ_0, ρ_0) in the xy-plane if and only if its image under the Hough transform intersects the point (χ_0, ψ_0, ρ_0) in $\chi\psi\rho$-space. More generally, the point $\mathbf{x}_i$ will lie on the curve $f(\mathbf{x}, \mathbf{p}_0) = 0$ in the domain of the image if and only if the curve $f(\mathbf{x}_i, \mathbf{p}) = 0$ intersects the point $\mathbf{p}_0$ in the parameter space. Therefore, the number of feature points in the domain of the image that lie on the curve $f(\mathbf{x}, \mathbf{p}_0) = 0$ can be counted by counting the number of elements in the range of the Hough transform that intersect $\mathbf{p}_0$.

As in the case of line detection, the parameter space must be quantized. The accumulator matrix is the representation of the quantized parameter space. For circle detection the accumulator $\mathbf{a}$ will be a three-dimensional matrix with all entries initially set to 0. The entry $\mathbf{a}(\chi_r, \psi_s, \rho_t)$ is incremented by 1 for every feature point (x_i, y_i) in the domain of the image whose conic surface in $\chi\psi\rho$-space passes through (χ_r, ψ_s, ρ_t). More precisely, $\mathbf{a}(\chi_r, \psi_s, \rho_t)$ is incremented provided

$$\left| (\chi_r - x_i)^2 + (\psi_s - y_i)^2 - \rho_t \right| < \epsilon,$$

where ϵ is used to compensate for digitization and quantization. Shapiro [19, 20, 21] discusses error analysis when using the Hough transform. If the above inequality holds, it implies that the conic surface $(\chi - x_i)^2 + (\psi - y_i)^2 - \rho = 0$ passes through the point (χ_r, ψ_s, ρ_t) (within a margin of error) in $\chi\psi\rho$ space. This means the point (x_i, y_i) lies on the circle $(x - \chi_r)^2 + (y - \psi_s)^2 - \rho_t = 0$ in the xy-plane, and thus the accumulator value $\mathbf{a}(\chi_r, \psi_s, \rho_t)$ should be incremented by 1.

Circles are among the most commonly found objects in images. Circles are special cases of ellipses, and circles viewed at an angle appear as ellipses. We extend circle detection to ellipse detection next. The method for ellipse detection that we will be discussing is taken from Ballard [22].

The equation for an ellipse centered at (χ, ψ) with axes parallel to the coordinate axes (Figure 9.7.3) is

$$\frac{(x - \chi)^2}{\alpha^2} + \frac{(y - \psi)^2}{\beta^2} = 1, \text{or}$$

$$f((x, y), (\chi, \psi, \alpha, \beta)) = \frac{(x - \chi)^2}{\alpha^2} + \frac{(y - \psi)^2}{\beta^2} - 1 = 0.$$

Differentiating with respect to x, the equation becomes

$$\frac{2(x - \chi)}{\alpha^2} + \frac{2(y - \psi)}{\beta^2}\frac{dy}{dx} = 0,$$

which gives us

$$(x - \chi)^2 = \left(\frac{\alpha^2}{\beta^2}\frac{dy}{dx}\right)^2 (y - \psi)^2.$$

Substituting into the original equation for the ellipse yields

$$\frac{(y - \psi)^2}{\beta^2}\left(1 + \frac{\alpha^2}{\beta^2}\left(\frac{dy}{dx}\right)^2\right) = 1.$$

Solving for ψ we get

$$\psi = y \pm \frac{\beta^2}{\sqrt{\left(1 + \frac{\alpha^2\left(\frac{dy}{dx}\right)^2}{\beta^2}\right)}}.$$

It then follows by substituting for ψ in the original equation for the ellipse that

$$\chi = x \pm \frac{\alpha^2}{\sqrt{\left(1 + \frac{\beta^2}{\alpha^2\left(\frac{dy}{dx}\right)^2}\right)}}.$$

For ellipse detection we will assume that the original image has been preprocessed by a direction edge detector and thresholded based on edge magnitude (Chapters 3 and 4). Therefore, we assume that an edge direction image $\mathbf{d} \in [0, 2\pi)^{\mathbf{X}}$ exist, where $\mathbf{X}$ is the domain of the original image. The direction $\mathbf{d}(x, y)$ is the direction of the gradient at the point (x, y) on the ellipse. The tangent to the ellipse at (x, y) is $\frac{dy}{dx}|_{(x,y)}$. Since the gradient is perpendicular to the tangent, the following holds:

$$\frac{dy}{dx}|_{(x,y)} = tan\left(\mathbf{d}(x, y) - \frac{\pi}{2}\right).$$

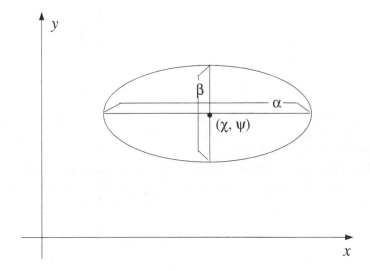

Figure 9.7.3. Parameters of an ellipse.

Recall that so far we have only been considering the equation for an ellipse whose axes are parallel to the axes of the coordinate system. Different orientations of the ellipse corresponding to rotations of an angle θ about (χ, ψ) can be handled by adding a fifth parameter θ to the descriptors of an ellipse. This rotation factor manifests itself in the expression for $\frac{dy}{dx}$, which becomes

$$\frac{dy}{dx}|_{(x,y)} = tan\left(\mathbf{d}(x,y) - \frac{\pi}{2} - \theta\right).$$

With this edge direction and orientation information we can write χ and ψ as

$$\chi = x \pm \frac{\alpha^2}{\sqrt{\left(1 + \frac{\beta^2}{\alpha^2\left(tan\left(\mathbf{d}(x,y) - \frac{\pi}{2} - \theta\right)\right)^2}\right)}}$$

and

$$\psi = y \pm \frac{\beta^2}{\sqrt{\left(1 + \frac{\alpha^2\left(tan\left(\mathbf{d}(x,y) - \frac{\pi}{2} - \theta\right)\right)^2}{\beta^2}\right)}},$$

respectively.

The accumulator array for ellipse detection will be a five-dimensional array $\mathbf{a}$. Every entry of $\mathbf{a}$ is initially set to zero. For every feature point (x, y) of the edge direction image, the accumulator cell $\mathbf{a}(\chi_r, \psi_s, \theta_t, \alpha_u, \beta_v)$ is incremented by 1 whenever

$$\left|\chi_r - \left(x \pm \frac{\alpha_u^2}{\sqrt{\left(1 + \frac{\beta_v^2}{\alpha_u^2\left(tan\left(\mathbf{d}(x,y) - \frac{\pi}{2} - \theta_t\right)\right)^2}\right)}}\right)\right| < \epsilon_1$$

and

$$\left|\psi_s - \left(y \pm \frac{\beta_v^2}{\sqrt{\left(1 + \frac{\alpha_u^2 \left(tan\left(\mathbf{d}(x,y) - \frac{\pi}{2} - \theta_t\right)\right)^2}{\beta_v^2}\right)}}\right)\right| < \epsilon_2.$$

Larger accumulator entry values are assumed to correspond to better ellipses. If an accumulator entry is judged large enough, its coordinates are deemed to be the parameters of an ellipse in the original image.

It is important to note that gradient information is used in the preceding description of an ellipse. As a consequence, gradient information is used in determining whether a point lies on an ellipse. Gradient information shows up as the term

$$\frac{dy}{dx}\Big|_{(x,y)} = tan\left(\mathbf{d}(x,y) - \frac{\pi}{2} - \theta\right)$$

in the equations that were derived above. The incorporation of gradient information improves the accuracy and computational efficiency of the algorithm. Our original example of circle detection did not use gradient information. However, circles are special cases of ellipses and circle detection using gradient information follows immediately from the description of the ellipse detection algorithm.

Image Algebra Formulation

The input image $\mathbf{b} = (\mathbf{c}, \mathbf{d})$ for the Hough algorithm is the result of preprocessing the original image by a directional edge detector and thresholding based on edge magnitude. The image $\mathbf{c} \in \{0, 1\}^{\mathbf{X}}$ is defined by

$$\mathbf{c}(x, y) = \begin{cases} 1 & \text{if edge magnitude at } (x, y) \\ & \quad \text{exceeds threshold} \\ 0 & \text{otherwise.} \end{cases}$$

The image $\mathbf{d} \in [0, 2\pi)^{\mathbf{X}}$ contains edge direction information.

Let $\mathbf{a} \in \mathbb{Z}^{\mathbf{Y}}$ be the accumulator image, where

$$\mathbf{Y} = \{(\chi_r, \psi_s, \theta_t, \alpha_u, \beta_v) : 0 \leq r < R, 0 \leq s < S, 0 \leq t < T,$$
$$0 \leq u < U, \text{and } 0 \leq v < V\}.$$

Let $C(x, y, \chi, \psi, \theta, \alpha, \beta, \epsilon_1, \epsilon_2)$ denote the condition

$$\left|\chi - \left(x \pm \frac{\alpha^2}{\sqrt{\left(1 + \frac{\beta^2}{\alpha^2 \left(tan\left(\mathbf{d}(x,y) - \frac{\pi}{2} - \theta\right)\right)^2}\right)}}\right)\right| < \epsilon_1 \text{ and}$$

$$\left|\psi - \left(y \pm \frac{\beta_v^2}{\sqrt{\left(1 + \frac{\alpha^2 \left(tan\left(\mathbf{d}(x,y) - \frac{\pi}{2} - \theta\right)\right)^2}{\beta^2}\right)}}\right)\right| < \epsilon_2.$$

Define the parameterized template $\mathbf{t}$ by

$$\mathbf{t}(\mathbf{b})_{(x,y)}(\chi,\psi,\theta,\alpha,\beta) = \begin{cases} 1 & \text{if condition } C(x,y,\chi,\psi,\theta,\alpha,\beta,\epsilon_1,\epsilon_2) \text{ is satisfied} \\ & \text{and } \mathbf{c}(x,y) = 1 \\ 0 & \text{otherwise.} \end{cases}$$

The accumulator array is constructed using the image algebra expression

$$\mathbf{a} := \sum \mathbf{t}(\mathbf{b}).$$

Similar to the implementation of the Hough transform for line detection, efficient incrementing of accumulator cells can be obtained by defining the neighborhood function $N : \mathbf{Y} \to 2^{\mathbf{X}}$ by

$$N(\chi_r,\psi_s,\theta_t,\alpha_u,\beta_v) = \left\{ (x,y) : \chi_r = \left[x \pm \frac{\alpha_u^2}{\sqrt{\left(1 + \frac{\beta_v^2}{\alpha_u^2\left(tan\left(\mathbf{d}(x,y)-\frac{\pi}{2}-\theta_t\right)\right)^2}\right)}} \right] \right.$$

$$\left. \text{and } \psi_s = \left[y \pm \frac{\beta_v^2}{\sqrt{\left(1 + \frac{\alpha_u^2\left(tan\left(\mathbf{d}(x,y)-\frac{\pi}{2}-\theta_t\right)\right)^2}{\beta_v^2}\right)}} \right] \right\}.$$

The accumulator array can now be computed by using the following image algebra pseudocode:

$$\mathbf{Z} := domain(\mathbf{c}\|_\mathbf{1})$$
$$\mathbf{b} := \mathbf{b}|_\mathbf{Z}$$
$$\mathbf{a} := \mathbf{b} \oplus N.$$

9.8.   Generalized Hough Algorithm for Shape Detection

In this section we show how the Hough algorithm can be generalized to detect non-analytic shapes. A non-analytic shape is one that does not have a representation of the form $f(\mathbf{x},\mathbf{p}) = 0$ (see Section 9.7). The lack of an analytic representation for the shape is compensated for by the use of a lookup table. This lookup table is often referred to as the shape's *R-table*. We begin our discussion by showing how to construct an R-table for an arbitrary shape. Next we show how the R-table is used in the generalized Hough algorithm. The origin of this technique can be found in Ballard [22].

Let S be an arbitrary shape in the domain of the image (see Figure 9.8.1). Choose any point (χ,ψ) off the boundary of S to serve as a reference point for the shape. This reference point will serve to parameterize the shape $S(\chi,\psi)$. For any point on the boundary of $S(\chi,\psi)$ let $\varphi(x,y)$ be the angle of the gradient at (x,y) relative to the xy coordinate axis. The vector $\vec{\mathbf{r}}(x,y)$ from (x,y) to (χ,ψ) can be expressed as a magnitude-direction pair $(r(x,y),\alpha(x,y))$, where $r(x,y)$ is the distance from (x,y) to (χ,ψ) and $\alpha(x,y)$ is the angle $\vec{\mathbf{r}}(x,y)$ makes with the positive x axis.

The R-table is indexed by values of gradient angles φ_i for points on the boundary of $S(\chi,\psi)$. For each gradient angle in the R-table, there corresponds a set $R(\varphi_i)$ of (r,α) magnitude-angle pairs. The pair (r,α) is an element of $R(\varphi_i)$ if and only if there exists a

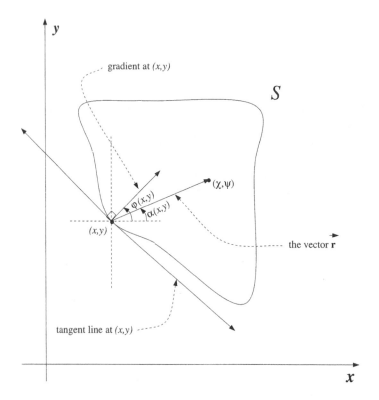

Figure 9.8.1. Non-analytic shape representation.

Table 9.8.1 The Structure of an R-Table

φ	$R(\varphi_i)$
φ_1	$\left\{\left(r_1^1, \alpha_1^1\right), \left(r_2^1, \alpha_2^1\right), \ldots, \left(r_{N_1}^1, \alpha_{N_1}^1\right)\right\}$
$\vdots$	$\vdots$
φ_i	$\left\{\left(r_1^i, \alpha_1^i\right), \left(r_2^i, \alpha_2^i\right), \ldots, \left(r_{N_i}^i, \alpha_{N_i}^i\right)\right\}$
$\vdots$	$\vdots$
φ_K	$\left\{\left(r_1^K, \alpha_1^K\right), \left(r_2^K, \alpha_2^K\right), \ldots, \left(r_{N_K}^K, \alpha_{N_K}^K\right)\right\}$

point (x, y) on the boundary of $S(\chi, \psi)$ whose gradient direction is φ_i and whose vector $\vec{\mathbf{r}}(x, y)$ has magnitude r and direction α. Thus,

$$R(\varphi_i) = \{(r, \alpha) : \exists (x, y) \in \partial S \ \ s.t. \, \varphi(x, y) = \varphi_i, \ \alpha(x, y) = \alpha, \text{ and } r(x, y) = r\},$$

which we will shorten for notational ease to

$$R(\varphi_i) = \left\{\left(r_j^i, \alpha_j^i\right) : 1 \leq j \leq N_i\right\},$$

where $N_i = card(R(\varphi_i))$. Table 9.8.1 illustrates the structure of an R-table. It provides a representation of the non-analytic shape $S(\chi, \psi)$ that can be used in the generalized Hough algorithm.

If $(x, y) \in \partial S(\chi, \psi)$ there is a φ_i, $1 \leq i \leq K$, in the R-table such that $\varphi_i = \varphi(x, y)$. For that φ_i there is a magnitude-angle pair (r_j^i, α_j^i) in $R(\varphi_i)$ such that

$$\chi = x + r_j^i \, cos(\alpha_j^i)$$

and

$$\psi = y + r_j^i \, sin(\alpha_j^i).$$

An important observation to make about the R-table is that by indexing the gradient direction, gradient information is incorporated into the shape's description. As in the case of ellipse detection (Section 9.7), this will make the Hough algorithm more accurate and more computationally efficient.

As it stands now, the shape $S(\chi, \psi)$ is parameterized only by the location of its reference point. The R-table for $S(\chi, \psi)$ can only be used to detect shapes that are the result of translations of $S(\chi, \psi)$ in the xy-plane.

A scaling factor can be accommodated by noting that a scaling of $S(\chi, \psi)$ by s is represented by scaling all the vectors in each of the $R(\varphi_i)$, $1 \leq i \leq K$, by s. If $R(\varphi_i) = \{(r_j^i, \alpha_j^i) : 1 \leq j \leq N_i\}$ is the set of vectors that correspond to φ_i for the original shape, then $R(\varphi_i, s) = \{(sr_j^i, \alpha_j^i) : 1 \leq j \leq N_i\}$ correspond to φ_i for the scaled shape. Thus, information from the R-table is easily adapted to take care of different scaling parameters.

A rotation of the shape through an angle θ is also easily represented through adaptation of R-table information. Let $R(\varphi_i, \theta)$ be the set of vectors that correspond to φ_i in an R-table for the rotated shape. The set $R(\varphi_i, \theta)$ is equal to the set of vectors in $R((\varphi_i - \theta) mod \, 2\pi)$ of the original shape rotated through the angle $\theta \, mod \, 2\pi$. More precisely, if $\varphi_j = (\varphi_i - \theta) mod \, 2\pi$ and $R(\varphi_j) = \{(r_k^j, \alpha_k^j) : 1 \leq k \leq N_j\}$, then $R(\varphi_i, \theta) = \{(r_k^j, (\alpha_k^j - \theta) mod \, 2\pi) : 1 \leq k \leq N_j\}$. Therefore, information from an R-table representing the shape $S(\chi, \psi)$ parameterized only by the location of its reference point can be adapted to represent a more general shape $S(\chi, \psi, s, \varphi)$ parameterized by reference point, scale, and orientation.

Having shown how an R-table is constructed, we can discuss how it is used in the generalized Hough algorithm. The input for the generalized Hough algorithm is an edge image with gradient information that has been thresholded based on edge magnitude. Let $\mathbf{d}(x, y)$ denote the direction of the gradient at edge point (x, y).

The entries of the accumulator are indexed by a quantization of the shape's $\chi\psi s\theta$ parameter space. Thus, the accumulator is a four-dimensional array with entries

$$\mathbf{a}(\chi_t, \psi_u, s_v, \theta_w), \text{where } 1 \leq t \leq T, 1 \leq u \leq U,$$
$$1 \leq v \leq V, \text{ and } 1 \leq w \leq W.$$

Entry $\mathbf{a}(\chi_s, \psi_t, s_u, \theta_w)$ is the counting bin for points in the xy-plane that lie on the boundary of the shape $S(\chi_s, \psi_t, s_u, \theta_w)$. Initially all the entries of the accumulator are 0.

For each feature edge pixel (x, y) add 1 to accumulator entry $\mathbf{a}(\chi_s, \psi_t, s_u, \theta_w)$ if the following condition, denoted by $C(x, y, \chi_s, \psi_t, s_u, \theta_w, \epsilon_1, \epsilon_2, \epsilon_3)$, holds. For positive numbers ϵ_1, ϵ_2, and ϵ_3, $C(x, y, \chi_s, \psi_t, s_u, \theta_w, \epsilon_1, \epsilon_2, \epsilon_3)$ is satisfied if there exists a $1 \leq k \leq N_j$ such that

$$|\chi_s - (x + s_u r_k^j cos((\alpha_k^j - \theta_w) mod \, 2\pi))| < \epsilon_1$$

and

$$|\psi_t - (y + s_u r_k^j sin((\alpha_k^j - \theta_w) mod \, 2\pi))| < \epsilon_2,$$

when $|(\mathbf{d}(x, y) - \theta_w) \bmod 2\pi - \varphi_j| < \epsilon_3$. The ϵ_1, ϵ_2, and ϵ_3 are error tolerances to allow for quantization and digitization.

Larger counts at an accumulator cell mean a higher probability that the indexes of the accumulator cell are the parameters of a shape in the image.

Image Algebra Formulation

The input image $\mathbf{b} = (\mathbf{c}, \mathbf{d})$ for the generalized Hough algorithm is the result of preprocessing the original image by a directional edge detector and thresholding based on edge magnitude. The image $\mathbf{c} \in \{0, 1\}^{\mathbf{X}}$ is defined by

$$\mathbf{c}(x, y) = \begin{cases} 1 & \text{if edge magnitude at } (x, y) \\ & \quad \text{exceeds threshold} \\ 0 & \text{otherwise.} \end{cases}$$

The image $\mathbf{d} \in [0, 2\pi)^{\mathbf{X}}$ contains edge gradient direction information. The accumulator $\mathbf{a}$ is defined over the quantized parameter space of the shape

$$\mathbf{Y} = \{(\chi_t, \psi_u, s_{v, \theta_w}) : 1 \leq t \leq T, 1 \leq u \leq U, 1 \leq v \leq U, 1 \leq w \leq W\}.$$

Define the template $\mathbf{t}$ by

$$\mathbf{t(b)}_{(x,y)}(\chi, \psi, s, \theta) = \begin{cases} 1 & \text{if } \mathbf{c}(x, y) = 1 \\ & \quad \text{and } C(x, y, \chi, \psi, s, \theta, \epsilon_1, \epsilon_2, \epsilon_3) \text{ is satisfied} \\ 0 & \text{otherwise.} \end{cases}$$

The set $R(\varphi_j)$ is the set from the R-table that corresponds to the gradient angle φ_j. The accumulator is constructed with the image algebra statement

$$\mathbf{a} := \sum \mathbf{t(b)}.$$

Again, in actual implementation, it will be more efficient to construct a neighborhood function $N : \mathbf{Y} \rightarrow 2^{\mathbf{X}}$ satisfying the above parameters in a similar fashion as the construction of the neighborhood function in the previous section (Section 9.7) and then using the three-line algorithm:

$$\mathbf{Z} := domain(\mathbf{c}\|_1)$$
$$\mathbf{b} := \mathbf{b}|_{\mathbf{Z}}$$
$$\mathbf{a} := \mathbf{b} \oplus N.$$

9.9.   Exercises

1.   Discuss similarities and differences between correlation and convolution. Write a relationship between them in image algebra (using some template $\mathbf{t}$).

2.   Write an image algebra expression of the autocorrelation of an image $\mathbf{a} \in \mathbb{R}^{\mathbf{X}}$ with itself. Then prove that the autocorrelation function is always even.

3.   Write an image algebra expression of the power spectral density function (power spectrum), which is defined as the Fourier transform of the autocorrelation function.

4.   Sections 9.4 and 9.5 present pattern matching methods which are invariant with respect to rotation and both rotation and scale, respectively. Formulate an image algebra method for scale-only invariant pattern matching.

5.   Sonar can determine the distance from a sound source by means of correlation. The system has two transducers separated by a distance. The transducers receive the signals $\mathbf{a}_t$ and the time-delayed $a\mathbf{a}_{t+\triangle t}$, respectively, where a is an attenuation factor which can be assumed to be 1 and $\triangle t$ is the time lag due to the different paths from the source to the transducers. Even in the presence of noise, the correlation of $\mathbf{a}_t$ and $a\mathbf{a}_{t+\triangle t}$ will have a peak at a distance from the origin corresponding to the delay $\triangle t$. Devise an image algebra scheme to determine the distance from a signal source, making any necessary assumptions.

6.   Suppose we have a fixed camera that captures images of a road at given time intervals. Devise an image algebra scheme that uses pattern matching and correlation (as in Exercise 4) to determine the speed of a vehicle in motion on the road.

7.   Find the Hough transform of lines enclosing an object with vertices $\mathbf{x}_1 = (10,0)$, $\mathbf{x}_2 = (10,10)$, and $\mathbf{x}_3 = (0,10)$. Sketch the modified object enclosed by lines obtained by replacing (ρ,θ) of the object lines by $(\rho^2,\theta + 90^\circ)$. Calculate the area of the modified object.

8.   Implement a Hough transform that will detect an ellipse.

9.   Develop a Hough transform that detects an ellipse in its rotated and scaled versions.

10. Generalize the two-dimensional Hough transform for line detection to a three-dimensional Hough transform that detects planes.

9.10. References

[1] R. Gonzalez and P. Wintz, *Digital Image Processing*. Reading, MA: Addison-Wesley, second ed., 1987.

[2] D. Ballard and C. Brown, *Computer Vision*. Englewood Cliffs, NJ: Prentice-Hall, 1982.

[3] W. Rudin, *Real and Complex Analysis*. New York, NY: McGraw-Hill, 1974.

[4] A. Oppenheim and J. Lim, "The importance of phase in signals," *IEEE Proceedings*, vol. 69, pp. 529–541, 1981.

[5] Q. Chen, M. Defrise, and F. Deconick, "Symmetric phase-only matched filtering of Fourier-Mellin transforms for image registration and recognition," *IEEE Transactions on Pattern Analysis and Machine Intelligence*, vol. 16, pp. 1156–1168, July 1994.

[6] V. K. P. Kumar and Z. Bahri, "Efficient algorithm for designing a ternary valued filter yielding maximum signal to noise ratio," *Applied Optics*, vol. 28, no. 10, 1989.

[7] O. K. Ersoy and M. Zeng, "Nonlinear matched filtering," *Journal of the Optical Society of America*, vol. 6, no. 5, pp. 636–648, 1989.

[8] J. Horver and P. Gianino, "Pattern recognition with binary phase-only filters," *Applied Optics*, vol. 24, pp. 609–611, 1985.

[9] D. Flannery, J. Loomis, and M. Milkovich, "Transform-ratio ternary phase-amplitude filter formulation for improved correlation discrimination," *Applied Optics*, vol. 27, no. 19, pp. 4079–4083, 1988.

[10] D. Casasent and D. Psaltis, "Position, rotation and scale-invariant optical correlation," *Applied Optics*, vol. 15, pp. 1793–1799, 1976.

[11] D. Casasent and D. Psaltis, "New optical transform for pattern recognition," *Proceedings of the IEEE*, vol. 65, pp. 77–84, 1977.

[12] Y. Hsu, H. Arsenault, and G. April, "Rotation-invariant digital pattern recognition using circular harmonic expansion," *Applied Optics*, vol. 21, no. 22, pp. 4012–4015, 1982.

[13] Y. Sheng and H. Arsenault, "Experiments on pattern recognition using invariant Fourier-Mellin descriptors," *Journal of the Optical Society of America*, vol. 3, no. 6, pp. 771–776, 1986.

[14] E. D. Castro and C. Morandi, "Registration of translated and rotated images using finite Fourier transforms," *IEEE Transactions on Pattern Analysis and Machine Intelligence*, vol. 9, no. 5, pp. 700–703, 1987.

[15] P. Hough, "Method and means for recognizing complex patterns," Dec. 18, 1962. U.S. Patent 3,069,654.

[16] R. Duda and P. Hart, "Use of the Hough transform to detect lines and curves in pictures," *Communications of the ACM*, vol. 15, no. 1, pp. 11–15, 1972.

[17] E. R. Davies, *Machine Vision: Theory, Algorithms, Practicalities*. New York: Academic Press, 1990.

[18] A. Rosenfeld, *Picture Processing by Computer*. New York: Academic Press, 1969.

[19] S. D. Shapiro, "Transformations for the noisy computer detection of curves in noisy pictures," *Computer Graphics and Image Processing*, vol. 4, pp. 328–338, 1975.

[20] S. D. Shapiro, "Properties of transforms for the detection of curves in noisy image," *Computer Graphics and Image Processing*, vol. 8, pp. 219–236, 1978.

[21] S. D. Shapiro, "Feature space transforms for curve detection," *Pattern Recognition*, vol. 10, pp. 129–143, 1978.

[22] D. Ballard, "Generalizing the Hough transform to detect arbitrary shapes," *Pattern Recognition*, vol. 13, no. 2, pp. 111–122, 1981.

CHAPTER 10
IMAGE FEATURES AND DESCRIPTORS

10.1. Introduction

The purpose of this chapter is to provide examples of image feature and descriptor extraction in an image algebra setting. Image features are useful extractable attributes of images or regions within an image. Examples of image features are the histogram of pixel values and the symmetry of a region of interest. Histograms of pixel values are considered a primitive characteristic, or low-level feature, while such geometric descriptors as symmetry or orientation of regions of interest provide examples of high-level features. Significant computational effort may be required to extract high-level features from a raw gray-value image.

Generally, image features provide ways of describing image properties, or properties of image components, that are derived from a number of different information domains. Some features, such as histograms and texture features, are statistical in nature, while others, such as the Euler number or boundary descriptors of regions, fall in the geometric domain. Geometric features are key in providing structural descriptors of images. Structural descriptions of images are of significant interest. They are an important component of high-level image analysis and may also provide storage-efficient representations of image data. The objective of high-level image analysis is to interpret image information. This involves relating image gray levels to a variety of non-image-related knowledge underlying scene objects and reasoning about contents and structures of complex scenes in images. Typically, high-level image analysis starts with structural descriptions of images, such as a coded representation of an image object or the relationships between objects in an image. The descriptive terms "inside of" and "adjacent to" denote relations between certain objects. Representation and manipulation of these kinds of relations is fundamental to image analysis. Here one assumes that a preprocessing algorithm has produced a regionally segmented image whose regions represent objects in the image, and considers extracting the relation involving the image regions from the regionally segmented image.

10.2. Area and Perimeter

Area and perimeter are commonly used descriptors for regions in the plane. In this section image algebra formulations for the calculation of area and perimeter are presented.

Image Algebra Formulation

Let $\mathbf{R}$ denote the region in $\mathbf{a} \in \{0,1\}^{Z_m \times Z_n}$ whose points have pixel value 1. One way to calculate the area A of $\mathbf{R}$ is simply to count the number of points in $\mathbf{R}$. This can be accomplished with the image algebra statement

$$A := \sum \mathbf{a}.$$

The perimeter P of $\mathbf{R}$ may be calculated by totaling the number of 0-valued 4-neighbors for each point in R. The image algebra statement for the perimeter of R using this definition is given by

$$P := 2 \cdot \sum (\chi_1(\mathbf{a} \oplus \mathbf{t}_1) + \chi_1(\mathbf{a} \oplus \mathbf{t}_2)),$$

where the templates t_1 and t_2 are defined pictorially as

$$t_1 = \boxed{\begin{array}{c|c} 2 & 1 \end{array}} \qquad t_2 = \boxed{\begin{array}{c} 2 \\ \hline 1 \end{array}}$$

Alternate Image Algebra Formulation

The formulas for area and perimeter that follow have been adapted from Pratt [1]. They provide more accurate measures for images of continuous objects that have been digitized. Both the formula for area and perimeter use the template t defined below.

$$t = \boxed{\begin{array}{c|c} 3 & 1 \\ \hline 1 & 3 \end{array}}$$

Let $b := a \oplus t$. The alternate formulations for area and perimeter are given by

$$A := \sum \{ \frac{1}{4} \cdot [\chi_1(\mathbf{b}) + \chi_3(\mathbf{b})] + \frac{1}{2} \cdot \chi_4(\mathbf{b}) + \frac{7}{8} \cdot [\chi_5(\mathbf{b}) + \chi_7(\mathbf{b})]$$
$$+ \chi_8(\mathbf{b}) + \frac{3}{4} \cdot [\chi_2(\mathbf{b}) + \chi_6(\mathbf{b})] \}$$

and

$$P := \sum \{ \chi_4(\mathbf{b}) + \frac{1}{\sqrt{2}} [\chi_1(\mathbf{b}) + \chi_3(\mathbf{b}) + \chi_5(\mathbf{b}) + \chi_7(\mathbf{b})$$
$$+ 2(\chi_2(\mathbf{b}) + \chi_6(\mathbf{b}))] \},$$

respectively.

10.3.   Euler Number

The Euler number is a topological descriptor for binary images. It is defined to be the number of connected components minus the number of holes inside the connected components [2].

There are two Euler numbers for a binary image $\mathbf{a}$, which we denote $e_4(\mathbf{a})$ and $e_8(\mathbf{a})$. Each is distinguished by the connectivity used to determine the number of feature pixel components and the connectivity used to determine the number of non-feature pixel holes contained within the feature pixel connected components. The Euler number of a binary image is referenced by the connectivity used to determine the number of feature pixel components.

The 4-connected Euler number, $e_4(\mathbf{a})$, is defined to be the number of 4-connected feature pixel components minus the number of 8-connected holes, that is

$$e_4(\mathbf{a}) = c_4(\mathbf{a}) - h_8(\mathbf{a}).$$

Here, $c_4(\mathbf{a})$ denotes the number of 4-connected feature components of $\mathbf{a}$ and $h_8(\mathbf{a})$ is the number of 8-connected holes within the feature components.

The 8-connected Euler number, $e_8(\mathbf{a})$, of $\mathbf{a}$ is defined by

$$e_8(\mathbf{a}) = c_8(\mathbf{a}) - h_4(\mathbf{a}),$$

where $c_8(\mathbf{a})$ denotes the number of 8-connected feature components of $\mathbf{a}$ and $h_4(\mathbf{a})$ is the number of 4-connected holes within the feature components.

Table 10.3.1 shows Euler numbers for some simple pixel configuration. Feature pixels are black.

Table 10.3.1 Examples of Pixel Configurations and Euler Numbers

No.	$\mathbf{a}$	$c_4(\mathbf{a})$	$c_8(\mathbf{a})$	$h_4(\mathbf{a})$	$h_8(\mathbf{a})$	$e_4(\mathbf{a})$	$e_8(\mathbf{a})$
1.		1	1	0	0	1	1
2.		5	1	0	0	5	1
3.		1	1	1	1	0	0
4.		4	1	1	0	4	0
5.		2	1	4	1	1	-3
6.		1	1	5	1	0	-4
7.		2	2	1	1	1	1

Image Algebra Formulation

Let $\mathbf{a} \in \{0,1\}^{\mathbf{X}}$ be a Boolean input image and let $\mathbf{b} := \mathbf{a} \oplus \mathbf{t}$, where $\mathbf{t} \in \left((\mathbb{Z}^{+})^{\mathbf{Z}^{2}} \right)^{\mathbf{Z}^{2}}$ is the template represented in Figure 10.3.1. The 4-connected Euler number is expressed as

$$e_4(\mathbf{a}) := \frac{1}{4} \sum [\chi_1(\mathbf{b}) + \chi_3(\mathbf{b}) - \chi_5(\mathbf{b}) - \chi_7(\mathbf{b}) + 2\chi_2(\mathbf{b}) + 2\chi_6(\mathbf{b})],$$

and the 8-connected Euler number is given by

$$e_8(\mathbf{a}) := \frac{1}{4} \sum [\chi_1(\mathbf{b}) + \chi_3(\mathbf{b}) - \chi_5(\mathbf{b}) - \chi_7(\mathbf{b}) - 2\chi_2(\mathbf{b}) - 2\chi_6(\mathbf{b})].$$

$$t = \begin{array}{|c|c|} \hline 3 & 1 \\ \hline 1 & 3 \\ \hline \end{array}$$

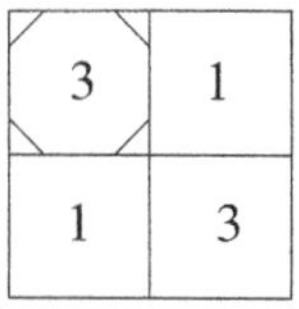

Figure 10.3.1. Pictorial representation of the
template used for determining the Euler number.

The image algebra expressions of the two types of Euler numbers were derived from the quad pattern formulations given in [1, 3].

Alternate Image Algebra Formulation

The formulation for the Euler numbers could also be expressed using lookup tables. Lookup tables may be more efficient. The lookup table for the 4-connected Euler number is

$$f(r) = \begin{cases} 1 & \text{if } r = 1 \text{ or } r = 3 \\ -1 & \text{if } r = 5 \text{ or } r = 7 \\ 2 & \text{if } r = 2 \text{ or } r = 6 \\ 0 & \text{otherwise}, \end{cases}$$

and the lookup table for the 8-connected Euler number is

$$g(\mathrm{r}) = \begin{cases} 1 & \text{if } r = 1 \text{ or } r = 3 \\ -1 & \text{if } r = 5 \text{ or } r = 7 \\ -2 & \text{if } r = 2 \text{ or } r = 6 \\ 0 & \text{otherwise}. \end{cases}$$

The 4-connected and 8-connected Euler numbers can then be evaluated using lookup tables via the expressions

$$e_4(\mathbf{a}) := \frac{1}{4} \sum f(\mathbf{a} \oplus \mathbf{t}) \text{ and } e_8(\mathbf{a}) := \frac{1}{4} \sum g(\mathbf{a} \oplus \mathbf{t}),$$

respectively.

10.4.   Chain Code Extraction and Correlation

The *chain code* provides a storage-efficient representation for the boundary of an object in a Boolean image. The chain code representation incorporates such pertinent information as the length of the boundary of the encoded object, its area, and moments [4, 5]. Additionally, chain codes are invertible in that an object can be reconstructed from its chain code representation.

The basic idea behind the chain code is that each boundary pixel of an object has an adjacent boundary pixel neighbor whose direction from the given boundary pixel can be specified by a unique number between 0 and 7. Given a pixel, consider its eight neighboring pixels. Each 8-neighbor can be assigned a number from 0 to 7 representing one of eight possible directions from the given pixel (see Figure 10.4.1). This is done with the same orientation throughout the entire image.

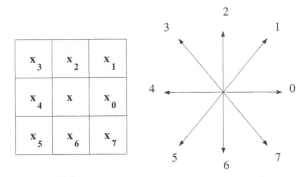

Figure 10.4.1. The 8-neighborhood of $\mathbf{x}$ and the associated eight directions.

The chain code for the boundary of a Boolean image is a sequence of integers, $\mathbf{c} = \{\mathbf{c}(0), \mathbf{c}(1), \ldots, \mathbf{c}(n-1)\}$, from the set $\{0, 1, \ldots, 7\}$; i.e., $\mathbf{c}(i) \in \mathbb{Z}_8$ for $i = 0, 1, \ldots, n-1$. The number of elements in the sequence $\mathbf{c}$ is called the length of the chain code. The elements $\mathbf{c}(0)$ and $\mathbf{c}(n-1)$ are called the *initial* and *terminal point* of the code, respectively. Starting at a given base point, the boundary of an object in a Boolean image can be traced out using the head-to-tail directions that the chain code provides. Figure 10.4.2 illustrates the process of tracing out the boundary of the SR71 by following direction vectors. The information of Figure 10.4.2 is then "flattened" to derive the chain code for its boundary. Suppose we choose the topmost left feature pixel of Figure 10.4.2 as the base point for the boundary encoding. The chain code for the boundary of the SR71 is the sequence

$$7, 6, 7, 7, 0, \ldots, 1, 1, 1.$$

Given the base point and the chain code, the boundary of the SR71 can be completely reconstructed. The chain code is an efficient way of storing boundary information because it requires only three bits ($2^3 = 8$) to determine any one of the eight directions.

The *chain code correlation function* provides a measure of similarity in shape and orientation between two objects via their chain codes. Let $\mathbf{c} \in (\mathbb{Z}_8)^{\mathbb{Z}_m}$ and $\mathbf{c}' \in (\mathbb{Z}_8)^{\mathbb{Z}_n}$ be two chains of length m and n, respectively, with $m \leq n$. The chain code correlation function $\rho_{\mathbf{c}\mathbf{c}'} : \mathbb{Z}_n \to [-1, 1]$ is defined by

$$\rho_{\mathbf{c}\,\mathbf{c}'}(j) = \frac{1}{m} \sum_{i=1}^{m} \cos\left[\frac{(\mathbf{c}(i) - \mathbf{c}'((i+j) \bmod n)) \cdot \pi}{4}\right].$$

The value of j at which $\rho_{\mathbf{c}\,\mathbf{c}'}$ takes on its maximum is the offset of the segment of $\mathbf{c}'$ that best matches $\mathbf{c}$. The closer the maximum is to 1 the better the match.

Image Algebra Formulation

Chain Code Extraction

In this section we present an image algebra formulation for extracting the chain code from an 8-connected object in a boolean image. This algorithm is capable of extracting the chain code from objects typically considered to be poor candidates for chain coding purposes. Specifically, this algorithm is capable of coping with *pinch points* and *feelers* (as shown in Figure 10.4.3).

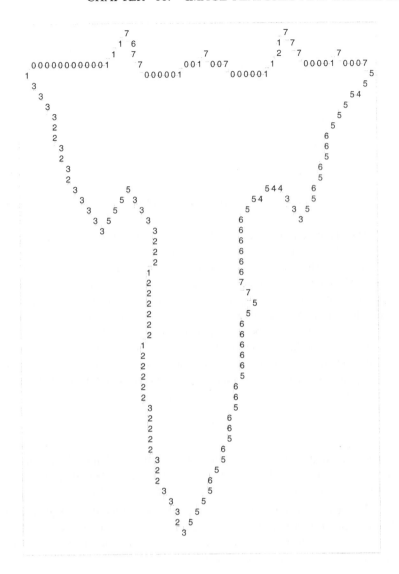

Figure 10.4.2. Chain code directions with associated direction numbers.

Let $\mathbf{a} \in \{0, 1\}^{\mathbf{X}}$ be the source image. The chain code extraction algorithm proceeds in three phases. The first phase uses the census template $\mathbf{t}$ shown in Figure 10.4.4 together with the linear image-template product operation $\bigoplus$ to assign each point $\mathbf{x} \in \mathbf{X}$ a number between 0 and 511. This number represents the configuration of the pixel values of $\mathbf{a}$ in the 8-neighborhood about $\mathbf{x}$. Neighborhood configuration information is stored in the image $\mathbf{n} \in (\mathbb{Z}_{512})^{\mathbf{X}}$, where

$$\mathbf{n} := \mathbf{a} \bigoplus \mathbf{t}.$$

This information will be used in the last phase of the algorithm to guide the selection of directions during the extraction of the chain code.

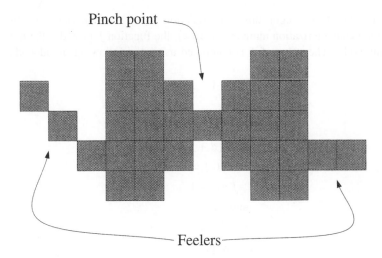

Figure 10.4.3. Boolean object with a pinch point and feelers.

$$\mathbf{t} = \begin{array}{|c|c|c|} \hline 8 & 4 & 2 \\ \hline 16 & 256 & 1 \\ \hline 32 & 64 & 128 \\ \hline \end{array}$$

Figure 10.4.4. Census template used for chain code extraction.

The next step of the algorithm extracts a starting point (the initial point) on the interior 8-boundary of the object and provides an initial direction. This is accomplished by the statement:

$$\mathbf{x}_0 := \bigwedge domain(\mathbf{a}\|_1).$$

Here we assume the lexicographical (row scanning) order. Hence, the starting point $\mathbf{x}_0$ is the lexicographical minimum of the point set $domain(\mathbf{a}\|_1)$. For a clockwise traversal of the object's boundary the starting direction d is initially set to 0.

There are three functions $b_k(i) : \mathbb{Z} \rightarrow \mathbb{Z}_2$, $\delta : \mathbb{Z}_8 \rightarrow \mathbb{Z}_{\pm 2}^2 \setminus (0,0)$, and $f : \mathbb{Z}_8 \times \mathbb{Z}_{512} \rightarrow \mathbb{Z}_8$ that are used in the final phase of the algorithm. The function $b_k(i)$ returns the kth bit in the binary representation of the integer i. For example, $b_2(6) = 1$, $b_1(6) = 1$, and $b_0(6) = 0$.

Given a direction, the function δ provides the increments required to move to the next point along the path of the chain code in $\mathbf{X}$. The values of δ are given by

$$\delta(0) = (0,1),$$
$$\delta(1) = (-1,1),$$
$$\delta(2) = (-1,0),$$
$$\delta(3) = (-1,-1),$$
$$\delta(4) = (0,-1),$$
$$\delta(5) = (1,-1),$$
$$\delta(6) = (1,0), \text{and}$$
$$\delta(7) = (1,1).$$

The key to the algorithm is the function f. Given a previous direction d and a neighborhood characterization number $c = \mathbf{n}(\mathbf{x})$, the function f provides the next direction in the chain code. The value of f is computed using the following pseudocode:

$$f := -1$$
$$i := 0$$
```
while f = -1 loop
```
$$f := (d + 2 - i) \bmod 8$$
```
    if b_f(c) = 0 then
```
$$f := -1$$
```
    end if
```
$$i := i + 1$$
```
end loop.
```

Note that f as defined above is designed for a clockwise traversal of the object's boundary. For a counterclockwise traversal, the line

$$f := (d + 2 - i) \bmod 8$$

should be replaced by

$$f := (d + 6 + i) \bmod 8.$$

Also, the starting direction d for the algorithm should be set to 5 for the counterclockwise traversal.

Let $\mathbf{a} \in \{0, 1\}^{\mathbf{X}}$ be the source image and let $\mathbf{c} \in (\mathbb{Z}_8)^{\mathbb{Z}_n}$ be the image variable used to store the chain code. Combining the three phases outlined above we arrive at the following image algebra pseudocode for chain code extraction from an 8-connected object contained in $\mathbf{a}$:

$$\mathbf{n} := \mathbf{a} \oplus \mathbf{t}$$
$$\mathbf{x} := \mathbf{x}_0$$
$$d := 0$$
$$d_0 := f(d, \mathbf{n}(\mathbf{x}_0))$$
$$i := 0$$
```
loop
```
$$d := f(d, \mathbf{n}(\mathbf{x}))$$
```
    if d = d_0 and x = x_0 then
        break
    end if
```
$$\mathbf{c}(i) := d$$
$$\mathbf{x} := \mathbf{x} + \delta(d)$$
$$i := i + 1$$
```
end loop.
```

In order to reconstruct the object from its chain code, the algorithm below can be used to generate the object's interior 8-boundary.

$$\mathbf{a} := 0$$

$$\mathbf{a}(\mathbf{x}_0) := 1$$

$$\mathbf{x} := \mathbf{x}_0 + \delta(\mathbf{c}(0))$$

for i in $1..n-1$ loop

$\quad \mathbf{a}(\mathbf{x}) := 1$

$\quad \mathbf{x} := \mathbf{x} + \delta(\mathbf{c}(i))$

end loop.

The region bounded by the 8–boundary can then be filled using one of the hole filling algorithms of Section 6.6.

Chain Code Correlation

Let $\mathbf{c} \in (\mathbb{Z}_8)^{\mathbb{Z}_m}$ and $\mathbf{c}' \in (\mathbb{Z}_8)^{\mathbb{Z}_n}$ be two chains where $m \leq n$. Let $\mathbf{c}'_j := \mathbf{c}' \circ f_j$ where $f_j(i) : \mathbb{Z} \to \mathbb{Z}_n$ is the function defined by $f_j(i) = (i+j) \bmod n$. The image $\mathbf{r} \in [-1, 1]^{\mathbb{Z}_n}$ that represents the chain code correlation between $\mathbf{c}$ and $\mathbf{c}'$ is given by

$$\mathbf{r}(j) := \frac{1}{m} \sum \cos\left[\frac{(\mathbf{c} - \mathbf{c}'_j) \cdot \pi}{4}\right].$$

Comments and Observations

The image algebra pseudocode algorithm presented here was developed by Robert H. Forsman at the Center for Computer Vision and Visualization of the University of Florida. To improve the performance of the chain code extraction algorithm, the function f used for chain code extraction should be implemented using a lookup table. Such an implementation is given in Ritter [6].

10.5.   Region Adjacency

A region, for the purposes of this section, is a subset of the domain of $\mathbf{a} \in \mathbb{Z}^{\mathbf{X}}$ whose points are all mapped to the same (or similar) pixel value by $\mathbf{a}$. Regions will be indexed by the pixel value of their members, that is, region $\mathbf{R}_i$ is the set $\{\mathbf{x} \in \mathbf{X} : \mathbf{a}(\mathbf{x}) = i\}$.

Adjacency of regions is an intuitive notion, which is formalized for $\mathbf{a} \in \mathbb{Z}^{\mathbf{X}}$ by the *adj* relation.

$$\mathbf{R}_i \; adj \; \mathbf{R}_j \Leftrightarrow \exists \mathbf{x}, \mathbf{y} \in \mathbf{X} \text{ with } d(\mathbf{x}, \mathbf{y}) = 1 \text{ s.t.} \begin{cases} 1. & \mathbf{a}(\mathbf{x}) = i \text{ and } \mathbf{a}(\mathbf{y}) = j \\ & or \\ 2. & \mathbf{a}(\mathbf{x}) = j \text{ and } \mathbf{a}(\mathbf{y}) = i. \end{cases}$$

Here d denotes the Euclidean distance between $\mathbf{x} = (x_1, x_2)$ and $\mathbf{y} = (y_1, y_2)$; i.e., $d(\mathbf{x}, \mathbf{y}) = \sqrt{(x_1 - y_1)^2 + (x_2 - y_2)^2}$. In other words, $\mathbf{R}_i$ and $\mathbf{R}_j$ are adjacent if and only if there is a point in $\mathbf{R}_i$ and a point in $\mathbf{R}_j$ that are a distance of one unit apart.

The adjacency relation, denoted by adj, can be represented by a set of ordered pairs, an image, or a graph. The ordered pair representation of the adj relation is the set

$$\{(i,j) : i,j \in \mathbb{Z} \text{ and } \mathbf{R}_i \ adj \ \mathbf{R}_j\}.$$

The adj relation is symmetric ($\mathbf{R}_i \ adj \ \mathbf{R}_j \Rightarrow \mathbf{R}_j \ adj \ \mathbf{R}_i$). Hence, the above set contains redundant information regarding the relation. This redundancy is eliminated by using the set $\{(i,j) : i > j \text{ and } \mathbf{R}_i \ adj \ \mathbf{R}_j\}$ to represent the adjacency relation.

The image $\mathbf{b} \in \{0,1\}^{\mathbf{Y}}$ defined by

$$\mathbf{b}(i,j) = \begin{cases} 1 & \text{if } \mathbf{R}_i \ adj \ \mathbf{R}_j \\ 0 & \text{otherwise}, \end{cases}$$

where $\mathbf{Y} = \{(i,j) : 2 \le i \le \vee \mathbf{a}, 1 \le j < i\}$ is the image representation of the adjacency relation among the regions in $\mathbf{X}$.

Region numbers form the vertices of the graph representation of the adj relation; edges are the pairs (i,j), where $\mathbf{R}_i \ adj \ \mathbf{R}_j$. Thus, there are three easy ways to represent the notion of region adjacency. As an example, we provide the three representations for the adjacency relation defined over the regions in Figure 10.5.1.

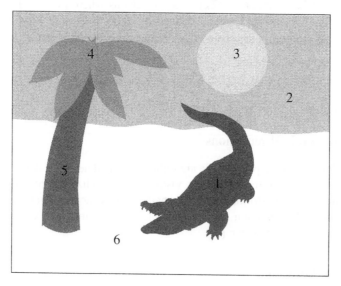

Figure 10.5.1. Albert the alligator (Region 1) and environs.

(a) Ordered pair representation —

$$adj = \{(2,1),(3,2),(4,2),(5,2)$$
$$(5,4),(6,1),(6,2),(6,5)\}$$

(b) Image representation —

$i \backslash j$	1	2	3	4	5	6
1						
2	1					
3	0	1				
4	0	1	0			
5	0	1	0	1		
6	1	1	0	0	1	

(c) Graph representation —

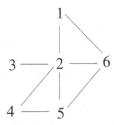

Image Algebra Formulation

Let $\mathbf{a} \in \mathbb{N}^{\mathbf{X}}$ be the input image, where $\mathbf{X}$ is an $m \times n$ array. In order for the adjacency relation to be meaningful, the image $\mathbf{a}$ in most applications will be the result of preprocessing an original image with some type of segmentation technique; e.g., component labeling (Sections 6.2 and 6.3).

The adjacency relation for regions in $\mathbf{X}$ will be represented by the image $\mathbf{b} \in \{0,1\}^{\mathbf{Y}}$, where $\mathbf{Y} = \{(i,j) : 2 \leq i \leq \vee \mathbf{a}, 1 \leq j < i\}$. The parameterized template $\mathbf{t}(\mathbf{a}) \in \left(\{0,1\}^{\mathbf{Y}}\right)^{\mathbf{X}}$ used to determine of adjacency is defined by

$$\mathbf{t}(\mathbf{a})_{(x,y)}(i,j) = \begin{cases} 1 & \text{if } (i,j) = (max(\mathbf{a}(x-1,y), \mathbf{a}(x,y)), min(\mathbf{a}(x-1,y), \mathbf{a}(x,y))) \\ & \text{or} \\ & (i,j) = (max(\mathbf{a}(x,y-1), \mathbf{a}(x,y)), min(\mathbf{a}(x,y-1), \mathbf{a}(x,y))) \\ 0 & \text{otherwise.} \end{cases}$$

The adjacency representation image is generated by the image algebra statement

$$\mathbf{b} := \bigvee \mathbf{t}(\mathbf{a}).$$

For a fixed point (x_0, y_0) of $\mathbf{X}$, $\mathbf{t}(\mathbf{a})_{(x_0,y_0)}$ is an image in $\{0,1\}^{\mathbf{Y}}$. If regions $\mathbf{R}_i$ and $\mathbf{R}_j$ "touch" at the point (x_0, y_0), $\mathbf{t}(\mathbf{a})_{(x_0,y_0)}(i,j)$ is equal to 1; otherwise it is 0. If regions $\mathbf{R}_i$ and $\mathbf{R}_j$ touch at any point in $\mathbf{X}$, the maximum

$$\bigvee_{(x,y)\in\mathbf{X}} \mathbf{t}(\mathbf{a})_{(x,y)}(i,j)$$

will take on value 1; otherwise it will be 0. Therefore,

$$\mathbf{b}(i,j) = \bigvee \mathbf{t}(\mathbf{a}) = \bigvee_{(x,y)\in\mathbf{X}} \mathbf{t}(\mathbf{a})_{(x,y)}(i,j)$$

is equal to 1 if $\mathbf{R}_i$ and $\mathbf{R}_j$ are adjacent and 0 otherwise. The ordered pair representation of the adjacency relation is the set

$$domain(\mathbf{b}\|_{=1}).$$

Comments and Observations

In the definition of adjacency above, two regions are adjacent if there is a point in one region and a point in the other region that are unit distance apart. That is, each point is in the 4-neighborhood of the other. Another definition of adjacency is that two regions are adjacent if they touch along a diagonal direction. With the second definition, two regions are adjacent if there is in each region a point which lies in the 8-neighborhood of the other. Another way to express the second adjacency relation is that there exist points in each region that are at distance $\sqrt{2}$ of each other.

If the second definition of adjacency is preferred, the template below should replace the template used in the original image algebra formulation.

$$\mathbf{t}(\mathbf{a})_{(x,y)}(i,j)=\begin{cases} 1 & \text{if } (i,j)=(max(\mathbf{a}(x-1,y),\mathbf{a}(x,y)), \\ & min(\mathbf{a}(x-1,y),\mathbf{a}(x,y))), \\ & (i,j)=(max(\mathbf{a}(x,y-1),\mathbf{a}(x,y)), \\ & min(\mathbf{a}(x,y-1),\mathbf{a}(x,y))), \\ & (i,j)=(max(\mathbf{a}(x-1,y-1),\mathbf{a}(x,y)), \\ & min(\mathbf{a}(x-1,y-1),\mathbf{a}(x,y))), \text{ or} \\ & (i,j)=(max(\mathbf{a}(x+1,y-1),\mathbf{a}(x,y)), \\ & min(\mathbf{a}(x+1,y-1),\mathbf{a}(x,y))) \\ 0 & \text{otherwise.} \end{cases}$$

The image algebra formulation of the algorithm presented above was first formulated in Ritter [6] and Shi and Ritter [7].

10.6. Inclusion Relation

The inclusion relation describes the hierarchical structure of regions in an image using the relation "inside of." Regions are defined as they were in Section 10.5. The inclusion relation can be represented by a set of ordered pairs, a directed graph, or an image. As an example, we present representations for the image shown in Figure 10.6.1.

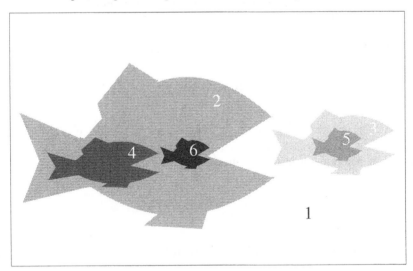

Figure 10.6.1. Inclusion of smaller fishes *inside of* a bigger fish.

(a) Ordered pairs — The ordered pair (i,j) is an element of the *inside of* relation iff region $\mathbf{R}_i$ is inside of region $\mathbf{R}_j$. For Figure 10.6.1 the *inside of* relation is represented by the set of ordered pairs

$$\{(6,2),(6,1),(5,3),(5,1),(4,2),(4,1),(3,1),$$
$$(2,1),(1,1)(2,2),(3,3),(4,4),(5,5),(6,6)\}.$$

(b)   Directed graph — The vertices of the graph representation of the inclusion relation are region numbers. The directed edge (i, j) with tail at i and head at j is an element of the graph iff $\mathbf{R}_i$ *inside of* $\mathbf{R}_j$. The directed graph representation of the inclusion relation is shown below. The *inside of* relation is reflexive,

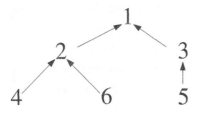

thus in the graph above each region is assumed to have an edge that points back to itself.

(c)   Image — The inclusion is represented by the image $\mathbf{a}$ defined by

$$\mathbf{a}(i, j) = \begin{cases} 1 & \text{if } \mathbf{R}_i \text{ } inside \text{ } of \text{ } \mathbf{R}_j \\ 0 & \text{otherwise.} \end{cases}$$

The following is the image representation of the inclusion relation for Figure 10.6.1.

$i \setminus j$	1	2	3	4	5	6
1	1	0	0	0	0	0
2	1	1	0	0	0	0
3	1	0	1	0	0	0
4	1	1	0	1	0	0
5	1	0	1	0	1	0
6	1	1	0	0	0	1

Note that regions are sets; however, the *inside of* relation and set inclusion $\subseteq$ relation are distinct concepts. Two algorithms used to describe the inclusion relation among regions in an image are presented next.

Image Algebra Formulation

The first image algebra formulation bases its method on the tenet that $\mathbf{R}_i$ *inside of* $\mathbf{R}_j$ iff the domain of $\mathbf{R}_i$ is contained ($\subseteq$) in the holes of $\mathbf{R}_j$. The algorithm begins by making binary images $\mathbf{r}_i = \chi_{=i}(\mathbf{a})$ and $\mathbf{r}_j = \chi_{=j}(\mathbf{a})$ to represent the regions $\mathbf{R}_i$ and $\mathbf{R}_j$, respectively, of the source image $\mathbf{a}$. Next, one of the hole filling algorithms of Section 6.6 is used to fill the holes in $\mathbf{r}_j$ to create the filled image $\bar{\mathbf{r}}_j$. Region $\mathbf{R}_i$ is inside of $\mathbf{R}_j$ iff $\mathbf{r}_i \vee \bar{\mathbf{r}}_j = \bar{\mathbf{r}}_j$.

Let $\mathbf{a} \in \mathbb{N}^{\mathbf{X}}$ be the source image, where $\mathbf{X} = \mathbb{Z}_m \times \mathbb{Z}_n$. Let $\mathbf{Y} = \mathbb{Z}_k \times \mathbb{Z}_k$, where $k = \vee \mathbf{a}$. We shall construct an image $\mathbf{b} \in \mathbb{N}^{\mathbf{Y}}$ that will represent the inclusion relation. Initially, all pixel values of $\mathbf{b}$ are set equal to 1. The image algebra pseudocode used to create the image representation $\mathbf{b}$ of the inclusion relation among the regions of

a is given by

$$\mathbf{b} := 1$$

```
for i in 1.. ∨ a loop
  for j in 1.. ∨ a loop
    if i ≠ j and b(i, j) ≠ 0 then
```
$$\mathbf{r}_i := \chi_{=i}(\mathbf{a})$$
$$\mathbf{r}_j := \chi_{=j}(\mathbf{a})$$
$$\bar{\mathbf{r}}_j := \mathtt{holefill}(\mathbf{r}_j)$$
```
      if rᵢ ∨ r̄ⱼ = r̄ⱼ then
```
$$\mathbf{b}(j, i) := 0$$
```
      else
```
$$\mathbf{b}(i, j) := 0$$
```
      end if
    end if
  end loop
end loop.
```

The subroutine `holefill` is an implementation of one of the hole filling algorithms from Section 6.6.

Alternate Image Algebra Formulation

The algorithm above requires $\frac{k(k-1)}{2}$ applications of the hole filling algorithm. Each application of the hole filling algorithm is, in itself, computationally very expensive. The alternate image algebra formulation assumes that regions at the same level of the *inside of* graph are not adjacent. This assumption will allow for greater computational efficiency at the cost of loss of generality.

The alternate image algebra formulation makes its determination of inclusion based on pixel value transitions along a row starting from a point (x_0, y_0) inside a region and moving in a direction toward the image's boundary. More specifically, if the image **a** is defined over an $m \times n$ grid, the number of transitions from region $\mathbf{R}_i$ to region $\mathbf{R}_j$ starting at (x_0, y_0) moving along a row toward the right side of the image is given by

$$card\{k : 1 \le k \le N - y_0 \text{ and } (i, j) = (\mathbf{a}(x_0, y_0 + k - 1), \mathbf{a}(x_0, y_0 + k))\}.$$

For example, in sequence

	(x_0, y_0)										(x_0, n)
$\mathbf{a}(x, y)$	3	4	4	6	6	4	4	6	6	7	7	1
k	1	2	3	4	5	6	7	8	9	10	11	12

there are two transitions from 4 to 6 and one transition from 6 to 4.

It follows from elementary plane topology that if $\mathbf{R}_i$ is *inside of* $\mathbf{R}_j$ then starting from a point in $\mathbf{R}_i$ the number of transitions from i to j is greater than the number of transitions from j to i. The image algebra formulation that follows bases its method on this fact.

Let $\mathbf{a} \in \mathbb{N}^{\mathbf{X}}$, where $\mathbf{X} = \mathbb{Z}_m \times \mathbb{Z}_n$ be the source image. Let I be the ordered pair representation of the inclusion relation. For $\mathbf{Y} = (\mathbb{Z}_{\vee \mathbf{a}})^2$, define the parameterized template $\mathbf{t} \in \left(\{0,1\}^{\mathbf{Y}}\right)^{\mathbb{Z}_n}$ by

$$\mathbf{t}(\mathbf{a}, x, y)_k(i, j) = \begin{cases} 1 & \text{if } (i,j) = (\mathbf{a}(x, y+k-1), \mathbf{a}(x, y+k)) \\ & \text{and } 1 \leq k \leq n - y \\ 0 & \text{otherwise.} \end{cases}$$

Let $f : \mathbb{Z}^2 \to \mathbb{Z}^2$ be the reflection $f(x, y) = (y, x)$. The alternate image algebra formulation used to generate the ordered pair representation of the inclusion relation for regions in $\mathbf{a}$ is

$$I := \{(1,1), (2,2), \ldots, (\vee \mathbf{a}, \vee \mathbf{a})\}$$

$$\text{for } l \text{ in } 1..\vee \mathbf{a} \text{ loop}$$

$$(x_0, y_0) := choice(domain(\mathbf{a}\|_l))$$

$$\mathbf{h} := \sum \mathbf{t}(\mathbf{a}, x_0, y_0)$$

$$I := I \cup domain((\mathbf{h} - \mathbf{h} \circ f)\|_{>0})$$

$$\text{end loop.}$$

Here $\mathbf{h} \circ f$ denotes the composition of $\mathbf{h}$ and f, and $(\mathbf{h} \circ f)(x, y) = \mathbf{h}(f(x,y)) = \mathbf{h}(y, x)$. A nice property of this algorithm is that one execution of the **for** loop finds all the regions that contain the region whose index is the current value of the loop variable. The first algorithm presented required nested **for** loops.

Note that for each point $k \in \mathbb{Z}$, $\mathbf{t}(\mathbf{a}, x_0, y_0)_k \in \{0, 1\}^{\mathbf{Y}}$ is an image whose value at (i, j) indicate whether the line crosses the boundary from $\mathbf{R}_i$ to $\mathbf{R}_j$ at $(x_0, y_0 + k - 1)$. Thus, $\mathbf{h}(i,j) = \sum_{k \in \mathbb{Z}} \mathbf{t}(\mathbf{a}, x_0, y_0)_k(i,j) = \sum_{k=1}^{N-y_0} \mathbf{t}(\mathbf{a}, x_0, y_0)_k(i,j)$ is the total number of boundary crossings from $\mathbf{R}_i$ to $\mathbf{R}_j$, and $\mathbf{h}(i,j) > \mathbf{h}(j,i)$ implies $\mathbf{R}_i$ *inside of* $\mathbf{R}_j$.

Comments and Observations

Both formulations, which were first presented in Shi and Ritter [7], produce a preordering on the set of regions. That is, the *inside of* relation is reflexive and transitive. The first formulation is not a partial ordering because it fails to be antisymmetric. To see this consider an image that consists of two regions, one of black pixels and the other of white pixels that together make a checkerboard pattern of one pixel squares. Under its assumption that regions at the same level are not adjacent, the second formulation does produce a partial ordering.

10.7. Quadtree Extraction

A quadtree is a hierarchical representation of a $2^n \times 2^n$ binary image based on successive partition of the image into quadrants. Quadtrees provide effective structural descriptions of binary images, and the operations on quadtrees are simple and elegant. Gargantini [8] introduced a new representation of a quadtree called linear quadtree. In a linear quadtree, he encodes each black node (foreground node) with a quaternary integer whose digits reflect successive quadrant subdivisions. He needs a special marker to encode big nodes with more than one pixel. Here, we encode a quadtree as a set of integers in base 5 as follows:

Each black node is encoded as an n-digit integer in base 5. Each successive digit represents the quadrant subdivision from which it originates. At the kth level, the NW quadrant is encoded with 1 at the kth digit, the NE with 2, the SW with 3, and the SE with 4. For a black node at the kth level, its last $n - k$ digits are encoded as 0's.

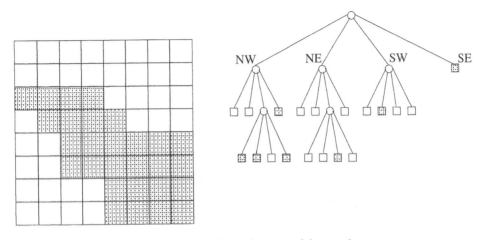

Figure 10.7.1. A binary image and its quadtree.

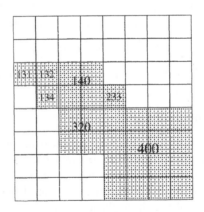

Figure 10.7.2. Codes of black pixels.

To obtain the set of quadtree codes for a $2^n \times 2^n$ binary image, we assign the individual black pixels of the image their corresponding quadtree codes, and work upwards to merge 4 codes representing 4 black nodes at a lower level into a code representing the black node consisting of those 4 black nodes if possible. The merging process can be performed as follows: whenever 4 black nodes at the $(k + 1)$th level are encoded as $c_1 c_2 \ldots c_k i \underbrace{0 \ldots 0}_{n-k-1}$, $i = 1, 2, 3, 4$, we merge those codes into $c_1 c_2 \ldots c_k \underbrace{0 \ldots 0}_{n-k}$ which represents a kth-level black node consisting of the 4 black nodes at the $(k + 1)$th level. This scheme provides an algorithm to transforms a $2^n \times 2^n$ binary image into its quadtree representation.

Image Algebra Formulation

Let the source image $\mathbf{a} \in \mathbb{Z}_2^{\mathbf{X}}$ be an image on $\mathbf{X} = \{(i,j) : 0 \leq i, j \leq 2^n - 1\}$. Define an image $\mathbf{d} \in \mathbb{Z}^{\mathbf{X}}$ by

$$\mathbf{d}(i,j) = \sum_{k=0}^{n-1} f(i_k, j_k) 5^k,$$

where

$$i = \sum_{k=0}^{n-1} i_k 2^k$$

$$j = \sum_{k=0}^{n-1} j_k 2^k$$

and

$$f(i_k, j_k) = \begin{cases} 1 & \text{if } i_k = 0 \text{ and } j_k = 0 \\ 2 & \text{if } i_k = 1 \text{ and } j_k = 0 \\ 3 & \text{if } i_k = 0 \text{ and } j_k = 1 \\ 4 & \text{if } i_k = 1 \text{ and } j_k = 1. \end{cases}$$

Next, define the template $\mathbf{t}(k)$ as follows:

$$\mathbf{t}(k) = \begin{array}{|c|c|} \hline {-5^{n-k}} & 0 \\ \hline 0 & 0 \\ \hline \end{array}$$

Finally, define two functions f_1 and f_2 by

$$f_1(y_1, y_2) = (2y_1, 2y_2)$$

and

$$f_2(x_1, x_2) = \left(\left\lfloor \frac{x_1}{2} \right\rfloor, \left\lfloor \frac{x_2}{2} \right\rfloor \right).$$

Now, if Q denotes the set of quadtree codes to be constructed, then Q can be obtained by using the following algorithm.

$$Q := \varnothing$$
$$\mathbf{a}_n := \mathbf{a} \cdot \mathbf{d}$$
$$k := n$$
$$\texttt{while } \mathbf{a}_k \not\leq 0 \texttt{ do}$$
$$\quad \mathbf{b}_k := (\mathbf{a}_k \boxtimes \mathbf{t}(k)) \circ f_1$$
$$\quad \mathbf{c}_k := (\chi_{\leq 0} \mathbf{b}_k) \circ f_2$$
$$\quad Q := Q \cup range((\mathbf{a}_k \cdot \mathbf{c}_k)\|_{>0})$$
$$\quad \mathbf{a}_{k-1} := \mathbf{b}_k$$
$$\quad k := k - 1.$$

In the above algorithm, which first appeared in Shi [7], $\mathbf{a}_k$ is a $2^k \times 2^k$ image in which any positive pixel encodes a kth-level black node which may be further merged to a big node at the $(k-1)$th level. The image $\mathbf{b}_k$ is a $2^{k-1} \times 2^{k-1}$ image in which any positive pixel encodes a black node merged from the corresponding four black nodes at the kth level. The image $\mathbf{c}_k$ is a $2^k \times 2^k$ binary image in which any pixel with value 1 indicates that the corresponding node cannot be further merged. Thus, a positive pixel of $\mathbf{a}_k \cdot \mathbf{c}_k$ encodes a kth level black node which cannot be further merged.

10.8.   Position, Orientation, and Symmetry

In this section, image algebra formulations for *position*, *orientation*, and *symmetry* are presented. Position, orientation, and symmetry are useful descriptors of objects within images [9, 10].

Let $\mathbf{a} \in \mathbb{R}^{\mathbb{R}^2}$ be an image that represents an object consisting of positive-valued pixels that is set against a background of 0-valued pixels. Position refers to the location of the object in the plane. The objects's *centroid* (or center of mass) is the point that is used to specify its position. The centroid is the point $(\overline{x}, \overline{y})$ whose coordinates are given by

$$\overline{x} = \frac{\int\limits_{\mathbb{R}^2} \int x \cdot \mathbf{a}(x, y) dx dy}{\int\limits_{\mathbb{R}^2} \int \mathbf{a}(x, y) dx dy}$$

$$\overline{y} = \frac{\int\limits_{\mathbb{R}^2} \int y \cdot \mathbf{a}(x, y) dx dy}{\int\limits_{\mathbb{R}^2} \int \mathbf{a}(x, y) dx dy}.$$

For the digital image $\mathbf{a} \in \mathbb{R}^{\mathbb{Z}_m \times \mathbb{Z}_n}$ the centroid's coordinates are given by

$$\overline{x} = \frac{\sum\limits_{x=1}^{m} \sum\limits_{y=1}^{n} x \cdot \mathbf{a}(x, y)}{\sum\limits_{x=1}^{m} \sum\limits_{y=1}^{n} \mathbf{a}(x, y)}$$

$$\overline{y} = \frac{\sum\limits_{x=1}^{m} \sum\limits_{y=1}^{n} y \cdot \mathbf{a}(x, y)}{\sum\limits_{x=1}^{m} \sum\limits_{y=1}^{n} \mathbf{a}(x, y)}.$$

Figure 10.8.1 shows the location of the centroid for an SR71 object.

Orientation refers to how the object lies in the plane. The object's *moment of inertia* is used to determine its angle of orientation. The moment of inertia about the line with slope $\tan \theta$ passing through $(\overline{x}, \overline{y})$ is defined as

$$M_\theta = \int\limits_{\mathbb{R}^2} \int [(x - \overline{x})\sin\theta - (y - \overline{y})\cos\theta]^2 \cdot \mathbf{a}(x, y) dx dy.$$

The angle θ_{min} that minimizes M_θ is the direction of the axis of least inertia for the object. If θ_{min} is unique, then the line in the direction θ_{min} is the *principal axis of inertia* for the object. For a binary image the principal axis of inertia is the axis about which the object appears elongated. The principal axis of inertia in the direction θ is shown for the SR71 object in Figure 10.8.1. The line in the direction θ_{max} is the axis about which the moment of inertia is greatest. In a binary image, the object is widest about this axis. The angles θ_{min} and θ_{max} are separated by 90°.

Let

$$m_{20} = \int\limits_{\mathbb{R}^2} \int (x - \overline{x})^2 \cdot \mathbf{a}(x, y) dx dy$$

$$m_{11} = \int\limits_{\mathbb{R}^2} \int (x - \overline{x})(y - \overline{y}) \cdot \mathbf{a}(x, y) dx dy$$

$$m_{02} = \int\limits_{\mathbb{R}^2} \int (y - \overline{y})^2 \cdot \mathbf{a}(x, y) dx dy.$$

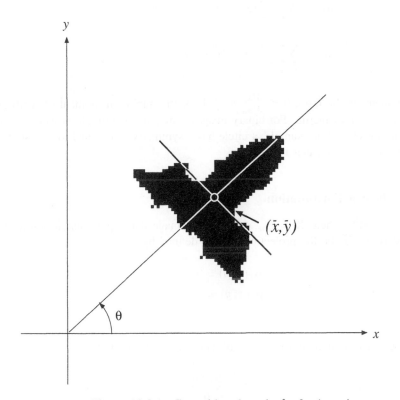

Figure 10.8.1. Centroid and angle θ of orientation.

The moment of inertia about the line in the direction θ can be written as

$$M_\theta = m_{20}sin^2\theta - 2m_{11}sin\theta\,cos\theta + m_{02}cos^2\theta.$$

The moment of inertia is minimized by solving

$$\frac{\partial M_\theta}{\partial \theta} = 0$$

for critical values of θ. Using the identity $tan\,2\theta = \frac{2\,tan\,\theta}{1-tan^2\theta}$ the search for critical values leads to the quadratic equation

$$tan^2\,\theta + \frac{m_{20} - m_{02}}{m_{11}}tan\,\theta - 1 = 0.$$

Solving the quadratic equations leads to two solutions for $tan\,\theta$, and hence two angles θ_{min} and θ_{max}. These are the angles of the axes about which the object has minimum and maximum moments of inertia, respectively. Determining which solution of the quadratic equation minimizes (or maximizes) the moment of inertia requires substitution back into the equation for M_θ.

For the digital image $\mathbf{a} \in \mathbb{R}^{\mathbb{Z}_m \times \mathbb{Z}_n}$ the central moments of order 2 are given by

$$m_{20} = \sum_{x=1}^{m} \sum_{y=1}^{n} (x - \overline{x})^2 \cdot \mathbf{a}(x, y)$$

$$m_{11} = \sum_{x=1}^{m} \sum_{y=1}^{n} (x - \overline{x}) \cdot (y - \overline{y}) \cdot \mathbf{a}(x, y)$$

$$m_{02} = \sum_{x=1}^{m} \sum_{y=1}^{n} (y - \overline{y})^2 \cdot \mathbf{a}(x, y).$$

Symmetry is the ratio, $0 \leq \frac{M_{\theta_{min}}}{M_{\theta_{max}}} \leq 1$, of the minimum moment of inertia to the maximum moment of inertia. For binary images symmetry is a rough measure of how "elongated" an object is. For example, a circle has a symmetry ratio equal to 1; a straight line has a symmetry ratio equal to 0.

Image Algebra Formulation

Let $\mathbf{a} \in \mathbb{R}^{\mathbf{X}}$, where $\mathbf{X} = \mathbb{Z}_m \times \mathbb{Z}_n$, be an image that represents an object. Let $\mathbf{p}_1 \in \mathbb{Z}_m^{\mathbf{X}}$ and $\mathbf{p}_2 \in \mathbb{Z}_n^{\mathbf{X}}$ be the projection images defined by

$$\mathbf{p}_1(x, y) = x$$
$$\mathbf{p}_2(x, y) = y.$$

The image algebra formulations for the coordinates of the centroid are

$$\overline{x} := \frac{\sum \mathbf{p}_1 \cdot \mathbf{a}}{\sum \mathbf{a}}$$

$$\overline{y} := \frac{\sum \mathbf{p}_2 \cdot \mathbf{a}}{\sum \mathbf{a}}.$$

The image algebra formulas for the central moments of order 2 used to compute the angle of orientation are given by

$$m_{20} := \sum (\mathbf{p}_1 - \overline{x})^2 \cdot \mathbf{a}$$
$$m_{11} := \sum (\mathbf{p}_1 - \overline{x}) \cdot (\mathbf{p}_2 - \overline{y}) \cdot \mathbf{a}$$
$$m_{02} := \sum (\mathbf{p}_2 - \overline{y})^2 \cdot \mathbf{a}.$$

10.9.   Region Description Using Moments

Moment invariants are image statistics that are independent of rotation, translation, and scale. Moment invariants are uniquely determined by an image and, conversely, uniquely determine the image (modulus rotation, translation, and scale). These properties of moment invariants facilitate pattern recognition in the visual field that is independent of size, position, and orientation. (See Hu [11] for experiments using moment invariants for pattern recognition.)

The moments invariants defined by Hu [11, 12] are derived from the definitions of moments, centralized moments, and normalized central moments. These statistics are defined as follows:

Let f be a continuous function defined over $\mathbb{R}^2$. The *moment of order* (p, q) of f is defined by

$$m_{pq} = \int\limits_{-\infty}^{\infty} \int\limits_{-\infty}^{\infty} x^p y^q f(x, y) \, dx \, dy,$$

where $p, q \in \{0, 1, 2, \ldots\}$. It has been shown [13] that if f is a piecewise continuous function with bounded support, then moments of all orders exist. Furthermore, under the same conditions on f, m_{pq} is uniquely determined by f, and f is uniquely determined by m_{pq}.

The *central moments* of f are defined by

$$\mu_{pq} = \int\limits_{-\infty}^{\infty} \int\limits_{-\infty}^{\infty} (x - \overline{x})^p (y - \overline{y})^q f(x, y) \, dx \, dy,$$

where

$$\overline{x} = \frac{m_{10}}{m_{00}} \quad \text{and} \quad \overline{y} = \frac{m_{01}}{m_{00}}.$$

The point $(\overline{x}, \overline{y})$ is called the image *centroid*. The center of gravity of an object is the physical analogue of the image centroid.

Let $\mathbf{a} \in \mathbb{R}^{\mathbf{X}}$, where $\mathbf{X} \subset \mathbb{Z}^2$ is an $m \times n$ array. The discrete counterpart of the centralized moment of order (p, q) is given by

$$\mu_{pq} = \sum_x \sum_y (x - \overline{x})^p (y - \overline{y})^q \mathbf{a}(x, y).$$

The *normalized central moment*, η_{pq}, is defined by

$$\eta_{pq} = \frac{\mu_{pq}}{\mu_{00}^{\gamma}}, \quad \text{where} \quad \gamma = \frac{p + q}{2} + 1.$$

We now present the seven, $\phi_1, \phi_2, \ldots, \phi_7$, *moment invariants* developed in Hu [11]. For the continuous case, these values are independent of rotation, translation, and scaling. In the discrete case, some aberrations may exists due to the digitization process.

$$\phi_1 = \eta_{20} + \eta_{02}$$
$$\phi_2 = (\eta_{20} - \eta_{02})^2 + 4\eta_{11}^2$$
$$\phi_3 = (\eta_{30} - 3\eta_{12})^2 + (3\eta_{21} - \eta_{03})^2$$
$$\phi_4 = (\eta_{30} + \eta_{12})^2 + (\eta_{21} + \eta_{03})^2$$
$$\phi_5 = (\eta_{30} - 3\eta_{12})(\eta_{30} + \eta_{12})\left[(\eta_{30} + \eta_{12})^2 - 3(\eta_{21} + \eta_{03})^2\right]$$
$$+ (3\eta_{21} - \eta_{03})(\eta_{21} + \eta_{03})\left[3(\eta_{30} + \eta_{12})^2 - (\eta_{21} + \eta_{03})^2\right]$$
$$\phi_6 = (\eta_{20} - \eta_{02})\left[(\eta_{30} + \eta_{12})^2 - (\eta_{21} + \eta_{03})^2\right]$$
$$+ 4\eta_{11}(\eta_{30} + \eta_{12})(\eta_{21} + \eta_{03})$$
$$\phi_7 = (3\eta_{21} - \eta_{30})(\eta_{30} + \eta_{12})\left[(\eta_{30} + \eta_{12})^2 - 3(\eta_{21} + \eta_{03})^2\right]$$
$$+ (3\eta_{12} - \eta_{30})(\eta_{21} + \eta_{03})\left[3(\eta_{30} + \eta_{12})^2 - (\eta_{21} + \eta_{03})^2\right].$$

Table 10.9.1 lists moment invariants for transformations applied to the binary image of the character "A." There is some discrepancy within the ϕ_i due to digitization.

Table 10.9.1 Moment Invariants of an Image under Rigid-Body Transformations

	ϕ_1	ϕ_1	ϕ_1	ϕ_1	ϕ_1	ϕ_1	ϕ_1
A	0.472542	0.001869	0.059536	0.005188	-0.000090	0.000223	0.000018
⊲	0.472452	0.001869	0.059536	0.005188	-0.000090	0.000223	0.000015
∀	0.472452	0.001869	0.059536	0.005188	-0.000090	0.000223	0.000018
A	0.321618	0.000519	0.016759	0.001812	-0.000009	0.000040	0.000004
⤸	0.453218	0.001731	0.058168	0.003495	-0.000047	0.000145	0.000031

Image Algebra Formulation

Let $\mathbf{a} \in \mathbb{R}^{\mathbf{X}}$, where $\mathbf{X} \subset \mathbb{Z}^2$ is an $m \times n$ array. Let $\mathbf{p}_i : \mathbf{X} \to \mathbb{Z}$ denote the ith coordinate projection. The image algebra formulation of the (p,q)th moment is

$$m_{pq} := \Sigma \mathbf{p}_1^p \cdot \mathbf{p}_2^q \cdot \mathbf{a}.$$

The (p,q)th central moment is given by

$$\mu_{pq} := \Sigma (\mathbf{p}_1 - \overline{x})^p \cdot (\mathbf{p}_2 - \overline{y})^q \cdot \mathbf{a},$$

where $\overline{x} = \frac{m_{10}}{m_{00}}$ and $\overline{y} = \frac{m_{01}}{m_{00}}$ are the coordinates of the image centroid. Computation of the invariant moments follow immediately from their mathematical formulations.

10.10.   Histogram

The histogram of an image is a function that provides the frequency of occurrence for each intensity level in the image. Image segmentation schemes often incorporate histogram information into their strategies. The histogram can also serve as a basis for measuring certain textural properties, and is the major statistical tool for normalizing and requantizing an image [12, 14].

Let $\mathbf{X}$ be a rectangular $m \times n$ array, $\mathbf{Y} = \{j \in \mathbb{N} : 0 \leq j \leq K\}$ for some fixed integer K, and $P = \{\mathbf{a} \in \mathbb{N}^{\mathbf{X}} : range(\mathbf{a}) \subset \mathbf{Y}\}$.

The histogram image, $\mathbf{h}(\mathbf{a}) \in \mathbb{N}^{\mathbf{Y}}$, that corresponds to $\mathbf{a} \in P$ is defined by

$$\mathbf{h}(\mathbf{a}, i) = card\{\mathbf{x} \in \mathbf{X} : \mathbf{a}(\mathbf{x}) = i\}.$$

That is, $\mathbf{h}(\mathbf{a}, i)$ corresponds to the number of elements in the domain of $\mathbf{a}$ that are mapped to i by $\mathbf{a}$. Figure 10.10.1 shows an image of a jet superimposed over a graphical representation of its histogram.

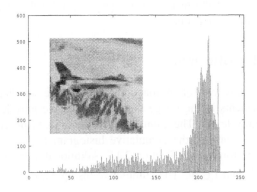

Figure 10.10.1. Image of a jet and its histogram.

The normalized histogram image $\overline{\mathbf{h}}(\mathbf{a}, i) \in [0, 1]^{\mathbf{Y}}$ of $\mathbf{a}$ is defined by

$$\overline{\mathbf{h}}(\mathbf{a}, i) = \frac{1}{card(\mathbf{X})} \cdot \mathbf{h}(\mathbf{a}, i).$$

It is clear that $\sum_{i \in \mathbf{Y}} \overline{\mathbf{h}}(\mathbf{a}, i) = 1$. The normalized histogram corresponds, in some ways, to a probability distribution function. Using the probability analogy, $\overline{\mathbf{h}}(\mathbf{a}, i)$ is viewed as the probability that $\mathbf{a}(\mathbf{x})$ takes on value i.

Image Algebra Formulation

Define a parameterized template

$$\mathbf{t} : P \to \left(\mathbb{N}^{\mathbf{Y}} \right)^{\mathbf{X}}$$

by defining for each $\mathbf{a} \in \mathbb{N}^{\mathbf{X}}$ the template $\mathbf{t}$ with parameter $\mathbf{a}$, i.e., $\mathbf{t}(\mathbf{a}) \in \left(\mathbb{N}^{\mathbf{Y}} \right)^{\mathbf{X}}$, by

$$\mathbf{t}(\mathbf{a})_{\mathbf{x}}(j) = \begin{cases} 1 & \text{if } \mathbf{a}(\mathbf{x}) = j \\ 0 & \text{otherwise.} \end{cases}$$

The image $\mathbf{h} \in \mathbb{R}^{\mathbf{Y}}$ obtained from the code

$$\mathbf{h} := \Sigma \mathbf{t}(\mathbf{a})$$

is the histogram of $\mathbf{a}$. This follows from the observation that since $\Sigma \mathbf{t}(\mathbf{a}) = \sum_{\mathbf{x} \in \mathbf{X}} \mathbf{t}(\mathbf{a})_{\mathbf{x}}$, $\mathbf{h}(j) = \sum_{\mathbf{x} \in \mathbf{X}} \mathbf{t}(\mathbf{a})_{\mathbf{x}}(j) =$ the number of pixels having value j.

If one is interested in the histogram of only one particular value j, then it would be more efficient to use the statement

$$n := \sum \chi_j(\mathbf{a})$$

since $\sum \chi_j(\mathbf{a})$ represents the number of pixels having value j, i.e., since $\sum \chi_j(\mathbf{a}) = \sum_{\mathbf{x} \in \mathbf{X}} \chi_j(\mathbf{a}(\mathbf{x}))$ and $\chi_j(\mathbf{a}(\mathbf{x})) = 1 \Leftrightarrow \mathbf{a}(\mathbf{x}) = j$.

Comments and Observations

Note that the template formulation above is designed for parallel implementation. For serial computers it is much more efficient to iterate over the source image's domain and increment the value of the histogram image at gray level g whenever the gray level g is encountered in the source image.

10.11.   Cumulative Histogram

For a given gray level of an image, the value of the cumulative histogram is equal to the number of elements in the image's domain that have gray level value less than or equal to the given gray level. The cumulative histogram can provide useful information about an image. For example, the cumulative histogram finds its way into each of the histogram specification formulas in Section 2.13. Additional uses can be found in [12, 15].

Let $\mathbf{a} \in \mathsf{N}^{\mathbf{X}}$ and $i \in \mathsf{N}$. The cumulative histogram $\mathbf{c}(\mathbf{a}) \in \mathsf{N}^{\mathsf{N}}$ is defined by

$$\mathbf{c}(\mathbf{a}, i) = card\{\mathbf{x} \in \mathbf{X} \, : \, \mathbf{a}(\mathbf{x}) \le i\}.$$

The normalized cumulative histogram $\overline{\mathbf{c}}(\mathbf{a}) \in [0, 1]^{\mathsf{N}}$ is defined by

$$\overline{\mathbf{c}}(\mathbf{a}, i) = \frac{1}{card(\mathbf{X})} \cdot \mathbf{c}(\mathbf{a}, i).$$

The normalized cumulative histogram is analogous to the cumulative probability function of statistics.

Figure 10.11.1 shows an image of a jet superimposed over a graphical representation of its cumulative histogram.

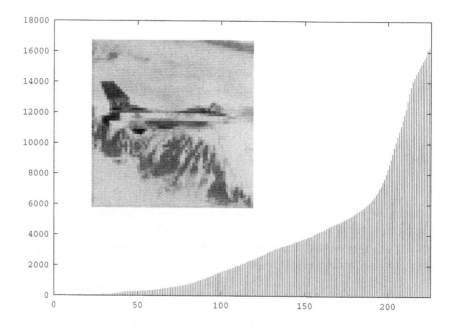

Figure 10.11.1. Image of a jet and its cumulative histogram.

Image Algebra Formulation

The image algebra formulation for the cumulative histogram is very similar to the formulation for the histogram (Section 10.10). Only the template $\mathbf{t}$ needs to be modified. The equality in the case statement of the template becomes an inequality so that the template $\mathbf{t}$ used for the cumulative histogram is

$$\mathbf{t(a)_x}(j) = \begin{cases} 1 & \text{if } \mathbf{a(x)} \leq j \\ 0 & \text{otherwise.} \end{cases}$$

The cumulative histogram is then represented by the image algebra statement

$$\mathbf{c} := \Sigma \mathbf{t(a)}.$$

Comments and Observations

As with the histogram in Section 10.10, the template formulation for the cumulative histogram is designed for parallel computers. For serial computers the cumulative histogram $\mathbf{c}$ should be calculated by iterating over the domain of the source image. Whenever the gray level g is encountered, the value of $\mathbf{c}$ at all gray levels less than or equal to g should be incremented.

10.12.   Texture Descriptors: Spatial Gray Level Dependence Statistics

Texture analysis is an important problem in image processing. Unfortunately, there is no precise definition of texture. In this section, the approach to texture analysis is based on the statistical properties of an image. In particular, statistical properties derived from spatial gray level dependence matrices of an image are formulated.

Spatial gray level dependence (SGLD) matrices have proven to be one of the most popular and effective sources of features in texture analysis. The SGLD approach computes an intermediate matrix of statistical measures from an image. It then defines features as functions of this matrix. These features relate to texture directionality, coarseness, contrast, and homogeneity on a perceptual level. The values of an SGLD matrix contains frequency information about the local spatial distribution of gray level pairs. Various statistics [16] derived from gray level spatial dependence matrices have been proposed for use in classifying image textures. We will only present a subset of these statistics. Our main purpose is to illustrate how their derivations are formulated in image algebra. Information on the interpretation of spatial dependence statistics and how they can be used for classifying textures can be found in various sources [16, 17, 18].

The statistics examined here are energy, entropy, correlation, inertia, and inverse difference moment. For a given image $\mathbf{a} \in \mathbb{Z}_G^{\mathbf{X}}$, where $\mathbf{X} = \mathbb{Z}_m \times \mathbb{Z}_n$ and G denotes the number of expected gray levels, all second-order statistical measures are completely specified by the joint probability

$$\mathbf{s}(\triangle x, \triangle y : i, j) = probability\{\mathbf{a}(x, y) = i \text{ and } \mathbf{a}(x + \triangle x, y + \triangle y) = j\},$$

where $\triangle x, \triangle y \in \mathbb{Z}$. Thus, $\mathbf{s}(\triangle x, \triangle y : i, j)$ is the probability that an arbitrary pixel location (x, y) has gray level i, while pixel location $(x + \triangle x, y + \triangle y)$ has gray level j.

By setting $\theta = tan^{-1}\left(\frac{\triangle y}{\triangle x}\right)$ and $d = \sqrt{(\triangle x)^2 + (\triangle y)^2}$, $\mathbf{s}(\triangle x, \triangle y : i, j)$ can be rewritten as

$$\mathbf{s}(d, \theta : i, j) = probability\{\mathbf{a}(x, y) = i \text{ and } \mathbf{a}(x + dcos\theta, y + dsin\theta) = j\}.$$

This provides for an alternate interpretation of $\mathbf{s}$; the four-variable function $\mathbf{s}(d, \theta : i, j)$ represents the probability that an arbitrary pixel location (x, y) has gray level i, while at an inter sample spacing distance d in angular direction θ the pixel location $(x + d\cos\theta, y + \sin\theta)$ has gray level j.

It is common practice to restrict θ to the angles $0\,^\circ$, 45°, 90°, and 135°, although different angles could be used. Also, distance measures other than the Euclidean distance are often employed. Many researchers prefer the chessboard distance, while others use the city block distance.

For each given distance d and angle θ the function $\mathbf{s}$ defines a $G \times G$ matrix $\mathbf{s}(d, \theta)$ whose (i, j)th entry $\mathbf{s}(d, \theta)_{ij}$ is given by $\mathbf{s}(d, \theta)_{ij} = \mathbf{s}(d, \theta : i, j)$. The matrix $\mathbf{s}(d, \theta)$ is called a *gray level spatial dependence matrix* (associated with $\mathbf{a}$). It is important to note that in contrast to regular matrix indexing, the indexing of $\mathbf{s}(d, \theta)$ starts at zero-zero; i.e., $0 \leq i, j \leq G - 1$.

The standard computation of the matrix $\mathbf{s}(\triangle x, \triangle y)$ is accomplished by setting

$$\mathbf{s}(\triangle x, \triangle y : i, j) = card\{(x, y) : \mathbf{a}(x, y) = i \; and \; \mathbf{a}(x + \triangle x, y + \triangle y) = j\,\}.$$

For illustration purposes, let $\mathbf{X}$ be a 4×4 grid and $\mathbf{a} \in \mathbb{Z}_3^{\mathbf{X}}$ be the image represented in Figure 10.12.1 below. In this figure, we assume the usual matrix ordering of $\mathbf{a}$ with $\mathbf{a}(0, 0)$ in the upper left-hand corner.

1	0	0	2
0	1	1	1
2	2	1	0
1	0	1	0

Figure 10.12.1. The 4×4 example image with gray values in $\mathbb{Z}_3$.

Spatial dependence matrices for $\mathbf{a}$, computed at various angles and chessboard distance $d = max\{|\triangle x|, |\triangle y|\}$, are presented below.

$$\mathbf{s}(1, 0^\circ) = \begin{pmatrix} 1 & 2 & 1 \\ 2 & 2 & 1 \\ 1 & 2 & 0 \end{pmatrix} \quad \mathbf{s}(2, 90^\circ) = \begin{pmatrix} 1 & 1 & 1 \\ 1 & 2 & 0 \\ 1 & 1 & 0 \end{pmatrix}$$

$$\mathbf{s}(1, 45^\circ) = \begin{pmatrix} 0 & 2 & 1 \\ 2 & 2 & 0 \\ 1 & 1 & 0 \end{pmatrix} \quad \mathbf{s}(1, 135^\circ) = \begin{pmatrix} 1 & 2 & 0 \\ 1 & 1 & 2 \\ 0 & 2 & 0 \end{pmatrix}$$

For purposes of texture classification, many researchers do not distinguish between $\mathbf{s}(d, \theta : i, j)$ and $\mathbf{s}(d, \theta : j, i)$; that is, they do not distinguish which one of the two pixels that are $(\triangle x, \triangle y)$ apart has gray value i and which one has gray value j. Therefore comparison between two textures is often based on the *texture co-occurence matrix*

$$\mathbf{c}(d, \theta) = \mathbf{s}(d, \theta) + \mathbf{s}'(d, \theta),$$

where $\mathbf{s}'$ denotes the transpose of $\mathbf{s}$. Using the SGLD matrices of the above example yields the following texture co-occurence matrices associated with the image shown in Figure

10.12.1:

$$c(1, 0°) = \begin{pmatrix} 2 & 4 & 2 \\ 4 & 4 & 3 \\ 2 & 3 & 0 \end{pmatrix} \quad c(2, 90°) = \begin{pmatrix} 2 & 2 & 2 \\ 2 & 4 & 1 \\ 2 & 1 & 0 \end{pmatrix}$$

$$c(1, 45°) = \begin{pmatrix} 0 & 4 & 2 \\ 4 & 4 & 1 \\ 2 & 1 & 0 \end{pmatrix} \quad c(1, 135°) = \begin{pmatrix} 2 & 3 & 0 \\ 3 & 2 & 4 \\ 0 & 4 & 0 \end{pmatrix}$$

Since co-occurence matrices have the property that $c(d, \theta) = c(d, \theta + \pi)$ — which can be ascertained from their symmetry — the rationale for choosing only angles between 0 and π becomes apparent.

Co-occurence matrices are used in defining the following more complex texture statistics. The marginal probability matrices $c_x(d, \theta : i)$ and $c_y(d, \theta : j)$ of $c(d, \theta)$ are defined by

$$c_x(d, \theta : i) = \sum_j c(d, \theta : i, j)$$

and

$$c_y(d, \theta : j) = \sum_i c(d, \theta : i, j),$$

respectively. The means and variances of c_x and c_y are

$$\mu_x(d, \theta) = \sum_i i c_x(d, \theta : i)$$

$$\mu_y(d, \theta) = \sum_j j c_y(d, \theta : j)$$

$$\sigma_x^2(d, \theta) = \sum_i (i - \mu_x(d, \theta))^2 c_x(d, \theta : i)$$

$$\sigma_y^2(d, \theta) = \sum_j (j - \mu_y(d, \theta))^2 c_y(d, \theta : j).$$

Five commonly used features for texture classification that we will reformulate in terms image algebra are defined below.

(a) *Energy:*

$$\sum_i \sum_j \{c(d, \theta : i, j)\}^2$$

(b) *Entropy:*

$$\sum_i \sum_j c(d, \theta : i, j) log(c(d, \theta : i, j))$$

(c) *Correlation:*

$$\frac{\sum_i \sum_j (i - \mu_x(d, \theta))(j - \mu_y(d, \theta)) c(d, \theta : i, j)}{\sigma_x(d, \theta) \sigma_y(d, \theta)}$$

(d) *Inverse Difference Moment:*

$$\sum_i \sum_j \frac{1}{1 + (i - j)^2} c(d, \theta : i, j)$$

(e) *Inertia:*

$$\sum_i \sum_j (i - j)^2 c(d, \theta : i, j)$$

Image Algebra Formulation

Let $\mathbf{a} \in \mathbb{Z}_G^{\mathbf{X}}$, where $\mathbf{X}$ is an $m \times n$ grid and G represents the number of gray levels in the image. First, we show how to derive the spatial dependence matrices $\mathbf{s}(d, \theta)$, where $d \in \mathbb{Z}^+$, and $\theta \in \{0°, 45°, 90°, 135°\}$.

By taking d to be the chessboard distance, $\mathbf{s}(d, \theta)$ can be computed in terms of the intersample spacings $\triangle x$ and $\triangle y$. In particular, let N denote the neighborhood function

$$N(x, y) = \{(x + \triangle x, y + \triangle y)\}.$$

The *image* $\mathbf{s}$ can now be computed using the image algebra following pseudocode:

> **for** $i := 0$ **to** $G - 1$ **loop**
>
> **for** $j := 0$ **to** $G - 1$ **loop**
>
> $\mathbf{s}(\triangle x, \triangle y)(i, j) := \Sigma \chi_2[\chi_i(\mathbf{a}) + \chi_j(\mathbf{a}) \oplus N]$
>
> **end loop**
>
> **end loop.**

Thus, for example, choosing $\triangle x = 1$ and $\triangle y = 0$ computes the image $\mathbf{s}(1, 0°)$, while choosing $\triangle x = 1$ and $\triangle y = 1$ computes $\mathbf{s}(1, 45°)$.

The co-occurence image $\mathbf{c}$ is given by the statement

$$\mathbf{c}(\triangle x, \triangle y) := \mathbf{s}(\triangle x, \triangle y) + \mathbf{s}'(\triangle x, \triangle y).$$

The co-occurence image can also be computed directly by using the parameterized template $\mathbf{t}(\mathbf{a}, \triangle x, \triangle y) \in \left(\mathbb{Z}^{\mathbb{Z}_G \times \mathbb{Z}_G}\right)^{\mathbf{X}}$ defined by

$$\mathbf{t}(\mathbf{a}, \triangle x, \triangle y)_{(x,y)}(i, j) = \begin{cases} 2 & \text{if } i = j \text{ and } \mathbf{a}(x, y) = \mathbf{a}(x + \triangle x, y + \triangle y) = i \\ 1 & \text{if } i \neq j \text{ and } ((\mathbf{a}(x, y) = i \text{ and } \mathbf{a}(x + \triangle x, y + \triangle y) = j) \\ & \quad \text{or } (\mathbf{a}(x, y) = j \text{ and } \mathbf{a}(x + \triangle x, y + \triangle y) = i)) \\ 0 & \text{otherwise.} \end{cases}$$

The computation of $\mathbf{c}(\triangle x, \triangle y)$ now reduces to the simple formula

$$\mathbf{c}(\triangle x, \triangle) := \sum \mathbf{t}(\mathbf{a}, \triangle x, \triangle).$$

To see how this formulation works, we compute $\mathbf{c}(\triangle x, \triangle y)(i, j)$ for $i \neq j$. Let

$$S(i, j) = \{(x, y) \in \mathbf{X} : (\mathbf{a}(x, y) = i \text{ and } \mathbf{a}(x + \triangle x, y + \triangle y) = j)$$
$$\text{or } (\mathbf{a}(x, y) = i \text{ and } \mathbf{a}(x + \triangle x, y + \triangle y) = j)\}.$$

By definition of template sum, we obtain

$$\mathbf{c}(\triangle x, \triangle y)(i, j) = \sum_{(x,y) \in \mathbf{X}} \mathbf{t}(\mathbf{a}, \triangle x, \triangle y)_{(x,y)}(i, j)$$
$$= \sum_{(x,y) \in S(i,j)} 1$$
$$= card(S(i, j))$$
$$= \mathbf{c}(\triangle x, \triangle y : i, j).$$

If $i = j$, then we obtain $\sum_{(x,y) \in S(i,j)} 2$ which corresponds to the doubling effect along the diagonal when adding the matrices $\mathbf{s} + \mathbf{s}'$. Although the template method of computing

$c(\triangle x, \triangle y)$ can be stated as a simple one-line formula, actual computation using this method
is very inefficient since the template is translation variant.

The marginal probability images $c_x(\triangle x, \triangle y)$ and $c_y(\triangle x, \triangle y)$ can be computed
as follows:

$$\text{for } i = 0 \text{ to } G - 1 \text{ loop}$$

$$c_x(\triangle x, \triangle y)(i) := \sum_{j=0}^{G-1} c(\triangle x, \triangle y)(i, j)$$

$$\text{end loop}$$

$$\text{for } j = 0 \text{ to } G - 1 \text{ loop}$$

$$c_y(\triangle x, \triangle y)(j) := \sum_{i=0}^{G-1} c(\triangle x, \triangle y)(i, j)$$

$$\text{end loop}.$$

The actual computation of these sums will probably be achieved by using the following
loops:

$$\text{for } i := 0 \text{ to } G - 1 \text{ loop}$$
$$\text{for } j := 0 \text{ to } G - 2 \text{ loop}$$
$$c_x(\triangle x, \triangle y)(i) := c(\triangle x, \triangle y)(i, j) + c(\triangle x, \triangle y)(i, j + 1)$$
$$\text{end loop}$$
$$\text{end loop}$$

and

$$\text{for } j := 0 \text{ to } G - 1 \text{ loop}$$
$$\text{for } i := 0 \text{ to } G - 2 \text{ loop}$$
$$c_y(\triangle x, \triangle y)(j) := c(\triangle x, \triangle y)(i, j) + c(\triangle x, \triangle y)(i + 1, j)$$
$$\text{end loop}$$
$$\text{end loop}.$$

Another version for computing $c_x(\triangle x, \triangle y)$ and $c_y(\triangle x, \triangle y)$, which does not
involve loops, is obtained by defining the neighborhood functions $N_x : \mathbb{Z}_G \to \mathbb{Z}_G \times \mathbb{Z}_G$
and $N_y : \mathbb{Z}_G \to \mathbb{Z}_G \times \mathbb{Z}_G$ by

$$N_x(i) = \{(i, j) : j = 0, 1, \ldots, G - 1\}$$

and

$$N_y(j) = \{(i, j) : i = 0, 1, \ldots, G - 1\},$$

respectively. The marginal probability images can then be computed using the following
statements:

$$c_x(\triangle x, \triangle y) := c(\triangle x, \triangle y) \oplus N_x$$
$$c_y(\triangle x, \triangle y) := c(\triangle x, \triangle y) \oplus N_y.$$

This method of computing $c_x(\triangle x, \triangle y)$ and $c_y(\triangle x, \triangle y)$ is preferable when using special
mesh-connected architectures.

Next, let $\mathbf{i} \in (\mathbb{Z}_G)^{\mathbb{Z}_G}$ be the identity image defined by

$$\mathbf{i}(i) = i, \quad i \in \{0, \ldots, G - 1\}.$$

The means and variances of $\mathbf{c}_x(\triangle x, \triangle y)$ and $\mathbf{c}_y(\triangle x, \triangle y)$ are given by

$$\mu_x(\triangle x, \triangle y) := \sum \mathbf{i} \cdot \mathbf{c}_x(\triangle x, \triangle y)$$

$$\mu_y(\triangle x, \triangle y) := \sum \mathbf{i} \cdot \mathbf{c}_y(\triangle x, \triangle y)$$

$$\sigma_x^2(\triangle x, \triangle y) := \sum (\mathbf{i} - \mu_x(\triangle x, \triangle y))^2 \cdot \mathbf{c}_x(\triangle x, \triangle y)$$

$$\sigma_y^2(\triangle x, \triangle y) := \sum (\mathbf{i} - \mu_y(\triangle x, \triangle y))^2 \cdot \mathbf{c}_y(\triangle x, \triangle y).$$

The image algebra pseudocode of the different statistical features derived from the spatial dependence image are

(a) Energy:
$$E := \sum [\mathbf{c}(\triangle x, \triangle y)]^2$$

(b) Entropy:
$$Entrp := \sum \mathbf{c}(\triangle x, \triangle y) \cdot log(\mathbf{c}(\triangle x, \triangle y))$$

(c) Correlation:
$$C := \frac{[\sum (\mathbf{p}_1 - \mu_x(\triangle x, \triangle y))(\mathbf{p}_2 - \mu_y(\triangle x, \triangle y))\mathbf{c}(\triangle x, \triangle y)]}{\sigma_x(\triangle x, \triangle y)\sigma_y(\triangle x, \triangle y)},$$

where the coordinate projections $\mathbf{p}_1, \mathbf{p}_2 \in (\mathbb{Z}_G)^{\mathbf{Y}}$ for the $G \times G$ grid $\mathbf{Y}$ are defined by

$$\mathbf{p}_1(i, j) = i \quad \text{and}$$
$$\mathbf{p}_2(i, j) = j.$$

(d) Inverse Difference Moment:
$$IDM := \sum \frac{\mathbf{c}(\triangle x, \triangle y)}{1 + (\mathbf{p}_1 - \mathbf{p}_2)^2}$$

(e) Inertia:
$$I := \sum (\mathbf{p}_1 - \mathbf{p}_2)^2 \cdot \mathbf{c}(\triangle x, \triangle y)$$

Comments and Observations

If the images to be analyzed are of different dimensions, it will be necessary to normalize the spatial dependence matrix so that meaningful comparisons among statistics can be made. To normalize, each entry should be divided by the number $c(d, \theta)$ of point pairs in the domain of the image that satisfy the relation of being at distance d of each other in the angular direction θ. The normalized spatial dependence image $\bar{\mathbf{c}}(d, \theta)$ is thus given by

$$\bar{\mathbf{c}}(d, \theta) := \frac{\mathbf{c}(d, \theta)}{R}.$$

Suppose $\mathbf{c}(1, 0°)$ is the spatial dependence matrix (at distance 1, direction $0°$) for an image defined over an $m \times n$ grid. There are $2n(m-1)$ nearest horizontal neighborhood pairs in the $m \times n$ grid. Thus, for this example

$$\bar{\mathbf{c}}(1, 0°) := \frac{\mathbf{c}(1, 0°)}{2n(m-1)}.$$

10.13. Exercises

1.   The ratio $4\pi \cdot \text{area}/(\text{perimeter})^2$ of a connected object in a binary image is known as its *measure of compactness*. Write an algorithm in image algebra that computes the compactness of objects in binary images. Implement your algorithm.

2.   Given a binary image **a** its horizontal and vertical signatures are defined as $p(x) = \int_y \mathbf{a}(x,y)$ and $p(y) = \int_x \mathbf{a}(x,.y)$, respectively (i.e., p projects **a** onto the x and y axes). Specify an image algebra algorithm that computes these projections.

3.   Projections are not information preserving transforms. However, given a sufficient number of projections of an objects allows for the reconstruction of the object to a high degree of accuracy. (This follows from the theory of computer assisted tomography.) Specify an image algebra algorithm that computes protection of a binary image onto lines having an angle of $\emptyset$ from the x-axis.

4.   Specify an algorithm in image algebra that will find the maximal diameter of an object in a binary image. Implement your algorithm.

5.   Specify an algorithm in image algebra that will find the smallest rectangle enclosing an object in a binary image. Note that the rectangle's sides may not be parallel to the x and y axis.

6.   Let $c = \{c(i) : i = 1,...,n\}$ denote the chain code of the boundary of an object in a binary image. An alternative formulation for the length of the perimeter of the object is given by the formulation

$$\text{perimeter} = \sum_{i=1}^{n} \Delta t_i,$$

where

$$\Delta t_i = 1 + \frac{(\sqrt{2}-1)}{2}\left[1-(-1)^{c(i)}\right].$$

Similarly, the area of the object can be computed using the formulation

$$\text{area} = \sum_{i=1}^{n} \Delta A_i$$

where

$$\Delta A_i = \Delta x_i \left(y_i + \frac{\Delta y_i}{2}\right),$$

$$\Delta x_i = sgn(6 - c(i))sgn(2 - c(i)),$$

$$\Delta y_i = sgn(4 - c(i))sgn(c(i)),$$

$$sgn(z) = \begin{cases} 1 & if\ z > 0 \\ 0 & if\ z = 0, \\ -1 & if\ z < 0 \end{cases}$$

and y_i denotes the y coordinate of the endpoint of $c(i)$. Specify an algorithm in image algebra that computes the perimeter and area of objects using the above formulations. Implement your algorithm and compare the results with those obtained using the formulations given in Section 10.2. Discuss the differences.

7. Suppose a binary image contains an 8–connected object with Euler number $-k$ $(k > 0)$. How many holes does the object contain? Suppose all holes are both 4–connected and 8–connected.

    a. Specify an algorithm in image algebra that computes the chain code of such objects.

    b. Extend your algorithm to also compute the total area and perimeter of this object using the specification as given in Exercise 6.

8. The image algebra formulation for computing region adjacency was specified in terms of parameterized templates which are often computationally inefficient. Provide an alternate image algebra formulation of region adjacency that avoids the use of parameterized templates.

9. Repeat Exercise 8 for region inclusion relationships.

10. In Section 10.10 it was noted that the parameterized template formulation for computing an image histogram was designed for parallel implementation. Explain how such implementation can be achieved using this particular specification.

11. A basic notion of pattern recognition is the feature vector. The feature vector $v = (v_1, \cdots, v_m)$ is used to condense the description of relevant properties of a textured image into a small subset of m-dimensional feature space. For instance, with each pixel $\mathbf{a}(i, j)$, one may associate a certain number of texture features $v(i, j) = (v_1(i, j), \cdots, v_m(i, j))$. Such feature vectors tend to form separate clusters in m-dimensional space according to the texture from which they were derived. Devise a method that uses spatial dependency statistics and feature vectors to segment an image of a natural scene containing possible regions of grasses, trees, and buildings.

10.14. References

[1] W. Pratt, *Digital Image Processing*. New York: John Wiley, 1978.

[2] G. Ritter, "Image algebra." Unpublished manuscript, available via anonymous ftp from `ftp://ftp.cise.ufl.edu/pub/src/ia/documents`, 1994.

[3] A. Rosenfeld and A. Kak, *Digital Picture Processing*. New York, NY: Academic Press, 2nd ed., 1982.

[4] H. Freeman, "On the encoding of arbitrary geometric configurations," *IRE Transactions on Electronic Computers*, vol. 10, 1961.

[5] H. Freeman, "Computer processing of line-drawing images," *Computing Surveys*, vol. 6, pp. 1–41, Mar. 1974.

[6] G. Ritter, "Recent developments in image algebra," in *Advances in Electronics and Electron Physics* (P. Hawkes, ed.), vol. 80, pp. 243–308, New York, NY: Academic Press, 1991.

[7] H. Shi and G. Ritter, "Structural description of images using image algebra," in *Image Algebra and Morphological Image Processing III*, vol. 1769 of *Proceedings of SPIE*, (San Diego, CA), pp. 368–375, July 1992.

[8] I. Gargantini, "An effective way to represent quadtrees," *Communications of the ACM*, vol. 25, pp. 905–910, Dec. 1982.

[9] A. Rosenfeld and A. Kak, *Digital Image Processing*. New York: Academic Press, 1976.

[10] J. Russ, *Computer-Assisted Microscopy: The Measurement and Analysis of Images.* New York, NY: Plenum Press, 1990.

[11] M. Hu, "Visual pattern recognition by moment invariants," *IRE Transactions on Information Theory*, vol. IT-8, pp. 179–187, 1962.

[12] R. Gonzalez and P. Wintz, *Digital Image Processing.* Reading, MA: Addison-Wesley, second ed., 1987.

[13] A. Papoulis, *Probability, Random Variables, and Stochastic Processes.* New York: McGraw Hill, 1965.

[14] E. Hall, "A survey of preprocessing and feature extraction techniques for radiographic images," *IEEE Transactions on Computers*, vol. C-20, no. 9, 1971.

[15] D. Ballard and C. Brown, *Computer Vision.* Englewood Cliffs, NJ: Prentice-Hall, 1982.

[16] R. Haralick, K. Shanmugan, and I. Dinstein, "Textural features for image classification," *IEEE Transactions on Systems, Man, and Cybernetics*, vol. SMC-3, pp. 610–621, Nov. 1973.

[17] R. Haralick, "Statistical and structural approaches to textures," *Proceedings of the IEEE*, vol. 67, pp. 786–804, May 1979.

[18] J. Weszka, C. Dyer, and A. Rosenfeld, "A comparative study of texture measures for terrain classification," *IEEE Transactions on Systems, Man, and Cybernetics*, vol. SMC-6, pp. 269–285, Apr. 1976.

CHAPTER 11
GEOMETRIC IMAGE TRANSFORMATIONS

11.1. Introduction

Geometric transformations change the spatial relationships among the objects in an image or change the geometry of an object in an image. Examples of geometric transformations are affine and perspective transforms. These types of transforms are commonly used in image registration and rectification. For example objects in magnetic resonance imagery (MRI) are displaced because of the warping effects of the field and raster-scanned satellite images of the earth exhibit the phenomenon that adjacent scan lines are offset slightly with respect to one another because the earth rotates as successive lines of an image are recorded.

Functions between spatial domains provide the underlying foundation for realizing naturally induced operations for spatial manipulation of image data. In particular, if $f : \mathbf{Y} \to \mathbf{X}$ and $\mathbf{a} \in \mathbb{F}^{\mathbf{X}}$, then we define the induced image $\mathbf{a} \circ f \in \mathbb{F}^{\mathbf{Y}}$ by

$$\mathbf{a} \circ f = \{(\mathbf{y}, \mathbf{a}(f(\mathbf{y}))) : \mathbf{y} \in \mathbf{Y}\}. \tag{11.1.1}$$

Thus, the operation defined by Eq. 11.1.1 transforms an $\mathbb{F}$–valued image defined over the space $\mathbf{X}$ into an $\mathbb{F}$–valued image defined over the space $\mathbf{Y}$. Figure 11.1.1 provides a visual interpretation of this composition operation.

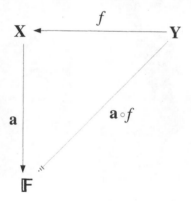

Figure 11.1.1. The spatial transform $\mathbf{a} \circ f$.

11.2. Image Reflection and Magnification

The basic transformations of the plane give rise to the basic geometric image transformations. Basic planar transformations are translations, rotations, reflections, contractions, and expansions. In the image domain, these correspond to shifting an image, rotating an image, reflecting an image across a straight line, and shrinking or magnifying an image. In this section we restrict our attention to the basic operations of image reflection and magnification. These operations are particularly easy if integral grid points map to integral grid points and lines of reflections are vertical or horizontal; i.e., of form $y = k$ or $x = k$, where k is an integer. In case of magnification, we have $\mathbf{X} \subset \mathbf{Y} \subset \mathbb{Z}^2$, and $f : \mathbf{Y} \to \mathbf{X}$. Thus, if $\mathbf{X}$ and $\mathbf{Y}$ are rectangular arrays, then f is a many-to-one map, meaning that in this particular case, magnification is achieved via replication of pixel values.

Image Algebra Formulation

We first consider the case of reflecting an image $\mathbf{a} \in \mathbb{R}^{\mathbf{X}}$ across a vertical line. Suppose $\mathbf{X} \subset \mathbb{Z}^2$ is a rectangular $m \times n$ array, $1 \leq k \leq \frac{m}{2}$, and $f : \mathbf{X} \to \mathbf{X}$ is defined as

$$f(x,y) = \begin{cases} (x,y) & \text{if } k \leq x \\ (2k - x, y) & \text{if } x < k \end{cases}.$$

The image $\mathbf{b}$ given by

$$\mathbf{b} := \mathbf{a} \circ f$$

represents the one-sided reflection of $\mathbf{a}$ across the line $x = k$. Figure 11.2.1 illustrates such a reflection on a 512×512 image across the line $k = 180$.

Figure 11.2.1. One-sided reflection across a line.

The image algebra formulation for magnifying an image by a factor of k, where k is a positive integer, by using simple replication of pixel values is also straight forward. For a given pair of real numbers $\mathbf{y} = (y_1, y_2)$, let $\lceil \mathbf{y} \rceil = (\lceil y_1 \rceil, \lceil y_2 \rceil)$. Suppose $\mathbf{a} \in \mathbb{R}^{\mathbf{X}}$ denotes the source image and $\mathbf{X}$ is a rectangular $m \times n$ array. Let $\mathbf{Y}$ be a $km \times kn$ array and define $f : \mathbf{Y} \to \mathbf{X}$ by $f(\mathbf{y}) = \lceil \frac{1}{k} \mathbf{y} \rceil$; i.e., $f(y_1, y_2) = (x_1, x_2)$, where $x_i = \lceil \frac{1}{k} y_i \rceil$. Then

$$\mathbf{b} := \mathbf{a} \circ f$$

represents the magnification of $\mathbf{a}$ by a factor of k. Figure 11.2.2 represents an example of a magnification by a factor of $k = 2$. Here each pixel was replicated four times, two times in the x direction and two times in the y direction.

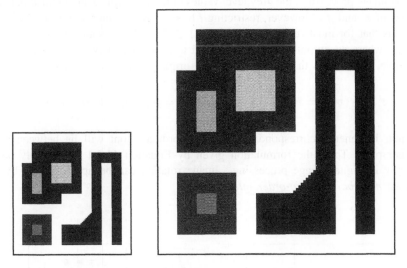

Figure 11.2.2. Magnification by replication.

Comments and Observations

As pointed out in the discussion above, image reflection and image magnification are easy as long as grid points are mapped to grid points. This is generally not the case as a simple reflection across a non-vertical or non-horizontal line illustrates. In such cases interpolation techniques are the usual tool for realizing geometric transforms. These techniques are discussed in subsequent sections. The induced image $\mathbf{a} \circ f$ given in either one of the above two examples could just as easily be obtained by use of an image-template operation. However, in various cases the statement $\mathbf{b} := \mathbf{a} \circ f$ provides for a more translucent expression and computationally more efficient method than an image-template convolution.

11.3. Nearest Neighbor Image Rotation

In this section we present a method for rotating an image about an arbitrary angle θ. The problem of rotating an image through an arbitrary degree is that, generally, grid points do not end up on grid points. Similar problems arise in a variety of geometric transformations. Note that in both image reflection and image magnification as described in the previous section, the function f can be viewed as a function on $\mathbb{Z}^2$ into $\mathbb{Z}^2$; that is, as a function preserving integral coordinates. In the strictest sense, however, there exist only a few geometric transformations of $\mathbb{Z}^2$, namely translations, reflections, and rotations through angles that are integral multiples of $90°$. Practical applications, however, force us to consider more general geometric transformations than these basic grid transforms. The general definition for geometric operations in the plane $\mathbb{R}^2$ is in terms of a function $f : \mathbb{R}^2 \to \mathbb{R}^2$ with

$$f(x,y) = (f_1(x,y), f_2(x,y)), \tag{11.3.1}$$

resulting in the expression

$$\mathbf{b}(x,y) = \mathbf{a}(x',y') = \mathbf{a}(f_1(x,y), f_2(x,y)), \tag{11.3.2}$$

where **a** denotes the input image, **b** the output image, $x' = f_1(x,y)$, and $y' = f_2(x,y)$. In digital image processing, the grey level values of the input image are defined only at integral values of x' and y'. However, restricting f to some rectangular subset **Y** of $\mathbb{Z}^2 \subset \mathbb{R}^2$, it is obvious that for most functions f, $range(f) \not\subset \mathbb{Z}^2$. This situation is illustrated in Figure 11.3.1, where an output location is mapped to a position between four input pixels. For example, if f denotes the rotation about the origin, then the coordinates

$$x' = f_1(x,y) = x cos\theta - y sin\theta \text{ and } y' = f_2(x,y) = x sin\theta + y cos\theta \qquad (11.3.3)$$

do not, in general, correspond to integral coordinates but will lie between an adjacent integer pair. Thus, the formulation given by Equation 11.1.1 can usually not be used directly in digital image processing. Image rotation can be approximated by using the *nearest neighbor*, or *zero-order interpolation*.

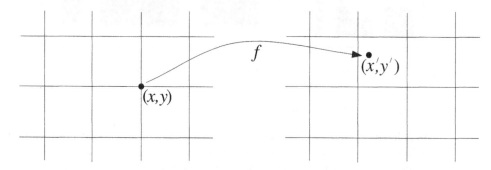

Figure 11.3.1. Mapping of integral to non-integral coordinates.

Image Algebra Formulation

Redefine the function f defined by Equation 11.3.1 as $\hat{f}(x,y) \equiv [f(x,y)]$. That is,

$$\hat{f}(x,y) = ([x'],[y']) \iff f(x,y) = (x',y'),$$

where $[r]$ denotes the rounding of r to the nearest integer. Then the image

$$\mathbf{b} := \mathbf{a} \circ \hat{f}$$

is obtained from **a** by use of nearest neighbor interpolation.

Comments and Observations

In nearest neighbor interpolation, the gray level of the output pixel is taken to be that of the input pixel nearest to the position to which it maps. This is computationally efficient and, in many cases, produces acceptable results. On the negative side, this simple interpolation scheme can introduce artifacts in images whose gray levels change significantly over one unit of pixel spacing. Figure 11.3.2 shows an example of rotating an image using nearest neighbor interpolation. The results show a sawtooth effect at the edges.

Additionally, due to rounding to the nearest neighbor, missing pixel values may also occur in the rotated image.

Figure 11.3.2. Rotation using nearest neighbor interpolation.

The image shown in Figure 11.3.2 is a very low resolution image. In larger high resolution images the sawtooth effect is not nearly as noticeable.

11.4. Image Rotation using Bilinear Interpolation

As mentioned in the previous section, rotating an image causes output pixels to be mapped to non-integral coordinate positions of the input image array. The new pixel locations are generally somewhere between four neighboring pixels of the input array. The simple nearest neighbor computation will, therefore, produce undesirable artifacts. Thus, higher order interpolation schemes are often necessary to determine the pixel values of the output image. The method of choice is *first-order* or *bilinear interpolation*. First-order interpolation produces more desirable results with only a modesr increase in computational complexity. Image rotation using bilinear interpolation requires the input image $\mathbf{a} \in \mathbb{R}^{\mathbf{X}}$ to be extended to an image $\mathbf{a}' \in (\mathbb{R})^{\mathbb{R}^2}$ and then resampled on an appropriate output array $\mathbf{Y}$.

Image Algebra Formulation

Suppose $\mathbf{a} \in \mathbb{R}^{\mathbf{X}}$, where $\mathbf{X}$ is an $m \times n$ array. Define an $m \times n$ rectangle

$$\mathbf{X}' = \left\{ (x_1', x_2') \ : \ 1 \leq x_1' \leq m, \ 1 \leq x_2' \leq n, \ (x_1', x_2') \in \mathbb{R}^2 \right\}$$

and extend $\mathbf{a}$ to a function $\mathbf{a}' \in (\mathbb{R})^{\mathbb{R}^2}$ as follows:
First set both $\mathbf{a}(x_1, x_2) = 0$ and $\mathbf{a}'(x_1, x_2) = 0$ whenever $(x_1, x_2) \in \mathbb{R}^2 \backslash \mathbf{X}'$. For $(x_1', x_2') \in \mathbf{X}'$, set

$$
\begin{aligned}
\mathbf{a}'(x_1', x_2') &= \mathbf{a}(x_1, x_2) + [\mathbf{a}(x_1 + 1, x_2) - \mathbf{a}(x_1, x_2)](x_1' - x_1) \\
&+ [\mathbf{a}(x_1, x_2 + 1) - \mathbf{a}(x_1, x_2)](x_2' - x_2) \\
&+ [\mathbf{a}(x_1 + 1, x_2 + 1) + \mathbf{a}(x_1, x_2) - \mathbf{a}(x_1, x_2 + 1) - \mathbf{a}(x_1 + 1, x_2)](x_1' - x_1)(x_2' - x_2),
\end{aligned}
$$

$$(11.4.1)$$

where $x_i = \lfloor x_i' \rfloor$. Note that $x_i \leq x_i' \leq x_i + 1$, and $\mathbf{a}'(x_1', x_2') = \mathbf{a}(x_1, x_2)$ whenever $x_1 = x_1'$ and $x_2 = x_2'$. Thus, $\mathbf{a}'$ is a continuous extension of $\mathbf{a}$ to $\mathbb{R}^2$ with the four corner values of $\mathbf{a}'$ on $\mathbf{X}$ agreeing with those of $\mathbf{a}$ as shown in Fiigure 11.4.1.

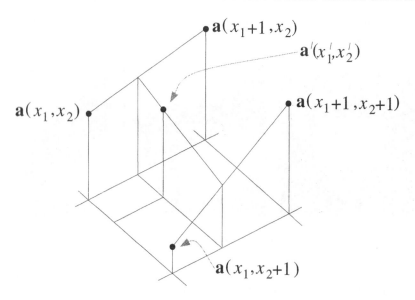

Figure 11.4.1. The bilinear interpolation of **a**.

Now suppose $f : \mathbb{R}^2 \to \mathbb{R}^2$, where

$$f(x, y) = (f_1(x, y), f_2(x, y)), \tag{11.4.2}$$

denotes the rotation

$$f_1(x, y) = x\cos\theta - y\sin\theta \quad \text{and} \quad f_2(x, y) = x\sin\theta + y\cos\theta. \tag{11.4.3}$$

If **Y** denotes the desired output array, then let $\hat{f} = f|_{\mathbf{Y}} : \mathbf{Y} \to \mathbb{R}^2$ and set

$$\mathbf{b} := \mathbf{a}' \circ \hat{f} \tag{11.4.4}$$

in order to obtain the desired spatial transformation of **a**. In particular, the values of x_i' for $i = 1, 2$ in Equation 11.4.1 are now replaced by the coordinate functions $f_i(y_1, y_2)$ of f. Figure 11.4.2 illustrates a rotation using first order interpolation. Note that in comparison to the rotation using zero-order interpolation (Figure 11.3.2), the boundary of the small interior rectangle has a smoother appearance; the sawtooth effect is visibly reduced. The *outer* boundary of the rotated image retains the sawtooth appearance since no interpolation occurs on points $(x_1', x_2') \in \mathbb{R}^2 \setminus \mathbf{X}'$.

Figure 11.4.2. Rotation using first-order interpolation.

Alternate Image Algebra Formulation

The computational complexity of the bilinear interpolation (Equation 11.4.1) can be improved if we first interpolate along one direction twice and then along the other direction once. Specifically, Equation 11.4.1 can be decomposed as follows. For a given point $(x_1', x_2') \in \mathbf{X}'$ compute

$$\mathbf{a}_1'(x_1', x_2) = \mathbf{a}(x_1, x_2) + [\mathbf{a}(x_1 + 1, x_2) - \mathbf{a}(x_1, x_2)](x_1' - x_1)$$

along the line segment with endpoints (x_1, x_2) and $(x_1 + 1, x_2)$, and

$$\mathbf{a}_2'(x_1', x_2 + 1) = \mathbf{a}(x_1, x_2 + 1) + [\mathbf{a}(x_1 + 1, x_2 + 1) - \mathbf{a}(x_1, x_2 + 1)](x_1' - x_1)$$

along the line segment with endpoints $(x_1, x_2 + 1)$ and $(x_1 + 1, x_2 + 1)$. Then set

$$\mathbf{a}'(x_1', x_2') = \mathbf{a}_1'(x_1', x_2) + [\mathbf{a}_2'(x_1', x_2 + 1) - \mathbf{a}_1'(x_1', x_2)](x_2' - x_2).$$

This reduces the four multiplications and eight additions or subtractions inherent in Equation 11.4.1 to only three multiplications and six additions/subtractions.

Although Equation 11.4.4 represents a functional specification of a spatial image transformation, it is somewhat deceiving; the image $\mathbf{a}'$ in the equation was derived using typical algorithmic notation (Equation 11.4.1). To obtain a functional specification for the interpolated image $\mathbf{a}'$ we can specify its values using three spatial transformations f_0, f_1, and f_2, mapping $\mathbf{X}' \to \mathbf{X}$, defined by

$$f_0(x_1', x_2') = (\lfloor x_1' \rfloor, \lfloor x_2' \rfloor)$$

$$f_1(x_1', x_2') = \begin{cases} (\lfloor x_1' \rfloor + 1, \lfloor x_2' \rfloor) & \text{if } \lfloor x_1' \rfloor < m \\ (\lfloor x_1' \rfloor, \lfloor x_2' \rfloor) & \text{otherwise} \end{cases}$$

and

$$f_2(x_1', x_2') = \begin{cases} (\lfloor x_1' \rfloor, \lfloor x_2' \rfloor + 1) & \text{if } \lfloor x_2' \rfloor < n \\ (\lfloor x_1' \rfloor, \lfloor x_2' \rfloor) & \text{otherwise} \end{cases},$$

and two real-valued images functions $\mathbf{w}_1, \mathbf{w}_2 \in \mathbb{R}^{\mathbf{X}'}$ defined by

$$\mathbf{w}_1(x_1', x_2') = x_1' - \lfloor x_1' \rfloor$$

and

$$\mathbf{w}_2(x_1', x_2') = x_2' - \lfloor x_2' \rfloor.$$

We now define

$$\mathbf{a}' = \mathbf{a} \circ f_0 + (\mathbf{a} \circ f_1 - \mathbf{a} \circ f_0) \cdot \mathbf{w}_1 + (\mathbf{a} \circ f_2 - \mathbf{a} \circ f_0) \cdot \mathbf{w}_2$$
$$+ (\mathbf{a} \circ f_1 \circ f_2 + \mathbf{a} \circ f_0 - \mathbf{a} \circ f_1 - \mathbf{a} \circ f_2) \cdot \mathbf{w}_1 \cdot \mathbf{w}_2.$$

A nice feature of this specification is that the interpolated image $\mathbf{a}'$ is only defined over the region of interest $\mathbf{X}'$ and not over all of $\mathbb{R}^2$.

Comments and Observations

Since $\hat{f} : \mathbf{Y} \to \mathbb{R}^2$, it is very likely that $\hat{f}(\mathbf{Y}) \not\subset \mathbf{X}'$. This means that the output image $\mathbf{b} = \mathbf{a}' \circ \hat{f}$ may contain many zero values, and — if $\mathbf{Y}$ is not properly chosen — not all values of $\mathbf{a}$ will be utilized in the computation of $\mathbf{b}$. The latter phenomenon is called *loss of information due to clipping*. A simple rotation of an image about its center provides an example of both, the introduction of zero values and loss of information due to clipping, if we choose $\mathbf{Y} = \mathbf{X}$. Figure 11.4.3 illustrates this case. Here the left image represents the input image $\mathbf{a}$ and the right image the output image $\mathbf{b} = \mathbf{a}' \circ \hat{f}$. Note that the value $\mathbf{b}(y_1, y_2)$ is zero since $\hat{f}(y_1, y_2) = (x_1', x_2') \notin \mathbf{X}'$. Also, the corner areas of $\mathbf{a}$ after rotation have been clipped since they do not fit into $\mathbf{Y}$. Of course, for rotations the problem clipping is easily resolved by choosing $\mathbf{Y}$ sufficiently large.

 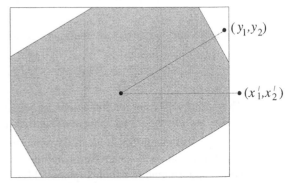

Figure 11.4.3. Rotation within the same array. The left image
is the input image and the right image is the output image.

The definition of the interpolated extension $\mathbf{a}'$ requires images to be specified as computational objects rather than enumerated objects (such as the input image $\mathbf{a}$). Once the spatial transform $f = (f_1, f_2)$ has been chosen, the dummy variables x_i and x_i' are replaced by $\lfloor f_i(y_i) \rfloor$ and $f_i(y_i)$, respectively. Each pixel value of $\mathbf{b} = \mathbf{a}' \circ \hat{f}$ can then be determined pixel by pixel, line by line, or in parallel.

The spatial transformation of a digital image as exhibited by Equation 11.4.4 represents a general scheme. It is not restricted to bilinear interpolation; any extension $\mathbf{a}'$ of $\mathbf{a}$ to a point set $\mathbf{X}'$ containing the range of f may be substituted. This is desirable even though bilinear interpolation and nearest neighbor approximation are the most widely used interpolation techniques. Similar to zero-order, first-order interpolation has its own drawbacks. The surface given by the graph of $\mathbf{a}'$ is not smooth; when adjacent four pixel neighborhoods are interpolated, the resulting surfaces match in amplitude at the boundaries but do not match in slope. The derivatives have, in general, discontinuities at the boundaries. In many applications, these discontinuities produce undesirable effects. In these cases, the extra additional computational cost of higher order interpolation schemes may be justified. Examples of higher order interpolation functions are cubic splines, Legendre interpolation, and $\frac{1}{x}\sin x$. Higher order interpolation is usually implemented by an image-template operation.

11.5.   Application of Image Rotation to the Computation
of Directional Edge Templates

In a continuous image, a sharp intensity transition between neighboring pixels as shown in Figure 11.5.1(a) would be considered to be an edge. Such steep changes in intensities can be detected by analyzing the derivatives of the signal function. In sampled waveforms such as shown in Figure 11.5.1(b), approximations to the derivative, such as finite difference methods, are used to detect the existence of edges. However, due to sampling, high frequency components are introduced, and every pair of pixels with different intensities could be considered an edge. For this reason, smoothing before edge enhancement followed by thresholding after enhancement are an important part of many edge detection schemes.

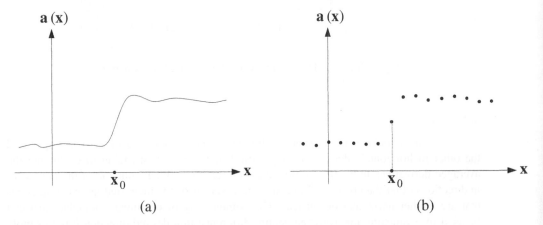

Figure 11.5.1. (a) Continuous image with edge phenomenon. (b) Sampled image function.

The gradient of an image $\mathbf{a}$ is defined in terms of direction oriented spatial derivatives as

$$\nabla \mathbf{a}(\mathbf{x}) = \frac{\partial \mathbf{a}(\mathbf{x})}{\partial \mathbf{x}} = \left(\begin{array}{c} \frac{\partial \mathbf{a}(x_1, x_2)}{\partial x_1} \\ \frac{\partial \mathbf{a}(x_1, x_2)}{\partial x_2} \end{array} \right).$$

One discrete approximation of the gradient is given in terms of the centered differences

$$d(x_1) = \frac{\mathbf{a}(x_1 + \Delta x_1, x_2) - \mathbf{a}(x_1 - \Delta x_1, x_2)}{2(\Delta x_1)}$$

and

$$d(x_2) = \frac{\mathbf{a}(x_1, x_2 + \Delta x_2) - \mathbf{a}(x_1, x_2 - \Delta x_2)}{2(\Delta x_2)}.$$

The centered differences $d(x_1)$ and $d(x_2)$ can be implemented using the templates show in Figure 11.5.2.

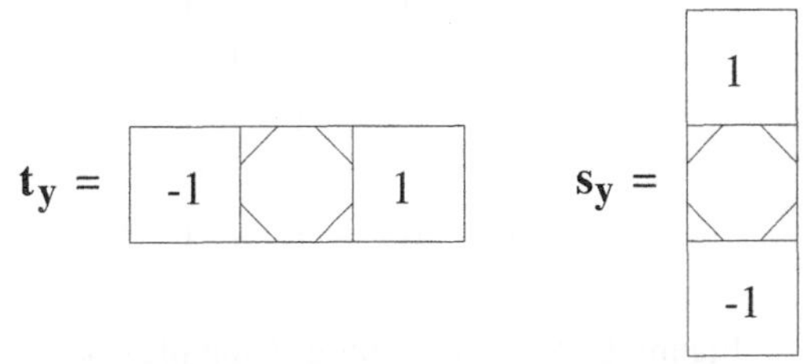

$$\mathbf{t_y} = \boxed{-1 \quad 1} \qquad \mathbf{s_y} = \boxed{\begin{array}{c} 1 \\ -1 \end{array}}$$

Figure 11.5.2. The centered difference templates $\mathbf{t}$ and $\mathbf{s}$.

The pixel values at location $\mathbf{x} = (x_1, x_2)$ of $\mathbf{a} \oplus \mathbf{t}$ and $\mathbf{a} \oplus \mathbf{s}$ are the centered differences $d(x_1)$ and $d(x_2)$, respectively. This concept forms the basis for extensions to various templates used for edge detection. Variants of the centered difference templates are the 3×3 templates shown in Figure 11.5.3 which form *smoothened* or *averaged* central

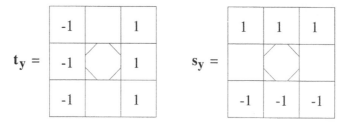

Figure 11.5.3. The averaged centered difference operators.

difference operators.

The templates **t** and **s** are orthogonal; one is sensitive to vertical edges and the other to horizontal edges. We may view **t** and **s** as corresponding to evaluating the averaged derivatives in the $0°$ and $90°$ direction, respectively. Evaluation of the derivative in directions other than $0°$ or $90°$ becomes necessary if one is interested in detecting edges that are neither horizontal or vertical. One simple way of obtaining a template that can be used for evaluating the averaged central difference in a desired direction θ is to simply rotate the image $\mathbf{t}_{(0,0)}$ through the angle θ and use the resulting image values for the weights of the new template. For example, applying a rotation f as defined in Section 11.4 to the image $\mathbf{t}_{(0,0)}$ with $\theta = 90°$ results in the image $\mathbf{t}_{(0,0)} \circ f$ which is identical to the image $\mathbf{s}_{(0,0)}$. Since **s** is translation invariant, the image $\mathbf{t}_{(0,0)} \circ f$ completely determines **s**. However, for angles θ other than $90°$ the resulting image $\mathbf{t}_{(0,0)} \circ f$ may not be a satisfactory approximation of the averaged central difference in that direction; for $\theta = 30°$ the image $\mathbf{t}_{(0,0)} \circ f$ is basically the same as $\mathbf{t}_{(0,0)}$. The reason for this is the small size of the support of $\mathbf{t}_{(0,0)}$ and associated interpolation errors. In order to obtain a more accurate representation, one scheme is to enlarge the image $\mathbf{t}_{(0,0)}$ to the image shown in Figure 11.5.4 and rotate

-1	-1	0	1	1
-1	-1	0	1	1
-1	-1	0	1	1
-1	-1	0	1	1
-1	-1	0	1	1

Figure 11.5.4. The enlarged template **t**.

this enlarged image using bilinear interpolation. For $\theta = 30°$ the resulting image $\mathbf{t}_{(0,0)} \circ f$

is of the form shown in Figure 11.5.5.

-0.73	0.13	1	1	1
-1	-0.37	0.5	1	1
-1	-0.87	0	0.87	1
-1	-1	-0.5	0.37	1
-1	-1	-1	-0.13	0.73

Figure 11.5.5. The enlarged template $\mathbf{t}$ rotated through angle $\theta = 30°$.

The final 3×3 template $\mathbf{r}$ is obtained by restricting its support to a 3×3 neighborhood of the center pixel location $\mathbf{y}$ (Figure 11.5.6).

$$\mathbf{r_y} = \begin{array}{|c|c|c|} \hline -0.37 & 0.5 & 1 \\ \hline -0.87 & 0 & 0.87 \\ \hline -1 & -0.5 & 0.37 \\ \hline \end{array}$$

Figure 11.5.6. The final $30°$ centered difference operator.

While $\mathbf{t}$ will result in maximal enhancement of vertical or $90°$ edges, $\mathbf{r}$ will be more sensitive to $120°$ edges. Likewise, rotation through an angle of $60°$ results in a template which is sensitive to $150°$ edges (Figure 11.5.7).

$$\mathbf{s_y} = \begin{array}{|c|c|c|} \hline 0.37 & 0.87 & 1 \\ \hline -0.5 & 0 & 0.5 \\ \hline -1 & -0.87 & -0.37 \\ \hline \end{array}$$

Figure 11.5.7. The averaged $60°$ centered difference operator.

Image Algebra Formulation

Let $\mathbf{X} = \{-2, \cdots, 2\} \times \{-2, \cdots, 2\}$. Define an ideal step image $\mathbf{a} \in \mathbb{R}^{\mathbf{X}}$ by

$$\mathbf{a}(x_1, x_2) = \begin{cases} -1 & \text{if } x_2 < 0 \\ 1 & \text{if } x_2 > 0 \\ 0 & \text{otherwise.} \end{cases}$$

Create the continuous extension of $\mathbf{a}$, namely $\mathbf{a}'$, as described in Section 11.4. Also define $f_1(x_1, x_2) = x_1 cos\theta - x_2 sin\theta$, $f_2(x_1, x_2) = x_1 sin\theta + x_2 cos\theta$, $f(x_1, x_2) =$

$(f_1(x_1, x_2), f_2(x_1, x_2))$. Let $\mathbf{Y} = \{-1, 0, 1\} \times \{-1, 0, 1\}$ and $\hat{f} = f|_\mathbf{Y} : \mathbf{Y} \to \mathbb{R}^2$. The translation invariant template $\mathbf{t}$, for calculating the derivative of an image in direction θ, is defined by specifying its image at the origin, $\mathbf{t}_{(0,0)} = \mathbf{a}' \circ \hat{f}$.

11.6.   General Affine Transforms

The general formulation of spatial transformation given by Equation 11.3.1 includes the class of affine transformations. A transformation $f : \mathbb{R}^2 \to \mathbb{R}^2$ of the form

$$f(y_1, y_2) = (ay_1 + by_2 + v_1, \; cy_1 + dy_2 + v_2),$$

where a, b, d, v_1, and v_2 are constants, is called a (2–dimensional) *affine transformation*. An equivalent definition of an affine transformation is given by

$$f(\mathbf{y}) = \mathbf{y} \cdot \mathrm{A} + \mathbf{v}, \qquad\qquad (11.6.1)$$

where

$$\mathbf{y} = (y_1, y_2), \quad \mathrm{A} = \begin{pmatrix} a & c \\ b & d \end{pmatrix}, \quad \text{and } \mathbf{v} = (v_1, v_2).$$

If $\mathrm{A} = \begin{pmatrix} \frac{1}{\alpha} & 0 \\ 0 & \frac{1}{\alpha} \end{pmatrix}$, $\mathbf{v} = \mathbf{0}$, and $\alpha > 1$, then f represents a magnification by a factor α. On the other hand, if $0 < \alpha < 1$, then f represents a contraction (shrinking) by the factor α.

The matrix A can always be written in the form

$$\begin{pmatrix} a & c \\ b & d \end{pmatrix} = \begin{pmatrix} r_1 \cos\theta_1 & r_1 \sin\theta_1 \\ -r_2 \sin\theta_2 & r_2 \cos\theta_2 \end{pmatrix},$$

where (r_1, θ_1) and $(r_2, \theta_2 + \frac{\pi}{2})$ correspond to the points (a, c) and (b, d) expressed in polar form. In particular, if $r_1 = r_2$, $\theta_1 = \theta_2$, and $\mathbf{v} = \mathbf{0}$, then Equation 11.6.1 corresponds to a rotation about the origin. If A is the identity matrix and $\mathbf{v} \neq \mathbf{0}$, then Equation 11.6.1 represents a translation.

In addition to simple image rotation, translation, magnification and contraction, affine transformations are used in such varied tasks as image compression, registration, analysis, and image generation [1, 2, 3, 4]. Various combinations of affine transformations can be used to produce highly complex patterns from simple patterns. In this section we provide examples of image construction using simple combinations of some of the affine transforms discussed above.

Image Algebra Formulation

Let $f_1(\mathbf{y}) = \mathbf{y} \cdot \mathrm{A}$, where $\mathrm{A} = \begin{pmatrix} \frac{1}{\alpha} & 0 \\ 0 & \frac{1}{\alpha} \end{pmatrix}$ and $\alpha = \frac{2}{3}$. The image $\mathbf{a} \circ f_1$ represents a contraction (shrinking) of the input image $\mathbf{a}$ by the factor of $\frac{2}{3}$. Figure 11.6.1 illustrates the contraction. Here the input image $\mathbf{a}$, shown on the left, contains a trapezoid of base length l and angle of inclination θ. The output image $\mathbf{a} \circ f_1$ is shown on the right of the figure, with pixels having zero values displayed in black.

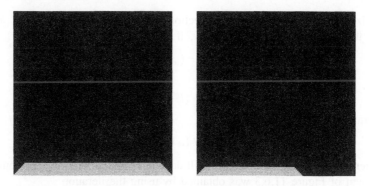

Figure 11.6.1. Image contraction using an affine map.

Suppose

$$f_2(\mathbf{y}) = \mathbf{y} \cdot \begin{pmatrix} 1 & 0 \\ 0 & 1 \end{pmatrix} + \left[(\alpha - 1)\frac{l}{2}, (\alpha - 1)\frac{l}{2}\tan\theta \right].$$

Then f_2 represents a shift in the direction θ. Composing $\mathbf{a} \circ f_1$ with f_2 results in the image $(\mathbf{a} \circ f_1) \circ f_2$ shown on the left of Figure 11.6.2.

The composition of two affine transformations is again an affine transformation. In particular, by setting $f = f_1 \circ f_2$, it is obvious that $(\mathbf{a} \circ f_1) \circ f_2$ could have been obtained from a single affine transformation, namely

$$(\mathbf{a} \circ f_1) \circ f_2 = \mathbf{a} \circ (f_1 \circ f_2) = \mathbf{a} \circ f.$$

Now iterating the process using the algorithm

$$\mathbf{b} := \mathbf{b} + \mathbf{b} \circ f$$

with initial image $\mathbf{b} = \mathbf{a}$, results in the *railroad to infinity* shown on the right of Figure 11.6.2.

Figure 11.6.2. A railroad to infinity.

Another example of iterating affine transformations in order to create geometric patterns from simple building blocks is the construction of a brick wall from a single brick. In this example, let w and l denote the width and length of the brick shown on the left of Figure 11.6.3. Suppose further that we want the cement layer between an adjacent pair of bricks to be of thickness t. A simple way of building the wall is to use two affine transformations f and g, where f is a horizontal shift by an amount $l + t$, and g is composed of a horizontal shift in the opposite direction of f by the amount $(l+t)/2$ and a vertical shift by the amount $w+t$. Specifically, if

$$f(y_1, y_2) = (y_1 - (l+t), y_2)$$

and

$$g(y_1, y_2) = (y_1 + (l+t)/2,\ y_2 - (w+t)),$$

then iterating the algorithm

$$\mathbf{a} := \mathbf{a} \vee (\mathbf{a} \circ f) \vee (\mathbf{a} \circ g)$$

will generate a brick wall whose size will depend on the number of iterations. The image on the right of Figure 11.6.3 was obtained by using the iteration

$$\text{repeat}$$

$$\mathbf{c} := \mathbf{a}$$

$$\mathbf{a} := \mathbf{a} \vee (\mathbf{a} \circ f) \vee (\mathbf{a} \circ g)$$

$$\text{until } \mathbf{c} = \mathbf{a}$$

Figure 11.6.3. Generation of a brick wall from a single brick.

11.7. Fractal Constructs

Although the name *fractal* was coined by B.B. Mandelbrot [5, 6], fractal objects were well known to mathematicians for at least a century and played an important role in such areas as topology, function theory, and dimension theory [7, 8]. Since these early works in fractal research, fractals have found wide-use applications in image analysis, image compression, image rendering, and computer graphics [9, 2, 1, 10].

A fractal object displays self-similarities, in a somewhat technical sense, on all scales. The object need not exhibit *exactly* the same structure at all scales, but the same *type* of structure must appear at all scales. Prototypical examples are the *Sirpinski curve*, or Sirpinski triangle, and the *von Koch snowflake*. The Sirpinski curve is formed by starting with a triangle and connecting the midpoints of the three edges, thus dividing the given triangle into four triangles. Removing the interior triangle (the one sharing no vertex with the original triangle) leaves three triangles, each sharing a vertex with the original triangle as shown in Figure 11.7.3. The process is then repeated on each of the triangles and continues ad infinitum.

The construction of the von Koch snowflake starts with an equilateral triangle. Each edge of the triangle is divided into three segments of equal length. Three new equilateral triangles are formed, one on each middle segment, resulting in a Star of David (Figure 11.7.5). The process is repeated on each of the *six* smaller triangles determining the boundary of the Star of David. Again, the iteration continues ad infinitum.

Image Algebra Formulation

In order to generate the Sirpinski curve image, let $\mathbf{X} = \mathbb{Z}_n^+ + \mathbb{Z}_n^+$, where $n = 2^k$, and let $\mathbf{a} \in \mathbb{Z}_2^{\mathbf{X}}$ be the input image containing the triangle defined by

$$\mathbf{a}(x, y) = \begin{cases} 1 & if \ 2x \geq y \ and \ 2(n - x) \geq y \\ 0 & otherwise. \end{cases}$$

Define an affine contraction map $f : \mathbb{R}^2 \to \mathbb{R}^2$ by $f(x, y) = (2x, 2y)$. The Sirpinski curve construction is now given by the following algorithm:

for $i := 1$ to k loop

  $\mathbf{b} := \mathbf{a} \circ f$

  $\mathbf{c} := \mathbf{b} + (n/2, 0)$

  $\mathbf{d} := \mathbf{b} + (n/2, n/2)$

  $\mathbf{a} := \mathbf{b} \vee \mathbf{c} \vee \mathbf{d}$

end loop

The input image $\mathbf{a}$ is shown on the left in Figure 11.7.1, while the contraction $\mathbf{a} \circ f$ is shown on the right of Figure 11.7.1. In Figure 11.7.2, the translated images $\mathbf{c}$ and $\mathbf{d}$ are shown on the left and right, respectively. Figure 11.7.3 shows the image $\mathbf{a} := \mathbf{b} \vee \mathbf{c} \vee \mathbf{d}$ after the first and second iteration while Figure 11.7.4 shows the fifth and kth iteration.

Figure 11.7.1. Input image $\mathbf{a}$ (left) and the contraction $\mathbf{a} \circ f$ (right).

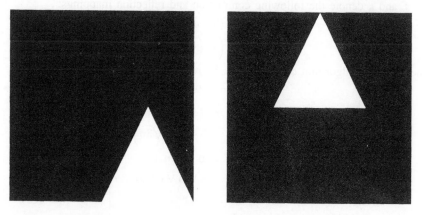

Figure 11.7.2. The translations $\mathbf{c}$ and $\mathbf{d}$ of the contraction $\mathbf{a} \circ f$.

Figure 11.7.3. Output $\mathbf{a} = \mathbf{b} \vee \mathbf{c} \vee \mathbf{d}$ after the first
iteration (left) and after the second iteration (right).

Figure 11.7.4. The Sirpinski curve after the fifth iteration (left) and the final iteration (right).

In order to construct the von Koch snowflake, let $\mathbf{X} = \mathbb{Z}_{3n}^+ + \mathbb{Z}_{4n}^+$, where
$n = 3^{k-1}$. Define the image $\mathbf{a} \in \mathbb{Z}_2^{\mathbf{X}}$ containing a solid equilateral triangle by

$$\mathbf{a}(xy) = \begin{cases} 1 & if\ y \geq n \\ and & y \leq \sqrt{3}x + n \\ and & y \leq -\sqrt{3}x + (3\sqrt{3}+1)n \\ 0 & otherwise \end{cases}$$

Let f and r denote the following contraction and reflection mappings:

$$f(x,y) = (3x, 3y - 2n)$$
$$r(x,y) = \left(x, \left(\sqrt{3}+2\right)n - y\right)$$

The following algorithm constructs the Snowflake:

$$\begin{aligned}
&\text{for } i := 1 \text{ to } k \text{ loop} \\
&\qquad \mathbf{b} := \mathbf{a} \circ f \\
&\qquad \mathbf{c} := \mathbf{b} + \left(n, \sqrt{3}n\right) \\
&\qquad \mathbf{d} := \mathbf{b} + (2n, 0) \\
&\qquad \mathbf{e} := \mathbf{a} \vee \mathbf{b} \vee \mathbf{c} \vee \mathbf{d} \\
&\qquad \mathbf{a} := \mathbf{e} \circ r \\
&\text{end loop}
\end{aligned}$$

Figures 11.7.5 and 11.7.6 show the image $\mathbf{a} = \mathbf{e} \circ r$ at various stages in the loop. The image on the left of Figure 11.7.5 shows the output image $\mathbf{a}$ after the first iteration while the image on the right shows the image after the second iteration. The third and kth iteration are illustrated in Figure 11.7.6.

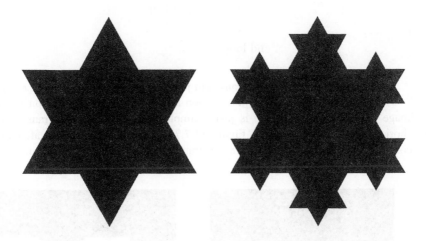

Figure 11.7.5.   The first and second iterations in the construction of the von Koch snowflake.

Figure 11.7.6.   The third and final iterations in the construction of the von Koch snowflake.

Alternate Image Algebra Formulation

There are various alternate image algebra formulations for constructing the Sirpinski curve or the von Koch snowflake. Different input images can also be used in order to obtain essentially the same fractal construct. Here we provide but one example. An alternate way of constructing the Sirpinski curve on $\mathbf{X} = \mathbb{Z}_n^+ + \mathbb{Z}_n^+$ is to start with the unit image $\mathbf{a} = \mathbf{1}$ on $\mathbf{X}$ and implement the following algorithm:

$$
\begin{aligned}
&\text{for } i := 1 \text{ to } k \text{ loop} \\
&\qquad \mathbf{b} := \mathbf{a} \circ f \\
&\qquad \mathbf{c} := \mathbf{b} + (0, n/2) \\
&\qquad \mathbf{d} := \mathbf{b} + (n/2, 0) \\
&\qquad \mathbf{a} := \mathbf{b} \vee \mathbf{c} \vee \mathbf{d} \\
&\text{end loop}
\end{aligned}
$$

In this algorithm, the affine transform is again given by $f(x, y) = (2x, 2y)$, and $n = 2^k$. The difference between the first version and the alternate formulation is that the input image in the alternate version is just a simple square and the two translations are simple horizontal and vertical shifts. Figure 11.7.7 shows the output image after the first and second iteration, while Figure 11.7.8 illustrates the fourth and kth iteration.

Figure 11.7.7. The output after the first and second iterations of the alternate Sirpinski curve construction.

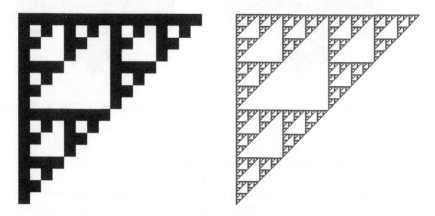

Figure 11.7.8. The output after the fourth and final iterations of the alternate Sirpinski curve construction.

Comments and Observations

Since digital images are finite (there are only $m \cdot n$ pixels in an $m \times n$ image) and a fractal object is obtained by an infinite iteration process, true fractals cannot be generated in digital image processing. Hence fractal objects in digital images are only first approximations of real fractal objects. This observation is sometimes lost in the engineering literature.

Another important consideration in digital fractal construction concerns the application of geometric contractions. In the transformation matrix $A = \begin{pmatrix} \frac{1}{\alpha} & 0 \\ 0 & \frac{1}{\alpha} \end{pmatrix}$ corresponding to $f(x, y) = (2x, 2y)$, $\alpha = \frac{1}{2} < 1$. It follows from the definition given in Section 11.6 that f is a contraction. This may seem a little odd as f actually *enlarges* or *dilates* objects in the plane by a factor of two. However, although $f : \mathbf{Y} \to \mathbf{X}$ enlarges spatial objects in $\mathbf{Y}$ to twice their size in $\mathbf{X}$, the input image $\mathbf{a}$ is an image on $\mathbf{X}$ while $\mathbf{a} \circ f$ is an image on $\mathbf{Y}$. Therefore, an object in image $\mathbf{a}$ will appear two times smaller in the image $\mathbf{a} \circ f$.

Although the affine map $f(x, y) = (2x, 2y)$ is a function from $\mathbb{R}^2$ to $\mathbb{R}^2$, in the underlying implementation of the algorithms specified in this as well as previous sections, the function is simply viewed as a map from the sub-array $\mathbf{Y} = \mathbb{Z}_{n/2}^+ + \mathbb{Z}_{n/2}^+$ of $\mathbf{X}$ into $\mathbf{X}$. Thus, strictly speaking, the resulting image $\mathbf{a} \circ f$ is an image on the array $\mathbf{Y}$. In our specification, we assume that $\mathbf{a} \circ f$ has been extended to the array $\mathbf{X}$ by assignment of zero values to the pixels in $\mathbf{X} \backslash \mathbf{Y}$.

Note that the triangle in the input image of the Sirpinski curve algorithm has base and height equal to $n = 2^k$, with the base equal to the bottom boundary of the image. In the von Koch snowflake construction, the starting triangle had to be moved up in the input image with its base sitting on the line $y = n$ in order to avoid clipping of objects when applying the reflection $\mathbf{a} = \mathbf{e} \circ r$. The line of reflection is parallel to the base of the input triangle at a distance corresponding to $\frac{1}{3}$ the height of the input triangle from the base. Also, the images are displayed in the standard (x, y)-coordinate system, with the origin in the bottom left-hand corner instead the usual upper right-hand corner as is common for digital images. In other words, the display images are upside down. This was done in order to order to avoid confusion and display the fractal objects as they usually appear in textbooks.

11.8. Iterated Function Systems

The rendering of images in the preceding two sections was accomplished by iteration of a given set of affine transforms. The notion of iterative function systems (IFS) was first formalized by M. Barnsley and has been used in a variety of applications such as image rendering, analysis, and compression [9, 2, 1, 10]. An IFS on $\mathbb{R}^2$ is a finite set of affine mappings $\{f_1, f_2, \ldots, f_n\}$ with each f_i being a contraction mapping. A contraction mapping $f : \mathbb{R}^2 \to \mathbb{R}^2$ has the property that

$$d(f(\mathbf{x}), f(\mathbf{y})) \leq s \cdot d(\mathbf{x}, \mathbf{y})$$

for some constant s with $0 < s < 1$ and $\forall \mathbf{x}, \mathbf{y} \in \mathbb{R}^2$. This property means that contraction mappings of compact sets must possess fixed points. For iterative function systems, these fixed points result in the *attractors* of the systems. More precisely, suppose $f : \mathbf{X} \to \mathbf{X}$ is a contraction, where $\mathbf{X}$ is some compact subset of $\mathbb{R}^2$. For $\mathbf{B} \subset \mathbf{X}$, define $f(\mathbf{B}) = \{f(\mathbf{x}) : \mathbf{x} \in \mathbf{B}\}$. If $\{f_1, f_2, \ldots, f_n\}$ is an IFS on $\mathbf{X}$, then the function $F : 2^{\mathbf{X}} \to 2^{\mathbf{X}}$, defined by

$$F(\mathbf{B}) = \bigcup_{i=1}^{n} f_i(\mathbf{B}),$$

is a contraction mapping. It can be shown that there is a unique set $\mathbf{A} \in 2^{\mathbf{X}}$ such that $\mathbf{A} = lim_{k \to \infty} F^k(\mathbf{B})$ for any compact set $\mathbf{B} \in 2^{\mathbf{X}}$. Here F^k denotes the kth iterate of F. The set $\mathbf{A}$ is called the *attractor* of the IFS $\{f_1, f_2, \dots, f_n\}$. The von Koch snowflake and the Sirpinski curve constructed in the previous section are examples of attractors of the respective iterative function systems used in their construction.

The algorithms for constructing the von Koch snowflake and the Sirpinski curve in the preceding section are deterministic algorithms. Far more interesting and complex images can be constructed by using random iterations of iterative function systems. In particular, given an IFS $\{f_1, f_2, \dots, f_n\}$ on $\mathbf{X}$, one can assign a probability p_i to each f_i with $\sum_{i=1}^{n} p_i = 1$. Choosing an initial point $\mathbf{x}_0 \in \mathbf{X}$ and then choosing recursively

$$\mathbf{x}_k \in \{f_1(\mathbf{x}_{k-1}), f_2(\mathbf{x}_{k-1}), \dots, f_n(\mathbf{x}_{k-1})\},$$

where the probability of the event $\mathbf{x}_k = f_i(\mathbf{x}_{k-1})$ is p_i, one obtains a sequence $\{\mathbf{x}_k : k = 0, 1, 2, \dots\} \subset \mathbf{X}$ that converges to the attractor of $\{f_1, f_2, \dots, f_n\}$. We illustrate the use of random iteration and attractors by providing an image algebra formulation for rendering the image of a fern.

Image Algebra Formulation

Let

$$f_1(x, y) = (0, \ 0.16y)$$
$$f_2(x, y) = (0.85x + 0.04y, \ -0.4x + 0.85y + 1.6)$$
$$f_3(x, y) = (0.2x - 0.26y, \ 0.23x + 0.22y + 1.6)$$
$$f_4(x, y) = (-0.15x + 0.28y, \ 0.26x + 0.24y + 0.44)$$

and $\mathbf{X} = \mathbb{Z}_n \times \mathbb{Z}_n$.

$$\mathbf{a} := 0 \in \mathbb{Z}_2^{\mathbf{X}}$$
$$(x, y) := (0, 0)$$

for in $1..m$ **loop**

$\quad k := choice\left(\mathbb{Z}_{100}^{+}\right)$

$\quad i := 1$ **if** $k = 1$

$\quad i := 2$ **if** $2 \le k \le 86$

$\quad i := 3$ **if** $87 \le k \le 93$

$\quad i := 4$ **if** $94 \le k \le 100$

$\quad (x, y) := f_i(x, y)$

$\quad \mathbf{a}([75x + 150, 48y]) := 1$

end loop

print $\mathbf{a}$

Applying this algorithm to a 500×500 array produces the attractor set shown in Figure 11.8.1.

Figure 11.8.1. Fern image after one, two, three, and nine iterations.

Comments and Observations

Observe that the probabilities for the event $\mathbf{x}_k = f_i(\mathbf{x}_{k-1})$, $i = 1, \ldots, 4$, in the above algorithm are given by $p_1 = 0.01$, $p_2 = 0.85$, $p_3 = 0.07$, and $p_4 = 0.07$. These probabilities are generally obtained by computing

$$p_i \approx \frac{|det\, A_i|}{\sum\limits_{i=1}^{n} |det\, A_i|} = \frac{|a_i d_i - b_i c_i|}{\sum\limits_{i=1}^{n} |a_i d_i - b_i c_i|} \, ,$$

where A_i denotes the transformation matrix corresponding to f_i and $\approx$ means approximately equal. If $det\, A_i = 0$, then p_i should be assigned a small positive number.

11.9.   Exercises

1. Construct a synthetic binary image $\mathbf{a}$ containing a disk D of some large radius r.

    a.   Implement a shrinking algorithm that shrinks D to a disk of radius $r - 1$.
    b.   Apply your algorithm successively until D consists of a single pixel. During successive applications, does D retain its circular structure?

2.

    a.   Implement the nearest neighbor image rotation algorithm.

    b.   Using a gray value image containing a scene as initial input, successively rotate the image through an angle of 30° using an input the image obtained from the previous rotation. Compute the output after twelve successive rotations with the original input image.

3.   Repeat Exercise 2 using bilinear interpolation.

4.   Specify an algorithm in image algebra that transforms an $m \times n$ rectangular image into a trapezoidal image with base n and height m

5.   Specify an algorithm in image algebra that transforms a horizontal line into a semi-circle. Implement your algorithm.

6.   In Section 11.6 we presented an algorithm that generated a rectangular brick wall from a single rectangular brick. Specify a similar algorithm that will generate a hexagonal brick wall from a single hexagonal brick. Your algorithm should leave a cement crack of thickness t between adjacent bricks. Implement your algorithms.

7.   Let $\mathbf{X} = \mathbb{Z}_n^+ \times \mathbb{Z}_n^+$, where $n = 2^k$ and $\mathbf{a} \in \mathbb{Z}_2^{\mathbf{X}}$ be defined by

$$\mathbf{a}(x, y) = \begin{cases} 1 & if\ x \leq y \\ 0 & otherwise \end{cases}.$$

Implement the Sirpinski curve algorithm using the image $\mathbf{a}$ as your input.

8.   Construct your own fractal construct by specifying affine transforms and an appropriate input image. Specify your algorithm in image algebra.

9.   Specify an alternative algorithm in image algebra that will create the von Koch snowflake.

10.   Specify an iterative function system for constructing the Sirpinski curve. Write an image algebra algorithm that will construct the Sirpinski curve using the specified iterative function system.

11.10. References

[1] M. Barnsley, *Fractals Everywhere*. Boston, MA: Academic Press, 1988.

[2] M. Barnsley, A. Jacquin, F. Malassenet, L. Reuter, and A. Sloan, "Harnessing chaos for image synthesis," *Computer Graphics*, vol. 22, pp. 131–137, Aug. 1988.

[3] D. Ballard and C. Brown, *Computer Vision*. Englewood Cliffs, NJ: Prentice-Hall, 1982.

[4] C. Bandt, "Self similar matrices and fractal tilings of r^n," *Proceedings of the American Mathematical Society*, vol. 112, no. 2, pp. 549–562, 1991.

[5] B. Mandelbrot, *Fractals: Form, Chance, and Dimension*. San Francisco: Freeman, 1977.

[6] B. Mandelbrot, *The Fractal Geometry of Nature*. San Francisco: Freeman, 1983.

[7] W. Hurewicz and H. Wallman, *Dimension Theory*. Princeton, NJ: Princeton University Press, 1948.

[8]   J. Hocking and G. Young, *Topology*. Reading, MA: Addison-Wesley, 1961.

[9]   A. Pentland, "Fractal-based description of natural scenes," *IEEE Transactions on Pattern Analysis and Machine Intelligence*, vol. 6, no. 6, 1984.

[10] A. Jacquin, "Fractal image coding: A review," *Proceedings of the IEEE*, vol. 81, pp. 1451–1465, Oct. 1993.

CHAPTER 12
NEURAL NETWORKS AND CELLULAR AUTOMATA

12.1.   Introduction

Artificial neural networks (ANNs) and cellular automata have been successfully employed to solve a variety of computer vision problems [1, 2]. One goal of this chapter is to demonstrate that image algebra provides an ideal environment for precisely expressing current popular neural network models and their computations. Artificial neural networks (ANNs) are systems of dense interconnected simple processing elements. There exist many different types of ANNs designed to address a wide range of problems in the primary areas of pattern recognition, signal processing, and robotics. The function of different types of ANNs is determined primarily by the processing elements' pattern of connectivity, the strengths (weights) of the connecting links, the processing elements' characteristics, and training or learning rules. These rules specify an initial set of weights and indicate how weights should be adapted during use to improve performance.

The theory and representation of the various ANNs is motivated by the functionality and representation of biological neural networks. For this reason, processing elements are usually referred to as neurons while interconnections are called axons and/or synaptic connections. Although representations and models may differ, all have the following basic components in common:

(a)   A finite set of neurons $a(1), a(2), \ldots a(n)$ with each neuron $a(i)$ having a specific neural value at time t which we will denote by $a_t(i)$.

(b)   A finite set of axons or neural connections $W = (w_{ij})$, where w_{ij} denotes the strength of the connection of neuron $a(i)$ with neuron $a(j)$.

(c)   A propagation rule $\tau_t(i) = \sum_{j=1}^{n} a_t(j) \cdot w_{ij}$.

(d)   An activation function f which takes τ as an input and produces the next state of the neuron

$$a_{t+1}(i) = f(\tau_t(i) - \theta),$$

where θ is a threshold and f a hard limiter, threshold logic, or sigmoidal function which introduces a nonlinearity into the network.

It is worthwhile noting that image algebra has suggested a more general concept of neural computation than that given by the classical theory [3, 4, 5, 6].

Cellular automata and artificial neural networks share a common framework in that the new or next stage of a neuron or cell depends on the states of other neurons or cells. However, there are major conceptual and physical differences between artificial neural networks and cellular automata. Specifically, an *n-dimensional cellular automaton* is a discrete set of cells (points, or sites) in $\mathbb{R}^n$. At any given time a cell is in one of a finite number of states. The arrangement of cells in the automaton form a regular array, e.g., a square or hexagonal grid.

333

As in ANNs, time is measured in discrete steps. The next state of a cell is determined by a spatially and temporally local rule. However, the new state of a cell depends only on the current and previous states of its neighbors. Also, the new state depends on the states of its neighbors only for a fixed number of steps back in time. The same update rule is applied to every cell of the automaton in synchrony.

Although the rules that govern the iteration locally among cells is very simple, the automaton as a whole can demonstrate very fascinating and complex behavior. Cellular automata are being studied as modeling tools in a wide range of scientific fields. As a discrete analogue to modeling with partial differential equations, cellular automata can be used to represent and study the behavior of natural dynamic systems. Cellular automata are also used as models of information processing.

In this chapter we present an example of a cellular automaton as well as an application of solving a problem using cellular automata. As it turns out, image algebra is well suited for representing cellular automata. The states of the automata (mapped to integers, if necessary) can be stored in an image variable $\mathbf{a} \in \mathbb{Z}_n^{\mathbf{X}}$ as pixel values. Template-image operations can be used to capture the state configuration of a cell's neighbors. The synchronization required for updating cell states is inherent in the parallelism of image algebra.

12.2.   Hopfield Neural Network

A pattern, in the context of the N node Hopfield neural network to be presented here, is an N-dimensional vector $\mathbf{p} = (p_1, p_2, \ldots, p_N)$ from the space $\mathbf{P} = \{-1, 1\}^N$. A special subset of $\mathbf{P}$ is the set of exemplar patterns $\mathbf{E} = \{\mathbf{e}^k : 1 \leq k \leq K\}$, where $\mathbf{e}^k = \left(e_1^k, e_2^k, \ldots, e_N^k\right)$. The Hopfield net associates a vector from $\mathbf{P}$ with an exemplar pattern in $\mathbf{E}$. In so doing, the net partitions $\mathbf{P}$ into classes whose members are in some way similar to the exemplar pattern that represents the class. For image processing applications the Hopfield net is best suited for binary image classification. Patterns were described as vectors, but they can just as easily be viewed as binary images and vice versa. A description of the components in a Hopfield net and how they interact follows next and is based on the description given by Lippmann [7]. An example will then be provided to illustrate image classification using a Hopfield net.

As mentioned in the introduction, neural networks have four common components. The specifics for the Hopfield net presented here are outlined below.

1. Neurons

The Hopfield neural network has a finite set of neurons $\mathbf{a}(i), 1 \leq i \leq N$, which serve as processing units. Each neuron has a value (or state) at time t denoted by $\mathbf{a}_t(i)$. A neuron in the Hopfield net can be in one of two states, either -1 or $+1$; i.e., $\mathbf{a}_t(i) \in \{-1, +1\}$.

2. Synaptic Connections

The permanent memory of a neural net resides within the interconnections between its neurons. For each pair of neurons, $\mathbf{a}(i)$ and $\mathbf{a}(j)$, there is a number w_{ij} called the strength of the synapse (or connection) between $\mathbf{a}(i)$ and $\mathbf{a}(j)$. The design specifications for this version of the Hopfield net require that $w_{ij} = w_{ji}$ and $w_{ii} = 0$ (see Figure 12.2.1).

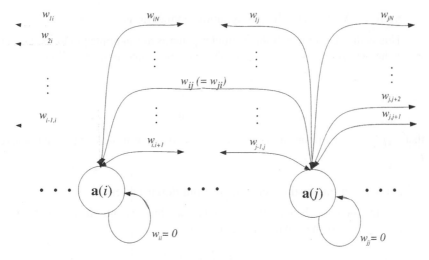

Figure 12.2.1. Synaptic connections for nodes $\mathbf{a}(i)$ and $\mathbf{a}(j)$ of the Hopfield neural network.

3. Propagation Rule

The propagation rule (Figure 12.2.2) defines how states and synaptic strengths combine as input to a neuron. The propagation rule $\tau_t(i)$ for the Hopfield net is defined by

$$\tau_t(i) = \sum_{j=1}^{N} \mathbf{a}_t(j) w_{ij}.$$

4. Activation Function

The activation function f determines the next state of the neuron $\mathbf{a}_{t+1}(i)$ based on the value $\tau_t(i)$ calculated by the propagation rule and the current neuron value $\mathbf{a}_t(i)$ (see Figure 12.2.2). The activation function for the Hopfield net is the hard limiter defined below.

$$\mathbf{a}_{t+1}(i) = f(\tau_t(i), \mathbf{a}_t(i)) = \begin{cases} \mathbf{a}_t(i) & \text{if } \tau_t(i) = 0 \\ 1 & \text{if } \tau_t(i) > 0 \\ -1 & \text{if } \tau_t(i) < 0. \end{cases}$$

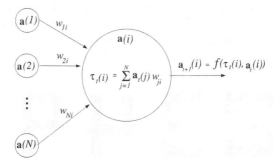

Figure 12.2.2. Propagation rule and activation function for the Hopfield network.

The patterns used for this version of the Hopfield net are N-dimensional vectors from the space $\mathbf{P} = \{-1, 1\}^N$. Let $\mathbf{e}^k = (e_1^k, e_2^k, \ldots, e_N^k)$ denote the kth exemplar pattern, where $1 \leq k \leq K$. The dimensionality of the pattern space determines the number of nodes in the net. In this case the net will have N nodes $\mathbf{a}(1), \mathbf{a}(2), \ldots, \mathbf{a}(N)$. The Hopfield net algorithm proceeds as outlined below.

Step 1. Assign weights to synaptic connections.

This is the step in which the exemplar patterns are imprinted onto the permanent memory of the net. Assign weights w_{ij} to the synaptic connections as follows:

$$
w_{ij} = \begin{cases} \sum\limits_{k=1}^{K} e_i^k e_j^k & \text{if } i \neq j \\ 0 & \text{if } i = j. \end{cases}
$$

Note that $w_{ij} = w_{ji}$, therefore it is only necessary to perform the computation above for $i < j$.

Step 2. Initialize the net with the unknown pattern.

At this step the pattern that is to be classified is introduced to the net. If $\mathbf{p} = (p_1, p_2, \ldots, p_N)$ is the unknown pattern, set

$$
\mathbf{a}_0(i) = p_i, 1 \leq i \leq N.
$$

Step 3. Iterate until convergence.

Calculate next state values for the neurons in the net using the propagation rule and activation function, that is,

$$
\mathbf{a}_{t+1}(i) = f\left(\sum_{j=1}^{N} \mathbf{a}_t(j) w_{ij}, \mathbf{a}_t(i) \right).
$$

Continue this process until further iteration produces no state change at any node. At convergence, the N-dimensional vector formed by the node states is the exemplar pattern that the net has associated with the input pattern.

Step 4. Continue the classification process.

To classify another pattern, repeat Steps 2 and 3.

Example

As an example, consider a communications system designed to transmit the six 12×10 binary images 1, 2, 3, 4, 9, and X (see Figure 12.2.3). Communications channels are subject to noise, and so an image may become garbled when it is transmitted. A Hopfield net can be used for error correction by matching the corrupted image with the exemplar pattern that it most resembles.

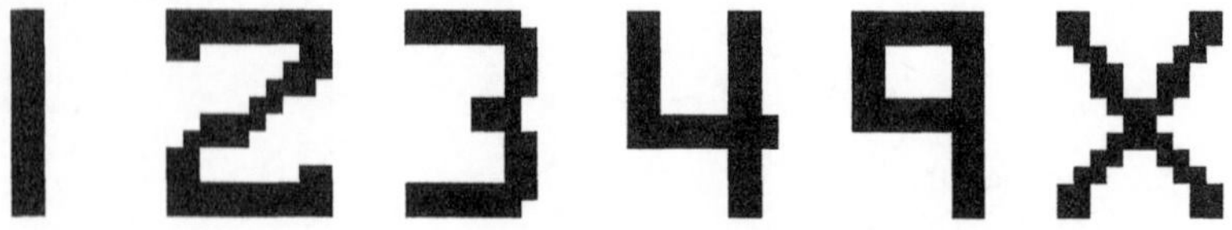

Figure 12.2.3. Exemplar patterns used to initialize synaptic connections w_{ij}.

A binary image $\mathbf{b} \in \{0,1\}^{\mathbf{X}}$, where $\mathbf{X} = \mathbb{Z}_{12} \times \mathbb{Z}_{10}$, can be translated into a pattern $\mathbf{p} = (p_1, p_2, \ldots, p_{120})$, and vice versa, using the relations

$$
p_{10(i-1)+j} = \begin{cases} -1 & \text{if } \mathbf{b}(i,j) = 0 \\ 1 & \text{if } \mathbf{b}(i,j) = 1 \end{cases}
$$

and

$$\mathbf{b}(i,j) = \begin{cases} 0 & \text{if } p_{10(i-1)+j} = -1 \\ 1 & \text{if } p_{10(i-1)+j} = 1. \end{cases}$$

The six exemplar images are translated into exemplar patterns. A 120-node Hopfield net is then created by assigning connection weights using these exemplar patterns as outlined earlier.

The corrupted image is translated into its pattern representation $\mathbf{p} = (p_1, p_2, \ldots, p_N)$ and introduced to the Hopfield net ($\mathbf{a}_0(i)$ is set equal to p_i, $1 \leq i \leq N$). The input pattern evolves through neuron state changes into the pattern $\hat{\mathbf{p}} = (\hat{p}_1, \hat{p}_2, \ldots, \hat{p}_N)$ of the neuron states at convergence ($\hat{p}_i = \mathbf{a}_c(i)$). If the net converges to an exemplar pattern, it is assumed that the exemplar pattern represents the true (uncorrupted) image that was transmitted. Figure 12.2.4 pictorially summarizes the use of a Hopfield net for binary pattern classification process.

Image Algebra Formulation

Let $\mathbf{a} \in \{-1, 1\}^{\mathbb{Z}_N}$ be the image used to represent neuron states. Initialize $\mathbf{a}$ with the unknown image pattern. The weights of the synaptic connections are represented by the template $\mathbf{t} \in \left(\mathbb{R}^{\mathbb{Z}_N}\right)^{\mathbb{Z}_N}$ which is defined by

$$\mathbf{t}_i(j) = \begin{cases} \displaystyle\sum_{k=1}^{K} e_j^k e_i^k & \text{if } i \neq j \\ 0 & \text{if } i = j, \end{cases}$$

where e_i^k is the ith element of the exemplar for class k. Let f be the hard limiter function defined earlier. The image algebra formulation for the Hopfield net algorithm is as follows:

$$\mathbf{b} := 0$$
$$\text{while } \mathbf{a} \neq \mathbf{b} \text{ loop}$$
$$\mathbf{b} := \mathbf{a}$$
$$\mathbf{a} := f(\mathbf{a} \oplus \mathbf{t}, \mathbf{a})$$
$$\text{end loop}.$$

Asynchronous updating of neural states is required to guarantee convergence of the net. The formulation above does not update neural states asynchronously. The implementation above does allow more parallelism and hence increased processing speed. The convergence properties using the parallel implementation may be acceptable; if so, the parallelism can be used to speed up processing.

Alternate Image Algebra Formulation

If asynchronous behavior is desired in order to achieve convergence, then either the template $\mathbf{t}$ needs to be parameterized so that at each application of $\mathbf{a} \oplus \mathbf{t}$ only one randomly chosen neuron changes state, or the following modification to the formulation

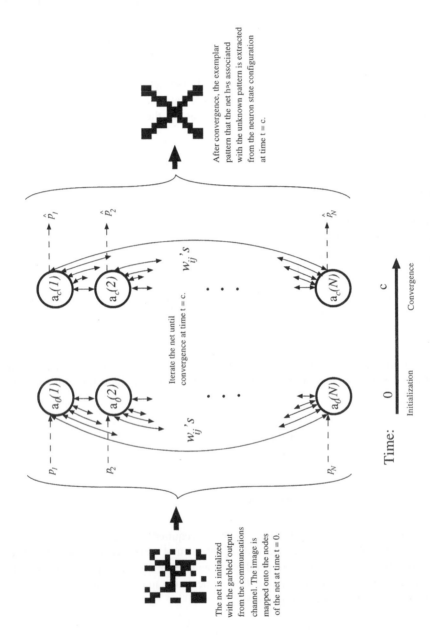

Figure 12.2.4. Example of a Hopfield network.

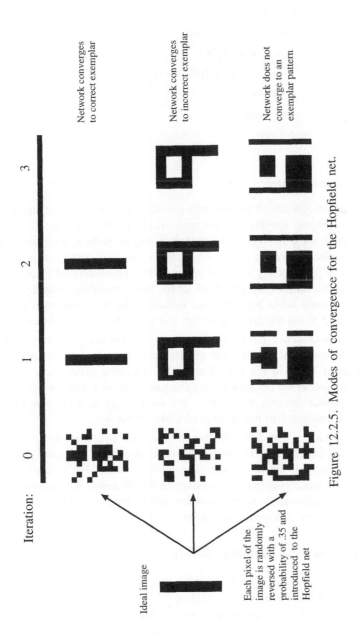

Figure 12.2.5. Modes of convergence for the Hopfield net.

above can be used.

$$\mathbf{b} := 0$$

```
while a ≠ b loop
    Y := domain(a)
    b := a
    while Y ≠ ∅ loop
        i := choice(Y)
        a(i) := f(a ⊕ t|ᵢ, a(i))
        Y := Y \ {i}
    end loop
end loop.
```

Comments and Observations

The Hopfield net is guaranteed to converge provided neuron states are updated asynchronously and the synaptic weights are assigned symmetrically, i.e., $w_{ij} = w_{ji}$. However, the network may not converge to the correct exemplar pattern, and may not even converge to an exemplar pattern. In Figure 12.2.5 three corrupted images have been created from the "1" image by reversing its pixel values with a probability of 0.35. The network is the same as in the earlier example. Each of the three corrupted images yields a different mode of convergence when input into the network. The first converges to the correct exemplar, the second to the incorrect exemplar, and the third to no exemplar.

The two major limitations of the Hopfield net manifest themselves in its convergence behavior. First, the number of patterns that can be stored and accurately recalled is a function of the number of nodes in the net. If too many exemplar patterns are stored (relative to the number of nodes) the net may converge to an arbitrary pattern which may not be any one of the exemplar patterns. Fortunately, this rarely happens if the number of exemplar patterns it small compared to the number of nodes in the net.

The second limitation is the difficulty that occurs when two exemplar patterns share too many pixel values in common. The symptom shows up when a corrupted pattern converges to an exemplar pattern, but to the wrong exemplar pattern. For example, the Hopfield net of the communication system tends to associate the "9" image with a corrupted "4" image. If the application allows, the second problem can be ameliorated by designing exemplar patterns that share few common pixel values.

12.3.   Bidirectional Associative Memory (BAM)

An *associative memory* is a vector space transform $\mathcal{T} : \mathbb{R}^M \rightarrow \mathbb{R}^N$ which may or may not be linear. The Hopfield net (Section 12.2) is an associative memory $\mathcal{H} : \mathbb{R}^M \rightarrow \mathbb{R}^M$. Ideally, the Hopfield net is designed so that it is equal to the identity transformation when restricted to its set of exemplars $\mathbf{E}$, that is,

$$\mathcal{H}|_{\mathbf{E}} = \mathcal{I}_{\mathbf{E}}.$$

The Hopfield net restricted to its set of exemplars can be represented by the set of ordered pairs

$$\mathcal{H}|_{\mathbf{E}} \equiv \left\{ (e^k, e^k) : e^k \in \mathbf{E} \right\}.$$

A properly designed Hopfield net should also take an input pattern from $\mathbb{R}^M$ that is not an exemplar pattern to the exemplar that best matches the input pattern.

A *bidirectional associative memory (BAM)* is a generalization of the Hopfield net [8, 9]. The domain and the range of the BAM transformation $\mathcal{B}$ need not be of the same dimension. A set of associations

$$\mathbf{A} = \left\{ \left(\mathbf{a}^k, \mathbf{b}^k \right) : \mathbf{a}^k \in \mathbb{R}^M \text{ and } \mathbf{b}^k \in \mathbb{R}^N, 1 \le k \le K \right\}$$

is imprinted onto the memory of the BAM so that

$$\mathcal{B}|_{\{\mathbf{a}^k : 1 \le k \le K\}} \equiv \mathbf{A}.$$

That is, $\mathcal{B}\left(\mathbf{a}^k \right) = \mathbf{b}^k$ for $1 \le k \le K$. For an input pattern $\mathbf{a}$ that is not an element of $\left\{ \mathbf{a}^k : 1 \le k \le K \right\}$ the BAM should converge to the association pair $\left(\mathbf{a}^k, \mathbf{b}^k \right)$ for the $\mathbf{a}^k$ that best matches $\mathbf{a}$.

The components of the BAM and how they interact will be discussed next. The bidirectional nature of the BAM will be seen from the description of the algorithm and the example provided.

The three major components of a BAM are given below.

1. Neurons

Unlike the Hopfield net, the domain and the range of the BAM need not have the same dimensionality. Therefore, it is necessary to have two sets of neurons $\mathbf{S_a}$ and $\mathbf{S_b}$ to serve as memory locations for the input and output of the BAM $\mathcal{B} : \mathbb{R}^M \to \mathbb{R}^N$:

$$\mathbf{S_a} = \{\mathbf{a}(m) : 1 \le m \le M\}$$
$$\mathbf{S_b} = \{\mathbf{b}(n) : 1 \le n \le N\}.$$

The state values of $\mathbf{a}(m)$ and $\mathbf{b}(n)$ at time t are denoted $\mathbf{a}_t(m)$ and $\mathbf{b}_t(n)$, respectively. To guarantee convergence, neuron state values will be either 1 or -1 for the BAM under consideration.

2. Synaptic Connection Matrix

The associations $\left(\mathbf{a}^k, \mathbf{b}^k \right), 1 \le k \le K$, where $\mathbf{a}^k = \left(a_1^k, a_2^k, \ldots, a_M^k \right)$ and $\mathbf{b}^k = \left(b_1^k, b_2^k, \ldots, b_N^k \right)$, are stored in the permanent memory of the BAM using an $M \times N$ weight (or synaptic connection) matrix w. The mnth entry of w is given by

$$w_{mn} = \sum_{k=1}^{K} a_m^k b_n^k.$$

Note the similarity in the way weights are assigned for the BAM and the Hopfield net.

3. Activation Function

The activation function f used for the BAM under consideration is the hard limiter defined by

$$f(x, y) = \begin{cases} y & \text{if } x = 0 \\ 1 & \text{if } x > 0 \\ -1 & \text{if } x < 0. \end{cases}$$

The next state of a neuron from $\mathbf{S_b}$ is given by

$$b_{t+1}(n) = f(\mathbf{a}_t w_{.n}, \mathbf{b}_t(n)) = \begin{cases} \mathbf{b}_t(n) & \text{if } \mathbf{a}_t w_{.n} = 0 \\ 1 & \text{if } \mathbf{a}_t w_{.n} > 0 \\ -1 & \text{if } \mathbf{a}_t w_{.n} < 0, \end{cases}$$

where $\mathbf{a}_t = (\mathbf{a}_t(1), \mathbf{a}_t(2), \ldots, \mathbf{a}_t(M))$ and $w_{.n}$ is the nth column of w.

The next state of $\mathbf{a}(m) \in \mathbf{S_a}$ is given by

$$a_{t+1}(m) = f(\mathbf{b}_t w'_{.m}, \mathbf{a}_t(m)) = \begin{cases} \mathbf{a}_t(m) & \text{if } \mathbf{a}_t w'_{.m} = 0 \\ 1 & \text{if } \mathbf{a}_t w'_{.m} > 0 \\ -1 & \text{if } \mathbf{a}_t w'_{.m} < 0, \end{cases}$$

where $\mathbf{b}_t = (\mathbf{b}_t(1), \mathbf{b}_t(2), \ldots, \mathbf{b}_t(N))$ and $w'_{.m}$ is the mth column of the transpose of w. The algorithm for the BAM is as outlined in the following four steps.

Step 1. Create the weight matrix w.

The first step in the BAM algorithm is to generate the weight matrix w using the formula

$$w_{mn} = \sum_{k=1}^{K} a_m^k b_n^k,$$

where $\mathbf{a}^k = (a_1^k, a_2^k, \ldots, a_M^k)$ and $\mathbf{b}^k = (b_1^k, b_2^k, \ldots, b_N^k)$ are from the association $(\mathbf{a}^k, \mathbf{b}^k)$.

Step 2. Initialize neurons.

For the unknown input pattern $\mathbf{p} = (p_1, p_2, \ldots, p_M)$ initialize the neurons of $\mathbf{S_a}$ as follows:

$$\mathbf{a}_0(m) = p_m, 1 \le m \le M.$$

The neurons in $\mathbf{S_b}$ should be assigned values randomly from the set $\{-1, 1\}$, i.e.,

$$\mathbf{b}_0(n) = choice(\{-1, 1\}), 1 \le n \le N.$$

Step 3. Iterate until convergence.

Calculate the next state values for the neurons in $\mathbf{S_b}$ using the formula

$$b_{t+1}(n) = f(\mathbf{a}_t w_{.n}, \mathbf{b}_t(n)), 1 \le n \le N,$$

then calculate the next state values for the neurons in $\mathbf{S_a}$ using

$$a_{t+1}(m) = f(\mathbf{b}_t w'_{.m}, \mathbf{a}_t), 1 \le m \le M.$$

The alternation between the sets $\mathbf{S_b}$ and $\mathbf{S_a}$ used for updating neuron values is why this type of associative memory neural net is referred to as "bidirectional." The forward feed (update of $\mathbf{S_b}$) followed by the feedback (update of $\mathbf{S_a}$) improves the recall accuracy of the net.

Continue updating the neurons in $\mathbf{S_b}$ followed by those in $\mathbf{S_a}$ until further iteration produces no state change for any neuron. At time of convergence t = c, the association that the BAM has recalled is $(\mathbf{a}, \mathbf{b})$, where $\mathbf{a} = (\mathbf{a}_c(1), \mathbf{a}_c(2), \ldots, \mathbf{a}_c(M))$ and $\mathbf{b} = (\mathbf{b}_c(1), \mathbf{b}_c(2), \ldots, \mathbf{b}_c(N))$.

Step 4. Continue classification.

To classify another pattern repeat Steps 2 and 3.

Example

Figure 12.3.1 shows the evolution of neuron states in a BAM from initialization to convergence. Three lowercase-uppercase character image associations (a, A), (b, B), and (c, C) were used to create the weight matrix w. The lowercase characters are 12×12 images and the uppercase characters are 16×16 images. The conversion from image to pattern and vice versa is done as in the example of Section 12.2. A corrupted "a" is input onto the $\mathbf{S_a}$ neurons of the net.

Image Algebra Formulation

Let $\mathbf{a} \in \{-1, 1\}^{Z_M}$ and $\mathbf{b} \in \{-1, 1\}^{Z_N}$ be the image variables used to represent the state for neurons in the sets $\mathbf{S_a}$ and $\mathbf{S_b}$, respectively. Initialize $\mathbf{a}$ with the unknown pattern. The neuron values of $\mathbf{S_b}$ are initialized to either -1 or 1 randomly. The weight matrix W is represented by the template $\mathbf{t} \in \left(\mathbb{R}^{Z_N}\right)^{Z_M}$ given by

$$\mathbf{t}_m(n) = \sum_{k=1}^{K} a_m^k b_n^k,$$

where a_m^k is the mth component of $\mathbf{a}^k$ and b_n^k is the nth component of $\mathbf{b}^k$ in the association pair $\left(\mathbf{a}^k, \mathbf{b}^k\right)$. The activation function is denoted f.

The image algebra pseudocode for the BAM is given by

```
c := 0
while a ≠ c loop
    c := a
    b := f(a ⊕ t, b)
    a := f(b ⊕ t', a)
end loop.
```

Comments and Observations

The BAM has the same limitations that the Hopfield net does. First, the number of associations that can be programmed into the memory and effectively recalled is limited. Second, the BAM may have a tendency to converge to the wrong association pair if components of two association pairs have too many pixel values in common. Figure 12.3.2 illustrates convergence behavior for the BAM.

The complement $\mathbf{d}^c$ of a pattern $\mathbf{d}$ is defined by

$$\mathbf{d}^c(i) = \begin{cases} 1 & \text{if } \mathbf{d}(i) = -1 \\ -1 & \text{if } \mathbf{d}(i) = 1. \end{cases}$$

Note that the weight matrix w is defined by

$$w_{mn} = \sum_{k=1}^{K} a_m^k b_n^k$$

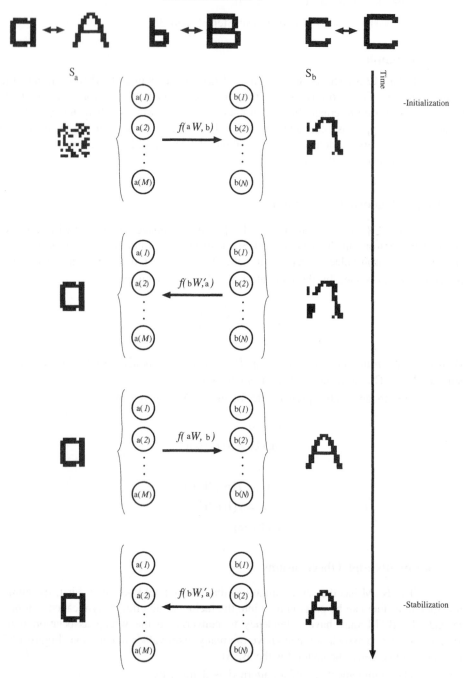

Figure 12.3.1. Bidirectional nature of the BAM.

which is equivalent to

$$w_{mn} = \sum_{k=1}^{K} \left(-a_m^k\right)\left(-b_n^k\right).$$

Therefore, imprinting the association pair $(\mathbf{a}, \mathbf{b})$ onto the memory of a BAM also imprints $(\mathbf{a}^c, \mathbf{b}^c)$ onto the memory. The effect of this is seen in Block 5 of Figure 12.3.2.

Associations encoded into the BAM

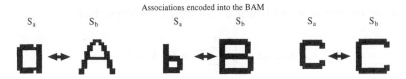

Five examples of the convergence behavior of the BAM

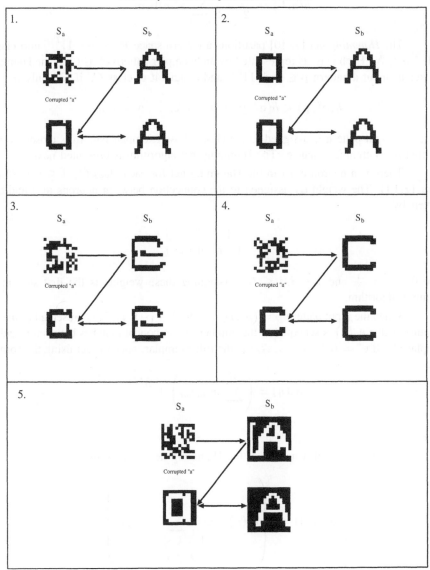

Figure 12.3.2. Modes of convergence for the BAM.

12.4. Hamming Net

The *Hamming distance* h between two patterns $\mathbf{a}, \mathbf{b} \in \{-1, 1\}^{Z_M}$ is equal to the number of components in **a** and **b** that do not match. More precisely,

$$h(\mathbf{a}, \mathbf{b}) = \frac{1}{2} \sum_{m=1}^{M} |\mathbf{a}(m) - \mathbf{b}(m)|,$$

or

$$h(\mathbf{a}, \mathbf{b}) = \frac{M}{2} - \frac{1}{2} \sum_{m=1}^{M} \mathbf{a}(m)\mathbf{b}(m).$$

The *Hamming net* [7, 10] partitions a pattern space $\mathbf{P} = \{-1, 1\}^{ZM}$ into classes $C_n, 1 \le n \le N$. Each class is represented by an exemplar pattern $e^n \in C_n$. The Hamming net takes as input a pattern $\mathbf{p} \in \{-1, 1\}^{ZM}$ and assigns it to class C_k if and only if

$$h(e^k, \mathbf{p}) < h(e^n, \mathbf{p}), \forall\, n = 1, 2, \ldots, N, k \neq n.$$

That is, the input pattern is assigned to the class whose exemplar pattern is closest to it as measured by Hamming distance. The Hamming net algorithm is presented next.

There is a neuron $\mathbf{a}(n)$ in the Hamming net for each class $C_n, 1 \le n \le N$ (see Figure 12.4.1). The weight t_{ij} assigned to the connection between neurons $\mathbf{a}(i)$ and $\mathbf{a}(j)$ is given by

$$t_{ij} = \begin{cases} 1 & \text{if } i = j \\ -\epsilon & \text{if } i \neq j, \end{cases}$$

where $0 < \epsilon < \frac{1}{N}$ and $1 \le i, j \le N$. Assigning these weights is the first step in the Hamming algorithm.

When the input pattern $\mathbf{p} = (p_1, p_2, \ldots, p_M) \in \{-1, 1\}^{ZM}$ is presented to the net, the neuron value $\mathbf{a}_0(n)$ is set equal to the number of component matches between $\mathbf{p}$ and the exemplar e^n. If $e^n = (e_1^n, e_2^n, \ldots, e_M^n)$ is the nth exemplar, $\mathbf{a}_0(n)$ is set using the formula

$$\mathbf{a}_0(n) = \left(\sum_{m=1}^{M} w_{mn} p_m \right) + \frac{M}{2},$$

where $w_{mn} = \frac{e_m^n}{2}$.

The next state of a neuron in the Hamming net is given by

$$\mathbf{a}_{t+1}(n) = f\left(\mathbf{a}_t(n) - \epsilon \sum_{\substack{1 \le k \le N \\ k \neq n}} \mathbf{a}_t(k) \right),$$

where f is the activation function defined by

$$f(x) = \begin{cases} x & \text{if } x > 0 \\ 0 & \text{otherwise.} \end{cases}$$

The next state of a neuron is calculated by first decreasing its current value by an amount proportional to the sum of the current values of all the other neuron values in the net. If the reduced value falls below zero then the new neuron value is set to 0; otherwise it assumes the reduced value. Eventually, the process of updating neuron values will lead to a state in which only one neuron has a value greater than zero. At that point the neuron with the non-zero value, say $a_c(k) \neq 0$, represents the class C_k that the net assigns to the input pattern. At time 0, the value of $a(k)$ was greater than all the other neuron values. Therefore, k is the number of the class whose exemplar had the most matches with the input pattern.

$$q_n > 0 \text{ iff } a_0(n) = max\{a_0(1), a_0(2),..., a_0(N)\}$$

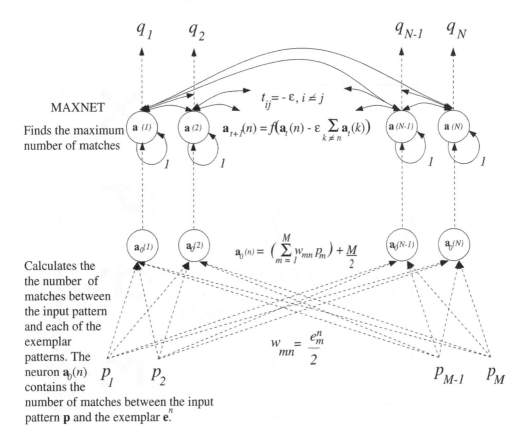

Figure 12.4.1. Hamming network.

Figure 12.4.2 shows a six-neuron Hamming net whose classes are represented by the exemplars 1, 2, 3, 4, 9, and X. The Hamming distance between the input pattern p and each of the exemplars is displayed at the bottom of the figure. The number of matches between exemplar e^n and p is given by the neuron states $a_0(n), 1 \leq n \leq 6$. At time of completion, only one neuron has a positive value, which is $a_c(5) = 37.5$. The Hamming net example assigns the unknown pattern to Class 5. That is, the input pattern had the most matches with the "9" exemplar which represents Class 5.

Image Algebra Formulation

Let $a \in \{-1, 1\}^{Z_n}$ be the image variable used to store neural states. The activation function f is as defined earlier. The template $s \in \left(\mathbb{R}^{Z_M}\right)^{Z_N}$ defined by

$$s_n(m) = \frac{e_m^n}{2},$$

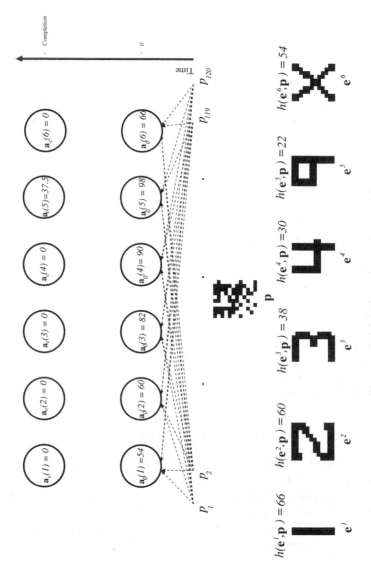

Figure 12.4.2. Hamming net example.

where $e^n = (e_1^n, e_2^n, \ldots, e_M^n)$ is the nth exemplar pattern, is used to initialize the net. The template $t \in \left(\mathbb{R}^{Z_N}\right)^{Z_N}$ defined by

$$t_j(i) = \begin{cases} 1 & \text{if } i = j \\ -\epsilon & \text{if } i \neq j \end{cases}$$

is used to implement the propagation rule for the net.

Given the input pattern $p \in \{-1, 1\}^{Z_M}$ the image algebra formulation for the Hamming net is as follows:

$$a := (p \oplus s) + \frac{M}{2}$$

```
while Σ(χ>0(a)) > 1 loop
    a := f(a ⊕ t)
end loop
```

$$c := domain(a\|_{>0}).$$

The variable c contains the number of the class that the net assigns to the input pattern.

Comments and Observations

The goal of our example was to find the number of the exemplar that best matches the input pattern as measured by Hamming distance. The formulation above demonstrated how to use image algebra to implement the Hamming neural network approach to the problem. A simpler image algebra formula that accomplished the same goal is given by the following two statements:

$$c(i) := \sum \left(\chi_{=1}(p \cdot e^i) \right)$$
$$c := domain(c\|_{\vee c}).$$

12.5. Single-Layer Perceptron (SLP)

A *single-layer perceptron* is used to classify a pattern $p = (p_1, p_2, \ldots, p_m) \in P \subset \mathbb{R}^m$ into one of two classes. An SLP consists of a set of weights,

$$\{w_0, w_1, \ldots, w_n\}, \ w_i \in \mathbb{R}, 0 \leq i \leq m,$$

and a limiter function $f : \mathbb{R}\backslash\{0\} \rightarrow \{0, 1\}$ defined by

$$f(x) = \begin{cases} 1 & \text{if } x > 0 \\ 0 & \text{if } x \leq 0. \end{cases}$$

Let $p = (p_1, p_2, \ldots, p_m) \in P$, in order to classify p the perceptron first calculates the sum of products

$$g(p) = w_0 + w_1 p_1 + \cdots + w_m p_m$$

and then applies the limiter function. If $f \circ g(p) < 0$, the perceptron assigns p to class C_0. The pattern p is assigned to class C_1 if $f \circ g(p) > 0$. The SLP is represented in Figure 12.5.1.

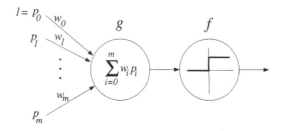

Figure 12.5.1. Single-layer perceptron.

The graph of $0 = g(\mathbf{x}) = w_0 + w_1 x_1 + \cdots + w_m x_m$ is a hyperplane that divides $\mathbb{R}^m$ into two regions. This hyperplane is called the perceptron's *decision surface*. Geometrically, the perceptron classifies a pattern based on which side the pattern (point) lies on. Patterns for which $g(\mathbf{p}) < 0$ lie on one side of the decision surface and are assigned to C_0. Patterns for which $g(\mathbf{p}) > 0$ lie on the opposite side of the decision surface and are assigned to C_1.

The perceptron operates in two modes — a classifying mode and a learning mode. The pattern classification mode is as described above. Before the perceptron can function as a classifier, the values of its weights must be determined. The learning mode of the perceptron is involved with the assignment of weights (or the determination of a decision surface).

If the application allows, i.e., if the decision surface is known *a priori*, weights can be assigned analytically. However, key to the concept of a perceptron is the ability to determine its own decision surface through a learning algorithm [7, 11, 12]. There are several algorithms for SLP learning. The one we present here can be found in Rosenblatt [11].

Let $\{(\mathbf{p}_k, y_k)\}_{k=1}^n$ be a training set, where $\mathbf{p}_k = (p_{k_1}, p_{k_2}, \ldots, p_{k_m}) \in \mathbf{P}$ and $y_k \in \{0, 1\}$ is the class number associated with $\mathbf{p}_k$. Let $w_i(t)$ be the value of the ith weight after the tth training pattern has been presented to the SLP. The learning algorithm is presented below.

Step 1. Set each $w_i(0)$, $0 \le i \le m$, equal to random real number.

Step 2. Present $\mathbf{p}_k$ to the SLP. Let y'_k denote the computed class number of $\mathbf{p}_k$, i.e.,

$$y'_k(t) = f\left(w_0(t-1) + \sum_{i=1}^m p_{k_i} \cdot w_i(t-1) \right).$$

Step 3. Adjust weights using the formula

$$w_i(t) = w_i(t-1) + \eta \cdot (y_k - y'_k(t)) \cdot p_{k_i}.$$

Step 4. Repeat Steps 2 and 3 for each element of the training set, recycling if necessary, until convergence or a predefined number of iterations.

The constant, $0 < \eta \le 1$, regulates the rate of weight adjustment. Small η results in slow learning. However, if η is too large the learning process may not be able to home in on a good set of weights.

Note that if the SLP classifies $\mathbf{p}_k$ correctly, then $y_k - y'_k(t) = 0$ and there is no adjustment made to the weights. If $y_k - y'_k(t) \ne 0$, then the change in the weight vector is proportional to the input pattern.

Figure 12.5.2 illustrates the learning process for distinguishing two classes in $\mathbb{R}^2$. Class 1 points have been plotted with diamonds and Class 0 points have been plotted with crosses. The decision surface for this example is a line. The lines plotted in the figure represent decision surfaces after $n = 0, 20, 40$, and 80 training patterns have been presented to the SLP.

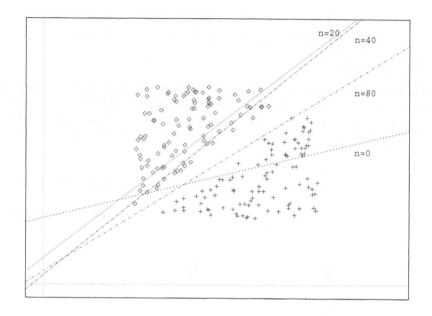

Figure 12.5.2. SLP learning — determination of the perceptron's decision surface.

Image Algebra Formulation

Let $\mathbf{w} \in \mathbb{R}^{m+1}$ be the image variable whose components represent the weights of the SLP. Let $\mathbf{p} \in \mathbb{R}^m$ be the pattern that is to be classified. Augment $\mathbf{p}$ with a 1 to form $\tilde{\mathbf{p}} = (1, p_1, \ldots, p_m) \in \{1\} \times \mathbb{R}^m$. The function f is the limiter defined earlier. The class number y of $\mathbf{p}$ is determined using the image algebra statement

$$y := f(\Sigma(\mathbf{w} \cdot \tilde{\mathbf{p}})).$$

To train the SLP, let $\{(\mathbf{p}_k, y_k)\}_{k=1}^{n}$ be a training set. Initially, each component of the weight image variable $\mathbf{w}$ is set to a random real number. The image algebra pseudocode for the SLP learning algorithm (iterating until convergence) is given by

$$\mathbf{v} := \mathbf{0}$$

```
while (v ≠ w) loop
    v := w
    for k in 1..n loop
        y′_k := f(Σ(p̃_k · w))
        w := w + η · (y_k − y′_k) · p̃_k
    end loop
end loop.
```

Comments and Observations

If the set of input patterns cannot be divided into two linear separable sets then an SLP will fail as a classifier. Consider the problem of designing an SLP for XOR classification. Such an SLP should be defined over the pattern space $\{(0,0),(0,1),(1,0),(1,1)\}$. The classes for the XOR problem are

$$C_0 = \{(0,0),(1,1)\} \quad (0\ XOR\ 0 = 0,\ 1\ XOR\ 1 = 0)$$
$$C_1 = \{(0,1),(1,0)\} \quad (0\ XOR\ 1 = 1,\ 1\ XOR\ 0 = 1).$$

The points in the domain of the problem are plotted in Figure 12.5.3. Points in Class 0 are plotted with open dots and points in Class 1 are plotted with solid dots. A decision surface in $\mathbb{R}^2$ is a line. There is no line in $\mathbb{R}^2$ that separates classes 0 and 1 for the XOR problem. Thus, an SLP is incapable of functioning as a classifier for the XOR problem.

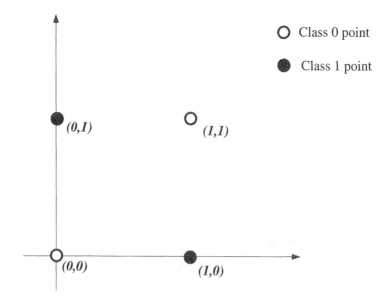

Figure 12.5.3. Representation of domain for XOR.

12.6. Multilayer Perceptron (MLP)

As seen in Section 12.5, a single-layer perceptron is not capable of functioning as an XOR classifier. This is because the two classes of points for the XOR problem are not linearly separable. A single-layer perceptron is only able to partition $\mathbb{R}^m$ into regions separated by a hyperplane. Fortunately, more complex regions in $\mathbb{R}^m$ can be specified by feeding the output from several SLPs into another SLP designed to serve as a multivariable AND gate. The design of a multilayer perceptron for the XOR problem is presented in the following example.

In Figure 12.6.1, the plane has been divided into two regions. The shaded region R_1 lies between the lines whose equations are $-1 + 2x + 2y = 0$ and $3 - 2x - 2y = 0$. Region R_0 consists of all points not in R_1. The points of class C_1 for the XOR problem lie in R_1 and the points in C_0 lie in R_0. Region R_1 is the intersection of the half-plane that lies above $-1 + 2x + 2y = 0$ and the half-plane that lies below $3 - 2x - 2y = 0$. Single-layer perceptrons SLP_1 and SLP_2 can be designed with decision surfaces $0 = -1 + 2x + 2y$ and $0 = 3 - 2x - 2y$, respectively. The intersection required to create R_1 is achieved by sending the output of SLP_1 and SLP_2 through another SLP that acts as an AND gate.

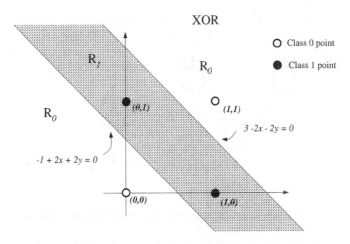

Figure 12.6.1. Multilayer perceptron solution strategy for XOR problem.

One SLP implementation of an AND gate and its decision surface is shown in Figure 12.6.2. All the components of a two-layer perceptron for XOR classification can be seen in Figure 12.6.3. The first layer consists of SLP_1 and SLP_2. An input pattern (x, y) is presented to both SLP_1 and SLP_2. Whether the point lies above $-1 + 2x + 2y = 0$ is determined by SLP_1; SLP_2 determines if it lies below $3 - 2x - 2y = 0$. The second layer of the XOR perceptron takes the output of SLP_1 and SLP_2 and determines if the input satisfies both the conditions of lying above $-1 + 2x + 2y = 0$ and below $3 - 2x - 2y = 0$.

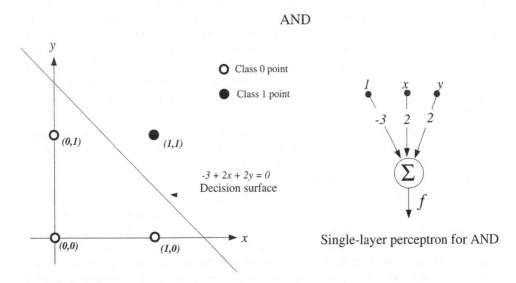

Figure 12.6.2. Single-layer perceptron for AND classification.

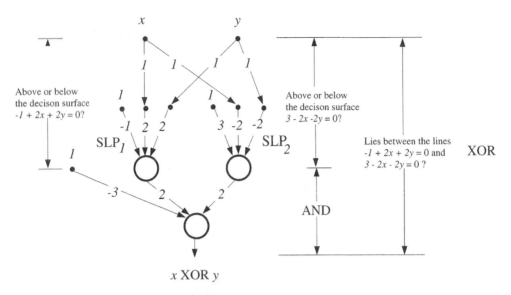

Figure 12.6.3. Two-layer perceptron for XOR classification.

By piping the output of SLPs through a multivariable AND gate, a two-layer perceptron can be designed for any class C_1 whose points lie in a region that can be constructed from the intersection of half-planes. Adding a third layer to a perceptron consisting of a multivariable OR gate allows for the creation of even more complex region for pattern classification. The outputs from the AND layer are fed to the OR layer which serves to union the regions created in the AND layer. An OR gate is easily implemented using an SLP.

Figure 12.6.4 shows pictorially how the XOR problem can be solved with a three-layer perceptron. The AND layer creates two quarter-planes from the half-planes created in the first layer. The third layer unions the two quarter-planes to create the desired classification region.

The example above is concerned with the analytic design of a multilayer perceptron. That is, it is assumed that the classification regions in $\mathbb{R}^2$ are known *a priori*. Analytic design is not accordant with the perceptron concept. The distinguishing feature of a perceptron is its ability to determine the proper classification regions, on its own, through a "learning by example" process. However, the preceding discussion does point out how the applicability of perceptrons can be extended by combining single-layer perceptrons. Also, by first approaching the problem analytically, insights will be gained into the design and operation of a "true" perceptron. The design of a *feedforward multilayer perceptron* is presented next.

A feedforward perceptron consists of an input layer of nodes, one or more hidden layers of nodes, and an output layer of nodes. We will focus on the two-layer perceptron of Figure 12.6.5. The algorithms for the two-layer perceptron are easily generalized to perceptrons of three or more layers.

A node in a hidden layer is connected to every node in the layer above and below it. In Figure 12.6.5 weight w_{ij} connects input node x_i to hidden node h_j and weight v_{jk} connects h_j to output node o_k. Classification begins by presenting a pattern to the input nodes $x_i, 1 \le i \le l$. From there data flows in one direction (as indicated by the arrows in Figure 12.6.5) through the perceptron until the output nodes $o_k, 1 \le k \le n$, are reached. Output nodes will have a value of either 0 or 1. Thus, the perceptron is capable of partitioning its pattern space into 2^n classes.

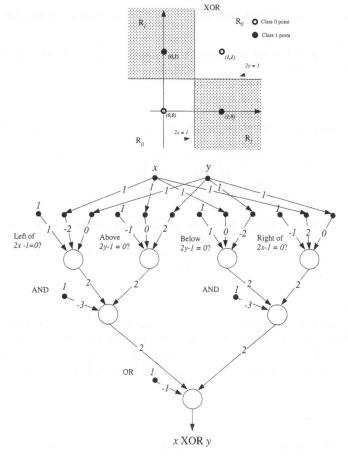

Figure 12.6.4. Three-layer perceptron implementation of XOR.

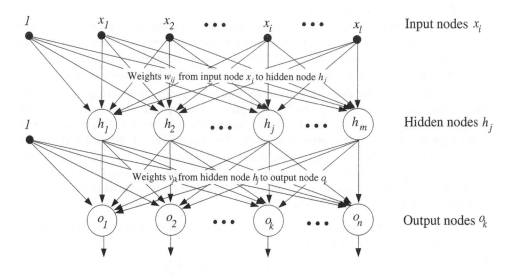

Figure 12.6.5. Two-layer perceptron.

The steps that govern data flow through the perceptron during classification are as follows:

Step 1. Present the pattern $\mathbf{p} = (p_1, p_2, \ldots, p_l) \in \mathbb{R}^l$ to the perceptron, i.e., set $x_i = p_i$ for $1 \leq i \leq l$.

Step 2. Compute the values of the hidden-layer nodes using the formula

$$h_j = \frac{1}{1 + e^{-\left(w_{0j} + \sum\limits_{i=1}^{l} w_{ij} x_i\right)}} \qquad 1 \leq j \leq m.$$

Step 3. Calculate the values of the output nodes using the formula

$$o_k = \frac{1}{1 + e^{-\left(v_{0k} + \sum\limits_{j=1}^{m} v_{jk} h_j\right)}} \qquad 1 \leq k \leq n.$$

Step 4. The class $\mathbf{c} = (c_1, c_2, \ldots, c_n)$ that the perceptron assigns $\mathbf{p}$ must be a binary vector which, when interpreted as a binary number, is the class number of $\mathbf{p}$. Therefore, the o_k must be thresholded at some level τ appropriate for the application. The class that $\mathbf{p}$ is assigned to is then $\mathbf{c} = (c_1, c_2, \ldots, c_n)$, where $c_k = \chi_{\geq \tau}(o_k)$.

Step 5. Repeat Steps 1, 2, 3, and 4 for each pattern that is to be classified.

Above, when describing the classification mode of the perceptron, it was assumed that the values of the weights between nodes had already been determined. Before the perceptron can serve as a classifier, it must undergo a learning process in which its weights are adjusted to suit the application.

The learning mode of the perceptron requires a training set $\{(\mathbf{p}^t, \mathbf{c}^t)\}_{t=1}^{s}$, where $\mathbf{p}^t \in \mathbb{R}^l$ is a pattern and $\mathbf{c}^t \in \{0, 1\}^n$ is a vector which represents the actual class number of $\mathbf{p}^t$. The perceptron learns (adjusts its weights) using elements of the training set as examples. The learning algorithm presented here is known as *backpropagation learning*.

For backpropagation learning, a forward pass and a backward pass are made through the perceptron. During the forward pass a training pattern is presented to the perceptron and classified. The backward pass recursively, level by level, determines error terms used to adjust the perceptron's weights. The error terms at the first level of the recursion are a function of $\mathbf{c}^t$ and output of the perceptron $(o_1, o_2, \ldots, o_n)$. After all the error terms have been computed, weights are adjusted using the error terms that correspond to their level. The backpropagation algorithm for the two-layer perceptron of Figure 12.6.5 is detailed in the steps which follow.

Step 1. Initialize the weights of the perceptron randomly with numbers between -0.1 and 0.1; i.e.,

$$w_{ij} = choice([-0.1, 0.1]) \qquad 0 \leq i \leq l, 1 \leq j \leq m$$

and

$$v_{jk} = choice([-0.1, 0.1]) \qquad 0 \leq j \leq m, 1 \leq k \leq n.$$

Step 2. Present $\mathbf{p}^t = (p_1^t, p_2^t, \ldots, p_l^t)$ from the training pair $(\mathbf{p}^t, \mathbf{c}^t)$ to the perceptron and apply Steps 1, 2, and 3 of the perceptron's classification algorithm outlined earlier. This completes the forward pass of the backpropagation algorithm.

Step 3. Compute the errors $\delta_{o_k}, 1 \leq k \leq n$, in the output layer using

$$\delta_{o_k} = o_k(1 - o_k)(c_k^t - o_k),$$

where $\mathbf{c}^t = (c_1^t, c_2^t, \ldots, c_n^t)$ represents the correct class of $\mathbf{p}^t$. The vector $(o_1, o_2, \ldots, o_n)$ represents the output of the perceptron.

Step 4. Compute the errors $\delta_{h_j}, 1 \leq j \leq m$, in the hidden-layer nodes using

$$\delta_{h_j} = h_j(1 - h_j) \sum_{k=1}^{n} \delta_{o_k} \cdot v_{jk}.$$

Step 5. Let $v_{jk}(t)$ denote the value of weight v_{jk} after the tth training pattern has been presented to the perceptron. Adjust the weights between the output layer and the hidden layer according to the formula

$$v_{jk}(t) = v_{jk}(t - 1) + \eta \cdot \delta_{o_k} \cdot h_j.$$

The parameter $0 < \eta \leq 1$ governs the learning rate of the perceptron.

Step 6. Adjust the weights between the hidden layer and the input layer according to

$$w_{ij}(t) = w_{ij}(t - 1) + \eta \cdot \delta_{h_j} \cdot p_i^t.$$

Step 7. Repeat Steps 2 through 6 for each element of the training set. One cycle through the training set is called an *epoch*. The performance that results from the network's training may be enhanced by repeating epochs.

Image Algebra Formulation

Let $\mathbf{h} \in \{1\} \times \mathbb{R}^m$ and $\mathbf{o} \in \mathbb{R}^n$ be the image variables used to store the values of the input layer and the output layer, respectively. The template $\mathbf{w} \in \left(\mathbb{R}^{\mathbb{Z}_{l+1}}\right)^{\mathbb{Z}_{m+1}}$ will be used to represent the weights between the input layer and the hidden layer. Initialize $\mathbf{w}$ as follows:

$$\mathbf{w}_0(0) = 1$$
$$\mathbf{w}_0(i) = 0 \quad 1 \leq i \leq l$$
$$\mathbf{w}_j(i) = choice([-0.1, 0.1]) \quad 0 \leq i \leq l, 1 \leq j \leq m.$$

The template $\mathbf{v} \in \left(\mathbb{R}^{\mathbb{Z}_{m+1}}\right)^{\mathbb{Z}_n}$ represents the weights between the hidden layer and the output layer. Initialize $\mathbf{v}$ by setting

$$\mathbf{v}_k(j) = choice([-0.1, 0.1]) \quad 0 \le j \le m, 1 \le k \le n.$$

The activation function for the perceptron is

$$f(x) = \frac{1}{1 + e^{-x}}.$$

Let $\{(\mathbf{p}^t, \mathbf{c}^t)\}_{t=1}^s$ be the training set for the two-layer perceptron, where $\mathbf{p}^t = (p_1^t, p_2^t, \dots, p_l^t) \in \mathbb{R}^l$ is a pattern and $\mathbf{c}^t \in \{0, 1\}^n$ is a vector which represents the actual class number of $\mathbf{p}^t$. Define $\tilde{\mathbf{p}}^t$ to be $(1, p_1^t, p_2^t, \dots, p_l^t)$. The parameterized templates $\mathbf{t}(\mathbf{d}, \mathbf{h}) \in \left(\mathbb{R}^{\mathbb{Z}_{m+1}}\right)^{\mathbb{Z}_n}$ and $\mathbf{u}(\mathbf{d}, \mathbf{p}) \in \left(\mathbb{R}^{\mathbb{Z}_{l+1}}\right)^{\mathbb{Z}_{m+1}}$ are defined by

$$\mathbf{t}(\mathbf{d}, \mathbf{h})_k(j) = \eta \cdot \mathbf{d}(k) \cdot \mathbf{h}(j) \quad 0 \le j \le m, 1 \le k \le n$$

and

$$\mathbf{u}(\mathbf{d}, \mathbf{p})_0(i) = 0 \quad 0 \le i \le l$$
$$\mathbf{u}(\mathbf{d}, \mathbf{p})_j(i) = \eta \cdot \mathbf{d}(j) \cdot \mathbf{p}(i) \quad 0 \le i \le l, 1 \le j \le m.$$

The image algebra formulation of the backpropagation learning algorithm for one epoch is

> **for** t **in** 1..s **loop**
> $\mathbf{h} := f(\tilde{\mathbf{p}}^t \oplus \mathbf{w})$
> $\mathbf{o} := f(\mathbf{h} \oplus \mathbf{v})$
> $\mathbf{d_v} := \mathbf{o}(1 - \mathbf{o})(\mathbf{c}^t - \mathbf{o})$
> $\mathbf{d_w} := \mathbf{h}(1 - \mathbf{h})(\mathbf{d_v} \oplus \mathbf{v}')$
> $\mathbf{v} := \mathbf{v} + \mathbf{t}(\mathbf{d_v}, \mathbf{h})$
> $\mathbf{w} := \mathbf{w} + \mathbf{u}(\mathbf{d_w}, \tilde{\mathbf{p}}^t)$
> **end loop.**

To classify $\mathbf{p} = (p_1, p_2, \dots, p_l)$, first augment it with a 1 to form $\tilde{\mathbf{p}} = (1, p_1, p_2, \dots, p_l)$. The pattern $\mathbf{p}$ is then assigned to the class represented by the image $\mathbf{c}$ using the image algebra code:

> $\mathbf{h} := f(\tilde{\mathbf{p}} \oplus \mathbf{w})$
> $\mathbf{o} := f(\mathbf{h} \oplus \mathbf{v})$
> $\mathbf{c} := \chi_{\ge \tau}(\mathbf{o}).$

Comments and Observations

The treatment of perceptrons presented here has been very cursory. Our purpose is to demonstrate techniques for the formulation of perceptron algorithms using image algebra. Other works present more comprehensive introductions to perceptrons [7, 10, 12, 13].

12.7. Cellular Automata and Life

Among the best known examples of cellular automata is John Conway's game *Life* [2, 14, 15]. The life processes of the synthetic organisms in this biosphere are presented next. Although this automaton is referred to as a game, it will provide an illustration of how easily the workings of cellular automata are formulated in image algebra.

Life evolves on a grid of points $X = \mathbb{Z}_k \times \mathbb{Z}_l$. A cell (point) is either alive or not alive. Life and its absence are denoted by state values 1 and 0, respectively. The game begins with any configuration of living and nonliving cells on the grid. Three rules govern the interactions among cells.

(a) Survivals — Every live cell with two or three live 8-neighbors survives for the next generation.

(b) Deaths — A live cell that has four or more 8-neighbors dies due to over-population. A live cell with one 8-neighbor or none dies from loneliness.

(c) Birth — A cell is brought to life if exactly three of its 8-neighbors are alive.

Figure 12.7.1 shows the life cycle of an organism. Grid 0 is the initial state of the life-form. After ten generations, the life-form cycles between the configurations in grids 11 and 12.

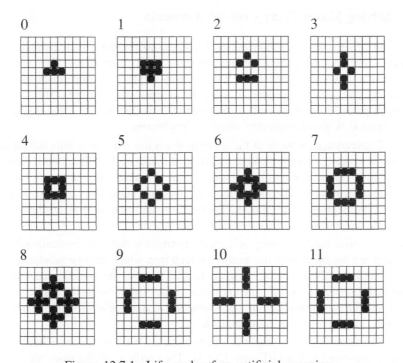

Figure 12.7.1. Life cycle of an artificial organism.

Depending on the initial configuration of live cells, the automaton will demonstrate one of four types of behavior [14]. The pattern of live cells may:

(a) vanish in time

(b) evolve to a fixed finite size

(c) grow indefinitely at a fixed speed

(d) enlarge and contract irregularly

Image Algebra Formulation

Let $\mathbf{a} \in \{0,1\}^{\mathbf{X}}$, where $\mathbf{X} = \mathbb{Z}_k \times \mathbb{Z}_l$, be the image variable that contains the state values of cells in the automata. The cells state at time t is denoted $\mathbf{a}_t$. The template used to capture the states of neighboring cell is shown below.

$$\mathbf{t} = \begin{array}{|c|c|c|} \hline 1 & 1 & 1 \\ \hline 1 & 9 & 1 \\ \hline 1 & 1 & 1 \\ \hline \end{array}$$

Let $\mathbf{a}_0$ be the initial configuration of cells. The next state of the automaton is given by the image algebra formula

$$\mathbf{a}_{t+1} := \chi_{\{3,11,12\}}(\mathbf{a}_t \oplus \mathbf{t}).$$

12.8.   Solving Mazes Using Cellular Automata

In this section an unconventional solution to finding a path through a maze is presented. The method uses a cellular automata (CA) approach. This approach has several advantageous characteristics. The memory required to implement the algorithm is essentially that which is needed to store the original maze image. The CA approach provides all possible solutions, and it can determine whether or not a solution exists [16]. The CA method is also remarkably simple to implement.

Conventional methods to the maze problem use only local information about the maze. As the mouse proceeds through the maze, it marks each intersection it passes. If the corridor that the mouse is currently exploring leads to a dead-end, the mouse backtracks to the last marker it placed. The mouse then tries another unexplored corridor leading from the marked intersection. This process is continued until the goal is reached. Thus, the conventional method uses a recursive depth-first search.

The price paid for using only local information with the conventional method is the memory needed for storing the mouse's search tree, which may be substantial. The CA approach requires only the memory that is needed to store the original maze image. The CA approach must have the information that a viewer overlooking the whole maze would have. Depending on the application the assumption of the availability of a global overview of the maze may or may not be reasonable.

The maze for this example is a binary image. Walls of the maze have pixel value 0 and are represented in black (Figure 12.8.1). The corridors of the maze are 4-connected and have pixel value 1. Corridors must be one pixel wide. Corridor points are white in Figure 12.8.1.

In some ways the CA approach to maze solving may be viewed as a poor thinning algorithm (Chapter 5). A desirable property of a thinning algorithm is the preservation of ends. To solve the maze the CA removes ends. Ends in the maze are corridor points that are at the terminus of a dead-end corridor. A corridor point at the end of a dead-end corridor is changed to a wall point by changing its pixel value from 1 to 0. At each iteration of the

Figure 12.8.1. Original maze.

algorithm, the terminal point of each dead-end corridor is converted into a wall point. This process is continued until no dead-end corridors exist, which will also be the time at which further iteration produces no change in the image. When all the dead-end corridors are removed, only solution paths will remain. The result of this process applied to the maze of Figure 12.8.1 is seen in Figure 12.8.2.

For this example, the next state of a point is function of the current states of the points in a von Neumann neighborhood of the target point. The next state rules that drive the maze solving CA are

    (a) A corridor point that is surrounded by three of four wall points becomes a wall point.

    (b) Wall points always remain wall points.

    (c) A corridor point surrounded by two or fewer wall points remains a corridor point.

Image Algebra Formulation

Let $\mathbf{a} \in \{0, 1\}^{\mathbf{X}}$ be the maze image. The template $\mathbf{t}$ that will be used to capture neighborhood configurations is defined pictorially as

$$
\mathbf{t} = \quad
\begin{array}{ccc}
 & 1 & \\
1 & 5 & 1 \\
 & 1 &
\end{array}
$$

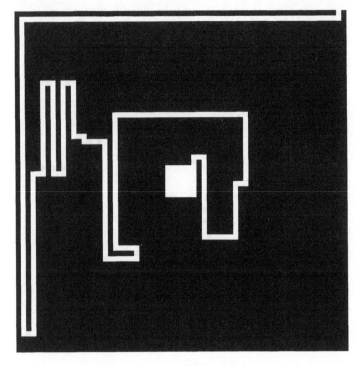

Figure 12.8.2. Maze solution.

The image algebra code for the maze solving cellular automaton is given by

$$\mathbf{b} := 0$$
$$\texttt{while } (\mathbf{a} \neq \mathbf{b}) \texttt{ loop}$$
$$\mathbf{b} := \mathbf{a}$$
$$\mathbf{a} := \chi_{\{7,8,9\}}(\mathbf{a} \oplus \mathbf{t})$$
$$\texttt{end loop}.$$

12.9.   Exercises

1.   Can a parallel implementation (i.e., all neurons fire simultaneously) guarantee convergence to a steady state of the Hopfield net? Explain your answer.

2.

    a.   Implement the Hopfield net using the patterns shown in Figure 12.2.3. Does the net provide perfect output for uncorrupted input of all patterns?

    b.   Add the four letter patterns shown in Figure 12.3.1 to the Hopfield memory established in 2.a. Does the expanded net still provide perfect output for uncorrupted input of all patterns?

3.   Implement the following variant of the Hopfield network:
Assign weights w_{ij} using the formula

$$w_{ij} = \bigwedge_{k=1}^{K} \left(e_i^k - e_j^k \right)$$

and use the propagation rule

$$\mathbf{a}_{t+1}(i) = f\left(\bigvee_{j=1}^{N} (\mathbf{a}_t(j) + w_{ij})\right).$$

Repeat Exercise 2 using this variation of the Hopfield net.

4.   Construct a BAM to establish the following three association:

a. $(1, 1, -1, -1) \leftrightarrow (1, 1)$,
b. $(1, 1, 1, 1) \leftrightarrow (1, -1)$,
c. $(-1, -1, 1, 1) \leftrightarrow (-1, 1)$.

What happens when a vector such as $(-1, -1, -1, -1)$ is presented to the BAM?

5.   There exists a wide variety of neural networks not discussed in this chapter. Consult your local science library or do a web search to find two networks not discussed here. Provide an image algebra description of the two networks.

6.   Design a 2–layer, 2–input, 1–output perceptron net that will solve the XOR problem. The net must work in the following way. The input to the net are the values +1, —1. The weights (which you determine, that is, you are not using the perceptron convergence procedure), should classify any input from the shaded region as a —1. Valid input points from the unshaded regions should be identified with value +1.

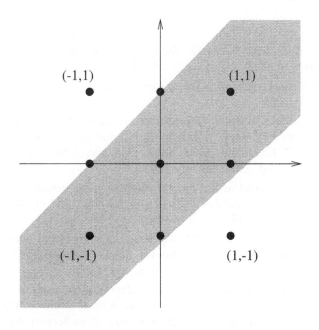

7.   Design a 2–layer, 2–input, 1–output perceptron net that will identify points within the region shaded in the figure below with -1 and points outside the region with +1. The weights for this net are fixed (determined by you) as in Exercise 6.

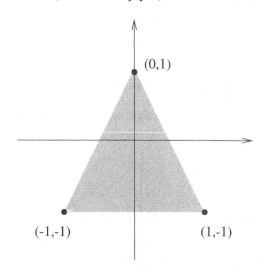

8.   Design a 3–layer, 2–input, 1–output feedforward perceptron net that will classify (-1,1) and (1,-1) as one class, and the line segment between (-1,-1) and (1,1) as another class.

9.   Design a multi-layer, 3–input, 1–output feedforward perceptron net that will classify (1,1,1) and (-1,-1,-1) as one class, but will classify (1,1,-1), (-1,1,1), (1,-1,1), (-1,-1,1), (-1,1,-1), and (1,-1,-1) as the other class.

12.10.   References

[1]   S. Chen, ed., *Neural and Stochastic Methods in Image and Signal Processing*, vol. 1766 of *Proceedings of SPIE*, (San Diego, CA), July 1992.

[2]   S. Wolfram, *Theory and Applications of Cellular Automata*. Singapore: World Scientific Publishing Co., 1986.

[3]   C. P. S. Araujo and G. Ritter, "Morphological neural networks and image algebra in artificial perception systems," in *Image Algebra and Morphological Image Processing III* (P. Gader, E. Dougherty, and J. Serra, eds.), vol. 1769 of *Proceedings of SPIE*, (San Diego, CA), pp. 128–142, July 1992.

[4]   J. Davidson, "Simulated annealing and morphological neural networks," in *Image Algebra and Morphological Image Processing III*, vol. 1769 of *Proceedings of SPIE*, (San Diego, CA), pp. 119–127, July 1992.

[5]   J. Davidson and R. Srivastava, "Fuzzy image algebra neural network for template identification," in *Second Annual Midwest Electro-Technology Conference*, (Ames, IA), pp. 68–71, IEEE Central Iowa Section, Apr. 1993.

[6]   J. Davidson and A. Talukder, "Template identification using simulated annealing in morphology neural networks," in *Second Annual Midwest Electro-Technology Conference*, (Ames, IA), pp. 64–67, IEEE Central Iowa Section, Apr. 1993.

[7]   R. Lippmann, "An introduction to computing with neural nets," *IEEE Transactions on Acoustics, Speech, and Signal Processing*, vol. ASSP-4, pp. 4–22, 1987.

[8]   B. Kosko, "Adaptive bidirectional associative memories," in *IEEE 16th Workshop on Applied Images and Pattern Recognition*, (Washington, D.C.), pp. 1–49, Oct. 1987.

[9]   J. A. Freeman and D. M. Skapura, *Neural Networks: Algorithms, Applications, and Programming Techniques*. Reading, MA: Addison Wesley, 1991.

[10] G. Ritter, D. Li, and J. Wilson, "Image algebra and its relationship to neural networks," in *Aerospace Pattern Recognition*, vol. 1098 of *Proceedings of SPIE*, (Orlando, FL), Mar. 1989.

[11] F. Rosenblatt, *Principles of Neurodynamics: Perceptrons and the Theory of Brain Mechanisms*. Washington, D.C.: Spartan Books, 1962.

[12] K. Knight, "Connectionist ideas and algorithms," *Communications of the ACM*, vol. 33, no. 11, pp. 59–74, 1990.

[13] R. Gonzalez and R. Woods, *Digital Image Processing*. Reading, MA: Addison-Wesley, 1992.

[14] S. Wolfram, "Cellular automata as models of complexity," *Nature*, vol. 311, pp. 419–424, Oct. 1884.

[15] M. Gardner, "The fantastic combinations of John Conway's new solitaire game "life"," *Scientific American*, vol. 223, pp. 120–123, Oct. 1970.

[16] B. Nayfeh, "Cellular automata for solving mazes," *Dr. Dobb's Journal*, pp. 32–38, Feb. 1993.

APPENDIX.
THE IMAGE ALGEBRA C++ LIBRARY

A.1. Introduction

In 1992, the U.S. Air Force sponsored work at the University of Florida to implement a C++ class library to support image algebra, iac++. This appendix gives a brief tour of the library then provides examples of programs implementing some of the algorithms provided in this text.

Current information on the most recent version of the iac++ class library is available via the Worldwide Web from URL

> http://www.cise.ufl.edu/research/IACC/.

The distribution and some documentation are available for anonymous ftp from

> ftp://ftp.cise.ufl.edu/pub/src/ia/iac++.

The class library has been developed in the C++ language [1, 2]. Several different platforms have been supported in the past. The current executable version of the library has been developed and tested for the following compiler/system combinations:

Table A.1.1 Currently tested iac++ platforms.

Compiler	System
g++ v 2.95.2	Solaris 2.7, 2.8 RedHat Linux 5.2

The library is likely to be portable to other systems if the g++ compiler is employed. Porting to other compilers might involve significant effort.

A.2. Classes in the **iac++** Class Library

The iac++ class library provides a number of classes and related functions to implement a subset of the image algebra. At the time of this writing the library contains the following major classes to support image algebra concepts:

(a) IA_Point<T>
(b) IA_Set<T>
(c) IA_Pixel<P,T>
(d) IA_Image<P,T>
(e) IA_Neighborhood<P,Q>
(f) IA_DDTemplate<I>

All names introduced into the global name space by iac++ are prefixed by the string IA_ to lessen the likelihood of conflict with programmer-chosen names.

Points

The IA_Point<*T*> class describes objects that are homogeneous points with coordinates of type *T*. The library provides instances of the following point classes:

(a) IA_Point<int>
(b) IA_Point<double>

The dimensionality of a point variable is determined at the time of creation or assignment, and may be changed by assigning a point value having different dimension to the variable. Users may create points directly with constructors. Functions and operations may also result in point values.

One must #include "ia/IntPoint.h" to have access to the class of points with int coordinates and its associated operations. One must #include "ia/DblPoint.h" to have access to the class of points with double coordinates and its associated operations.

Point operations described in the sections that follow correspond to those operations presented in Section 1.2 of this book.

Point Constructors and Assignment

Let i, $i1$, $i2$, ..., represent int-valued expressions. Let $iarray$ represent an array of int values with n elements. Let d, $d1$, $d2$, ..., represent double-valued expressions. Let $darray$ represent an array of double values with n elements. And let ip and dp represent points with int and double coordinate values, respectively.

Table A.2.1 Point Constructors and Assignment

construct a copy	IA_Point<int>(ip) IA_Point<double> (dp)
construct from coordinates	IA_Point<int> (i) IA_Point<int> ($i1,i2$) IA_Point<int> ($i1,i2,i3$) IA_Point<int> ($i1,i2,i3,i4$) IA_Point<int> ($i1,i2,i3,i4,i5$) IA_Point<double> (d) IA_Point<double> ($d1,d2$) IA_Point<double> ($d1,d2,d3$) IA_Point<double> ($d1,d2,d3,d4$) IA_Point<double> ($d1,d2,d3,d4,d5$)
construct from array of coordinates	IA_Point<int>($iarray$, n) IA_Point<double>($darray$, n)
assign a point	$ip1 = ip2$ $dp1 = dp2$

Binary Operations on Points

Let $p1$ and $p2$ represent two expressions both of type IA_Point<int> or IA_Point<double>.

Table A.2.2 Binary Operations on Points

addition	p1 + p2
subtraction	p1 - p2
multiplication	p1 * p2
division	p1 / p2
supremum	sup (p1, p2)
infimum	inf (p1, p2)
dot product	dot (p1, p2)
cross product	cross (p1, p2)
concatenation	concat (p1, p2)

Unary Operations on Points

Let p represent an expression of type IA_Point<int> or IA_Point<double>.

Table A.2.3 Unary Operations on Points

negation	-p
ceiling†	ceil (p)
floor†	float (p)
rounding†	rint (p)
projection (subscripting)	p(i) p[i]‡
sum	sum (p)
product	product (p)
maximum	max (p)
minimum	min (p)
Euclidean norm	enorm (p)
L^1 *norm*	mnorm (p)
L^∞ *norm*	inorm (p)
dimension	p.dim()

†This function applies only to points of type IA_Point<double>.

‡Subscripting with () yields a coordinate value, but subscripting with [] yields a coordinate *reference* to which one may assign a value, thus changing a point's value.

Relations on Points

Let *p1* and *p2* represent two expressions both of type IA_Point<int> or IA_Point<double>. Note that a relation on points is said to be *strict* if it must be satisfied on all of the corresponding coordinates of two points to be satisfied on the points themselves.

Table A.2.4 Relations on Points

less than (strict)	$p1 < p2$
less than or equal to (strict)	$p1 <= p2$
greater than (strict)	$p1 > p2$
greater than or equal to (strict)	$p1 >= p2$
equal to (strict)	$p1 == p2$
not equal to (complement of ==)	$p1 != p2$
lexicographic comparison	`pointcmp` $(p1, p2)$

Examples of Point Code Fragments

One can declare points using constructors.

```
IA_Point<int> point1 (0,0,0);
    // point1 == origin in 3D integral Cartesian space
IA_Point<double> dpoint (1.3,2.7);
    // dpoint is a point in 2D real Cartesian space
```

One may subscript the coordinates of a point. (Note that point coordinates use zero-based addressing in keeping with C vectors.)

```
point1[0] = point1[1] = 3;
    // point1 == IA_Point<int> (3,3,0)
```

One can manipulate such points using arithmetic operations.

```
point1 = point1 + IA_Point<int> (1,2,3);
    // point1 == IA_Point<int> (4,5,3)
```

And one may apply various functions to points.

```
point1 = floor (dpoint);
// point1 == IA_Point<int> (1,2)
```

Sets

The C++ template class `IA_Set<`*T*`>` provides an implementation of sets of elements of type `T`. The following instances of `IA_Set` are provided:

(a) `IA_Set<IA_Point<int> >`
(b) `IA_Set<IA_Point<double> >`
(c) `IA_Set<bool>`
(d) `IA_Set<unsigned char>`
(e) `IA_Set<int>`
(f) `IA_Set<float>`
(g) `IA_Set<IA_complex>`
(h) `IA_Set<IA_RGB>`

`IA_Set<IA_Point<int> >` and `IA_Set<IA_Point<double> >`, provide some capabilities beyond those provided for other sets.

Sets of Points and Their Iterators

The point set classes `IA_Set<IA_Point<int> >` and `IA_Set<IA_Point<double> >` provide the programmer with the ability to define and manipulate sets of point objects all having the same type and dimension.

One must `#include "ia/IntPS.h"` to have access to the class of sets of int points and its associated operations. One must `#include "ia/DblPS.h"` to have access to the class of sets of double points and its associated operations. To gain access to point set iterator classes and their associated operations, one must `#include "ia/PSIter.h"`.

Point set operations described in the sections that follow correspond to those operations presented in Section 1.2 of this book.

Point Set Constructors and Assignment

In Table A.2.5, let *ip*, *ip1*, *ip2*, ..., represent `IA_Point<int>`-valued expressions. Let *iparray* represent an array of `IA_Point<int>` values, each of dimension *d*, with *n* elements. Let *dp*, *dp1*, *dp2*, ..., represent `IA_Point<double>`-valued expressions. Let *dparray* represent an array of double values, each of dimension *d*, with *n* elements. And let *ips* and *dps* represent sets with `IA_Point<int>` and `IA_Point<double>` elements, respectively.

The function `IA_boxy_pset` creates the set containing all points bounded by the infimum and supremum of its two point arguments. The type of set is determined by the point types of the arguments. The `IA_universal_ipset` and `IA_empty_ipset` create universal or empty sets of type `IA_Point<int>` of dimension specified by the argument. Sets of type `IA_Point<double>` are created in a corresponding fashion by `IA_universal_dpset` and `IA_empty_dpset`.

Table A.2.5 Point Set Constructors and Assignment

construct a copy	`IA_Set<IA_Point<int> > (`*ips*`)` `IA_Set<IA_Point<double> > (`*dps*`)`

construct from points	`IA_Set<IA_Point<int> > (`*ip*`)`
	`IA_Set<IA_Point<int> > (`*ip1*`,`*ip2*`)`
	`IA_Set<IA_Point<int> > (`*ip1*`,`*ip2*`,`*ip3*`)`
	`IA_Set<IA_Point<int> > (`*ip1*`,`*ip2*`,`*ip3*`,`
	  ip4`)`
	`IA_Set<IA_Point<int> > (`*ip1*`,`*ip2*`,`*ip3*`,`
	  ip4`,`*ip5*`)`
	`IA_Set<IA_Point<double> > (`*dp*`)`
	`IA_Set<IA_Point<double> > (`*dp1*`,`*dp2*`)`
	`IA_Set<IA_Point<double> > (`*dp1*`,`*dp2*`,`
	  dp3`)`
	`IA_Set<IA_Point<double> > (`*dp1*`,`*dp2*`,`
	  dp3`,`*dp4*`)`
	`IA_Set<IA_Point<double> > (`*dp1*`,`*dp2*`,`
	  dp3`,`*dp4*`,`*dp5*`)`

construct from array of *points*	`IA_Set<IA_Point<int> > (`*d*`,`*iparray*`,`*n*`)` `IA_Set<IA_Point<double> > (`*d*`,`*dparray* *n*`)`
assign a point set	*ips1* = *ips2* *dps1* = *dps2*

functions returning
point sets
	`IA_boxy_pset (`*ip1*`,  `*ip2*`)`
	`IA_boxy_pset (`*dp1*`,  `*dp2*`)`
	`IA_universal_ipset (`*dim*`)`
	`IA_universal_dpset (`*dim*`)`
	`IA_empty_ipset (`*dim*`)`
	`IA_empty_dpset (`*dim*`)`

Binary Operations on Point Sets

In Table A.2.6, let *ps*, *ps1* and *ps2* represent expressions all of type
`IA_Set<IA_Point<int> >` or `IA_Set<IA_Point<double> >`. Let *p* represent
an expression having the same type as the elements of *ps*.

Table A.2.6 Binary Operations on Point Sets

addition	`ps1 + ps2`
subtraction	`ps1 - ps2`
point addition	`ps + p` `p + ps`
point subtraction	`ps - p` `p - ps`
union	`ps1	ps2`
intersection	`ps1 & ps2`
set difference	`ps1 / ps2`
symmetric difference	`ps1 ^ ps2`

Unary Operations on Point Sets

Let *ps* represent an expression of type `IA_Set<IA_Point<int> >` or `IA_Set<IA_Point<double> >`.

Table A.2.7 Unary Operations on Point Sets

negation	`-ps`
complementation	`~ps`
supremum	`sup (ps)`
infimum	`inf (ps)`
choice function	`ps.choice()`
cardinality	`ps.card()`

Relations on Point Sets

Let *ps*, *ps1*, and *ps2* represent expressions all of type `IA_Set<IA_Point<int> >` or `IA_Set<IA_Point<double> >`. Let *p* represent an expression having the same type as the elements of *ps*.

Table A.2.8 Relations on Point Sets

containment test	`ps.contains(p)`
proper subset	`ps1 < ps2`
(improper) subset	`ps1 <= ps2`
proper superset	`ps1 > ps2`
(improper) superset	`ps1 >= ps2`
equality	`ps1 == ps2`
inequality	`ps1 != ps2`
emptiness test	`ps.empty()`

Point Set Iterators

The class `IA_PSIter<P>` supports iteration over the elements of sets of type `IA_Set<IA_Point<int> >` and `IA_Set<IA_Point<double> >`. This provides a single operation, namely function application (the parentheses operator), taking a reference argument of the associated point type and returning a `bool`. The first time an iterator object is applied as a function, it attempts to assign its argument the initial point in the set (according to an implementation specified ordering). If the set is empty, the iterator returns a false value to indicate that it failed; otherwise it returns a true value. Each subsequent call assigns to the argument the value following that which was assigned on the immediately preceding call. If it fails (due to having no more points) the iterator returns a false value; otherwise it returns a true value.

Examples of Point Set Code Fragments

One can declare point sets using a variety of constructors.

```
IA_Set<IA_Point<int> > ps1;
    // uninitialized point set

IA_Set<IA_Point<int> > ps2(IA_IntPoint(0,0,0));
    // point set with a single point

IA_Set<IA_Point<int> > ps3(IA_boxy_pset(IA_Point<int>(-1,-1),
                                        IA_Point<int>(9,9)));
    // contains all points in the rectangle
    // with corners (-1,-1) and (9,9)

// Introduce a predicate function
int pred(IA_IntPoint p) { return (p[0] >= 0)? 1 : 0; }

IA_Set<IA_Point<int> > ps4(2, &pred);
    // the half plane of 2D space having nonnegative 0th coord.

IA_Set<IA_Point<int> > ps5(3, &pred);
    // the half space of 3D space having nonnegative 0th coord.
```

One can operate upon those sets.

```
ps1 = ps3 & ps4;
    // intersection of ps3 and ps4 contains all points
    // in the rectangle with corners (0,0) and (9,9)

ps1 = ps3 | ps4;
    // union of ps3 and ps4

ps1 = ps3 + IA_Point<int> (5,5);
    // pointset translation

ps1 = ps3 + ps3;
    // Minkowski (pairwise) addition

// relational operations and predicates
if ((ps2 <= ps3) || ps2.empty())
    ps2 = ~ps2;  // complementation
```

And one can iterate over the elements in those sets.

```
IA_PSIter<IA_Point<int> > iter(ps3);
    // iter will be an iterator over the points
    // in set ps3
IA_Point<int> p;

// the while loop below iterates over all the points in the
// set ps3 and writes them to the standard output device
while(iter(p)) { cout << p << "\n"; }
```

Sets of Values and Their Iterators

In addition to point sets, the user may employ the C++ template class $IA\_Set<T>$ to construct and manipulate extensive sets of values of any C++ type T for which there is an ordering. Instances of sets with bool, unsigned char, int, float, IA_complex, and IA_RGB elements are instantiated in the iac++ library. The class $IA\_Set<T>$ imposes a total ordering on the elements of type T in the set, and supports operations max and min. If the value type does not provide underlying definitions for max and min (as with IA_complex), then the library chooses some arbitrary (but consistent) definition for them.

One must #include "ia/Set.h" to have access to instances of the value set class and associated operations. To gain access to the iterators and associated operations, one must #include "ia/SetIter.h".

Value Set Constructors and Assignment

Let v, $v1$, $v2$, ..., represent T-valued expressions. Let $varray$ represent an array of T values with n elements, and let vs, $vs1$, and $vs2$ represent sets with T-valued elements.

Table A.2.9 Value Set Constructors and Assignment

construct a copy	IA_Set<T>(vs)
construct from values	IA_Set<T> (v) IA_Set<T> (v1, v2) IA_Set<T> (v1, v2, v3) IA_Set<T> (v1, v2, v3, v4) IA_Set<T> (v1, v2, v3, v4, v5)
construct from array of values	IA_Set<T>(varray, n)
assign a value set	vs1 = vs2

Binary Operations on Value Sets

Let $vs1$ and $vs2$ represent two expressions, both of type $IA\_Set<T>$.

Table A.2.10 Binary Operations on Value Sets

union	vs1 | vs2
intersection	vs1 & vs2
set difference	vs1 / vs2
symmetric difference	vs1 ^ vs2

Unary Operations on Value Sets

Let *vs* represent an expressions of type IA_Set<*T*>.

Table A.2.11 Unary Operations on Value Sets

maximum	max (vs)
Minimum	min (vs)
choice function	vs.choice()
cardinality	vs.card()

Relations on Value Sets

Let *vs*, *vs1* and *vs2* represent expressions all of type IA_Set<T>.

Table A.2.12 Relations on Value Sets

containment test	vs.contains(v)
proper subset	vs1 < vs2
(improper) subset	vs1 <= vs2
proper superset	vs1 > vs2
(improper) superset	vs1 >= vs2
equality	vs1 == vs2
inequality	vs1 != vs2
emptiness test	vs.empty()

Value Set Iterators

The class IA_SetIter<*T*> supports iteration over the elements of an IA_Set<*T*>. This provides a single operation, namely function application (the parentheses operator), taking a reference argument of type *T* and returning a bool. The first time an iterator object is applied as a function, it attempts to assign its argument the initial point in the set (according to an implementation-specified ordering). If the set is empty, the iterator returns a false value to indicate that it failed; otherwise it returns a true value. Each subsequent call assigns to the argument the value following that which was assigned on the immediately preceding call. If it fails (due to having no more points) the iterator returns a false value; otherwise it returns a true value.

Examples of Value Set Code Fragments

Value sets can be constructed by explicitly listing the contained values.

```
IA_Set<int> v1(1,2,3);
    // constructs a set containing three integers.
    // sets of up to 5 elements can be
    // constructed in this way

float vec[] = { 1.0, 2.0, 3.0,
                1.0, 2.0, 3.0,
                3.0, 2.0, 1.0 };
IA_Set<float> v2( vec, 9);
    // constructs a set containing the three
    // unique values 1.0, 2.0, and 3.0, that
    // are contained in the 9 element vector vec
```

One can apply operations to those value sets

```
v2 = v2 & IA_Set<float>(2.0, 3.0);
    // the intersection of v2 and the
    // specified set is assigned to v2
v2 = v2 | 6.0;
    // the single element 6.0 is united with set v2

cout << v2.card() << "\n";
    // writes the cardinality of set

if (IA_Set<int>(1, 2) <= v1)     // test for subset
    cout << v1 << "\n";          // write the set
```

One can iterate over the values in those sets.

```
IA_SetIter<int> iter(v1);
int i;

// write all the elements of a set
while (iter(i)) { cout << i << "\n"; }
```

Images, Pixels, and Iterators Over Image Pixels and Values

The image classes defined in the iac++ library are instances of the template class IA_Image<P, T>, comprising the images defined over sets of points of type P having values of type T. The following instances of the IA_Image class are provided in the library:

```
(a)  IA_Image<IA_Point<int>, bool>
(b)  IA_Image<IA_Point<int>, unsigned char>
(c)  IA_Image<IA_Point<int>, int>
(d)  IA_Image<IA_Point<int>, float>
(e)  IA_Image<IA_Point<int>, IA_complex>
(f)  IA_Image<IA_Point<int>, IA_RGB>
(g)  IA_Image<IA_Point<double>, float>
(h)  IA_Image<IA_Point<double>, IA_complex>
```

To gain access to the class definitions and associated operations for each of these image types, one must #include the associated header file. The basic unary and binary operations on the supported image types are included in the following files, respectively:

(a) `ia/BoolDI.h`
(b) `ia/UcharDI.h`
(c) `ia/IntDI.h`
(d) `ia/FloatDI.h`
(e) `ia/CplxDI.h`
(f) `ia/RGBDI.h`
(g) `ia/FloatCI.h`
(h) `ia/CplxCI.h`

In these file names, the `DI` designation (discrete image) denotes images over sets of point with `int` coordinates and `CI` (continuous image) denotes images over sets of points with `double` coordinates.

A pixel is a point together with a value. Pixels drawn from images mapping points of type P into values of type T are supported with the class `IA_Pixel<P, T>` which can be instantiated for any combination of P and T. Objects belonging to this class have the two publicly accessible fields `point` and `value`. An `IA_Pixel` object, while it does bring together a point and a value, is not in any way associated with a specific image. Thus, assigning to the `value` field of an `IA_Pixel` object does not change the actual value associated with a point in any image. The definition of the `IA_Pixel` class is contained in the header file `ia/Pixel.h`.

Image Constructors and Assignment

Given a point type P and a value type T, one can create and assign images mapping points of type P into values of type T. In the following table *img*, *img1*, and *img2* denote objects of type `IA_Image<P, T>`, *p* denotes an object of type `IA_Set<P>`, *t* denotes a value of type T, *pixarray* denotes an array of n objects of type `IA_Pixel<P, T>`, and *f* denotes a function with signature `T f(P)` or `T f(const P&)`.

Table A.2.13 Image Constructors and Assignment

construct an empty image	`IA_Image<P, T> ()`
construct a copy	`IA_Image<P, T> (img)`
construct a constant-valued image	`IA_Image<P, T> (ps, t)`
construct from an array of pixels	`IA_Image<P, T> (pixarray, n)`
construct from a point-to-value function	`IA_Image<P, T> (ps, f)`
assign an image	`img1 = img2`
assign each pixel a constant value	`img = t`
assign to img1 the overlapping parts of img2	`img1.restrict_assign(img2)`

Binary Operations on Images

Let *img1* and *img2* represent two expressions both of type
`IA_Image<IA_Point<int>,` *T*`>`.

Table A.2.14 Binary Operations on Images

addition	`img1 + img2`
pointwise maximum	`max (img1, img2)`
subtraction	`img1 - img2`
multiplication	`img1 * img2`
division	`img1 / img2`
pseudo-division	`pseudo_div (img1, img2)`
modulus†	`img1 % img2`
binary and†	`img1 & img2`
binary or†	`img1	img2`
binary exclusive or†	`img1 ^ img2`
logical and	`img1 && img2`
logical or	`img1 || img2`
left arithmetic shift†	`img1 << img2`
right arithmetic shift†	`img1 >> img2`
pointwise minimum	`min (img1, img2)`
characteristic less than	`chi_lt (img1, img2)`
characteristic less than or equal to	`chi_le (img1, img2)`
characteristic equal	`chi_eq (img1, img2)`
characteristic greater than	`chi_gt (img1, img2)`
characteristic greater than or equal	`chi_ge (img1, img2)`
characteristic value set containment	`chi_contains (img1, vset)`

† Available only for images for which *T* is an integer type (`bool`, `unsigned char`, or `int`).

Unary Operations on Images

Let *img* represent an expressions of type `IA_Image<IA_Point<int>,` *T*`>`.

Table A.2.15 Unary Operations on Images

projection	`img(p)`
	`img[p]`†
negation	`-img`
pointwise one's complement	`~img`
pointwise logical not	`!img`
cardinality	`img.card()`
domain extraction	`img.domain()`
range extraction	`img.range()`
sum	`sum (img)`
maximum	`max (img)`
minimum	`min (img)`
product	`prod (img)`
absolute value	`abs (img)`
ceiling	`ceil (img)`
floor	`floor (img)`
exponential	`exp (img)`
natural logarithm	`log (img)`
cosine	`cos (img)`
sine	`sin (img)`
tangent	`tan (img)`
complex magnitude	`abs_f (img)`
complex angle	`arg_f (img)`
complex real part	`real_f (img)`
complex imaginary part	`imag_f (img)`
sqare root	`sqrt (img)`
integer square root	`isqrt (img)`
square	`sqr (img)`

†Subscripting with `( )` yields the value associated with the point p, but subscripting with `[ ]` yields a *reference* to the value associated with p. One may assign such a reference a new value, thus changing the image. This operation is potentially expensive. This is discussed at greater length in the section providing example code fragments below.

Domain Transformation Operations on Images

Let *img*, *img1*, and *img2* represent expressions of type `IA_Image<P, T>`. Let *tset* represent an expression of type `IA_Set<T>`. Let *pset* represent an expression of type `IA_Set<P>` and let p represent an element of such a point set.

Table A.2.16 Domain Transformation Operations on Images

translation	`translate (`*`img, p`*`)`
restriction of domain to a point set	`restrict (`*`img, pset`*`)`
restriction of domain by a value set	`restrict (`*`img, tset`*`)`
extension of one image by another	`extend (`*`img1, img2`*`)`

Relations on Images

Let *img1* and *img2* represent two expressions both of type `IA_Image<`*`P, T`*`>`.

Table A.2.17 Relations on Images

less than	*img1* `<` *img2*
less than or equal to	*img1* `<=` *img2*
equal to	*img1* `==` *img2*
greater than	*img1* `>` *img2*
greater than or equal to	*img1* `>=` *img2*
not equal to (complement of equal to)	*img1* `!=` *img2*
strictly not equal to	`strict_ne (`*`img1, img2`*`)`

Input/Output Operations on Images

The `iac++` library supports the reading and writing of images in extended portable bitmap format (EPBM) [3]. This format supports the commonly used pbm, pgm, and ppm formats. The EPBM supports only two-dimensional rectangular images and does not represent an image's point set, thus not all images representable in the `iac++` library can be directly read or written. When an EPBM image is read, it is assigned an `int` coordinate point set spanning from the origin to the point whose coordinates are one less than the number of rows and number of columns in the image. To insure that we write only rectangular images, any image to be written is extended to the smallest enclosing rectangular domain with the value 0 used wherever the image was undefined.

Let *istr* represent an `istream` and let *in_file_name* represent a character string containing an input file name. Let *max_ptr* be a pointer to an `int` variable that will receive as its value the maximum value in the image read.

Table A.2.18 Input Operations on Images

read a bit image	read_PBM (*in_file_name*) read_PBM (*istr*)
read an unsigned char image	read_uchar_PGM (*in_file_name*) read_uchar_PGM (*in_file_name, max_ptr*) read_uchar_PGM (*istr*) read_uchar_PGM (*istr, max_ptr*)
read an integer image	read_int_PGM (*in_file_name*) read_int_PGM (*in_file_name, max_ptr*) read_int_PGM (*istr*) read_int_PGM (*istr, max_ptr*)
read an RGB image	read_PPM (*in_file_name*) read_PPM (*in_file_name, max_ptr*) read_PPM (*istr*) read_PPM (*istr, max_ptr*)

One may write either to a named file or to an ostream. If writing to a named file, the image output function returns no value. If writing to an ostream, the image output function returns the ostream. The display function displays an image to the currently selected display. The IMAGE_DISPLAY environment variable should map to a program that will display the pgm file which is the program's argument. The default value for IMAGE_DISPLAY is xv.

Let *ostr* represent an ostream and let *out_file_name* represent a character string containing an output file name. Let *bit_img* denote an image with bool values, *uchar_img* denote an image with unsigned char values, *int_img* denote an image with int values, and *rgb_img* denote an image with IA_RGB values. Let *maxval* be an unsigned int representing the maximum value in an image.

Image Iterators

The class IA_Pixel<P,T> supports storing of a point and value as a unit and provides the two field selectors point and value. Instances of the IA_Pixel class are provided for the same combinations of P and T on which IA_Image instances are provided.

Iterators over either the values or the pixels of an image can be constructed and used. The class IA_IVIter<P, T> supports image value iteration and class IA_IPIter<P, T> supports pixel iteration. The value iterator provides an overloading of function call with a reference argument of the image's range type. The pixel iterator function call overloading takes a reference argument of type IA_Pixel<P, T>. To use image iterators, one must #include "ia/ImageIter.h".

The first time an image value (pixel) iterator object is applied as a function, it attempts to assign its argument the value (pixel) that is associated with the first point in the image's domain point set. If the image is empty, a false value is returned to indicate that it failed, otherwise a true value is returned. Each subsequent call assigns to the argument the value (pixel) associated with the next point in the image's domain. If the iterator fails to find such a value (pixel) because it has exhausted the image's domain point set, the iterator returns a false value, otherwise it returns a true value.

Table A.2.19 Output Operations on Images

write a bit image	write_PBM (*bit_img*, *out_file_name*) write_PBM (*bit_img*, *ostr*)
write an unsigned char image	write_PGM (*uchar_img*, *out_file_name*) write_PGM (*uchar_img*, *out_file_name*, *maxval*) write_PGM (*uchar_img*, *ostr*) write_PGM (*uchar_img*, *ostr*, *maxval*)
display an unsigned char image	display (*uchar_img*)
write an integer image	write_PGM (*int_img*, *out_file_name*) write_PGM (*int_img*, *out_file_name*, *maxval*) write_PGM (*int_img*, *ostr*) write_PGM (*int_img*, *ostr*, *maxval*)
write an RGB image	write_PPM (*rgb_img*, *out_file_name*) write_PPM (*rgb_img*, *out_file_name*, *maxval*) write_PPM (*rgb_img*, *ostr*) write_PPM (*rgb_img*, *ostr*, *maxval*)

Image Composition Operations

Two kinds of composition operations are supported for images. Since an image maps points to values, one may generate a new image by composing a value-to-value mapping with an image or by composing an image with a spatial transform (or point-to-point mapping). To use composition of a value-to-value function with an image, one must #include "ia/ImgComp.h".

Let *img* be an image of type IA_Image<P, T>, let *value_map* be a function with signature T *value_map*(T), let *point_map* be a function pointer with signature P *point_map*(const P&), and let *result_pset* be the set of type IA_Set<P> over which the result of the composition is to be defined. The result of composing the point-to-point mapping function with the image at any point *p* in the *result_pset* is equal to *img*(*point_map*(*p*)). The result of composing the value-to-value mapping function with the image is *value_map* (*img*(*p*)).

Table A.2.20 Image Composition Operations

compose an image with a point-to-point function	compose (*img*, *point_map*, *result_pset*)
compose a value-to-value function with an image	compose (*value_map*, *img*)

Note that the composition of an image with a point-to-point mapping function takes a third argument, namely the point set over which the result is to be defined. This is necessary because determining the set of points over which the result image defines values

would require us to solve the generally unsolvable problem of computing the inverse of the
point_map function. Thus, we require the user to specify the domain of the result.

Examples of Image Code Fragments

One may construct images.

```
IA_Image <IA_Point<int>, int> i1;
    // uninitialized image
IA_Image<IA_Point<int>, int> i2(ps3, 5);
    // a constant image over point set ps3
    // having value 5 at each point

// we may use a function such as p0 below to create an
// image with value at each point specified by a function
int p0(IA_Point<int> p) { return p[0]; }
IA_Image<IA_Point<int>, int> i2(ps3, &p0);
```

It is important to note the significant difference in application of an image as a
function (as in i2 (...)) versus subscripting of an image (as in i2 [...]). In the first
case, the value of the image at a point is returned. In the second case, a reference to a
pixel, the specific value mapping of a point within an image, is returned. An image pixel
can have its value assigned through such a reference, thus modifying the image. Assigning
an image to a variable can be quite time consuming. The iac++ library tries to avoid
such copying wherever possible. Assigning an image to a variable usually does not copy
storage. Instead, it uses a reference counting technique to keep track of multiple uses of
a single image representation. If an image having multiple readers is subscripted (with
operation [ ]), a unique version of the image is created so that the subscripter (writer) will
get a unique copy of the potentially modified image. While this insulates other readers
of that image from future changes, it also imposes a significant penalty in both time and
space. Thus, it is preferable to use function application notation rather than subscripting
wherever possible.

```
i1 = i2;      // i1 and i2 now share the same value map

// write a point of image i2 (same as i1)
cout << i2(IA_Point<int>(3,5));

i1[IA_Point<int>(2,4)] = 10;
    // Subscripting a pixel of i1 causes new
    // storage for its value mapping to be allocated
    // whether or not an assignment is actually made
```

The iac++ library also supports assignment to a subregion of an image with the re-
strict_assign member function.

```
i2.restrict_assign (i1);
    // For each point p in both i1 and i2's domains
    // This performs assignment
    //     i2[p] = i1(p);
```

One may restrict the value of an image to a specified set of points or to the set of points
containing values in a specified set, and one may extend an image to the domain of another
image.

```
i1 = restrict(i2, ps4);
    // i1's domain will be the intersection of
    // ps4 and i2.domain()

i1 = restrict(i2, IA_Set<int>(1, 2, 3));
    // i1's domain will be all those
    // points in i2's domain associated with value
    // 1, 2, or 3 by i2.

i1 = extend(i2, IA_Image<IA_Point<int>, int>(ps4, 1));
    // i1's domain will be the union of i2.domain() and ps4
    // and value 1 will be associated with those
    // points of ps4 not in i2.domain().
```

One may reduce an image to a single value with a binary operations on the range type if the image has an extensive domain.

```
// binary operation defined by a function
int add(int i, int j) { return i + j; }

cout << i1.reduce(&add, 0) << "\n";
    // writes the result of adding
    // all the pixel values of i1
```

One can compose a C++ function of one parameter with an image having elements of the same type as the function's parameter. This lets one efficiently apply a user-defined function to each element of an image.

```
int my_function(int i) { ... }
i1 = compose (my_function, i2);
    // This is equivalent to assigning i1 = i2
    // and for each point p in i1.domain() executing
    // i1[p] = my_function(i1(p))
```

Likewise, since an image is conceptually a function composition of an image with a function mapping points to points will yield a new image. The point set of such an image cannot be effectively computed, thus one must specify the set of points in the resulting image.

```
IA_Point<int>
reflect_through_origin(const IA_Point<int> &p)
{
    return -p;
}

i1 = compose (i2,
              reflect_through_origin,
              IA_boxy_pset (-max(i2.domain()),
                            -min(i2.domain())));
    // After executing this,
    // i1(p) == i2(reflect_through_origin(p)) = i2(-p).
```

Input/output, arithmetic, bitwise, relational, and type conversion operations are provided on the image classes.

```
IA_Image<IA_Point<int>, int> i3, i4;
IA_Image<IA_Point<int>, float> f1, f2;


// image I/O is supported for pbm, pgm, and ppm image formats
i3 = read_int_PGM ("image1.pgm");
i4 = read_int_PGM ("image3.pgm");

// conversions mimic the behavior of C++ casts
f1 = to_float (i3*i3 + i4*i4);
f2 = sqrt (f1);

// All relational operations between images, with the
// exception of !=, return a true boolean value
// if the relation holds at every pixel.
// != is the complement of the == relation.
if (i3 < i4)
    write_PGM (max(i3) - i3);
```

Iteration over the values or pixels in an image is supported by the library.

```
IA_IVIter<IA_Point<int>, int> v_iter(i1);
    // declares v_iter to be an iterator over the
    // values of image i1

IA_IPIter<IA_Point<int>, int> p_iter(i1);
    // declares p_iter to be an iterator over the
    // pixels of image i1

int i, sum = 0;

// iterate over values in i1 and collect sum
while (v_iter(i)){ sum += i; }
cout << sum << "\n";
    // prints the sum

IA_Pixel<IA_Point<int>, int> pix;
IA_Point<int> psum = extend_to_point(0, i1.domain().dim());

// sum together the value-weighted pixel locations in i1
while (p_iter(pix))
    psum += pix.point * pix.value;

cout << (psum / i1.card()) << "\n";
    // prints i1's centroid.
```

Neighborhoods and Image-Neighborhood Operations

A *neighborhood* is a mapping from a point to a set of points. The iac++ library template class IA_Neighborhood<P, Q> provides the functionality of image algebra neighborhoods. To specify an instance of this class, one gives the type of point in the neighborhood's domain and the type of point which is an element of the range. The library contains the single instance IA_Neighborhood <IA_Point<int>,

`IA_Point<int> >`, mapping from points with integral coordinates to sets of points with integral coordinates. The library provides the following kinds of operations:

(a) constructors and assignment,
(b) function application of the neighborhood to a point,
(c) domain extraction, and
(d) image-neighborhood reduction operations.

Neighborhood Constructors and Assignment

Let Q be the argument point type of a neighborhood and P be the type of elements in the result point set. Let *nbh*, *nbh1*, and *nbh2* denote objects of type `IA_Neighborhood<P,Q>`. Let *result_set* be a set of type `IA_Set<P>` containing points with dimensionality *dim* that is to be associated with the origin of a translation invariant neighborhood. Let *domain_set* be a set of type `IA_Set<Q>` over which a neighborhood shall be defined. Let *point_to_set_map* be a function pointer with prototype `IA_Set<P> point_to_set_map (Q)` or with prototype `IA_Set<P> point_to_set_map(const Q&)`.

Table A.2.21 Neighborhood Constructors and Assignment

construct an uninitialized neighborhood	`IA_Neighborhood<P, Q> ()`
construct a copy of a neighborhood	`IA_Neighborhood<P, Q> (nbh)`
construct a translation invariant neighborhood	`IA_Neighborhood<P, Q> (dim, result_set)`
construct a translation variant neighborhood from a function	`IA_Neighborhood<P, Q> (domain_set, point_to_set_map)`
assign a neighborhood value	`nbh1 = nbh2`

Neighborhood operations presented in the sections that follow are introduced in Section 1.7.

Image-Neighborhood Reduction Operations

The definitions for the neighborhood reduction operations are provided in the following header files:

(a) `ia/BoolNOps.h` for images of type `IA_Image<P, bool>`
(b) `ia/UcharNOps.h` for images of type `IA_Image<P, unsigned char>`
(c) `ia/IntNOps.h` for image of type `IA_Image<P, int>`
(d) `ia/FloatNOps.h` for images of type `IA_Image<P, float>`
(e) `ia/CplxNOps.h` for image of type `IA_Image<P, IA_Complex>`

In the table following, *img* denotes an image object, *nbh* denotes a neighborhood, and `result_pset` denotes the point set over which the result image is to be defined. If no result point set is specified, the resulting image has the same domain as *img*. For the generic neighborhood reduction operation, the function with prototype T `binary_reduce` (T, T) is a commutative function with identity `zero` used to reduce all the elements of a neighborhood. Alternatively, the function T `n_ary_reduce` (T^*, `unsigned` n), which takes an array of n values of type T and reduces them to a single value, can be used to specify a neighborhood reduction.

Table A.2.22 Image-Neighborhood Reduction Operations

calculate the right image-neighborhood sum	`sum (`*img*`, `*nbhd*`, `*result_pset*`)` `sum (`*img*`, `*nbhd*`)`
calculate the right image-neighborhood product	`prod (`*img*`, `*nbhd*`, `*result_pset*`)` `prod (`*img*`, `*nbhd*`)`
calculate the right image-neighborhood maximum	`max (`*img*`, `*nbhd*`, `*result_pset*`)` `max (`*img*`, `*nbhd*`)`
calculate the right image-neighborhood minimum	`min (`*img*`, `*nbhd*`, `*result_pset*`)` `min (`*img*`, `*nbhd*`)`

calculate a right image-neighborhood product with a programmer-specified reduction function	`neighborhood_reduction (`*img*`, `*nbhd*`,`
	` `*result_pset*`, `*binary_reduce*`, `*zero*`)`
	`neighborhood_reduction (`*img*`, `*nbhd*`,`
	` `*result_pset*`, `*n_ary_reduce*`)`

Examples of Neighborhood Code Fragments

The neighborhood constructors provided in the library support both translation invariant and variant neighborhoods. A translation invariant neighborhood is specified by giving the set the neighborhood associates with the origin of its domain.

```
IA_Neighborhood<IA_Point<int>, IA_Point<int> >
    n1 (2,
        IA_boxy_pset (IA_Point<int>(-1,-1),
                      IA_Point<int>( 1, 1)));
    // n1 is a neighborhood defined for all
    // 2 dimensional points,
    //   associating a 3x3 neighborhood with each point.

// A neighborhood mapping function.
IA_Set<IA_Point<int> > nfunc (IA_Point<int> p) { ... }

// A neighborhood constructed from a mapping function.
IA_Neighborhood<IA_Point<int>, IA_Point<int> >
    n2 (IA_boxy_pset(IA_Point<int>(0,0) IA_Point<int>(511, 511),
        nfunc)
```

Neighborhood reduction functions map an image and a neighborhood to an image by reducing the neighbors of a pixel to a single value. The built-in neighborhood reduction functions max, min, product, and sum are provided, together with a functions supporting a user-specified reduction function.

```
IA_Image<IA_Point<int>, int> i1, i2;

i1 = max (i2, n1);
      // Finds the maximum value in the neighbors of each pixel
      // in image i1 and assigns the resultant image to i2.

// Specify a reduction function.
int or (int x, int y) { return (x | y); }

// Reduce an image by giving the image, neighborhood,
// reduction function, and the identity of the reduction
// functions as arguments to the generic reduction functional.

i1 = neighborhood_reduction (i2, n1, i1.domain(), or, 0);
```

Templates and Image-Template Product Operations

An image algebra template can be thought of as an image which has image values. Thus, image algebra templates concern themselves with two different point sets, the domain of the template itself, and the domain of the images in the template's range. The iac++ library currently supports the creation of templates — in which both of these point sets are discrete — with the C++ template class IA_DDTemplate<I> which takes as a class parameter an image type. The iac++ class library provides the following instances of IA_DDTemplate:

(a) IA_DDTemplate<IA_Image<IA_Point<int>,bool> >
(b) IA_DDTemplate<IA_Image<IA_Point<int>,unsigned char> >
(c) IA_DDTemplate<IA_Image<IA_Point<int>,int> >
(d) IA_DDTemplate<IA_Image<IA_Point<int>,float> >
(e) IA_DDTemplate<IA_Image<IA_Point<int>, IA_Complex> >

The operations defined upon these IA_DDTemplate instances are defined in the following include files:

(a) ia/BoolProd.h
(b) ia/UcharProd.h
(c) ia/IntProd.h
(d) ia/FloatProd.h
(e) ia/CplxProd.h

Template Constructors

Let the argument image type I in the IA_DDTemplate class be IA_Image<P,T>. Let $templ$, $templ1$, and $templ2$ denote objects of type IA_DDTemplate<I>. Let img denote the image to be associated with the origin by an image algebra template. Let dim denote the dimensionality of the domain of img. Let $domain\_pset$ denote the point set over which an image algebra template is to be defined. Let the function I $templ\_func$ (IA_Point<int>) be the point-to-image mapping function of a translation variant template.

Table A.2.23 Template Constructors and Assignment

construct an uninitialized template	`IA_DDTemplate<I> ()`
construct a copy of a template	`IA_DDTemplate<I> (templ)`
construct a translation invariant template	`IA_DDTemplate<I> (dim, img)` `IA_DDTemplate<I> (domain_pset, img)`
construct a translation variant template from a function	`IA_DDTemplate<I> (domain_pset, templ_func)`
assign a template value	`templ1 = templ2`

Template operations presented in the sections that follow are introduced in Section 1.5.

Image-Template Product Operations

In Table A.2.24, let *img* denotes an image object of type *I* which is `IA_Image<P, T>`. Let *templ* denote an object of type `IA_DDTempl<I>`, and *result_pset* denotes the point set over which the result image is to be defined. If no result point set is specified, the resulting image has the same domain as *img*. For the generic template reduction operation, the function with prototype `T pwise_op (T, T)` is a commutative function with identity *pwise_zero* used as a pointwise operation on the source image and template image and the function with prototype `T binary_reduce (T, T)` is a commutative function with identity *reduce_zero* used to reduce the image resulting from the pointwise combining. Alternatively, the function `T n_ary_reduce (T*, unsigned n)` which takes an array of *n* values of type *T* and reduces them to a single value, can be used to specify a template reduction.

In the product operations, the following correspondences may be drawn between `iac++` library function names and the image algebra symbols used to denote them:

(a) `linear_product` corresponds to $\oplus$,
(b) `additive_maximum` corresponds to $\boxdot$,
(c) `additive_minimum` corresponds to $\boxdot$,
(d) `multiplicative_maximum` corresponds to $\bigvee$,
(e) `multiplicative_minimum` corresponds to $\bigwedge$, and
(f) `generic_product` corresponds to $\bigcirc\!\!\!\!\gamma$ with *pwise_op* corresponding to $\bigcirc$, *binary_reduce* corresponding to γ, and *n_ary_reduce* corresponding to Γ.

Examples of Template Code Fragments

Constructors for `IA_DDTemplate` support both translation invariant and variant templates. The translation invariant templates are specified by giving the image result of applying the template function at the origin.

Table A.2.24 Image-Neighborhood Reduction Operations

calculate the right image-template linear product	`linear_product` (*img, templ, result_pset*) `linear_product` (*img, templ*)
calculate the right image-neighborhood additive maximum	`addmax_product` (*img, templ, result_pset*) `addmax_product` (*img, templ*)
calculate the right image-neighborhood additive minimum	`addmin_product` (*img, templ, result_pset*) `addmin_product` (*img, templ*)
calculate the right image-neighborhood multiplicative maximum	`multmax_product` (*img, templ, result_pset*) `multmax_product` (*img, templ*)
calculate the right image-neighborhood multiplicative minimum	`multmin_product` (*img, templ, result_pset*) `multmin_product` (*img, templ*)
calculate a right image-neighborhood product with a programmer-specified reduction function	`generic_product` (*img, templ, result_pset, pwise_op, binary_reduce, reduce_zero, pwise_zero*) `generic_product` (*img, templ, result_pset, pwise_op, n_ary_reduce, pwise_zero*)
calculate the left image-template linear product	`linear_product` (*templ, img, result_pset*) `linear_product` (*templ, img*)
calculate the left image-neighborhood additive maximum	`addmax_product` (*templ, img, result_pset*) `addmax_product` (*templ, img*)
calculate the left image-neighborhood additive minimum	`addmin_product` (*templ, img, result_pset*) `addmin_product` *templ, img*)
calculate the left image-neighborhood multiplicative maximum	`multmax_product` (*templ, img, result_pset*) `multmax_product` (*img, templ, img*)
calculate the left image-neighborhood multiplicative minimum	`multmin_product` (*templ, img, result_pset*) `multmin_product` (*templ, img*)
calculate a left image-neighborhood product with a programmer-specified reduction function	`generic_product` (*templ, img, result_pset, pwise_op, binary_reduce, reduce_zero, pwise_zero*) `generic_product` (*templ, img, result_pset, pwise_op, n_ary_reduce, pwise_zero*)

```
IA_Image<IA_Point<int>, int> i1, i2;
IA_DDTemplate<IA_Image<IA_Point<int>, int> > t1, t2;

i1 = read_int_PGM ("temp-file.pgm");
    // read the template image
i2 = read_int_PGM ("image.pgm");
    // read an image to process

t1 = IA_DTemplate<IA_Image<IA_Point<int>,int> > (i2.domain(),
                                                  i1);
    // Define template t1 over the same domain as image i2.

t2 = IA_DDTemplate<IA_Image<IA_Point<int>, int> (2, i1);
    // Define template t2to apply anywhere
    // in 2D integer Cartesian space.
```

Templates can be defined by giving a function mapping points into their image values.

```
IA_DDTemplate<IA_Image<IA_Point<int>, int> > tv;
const IA_Set<IA_Point<int> >
    ps (IA_boxy_pset (IA_Point<int>(0,0),
                      IA_Point<int>(511,511)));

// function from point to image
IA_Image<IA_Point<int>, int> f(const IA_IntPoint &p)
{ return IA_Image<IA_Point<int>, int>(ps, p[0]); }

tv = IA_DDTemplate<IA_Image<IA_Point<int>, int> >(ps, &f);
    // tv applied to any point p, yields the image
    // f(p) as its value.
```

Finally, any of these types of templates may be used with image-template product operations. The operations linear_product, addmax_product, addmin_product, multmax_product, and multmin_product are provided by the iac++ library together with a generic image-template product operation supporting any user specifiable image-template product operations.

```
IA_DDTemplate<IA_Image<IA_Point<int>, int> > templ;
IA_Image<IA_Point<int>, int> source, result;
IA_Set<IA_Point<int> > ps;

// ... code to initialize source and templ

result = linear_product(source, templ, ps);
    // result will be the linear product of source and
    // templ defined over point set ps.

result = addmax_product(source, templ);
    // result will be the add_max product of source and
    // templ defined over default point
    // set source.domain()

result = multmin_product(source, templ);
```

A.3. Examples of Programs Using `iac++`

The examples that follow present brief programs employing the iac++ library. The first example shows how to perform averaging of multiple images. This example involves the use of image input and output operations and binary operations upon images. The second example shows how to compose an image to a point-to-point function to yield a spatial/geometric modification of the original image. The third example presents local averaging using a neighborhood operation in two ways: using the sum reduction operation (and incurring an edge effect) and using an n-ary reduction operation. The fourth and final example presents the Hough transform.

Example 1. Averaging of Multiple Images

```
//
// example1.c -- Averaging Multiple Images
//
// usage: example1 file-1.pgm [file-2.pgm ... file-n.pgm]
//
// Read a sequence of images from files and average them.
// Display the result.
// Assumes input files are pgm images all having the same pointset
// and containing unsigned char values.
//

#include "ia/UcharDI.h"
#include "ia/FloatDI.h"

int main(int argc, char **argv)
{
    IA_Image<IA_Point<int>, float>      accumulator;
    IA_Image<IA_Point<int>, u_char>     result;

    if (argc < 2) {
        cerr << "usage: "
            << argv[0]
            << " file-1.pgm [file-2.pgm ... file-n.pgm]"
            << endl;
        abort();
    }

    // Get the first image
    accumulator = to_float (read_uchar_PGM (argv[1]));

    // Sum the first image together with the rest
    for (int i = 1; i < argc - 1; i++) {
        cout << "Reading " << argv[i+1] << endl;
        accumulator += to_float (read_uchar_PGM (argv[i+1]));
    }

    result = to_uchar (accumulator / float (i));
    display (result);
}
```

Example 2. Composing an Image with a Point-to-Point Function

```
//
// example2.c -- Composition of an Image with a Function
//

#include <iostream.h>
#include "ia/UcharDI.h"

//
// point-to-point mapping function
//

IA_Point<int>
reflect_through_origin (const IA_Point<int> &p)
{
    return -p;
}

int
main ()
{
    // read in image from cin
    IA_Image<IA_Point<int>, u_char> img = read_uchar_PGM (cin);

    IA_Image<IA_Point<int>, u_char> result;

    result = compose (img,
                      reflect_through_origin,
                      IA_boxy_pset (-max (img.domain()),
                                    -min (img.domain()))));

    display (result);

    return 0;
}
```

Example 3. Using Neighborhood Reductions for Local Averaging

```
//
// example3.c -- Local Averaging
//
// usage: example3 < file.pgm
//
// Read an image from cin and average it locally
// with a 3x3 neighborhood.
//

#include "math.h"
#include "ia/UcharDI.h"
#include "ia/IntDI.h"
#include "ia/Nbh.h"
#include "ia/UcharNOps.h"
```

```
u_char
average (u_char *uchar_vector, unsigned num)
{
    if (0 == num) {
        return 0;
    } else {
        int sum = 0;
        for (int i = 0; i < num; i++) {
            sum += uchar_vector[i];
        }
        return u_char (irint (float (sum) / num));
    }
}

int main(int argc, char **argv)
{
    IA_Image<IA_Point<int>, u_char>
        source_image = read_uchar_PGM (cin);

    IA_Neighborhood <IA_Point<int>, IA_Point<int> >
        box (2, IA_boxy_pset (IA_Point<int> (-1, -1),
                              IA_Point<int> ( 1,  1)));

    //
    // Reduction with sum and division by nine yields a boundary
    // effect at the limits of the image point set due to the
    // lack of nine neighbors.
    //
    display (to_uchar (sum (source_image, box) / 9));

    //
    // Reduction with the n-ary average function correctly
    // generates average values even at the boundary.
    //
    display (neighborhood_reduction (source_image,
                                     box,
                                     source_image.domain(),
                                     average));
}
```

Example 4. Hough Transform

main.c

```
//
// example4.c -- Averaging Multiple Images
//
// usage: example4 file-1.pgm [file-2.pgm ... file-n.pgm]
//
// Read a sequence of images from files and average them.
// Display the result.
// Assumes input files are pgm images all having the same pointset
```

```cpp
// and containing unsigned char values.
//

#include "ia/IntDI.h"    // integer valued images
#include "ia/UcharDI.h"  // unsigned character images
#include "ia/IntNOps.h"  // neighborhood operations

#include "hough.h"       // Hough transform functions

int main(int argc, char **argv)
{
    IA_Image<IA_Point<int>, int>
        source = read_int_PGM (argv[1]);

    IA_Image<IA_Point<int>, int>
        result;

    IA_Set<IA_Point<int> >
        accumulator_domain = source.domain();

    //
    // Initialize parameters for the Hough Neighborhood
    //
    hough_initialize (source.domain(), accumulator_domain);

    IA_Neighborhood<IA_Point<int>, IA_Point<int> >
        hough_nbh (accumulator_domain, hough_function);

    IA_Image<IA_Point<int>, int> accumulator;

    display (to_uchar (source * 255 / max(source)));

    //
    // Map feature points to corresponding locations
    // in the accumulator array.
    //
    accumulator = sum (source, hough_nbh, accumulator_domain);

    display (to_uchar(accumulator * 255 / max(accumulator)));

    //
    // Threshold the accumulator
    //

    accumulator = to_int(chi_eq (accumulator, max(accumulator)));

    display (to_uchar(accumulator*255));

    restrict (accumulator, IA_Set<int>(1));

    //
    // Map back to see the corresponding lines in the
    // source domain.
    //
```

```
    result = sum (hough_nbh, accumulator, source.domain());

    display (to_uchar(result * 255 / max(result)));
}
```

hough.h

```
// hough.h
//
// Copyright 1995, Center for Computer Vision and Visualization,
// University of Florida.  All rights reserved.

#ifndef _hough_h_
#define _hough_h_

#include "ia/Nbh.h"
#include "ia/IntPS.h"

void
hough_initialize (IA_Set<IA_Point<int> > image_domain,
                  IA_Set<IA_Point<int> > accumulator_domain);

IA_Set<IA_Point<int> >
hough_function (const IA_Point<int> &r_t);

// Accomplish the Hough Transform as follows:
//
// Given a binary image 'Source' containing linear features
//
// Call hough_initialize with first argument Source.domain()
//     and with second argument being the accumulator domain
//     (an r by t set of points with r equal to the number of
//      accumulator cells for rho and t equal to the number of
//      accumulator sells for theta).
//
// Create an IA_Neighborhood as follows:
//     IA_Neighborhood<IA_Point<int>, IA_Point<int> >
//         HoughNbh (AccumulatorDomain, hough_function);
//
// Then calculate the Hough Transform of a (binary) image
// as follows:
//
//     Accumulator = sum (Source, HoughNbh);

#endif
```

hough.c

```
// hough.c
//
// Copyright 1995, Center for Computer Vision and Visualization,
// University of Florida.  All rights reserved.
```

```
#include "hough.h"
#include "math.h"
#include <iostream.h>

static const double HoughPi = atan(1.0)*4.0;
static double *HoughCos;
static double *HoughSin;

static int RhoCells;
static int ThetaCells;
static IA_Point<int> IterationMin;
static IA_Point<int> IterationMax;
static IA_Point<int> ImageSize;
static IA_Point<int> Delta;

//
// hough_initialize:
//
// Initializes parameters for HoughFunction to allow a neighborhood
// to be created for given image and accumulator image point sets.
//

void
hough_initialize (IA_Set<IA_Point<int> > image_domain,
                  IA_Set<IA_Point<int> > accumulator_domain)
{
    //
    // Check to make sure the image domain and accumulator domain
    // are both 2 dimensional rectangular point sets.
    //

    if  (!image_domain.boxy() ||
         !accumulator_domain.boxy() ||
         image_domain.dim() != 2 ||
         accumulator_domain.dim() != 2) {

        cerr << "Hough transform needs 2-D rectangular domains."
            << endl;
    }

    //
    // Record data necessary to carry out rho,theta to
    // source pointset transformation.
    //

    ImageSize = image_domain.sup() - image_domain.inf() + 1;
    Delta = ImageSize / 2;
    IterationMin = image_domain.inf() - Delta;
    IterationMax = image_domain.sup() - Delta;

    RhoCells = accumulator_domain.sup()(0) -
        accumulator_domain.inf()(0) + 1;
    ThetaCells = accumulator_domain.sup()(1) -
        accumulator_domain.inf()(1) + 1;
```

```
    //
    // Create sine and cosine lookup tables for specified
    // accumulator image domain.
    //

    HoughSin = new double [ThetaCells];
    HoughCos = new double [ThetaCells];

    double t = HoughPi / ThetaCells;
    for (int i = accumulator_domain.inf()(1);
         i <= accumulator_domain.sup()(1);
         i++ ) {
        HoughSin[i] = sin (t * i);
        HoughCos[i] = cos (t * i);
    }
}


//
// hough_function
//
// This function is used to construct a Hough transform
// neighborhood.  It maps a single accumulator cell location
// into a corresponding set of (x,y) coordinates in the
// source image.
//

IA_Set<IA_Point<int> >
hough_function (const IA_Point<int> &r_t)
{
    double theta, rho;

    //
    // Convert accumulator image pixel location to
    // correct (rho, theta) location.
    //

    rho = double (r_t(0) - RhoCells/2) *
        enorm(ImageSize)/RhoCells;
    theta = r_t(1)* HoughPi / ThetaCells;

    IA_Point<int> *p_ptr, *pp;
    int coord, i;
    int num_points;

    //
    // Construct vector of (x,y) points associated with (rho,theta)
    // We check theta to determine whether we should make
    // x a function of y or vice versa.
    //

    if (theta > HoughPi/4.0 && theta < 3.0*HoughPi/4.0) {
        //
        // Scan across all 0th coordinate indices
        // assigning corresponding 1st coordinate indices.
```

```
        //
        num_points = ImageSize(0);
        p_ptr = new IA_Point<int> [num_points];

        for (coord = IterationMin(0), pp = p_ptr;
             coord <= IterationMax(0);
             coord++, pp++) {

            *pp = Delta +
                IA_Point<int> (coord,
                               nint (rho -
                                    (coord * HoughCos [r_t(1)]
                                    / HoughSin [r_t(1)])));
        }
    } else {
        //
        // Scan across all 1st coordinate indices
        // assigning corresponding 0th coordinate indices.
        //
        num_points = ImageSize(1);
        p_ptr = new IA_Point<int> [num_points];

        for (coord = IterationMin(1), pp = p_ptr;
             coord <= IterationMax(1);
             coord++, pp++) {

            *pp = Delta +
                IA_Point<int> (nint (rho -
                                    (coord * HoughSin [r_t(1)] /
                                    HoughCos [r_t(1)])),
                               coord);
        }
    }

    //
    // Turn vector of points into a point set
    //

    IA_Set<IA_Point<int> > result (2, p_ptr,  num_points);
    delete [] p_ptr;
    return result;
}
```

 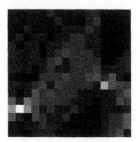

Figure A.3.1. Source image (left) and associated accumulator array image (right).

Figure A.3.2. Binarized accumulator image (left) and
the line associated with the identified cell (right).

Example 5. Canny Edge Detection

canny_edge.c

```c
#include "canny_params.h"

#include "ia/FloatDI.h"
#include "ia/FloatCI.h"
#include "ia/UcharDI.h"
#include "ia/BitDI.h"
#include "ia/Pixel.h"
#include "ia/DDTempl.h"
#include "ia/FloatProd.h"
#include "ia/Nbh.h"

#include "ia/chistogram.h"
#include "ia/gauss_templates.h"
#include "ia/BNops.h"
#include "ia/scale_to_uchar.h"

#include "edge_templates.h"
#include "local_maximum_templates.h"
#include "connector_neighborhoods.h"
#include "iostream.h"

float SIGMA = 1.0;

main(int argc, char **argv) {

  IA_Image<IA_Point<int>, float>
    source_image = to_float (read_int_PGM(argv[1]));

  if (argc > 2) {
    SIGMA = atof (argv[2]);
  }

  //
  // Perform Gaussian smoothing on the source image.
  //
```

```
IA_DDTemplate<IA_Image<IA_Point<int>, float> > tgx=gauss_x(SIGMA);
IA_DDTemplate<IA_Image<IA_Point<int>, float> > tgy=gauss_y(SIGMA);

IA_Image<IA_Point<int>, float>
  smoothed_image =
    linear_product(linear_product(source_image, tgx),
      tgy);

IA_Image<IA_Point<int>, float>
  edge_magnitude(smoothed_image.domain(), 0.0);
IA_Image<IA_Point<int>, int>
  direction_image(smoothed_image.domain(), -1);

//
// Calculate edge magnitude in each of the chosen directions and
// maintain a map of the edge direction associated with each pixel.
//

for (int i = 0; i < NUM_DIRECTIONS; i++) {

  IA_Image<IA_Point<int>, float>
    dir_i_edge(abs(linear_product (smoothed_image,
      edge_templates[i])));

  //
  // Note where the current edge point dominates in the edge image
  //
  direction_image =
    max (i*to_int(chi_gt(dir_i_edge, edge_magnitude)),
    direction_image);

  //
  // Store edge magnitude of dominators in edge_magnitude
  //
  edge_magnitude = max(edge_magnitude, dir_i_edge);
}



//
// Apply local-maximum finding
//

IA_Image<IA_Point<int>, bool>
  edges(edge_magnitude.domain(), false);

for (int i=0; i < NUM_DIRECTIONS; i++) {

 edges |=
  chi_eq(i,direction_image)
  &chi_ge(edge_magnitude,
    linear_product(edge_magnitude,
    local_maximum_templates[i]))
   &chi_ge(edge_magnitude,
    linear_product(edge_magnitude,
```

```
        local_maximum_templates[i+
          NUM_DIRECTIONS]));
}

//
// Get rid of pixels displaying boundary effect
//
IA_Set<IA_Point<int> >
  interior = IA_boxy_pset(source_image.domain().inf() +
      IA_Point<int>(1,1),
    source_image.domain().sup() -
      IA_Point<int>(1,1));

edge_magnitude = restrict(edge_magnitude, interior);
edges = restrict(edges, interior);


display(scale_to_uchar(edge_magnitude));
display(to_uchar(edges));

//
// Normalize edge values to range 0 .. 255 for creation of
// cumulative histogram.
//
float edge_max = max(edge_magnitude);

IA_Image<IA_Point<int>, u_char>
  histogrammable_edge = to_uchar(255 / edge_max *
      edge_magnitude);

CumulativeHistogram<u_char>
  edge_cumulative_histo(histogrammable_edge);

//
// Canny recommends using the 80% percentile edge value as
// a good threshold to keep noise out of the edge map.
//
int t_low;
int t_high;

for (t_low=0;
     t_low < 256 && edge_cumulative_histo(t_low) <= 0.4;
     t_low++);

for(t_high=t_low++;
    t_high < 256 && edge_cumulative_histo(t_high) <= 0.8;
    t_high++);

//
// Restore edge threshold to correct range (from 0 .. 255)
//
t_low = int(t_low * edge_max / 255.0);
t_high = int(t_high * edge_max / 255.0);

//
```

```
// Start by identifying edges_high points as edge points.
//
IA_Image<IA_Point<int>, bool>
  new_edges = edges &&
    chi_gt(edge_magnitude, t_high );

//
// Also find points in edge image satisfying a low threshold
//
IA_Image<IA_Point<int>, bool>
  edges_low = edges &&
    chi_gt(edge_magnitude, t_low );

display(to_uchar(new_edges));
display(to_uchar(edges_low));


IA_Image<IA_Point<int>, bool>
  initial_edges = new_edges;

//
// Repeatedly search to see if a point is
//  1. in the edges_low image,
//  2. has a neighboring point that is a previously identified
//     edge point, and
//  3. lies in that neighboring point's edge direction
// If a point satisfies these criteria, identify it as an edge
// point.
//
direction_image = restrict(direction_image, interior);

do {

  // Include previously identified points
  edges = new_edges;

  //
  // Check each edge direction
  //
  for (int i = 0; i < NUM_DIRECTIONS; i++) {
    new_edges |=
        edges_low
    & max (edges & chi_eq (i, direction_image),
    connector_neighborhood[i]);
  }

  //
  // Continue until no new edge points are added
  //
} while (new_edges != edges);

display (to_uchar (edges ^ initial_edges));
display (to_uchar (edges));
}
```

canny_params.h

```
#ifndef canny_params_h_
#define canny_params_h_

#define NUM_DIRECTIONS 6

#endif
```

edge_templates.h

```
#ifndef edge_templates_h_
#define edge_templates_h_
#include "ia/Pixel.h"
#include "ia/DDTempl.h"
#include "ia/FloatDI.h"

#include "canny_params.h"

extern
IA_DDTemplate<IA_Image<IA_Point<int>, float> >
edge_templates[NUM_DIRECTIONS];

#endif;
```

edge_templates.c

```
#include "edge_templates.h"

static
IA_Pixel<IA_Point<int>, float>
e0[] = { IA_Pixel<IA_Point<int>, float>(IA_Point<int>(-1,-1),
      -1),
   IA_Pixel<IA_Point<int>, float>(IA_Point<int>(-1,1),
      1),
   IA_Pixel<IA_Point<int>, float>(IA_Point<int>(0,-1),
      -1),
   IA_Pixel<IA_Point<int>, float>(IA_Point<int>(0,1),
      1),
   IA_Pixel<IA_Point<int>, float>(IA_Point<int>(1,-1),
      -1),
   IA_Pixel<IA_Point<int>, float>(IA_Point<int>(1,1),
      1)};
static const int e0n = 6;
```

```
static
IA_Pixel<IA_Point<int>, float>
e1[] = {IA_Pixel<IA_Point<int>, float>(IA_Point<int>(-1,-1),
          -0.366025),
  IA_Pixel<IA_Point<int>, float>(IA_Point<int>(-1,0),
          0.5),
  IA_Pixel<IA_Point<int>, float>(IA_Point<int>(-1,1),
          1),
  IA_Pixel<IA_Point<int>, float>(IA_Point<int>(0,-1),
          -0.866025),
  IA_Pixel<IA_Point<int>, float>(IA_Point<int>(0,1),
          0.866025),
  IA_Pixel<IA_Point<int>, float>(IA_Point<int>(1,-1),
          -1),
  IA_Pixel<IA_Point<int>, float>(IA_Point<int>(1,0),
          -0.5),
  IA_Pixel<IA_Point<int>, float>(IA_Point<int>(1,1),
          0.366025)};
static const int e1n = 8;

static
IA_Pixel<IA_Point<int>, float>
e2[] = {IA_Pixel<IA_Point<int>, float>(IA_Point<int>(-1,-1),
          0.366025),
  IA_Pixel<IA_Point<int>, float>(IA_Point<int>(-1,0),
          0.866025),
  IA_Pixel<IA_Point<int>, float>(IA_Point<int>(-1,1),
          1),
  IA_Pixel<IA_Point<int>, float>(IA_Point<int>(0,-1),
          -0.5),
  IA_Pixel<IA_Point<int>, float>(IA_Point<int>(0,1),
          0.5),
  IA_Pixel<IA_Point<int>, float>(IA_Point<int>(1,-1),
          -1),
  IA_Pixel<IA_Point<int>, float>(IA_Point<int>(1,0),
          -0.866025),
  IA_Pixel<IA_Point<int>, float>(IA_Point<int>(1,1),
          -0.366025)};
static const int e2n = 8;

static
IA_Pixel<IA_Point<int>, float>
e3[] = {IA_Pixel<IA_Point<int>, float>(IA_Point<int>(-1,-1),
          1),
  IA_Pixel<IA_Point<int>, float>(IA_Point<int>(-1,0),
          1),
  IA_Pixel<IA_Point<int>, float>(IA_Point<int>(-1,1),
          1),
  IA_Pixel<IA_Point<int>, float>(IA_Point<int>(1,-1),
          -1),
  IA_Pixel<IA_Point<int>, float>(IA_Point<int>(1,0),
          -1),
  IA_Pixel<IA_Point<int>, float>(IA_Point<int>(1,1),
          -1)};
static const int e3n = 6;
```

```
static
IA_Pixel<IA_Point<int>, float>
e4[] = {IA_Pixel<IA_Point<int>, float>(IA_Point<int>(-1,-1),
           1),
 IA_Pixel<IA_Point<int>, float>(IA_Point<int>(-1,0),
           0.866025),
 IA_Pixel<IA_Point<int>, float>(IA_Point<int>(-1,1),
           0.366025),
 IA_Pixel<IA_Point<int>, float>(IA_Point<int>(0,-1),
           0.5),
 IA_Pixel<IA_Point<int>, float>(IA_Point<int>(0,1),
           -0.5),
 IA_Pixel<IA_Point<int>, float>(IA_Point<int>(1,-1),
           -0.366025),
 IA_Pixel<IA_Point<int>, float>(IA_Point<int>(1,0),
           -0.866025),
 IA_Pixel<IA_Point<int>, float>(IA_Point<int>(1,1),
           -1)};
static const int e4n = 8;

static
IA_Pixel<IA_Point<int>, float>
e5[] = {IA_Pixel<IA_Point<int>, float>(IA_Point<int>(-1,-1),
           1),
 IA_Pixel<IA_Point<int>, float>(IA_Point<int>(-1,0),
           0.5),
 IA_Pixel<IA_Point<int>, float>(IA_Point<int>(-1,1),
           -0.366025),
 IA_Pixel<IA_Point<int>, float>(IA_Point<int>(0,-1),
           0.866025),
 IA_Pixel<IA_Point<int>, float>(IA_Point<int>(0,1),
           -0.866025),
 IA_Pixel<IA_Point<int>, float>(IA_Point<int>(1,-1),
           0.366025),
 IA_Pixel<IA_Point<int>, float>(IA_Point<int>(1,0),
           -0.5),
 IA_Pixel<IA_Point<int>, float>(IA_Point<int>(1,1),
           -1)};
static const int e5n = 8;

IA_DDTemplate<IA_Image<IA_Point<int>, float> >
edge_templates[NUM_DIRECTIONS] = {
  IA_DDTemplate<IA_Image<IA_Point<int>, float> >
    (2, IA_Image<IA_Point<int>, float>( e0, e0n)),
  IA_DDTemplate<IA_Image<IA_Point<int>, float> >
    (2, IA_Image<IA_Point<int>, float>( e1, e1n)),
  IA_DDTemplate<IA_Image<IA_Point<int>, float> >
    (2, IA_Image<IA_Point<int>, float>( e2, e2n)),
  IA_DDTemplate<IA_Image<IA_Point<int>, float> >
    (2, IA_Image<IA_Point<int>, float>( e3, e3n)),
  IA_DDTemplate<IA_Image<IA_Point<int>, float> >
    (2, IA_Image<IA_Point<int>, float>( e4, e4n)),
  IA_DDTemplate<IA_Image<IA_Point<int>, float> >
    (2, IA_Image<IA_Point<int>, float>( e5, e5n))};
```

local_maximum_templates.h

```
#ifndef local_maximum_templates_h_
#define local_maximum_templates_h_
#include "ia/Pixel.h"
#include "ia/DDTempl.h"
#include "ia/FloatDI.h"

#include "canny_params.h"

extern
IA_DDTemplate<IA_Image<IA_Point<int>, float> >
local_maximum_templates[2*NUM_DIRECTIONS];

#endif;
```

local_maximum_templates.c

```
#include "local_maximum_templates.h"

static
IA_Pixel<IA_Point<int>, float>
nm0 [] = {
  IA_Pixel<IA_Point<int>, float>(IA_Point<int>(0, -1),1)};
static const int nm0n = 1;

static
IA_Pixel<IA_Point<int>, float>
nm1 [] = {
  IA_Pixel<IA_Point<int>, float>(IA_Point<int>(0, -1),0.381854),
  IA_Pixel<IA_Point<int>, float>(IA_Point<int>(1, -1),0.559073),
  IA_Pixel<IA_Point<int>, float>(IA_Point<int>(1, 0),0.059073)};
static const int nm1n = 3;

static
IA_Pixel<IA_Point<int>, float>
nm2 [] = {
  IA_Pixel<IA_Point<int>, float>(IA_Point<int>(0, -1),0.059073),
  IA_Pixel<IA_Point<int>, float>(IA_Point<int>(1, -1),0.559073),
  IA_Pixel<IA_Point<int>, float>(IA_Point<int>(1, 0),0.381854)};
static const int nm2n = 3;

static
IA_Pixel<IA_Point<int>, float>
nm3 [] = {
  IA_Pixel<IA_Point<int>, float>(IA_Point<int>(1, 0),1)};
static const int nm3n = 1;
```

```
static
IA_Pixel<IA_Point<int>, float>
nm4 [] = {
  IA_Pixel<IA_Point<int>, float>(IA_Point<int>(0, 1),0.059073),
  IA_Pixel<IA_Point<int>, float>(IA_Point<int>(1, 0),0.381854),
  IA_Pixel<IA_Point<int>, float>(IA_Point<int>(1, 1),0.559073)};
static const int nm4n = 3;

static
IA_Pixel<IA_Point<int>, float>
nm5 [] = {
  IA_Pixel<IA_Point<int>, float>(IA_Point<int>(0, 1),0.381854),
  IA_Pixel<IA_Point<int>, float>(IA_Point<int>(1, 0),0.059073),
  IA_Pixel<IA_Point<int>, float>(IA_Point<int>(1, 1),0.559073)};
static const int nm5n = 3;

static
IA_Pixel<IA_Point<int>, float>
nm6 [] = {
  IA_Pixel<IA_Point<int>, float>(IA_Point<int>(0, 1),1)};
static const int nm6n = 1;

static
IA_Pixel<IA_Point<int>, float>
nm7 [] = {
  IA_Pixel<IA_Point<int>, float>(IA_Point<int>(-1, 0),0.059073),
  IA_Pixel<IA_Point<int>, float>(IA_Point<int>(-1, 1),0.559073),
  IA_Pixel<IA_Point<int>, float>(IA_Point<int>(0, 1),0.381854)};
static const int nm7n = 3;

static
IA_Pixel<IA_Point<int>, float>
nm8 [] = {
  IA_Pixel<IA_Point<int>, float>(IA_Point<int>(-1, 0),0.381854),
  IA_Pixel<IA_Point<int>, float>(IA_Point<int>(-1, 1),0.559073),
  IA_Pixel<IA_Point<int>, float>(IA_Point<int>(0, 1),0.059073)};
static const int nm8n = 3;

static
IA_Pixel<IA_Point<int>, float>
nm9 [] = {
  IA_Pixel<IA_Point<int>, float>(IA_Point<int>(-1, 0),1)};
static const int nm9n = 1;

static
IA_Pixel<IA_Point<int>, float>
nm10 [] = {
  IA_Pixel<IA_Point<int>, float>(IA_Point<int>(-1, -1),0.559073),
  IA_Pixel<IA_Point<int>, float>(IA_Point<int>(-1, 0),0.381854),
  IA_Pixel<IA_Point<int>, float>(IA_Point<int>(0, -1),0.059073)};
static const int nm10n = 3;

static
IA_Pixel<IA_Point<int>, float>
nm11 [] = {
```

```
  IA_Pixel<IA_Point<int>, float>(IA_Point<int>(-1, -1),0.559073),
  IA_Pixel<IA_Point<int>, float>(IA_Point<int>(-1, 0),0.059073),
  IA_Pixel<IA_Point<int>, float>(IA_Point<int>(0, -1),0.381854)};
static const int nm11n = 3;

IA_DDTemplate<IA_Image<IA_Point<int>, float> >
local_maximum_templates[2*NUM_DIRECTIONS] = {
  IA_DDTemplate<IA_Image<IA_Point<int>, float> >
    (2, IA_Image<IA_Point<int>, float>( nm0, nm0n)),
  IA_DDTemplate<IA_Image<IA_Point<int>, float> >
    (2, IA_Image<IA_Point<int>, float>( nm1, nm1n)),
  IA_DDTemplate<IA_Image<IA_Point<int>, float> >
    (2, IA_Image<IA_Point<int>, float>( nm2, nm2n)),
  IA_DDTemplate<IA_Image<IA_Point<int>, float> >
    (2, IA_Image<IA_Point<int>, float>( nm3, nm3n)),
  IA_DDTemplate<IA_Image<IA_Point<int>, float> >
    (2, IA_Image<IA_Point<int>, float>( nm4, nm4n)),
  IA_DDTemplate<IA_Image<IA_Point<int>, float> >
    (2, IA_Image<IA_Point<int>, float>( nm5, nm5n)),
  IA_DDTemplate<IA_Image<IA_Point<int>, float> >
    (2, IA_Image<IA_Point<int>, float>( nm6, nm6n)),
  IA_DDTemplate<IA_Image<IA_Point<int>, float> >
    (2, IA_Image<IA_Point<int>, float>( nm7, nm7n)),
  IA_DDTemplate<IA_Image<IA_Point<int>, float> >
    (2, IA_Image<IA_Point<int>, float>( nm8, nm8n)),
  IA_DDTemplate<IA_Image<IA_Point<int>, float> >
    (2, IA_Image<IA_Point<int>, float>( nm9, nm9n)),
  IA_DDTemplate<IA_Image<IA_Point<int>, float> >
    (2, IA_Image<IA_Point<int>, float>( nm10, nm10n)),
  IA_DDTemplate<IA_Image<IA_Point<int>, float> >
    (2, IA_Image<IA_Point<int>, float>( nm11, nm11n))};
```

connector_neighborhoods.h

```
#ifndef connector_templates_h_
#define connector_templates_h_

#include "ia/Nbh.h"
#include "canny_params.h"

extern
IA_Neighborhood<IA_Point<int>, IA_Point<int> >
connector_neighborhood[NUM_DIRECTIONS];

#endif
```

connector_neighborhoods.c

```
#include "connector_neighborhoods.h"
```

```
static
IA_Set<IA_Point<int> >
s0 = IA_Set<IA_Point<int> >(
        IA_Point<int>(-1,   0),
        IA_Point<int>( 0,   0),
        IA_Point<int>( 1,   0));

static
IA_Set<IA_Point<int> >
s1 = IA_Set<IA_Point<int> >(
        IA_Point<int>(-1, -1),
        IA_Point<int>( 0,   0),
        IA_Point<int>( 1,   1));

static
IA_Set<IA_Point<int> >
s2 = IA_Set<IA_Point<int> >(
                        IA_Point<int>(-1, -1),
                        IA_Point<int>( 0,   0),
                        IA_Point<int>( 1,   1));

static
IA_Set<IA_Point<int> >
s3 = IA_Set<IA_Point<int> >(
                        IA_Point<int>(0, -1),
                        IA_Point<int>(0,   0),
                        IA_Point<int>(0,   1));

static
IA_Set<IA_Point<int> >
s4 = IA_Set<IA_Point<int> >(
                        IA_Point<int>(-1,   1),
                        IA_Point<int>( 0,   0),
                        IA_Point<int>( 1, -1));

static
IA_Set<IA_Point<int> >
s5 = IA_Set<IA_Point<int> >(
                        IA_Point<int>(-1,   1),
                        IA_Point<int>( 0,   0),
                        IA_Point<int>( 1, -1));

static int numpoints = 3;

IA_Neighborhood<IA_Point<int>, IA_Point<int> >
connector_neighborhood [NUM_DIRECTIONS] = {
  IA_Neighborhood<IA_Point<int>, IA_Point<int> > (2, s0),
  IA_Neighborhood<IA_Point<int>, IA_Point<int> > (2, s1),
  IA_Neighborhood<IA_Point<int>, IA_Point<int> > (2, s2),
  IA_Neighborhood<IA_Point<int>, IA_Point<int> > (2, s3),
  IA_Neighborhood<IA_Point<int>, IA_Point<int> > (2, s4),
  IA_Neighborhood<IA_Point<int>, IA_Point<int> > (2, s5)};
```

A.4. References

[1]   B. Stroustrup, *The C++ Programming Language Second Edition*. Reading, MA:
      Addison-Wesley, 1991.

[2]   M. Ellis and B. Stroustrup, *The Annotated C++ Reference Manual*. Reading, MA:
      Addison-Wesley, 1990.

[3]   J. Poskanzer, "Pbmplus: Extended portable bitmap toolkit." Internet distribution, Dec.
      1991.

INDEX

T - #0317 - 101024 - C0 - 254/178/24 [26] - CB - 9780849300752 - Gloss Lamination